Life Sciences

Until the nineteenth century, the various subjects now known as the life sciences were regarded either as arcane studies which had little impact on ordinary daily life, or as a genteel hobby for the leisured classes. The increasing academic rigour and systematisation brought to the study of botany, zoology and other disciplines, and their adoption in university curricula, are reflected in the books reissued in this series.

The Animal Kingdom

Georges Cuvier (1769–1832), made a peer of France in 1819 in recognition of his work, was perhaps the most important European scientist of his day. His most famous work, Le Règne Animal, was published in French in 1817; Edward Griffith (1790–1858), a solicitor and amateur naturalist, embarked in 1824, with a team of colleagues, on an English version which resulted in this illustrated sixteen-volume edition with additional material, published between 1827 and 1835. Cuvier was the first biologist to compare the anatomy of fossil animals with living species, and he named the now familiar 'mastodon' and 'megatherium'. However, his studies convinced him that the evolutionary theories of Lamarck and St Hilaire were wrong, and his influence on the scientific world was such that the possibility of evolution was widely discounted by many scholars both before and after Darwin. Volume 15 is the second of two covering insects.

The Animal Kingdom

Volume 15:
The Class Insecta 2

Georges Cuvier
Edited and translated by
Edward Griffith
with Edward Pidgeon

CAMBRIDGE UNIVERSITY PRESS

Cambridge, New York, Melbourne, Madrid, Cape Town,
Singapore, São Paolo, Delhi, Mexico City

Published in the United States of America by Cambridge University Press, New York

www.cambridge.org
Information on this title: www.cambridge.org/9781108049689

This edition first published 1832
This digitally printed version 2012

ISBN 978-1-108-04968-9 Paperback

The original edition of this book contains a number of colour plates,
which have been reproduced in black and white. Colour versions of these
images can be found online at www.cambridge.org/9781108049689

THE

ANIMAL KINGDOM

ARRANGED IN CONFORMITY WITH ITS ORGANIZATION,

BY THE BARON CUVIER,

MEMBER OF THE INSTITUTE OF FRANCE, &c. &c. &c.

WITH

SUPPLEMENTARY ADDITIONS TO EACH ORDER,

BY

EDWARD GRIFFITH, F.L.S., A.S.,

CORRESPONDING MEMBER OF THE ACADEMY OF NATURAL SCIENCES OF PHILADELPHIA, &c.

AND OTHERS.

VOLUME THE FIFTEENTH.

LONDON:

PRINTED FOR WHITTAKER, TREACHER, AND CO.

AVE-MARIA-LANE.

MDCCCXXXII.

THE

CLASS INSECTA

ARRANGED BY THE

BARON CUVIER,

WITH

SUPPLEMENTARY ADDITIONS TO EACH ORDER

BY

EDWARD GRIFFITH, F.L.S., A.S., &c.

AND

EDWARD PIDGEON, Esq.

AND NOTICES OF NEW GENERA AND SPECIES

BY

GEORGE GRAY, Esq.

VOLUME THE SECOND.

LONDON:

PRINTED FOR WHITTAKER, TREACHER, AND CO.

AVE-MARIA-LANE.

MDCCCXXXII.

THE

ANIMAL KINGDOM.

THIRD CLASS OF ARTICULATED ANIMALS,

AND PROVIDED WITH ARTICULATED FEET.

THE INSECTS.

SECOND FAMILY

OF THE

HETEROMEROUS COLEOPTERA.

THE TAXICORNES

HAVE no corneous claw at the internal side of the jaws, and are all winged; their body is for the most part square, with the corslet trapezoid, or semi-circular, and concealing or receiving the head; in some, the antennæ, usually inserted under a marginal projection of the sides of the head, are short, more or less perfoliate or grained, thicken insensibly, or terminate in a knob. The feet are proper only for running, and all the articulations of the tarsi are entire, and terminated by simple hooks. The anterior legs are often broad and triangular. Many males have the head provided with horns. The majority of these heteromera are found in the mushrooms of trees, or under the bark. Some others live on the ground under stones. M. Leon Dufour has observed in some sub-genera of this family, such as the hypophlœi, diaperis proper, and the eledonæ or boletophagus, an apparatus for excrementitious secretion, and in the second, some salivary vessels. The chylific ventricle of these heteromera is bristling with small papillæ in the form of hairs. These characters, and the conformation moreover of the organs of generation, indicate to us that this family is connected with the preceding.

Some have the head uncovered, and never entirely engaged in a deep and anterior notch of the corslet. This last part of the body is sometimes trapezoid or square, sometimes almost cylindrical; its sides, as well as those of the elytra, do not out-edge the body remarkably.

This division will form the tribe of DIAPERALES, having for type the genus

DIAPERIS.

Sometimes the antennæ are generally thick, almost straight, for the greater part perfoliate, or terminating abruptly in a thick knob. The body is smooth, or but slightly striated on the elytra. The two anterior feet are triangular, and dilated externally at their extremity, in a great number.

Sometimes the antennæ thicken insensibly, or at least do not terminate abruptly in an oval or ovoid knob, and the majority of the articulations are much larger than the preceding.

Some, and these form the greatest number, have the body oval or ovoid, sometimes even hemispherical, with the corslet almost square or trapezoid, most frequently transverse, but never long and narrow.

PHALERIA, Lat. *Uloma, Phaleria,* Dej., have the last articulation of the maxillary palpi larger, in the form of a reversed triangle or hatchet, and the anterior legs broader, dilated in the manner of a reversed triangle, and often denticulated, or provided with small spurs on one of its sides.

Some, from their elongated form, approach Tenebrio. The intermediate articulations of the antennæ are almost abconical, and the last four form a perfoliated knob. The head of the males is horned.

Others have the body oval, depressed, and the antennæ very much perfoliated. Such are the tenebrio *culinaris*, *retusus*, *chrysomelinus*, *impressus*, *nitidulus*, of this author.

The species of these two divisions form the genus *Uloma*, of MM. Megerle and Dejean. Those whose body is shorter, and more rounded, in the form of a short or even hemispherical ellipsoid, in which the six or seven articulations of the antennæ are almost globular, are, with M. Dejean, Phaleria. The tenebrio *cadaverinus* of Fabricius is of this number.

One species (*bicolor*) of the Cape of Good Hope, and of this division, is distinguished from the preceding by the maxillary palpi terminating in an articulation proportionally larger, in the form of a hatchet, and by its antennæ, the last four articulations of which alone are globular.

Another (*peltoides*), proper to Senegal, approaches the *peltis* of Fabricius, and the *Cossyphus* in its flatted form. Its antennæ are not almost perfoliate in the majority of their articulations, and even the last, being in the form of an inverted cone.

Diaperis (proper), Geoff. Fab., whose maxillary palpi terminate in an articulation scarcely thicker than the preceding, almost cylindrical, and whose anterior legs, not at all or but little broader than the following, are narrow, almost linear, and slightly dilated at their extremity.

Some others, but in which the last five articulations alone are perfoliate, and form a small knob, compose also a genus proper, that of *Pentaphyllus*.

Other insects of this tribe, whose antennæ proceed thickening, and are almost entirely perfoliate, are distinguished from Diaperis and Phaleria, by the linear form of their body and their corslet, in a long or almost cylindrical square. Such are, HYPOPHLÆUS, Fab.—*Ips.* Oliv. They are found under the barks of trees.

In others, the antennæ, the insertion of which is uncovered, or very little concealed, terminate abruptly in a large oval or ovoid knob, perfoliate, of four articulations at least, the second of which, in those in which it is formed of five, is very small. The body is ovoid or almost hemispherical and convex.

TRACHYSCELIS, Latr. Dej., have the antennæ but little longer than the head, terminating in an ovoid knob of six articulations. All the legs are broad and triangular, and proper for digging, and the body is short, and most frequently hemispherical. They bury themselves in the sand of the sea-shore.

LEIODES, Latr. *Anisotoma*, Illig. Fab., whose body is likewise short and gibbous, but whose antennæ, of the length of the head and corslet, terminate in an oval knob of five articulations, the second of which is smaller. The legs are narrow, elongated, or but little dilated. The four anterior ones at least are spinous.

TETRATOMA, Herbst. Fab., have the body a little more elongated than the preceding, ovoid and less raised above; all the legs are narrow and without spines, and the antennæ of the length of the head and corslet, terminating in an ovaliform club of four articulations.

Sometimes the antennæ always terminating in a perfoliate knob of five or six articulations, the preceding of which are almost in the form of an inverted cone, or a little dilated at the internal side in the manner of a tooth, are arched or a little curved. The body is ovoid, very unequal above, or deeply punctuated and striated on the elytra. The corslet is depressed laterally, and the edges of this marginal border are denticulated. It is separated posteriorly on each side by a remarkable vacancy. The palpi are filiform, or but slightly thicker at their extremity, as it is in Phaleria and Diaperis. The head of the males is often horned. They are also found in the mushrooms of trees. They form the genus

ELEDONA of Latr., or BOLETOPHAGUS of Fabricius, and of most other naturalists.

M. Ziegler, and after him, M. le Comte Dejean, comprehend here only the species whose antennæ have a knob formed by the last five articulations, the preceding to which are a little serrated.

Those in which the last three articulations alone form the knob, while the preceding are almost in the form of an inverted cone without internal projection, compose the genus COXELUS. The *cis* appear in a natural order to approximate to these insects.

Our second tribe of Taxicornes, the COSSYPHENES, is formed of insects analogous in the general form of the body to the *peltis* of Fabricius, and to many nitidulæ and cassidæ. It is ovoid or sub-hemispherical, out-edged in its circumference by the sides, dilated and flatted in the manner of a border, or margin, of the corslet and the elytra. The head is sometimes entirely concealed under this corslet, sometimes received, or as it were, emboxed, in an anterior emargination of this part of the body. The final articulation of the maxillary palpi is larger than the preceding, and formed like an axe.

This tribe is composed of the genus

COSSYPHUS, Oliv. Fab.

Some have the body flatted, in the form of a buckler, of a solid consistence, and the antennæ terminating in a knob of from four to five articulations. They are proper to the ancient continent, or at least to New Holland. Such are,

Cossyphus (proper), Oliv. Fab., whose corslet, almost semicircular, does not present anteriorly any emargination, and entirely conceals the head; whose antennæ are short, and terminate abruptly in an oval knob of four articulations, the most part transverse. The second of all, and the following, are almost identical.

Helœus, Latr. Kirb., have the head engaged in a deep emargination, or in a middle aperture of the corslet, and uncovered, or at least, partly so, above. The antennæ of the length at least of these two parts of the body united, terminate almost gradually in a narrow knob, elongated and formed by the last five articulations, the last of which is ovoid, and the preceding in the form of a top. The second of all is shorter than the third. These insects are proper to Australasia. (*Helæus Brownii*, Kirb. Linn. Trans. XII. xxiii. 8.)

The others, whose head is always uncovered, and received in a deep emargination of the corslet, have the body almost hemispherical, gibbous, soft, or of not a very solid consistence; the corslet very short, and the antennæ almost of the same thickness throughout, and grained. They are proper to South America, and resemble, at the first glance, the coccinellæ, and divers species of Erotylus. Such are, Nilio, Latr.

SUPPLEMENT

ON THE

TAXICORNES.

THE insects of the first tribe of this family, the DIAPERALES, live for the most part in mushrooms, or under the barks of trees. The insects of the first sub-genus of this family, PHALERIA, entomologists, deceived by general similitudes of form and colour, have confounded with tenebrio; but a deep, and comparative study of the antennæ and the parts of the mouth in these heteromera, will speedily evince the viciousness of this classification. The phaleriæ have no scaly claws at the internal sides of their jaws, and their antennæ are more or less perfoliate at their extremity. These coleoptera also appear to be removed from tenebrio in their habits. They remain under the barks of trees, and in the sands on maritime coasts. They are much more near, in a natural order, to Diaperis, but they differ from them by the characters which the reader will find in the text. As well as in this last mentioned genus, many species of *Phaleria* present very remarkable sexual differences. Their males have sometimes two eminences in the form of horns upon the head, and sometimes an excavation on the corslet.

The genus DIAPERIS was first distinguished by Geoffroy, who gave it this name in consequence of the singular form of its antennæ, composed of lenticular rings, threaded from their centre, one following the other. Linnæus had placed it with chrysomela, the only species with which he was acquainted. Degeer placed it with tenebrio. Fabricius had also ranged among the chrysomelæ, the species of Linnæus, and two other species among the *hispæ*. But in his latter works he adopted the genus Diaperis.

These insects are found in agaris, &c. which they gnaw, as well in their final form as in the larva state. Many species are remarkable for two horns, more or less long, which the male carries above his head.

The larva of *Diaperis boleti* has the body soft, smooth, and divided into twelve distinct rings. The head is scaly, a little flatted, provided with two small antennæ, divided into three or four articulations. These larvæ are usually found in great numbers in the agaries, which are on the point of being decomposed. When they are desirous of changing into nymphs, they construct a shell, from which they issue forth under the form of the perfect insect.

We insert a new species of Diaperis, named *nigro punctatus*. Ferruginous, with some black spots on the thorax; elytra, with the breast and legs black. Inhabits Brazil.

Many of the insects of the genus Hypophlœus, had been put along with the *Ips* of Fabricius. This author has formed them into a separate genus under the former title.

The hypophlœi have such great relations with Diaperis, and Phaleria, that, on the first glance, they seem to differ only in the linear or cylindrical form of their bodies. They are found in spring and summer under the barks of different trees. They are tolerably agile. The larvæ of these insects are not known; but it may be presumed that they live in the trunks of carious trees.

Fabricius, who established the genus Tetratoma, mentions four species of it, but the last of which must be separated from it, and perhaps, the second and third. The

Tetratoma fungorum is found in mushrooms, in different parts of Europe.

The insects of the genus Eledona, approximate very much to *Diaperis* in the characters taken from the number of articulations in the tarsi, and in those of the different parts of the mouth, and in the habits.

This genus, formed by Latreille, is composed of several species taken from the genus *Opatrum.* Illiger, in recognizing the existence of this genus, has given it the name of *Bolitophagus,* which all the German naturalists have adopted.

These insects are in general small, and of an obscure colour. They are found in rotten mushrooms, and appear to feed upon their substance. Their larva is unknown. One species (*Bolitophagus cornutus*) was brought from Carolina by M. Bosc, who discovered it in mushrooms.

Cossyphus depressus, the type of the genus Cossyphus, had been confounded by Fabricius with Lampyris. But the number of articulations of the tarsi, the antennæ, and other characters, remove it from Lampyris, and Olivier has justly formed it into a separate genus. These insects are found in the most southern parts of Europe, in Bombay, in Egypt, and the East Indies.

Helæus is peculiar to the Australian islands, and Nilio to South America. Their habits are unknown.

THE THIRD FAMILY

OF THE

HETEROMEROUS COLEOPTERA.

THE STENELYTRA

DOES not differ from the preceding but in the antennæ, which are neither grained nor perfoliate, and the extremity of which, in the greater number, is not thickened. The body is most frequently oblong, arched above, with the feet elongated, as well as in many other insects. The males, with the exception of the antennæ, and of size, resemble the females. Linnæus has referred the greater number of them to his genus *Tenebrio*. He has disposed the others in those of *Necydalis*, *Chrysomela*, *Cerambyx*, and *Cantharis*. In the first edition of this work we had united these heteromera into a single genus, that of HELOPS. But anatomy, both internal and external, indicated to us the necessity of dividing this family into five tribes, attached to as many genera, namely, the Helops of Fabricius, his Cistela, his Dircœa, the Œdemera and Mycterus of Olivier. We know from M. Dufour, that respecting the biliary vessels, (the insertion of which is conical, or at least that of the posterior ones,) this insertion is not effectuated in the last two genera as in the first, and in the other preceding heteromera, by a common trunk, but by three conduits, one of which is simple, the second bifid, and the third with three branches. The œdemeræ present some salivary vessels. Their head is more or less narrowed, and prolonged anteriorly in the form of a muzzle, and the penultimate articulation of the tarsi is always bilobate, characters which seem to approximate these insects to the rhyncophorous coleoptera. In the relation of the digestive canal, and many other considerations, the helops and cistelæ border on tenebrio, but the cistelæ have the chylific ventricle smooth, the mandibles entire, and generally live on flowers and leaves, which peculiarities distinguish them from helops. Most of the Dirceæ possess the faculty of jumping, and the penultimate articulation of their tarsi, or at least of some of them, is bifid; some of them live in mushrooms, others in old wood. These insects are connected on one side with the helops, and on the other with the œdemera, and still more with nothus, a sub-genus of the same tribe. Such are the principles by which we have been directed in the sub-division of this family.

Some have the antennæ approximating to the eyes, and the head not prolonged in the manner of a proboscis, but, at the most, terminated by a very short muzzle. They will compose our first four tribes.

Those of the first, or the Helopii, have the antennæ covered at their insertion, by the edges of the head, almost filiform, or a little thicker towards their extremity, generally composed of articulations almost cylindrical, attenuated towards their base, the last but one of which is often a little shorter, in the form of a reversed cone, and the terminal one is usually almost ovoid; the third is always elongated. The extremity of the mandibles is bifid, the last articulation of the maxillary palpi is larger, in the form of a reversed triangle or hatchet. The eyes are oblong, kidney-formed, or emarginated. None of the feet are adapted for jumping. The penult articulation, or at least of the last, is always entire, or not deeply bilobate. The hooks of the end are simple, or without fissures or denticulations; the body is most frequently arched above, and always of a solid consistence.

The larvæ which are known to us are filiform, smooth and shining, with very short feet, like those of tenebrio. They are found in old wood. It is also under the old barks of trees that the perfect insect is found.

This tribe corresponds in a great measure to the genus

HELOPS, of Fabricius.

Some have the body nearly elliptical, much arched, or very convex above, with the antennæ of the length, at most, of the corslet, compressed and dilated like the teeth of a saw towards the extremity, the corslet transversal, flat above, either trapezoid and enlarged posteriorly, or nearly square, and the elytra frequently ending in a point or tooth. The posterior end of the presternum has a slight pointed projection, which is received in a fork-formed emargination of the mesosternum.

In these the mentum is large, and conceals the origin of the jaws. The middle of the hind extremity of the corslet is advanced on the side of the scutellum like an angle. Such are EPITRAGUS, Lat.

In the others the mentum does not cover the base of the jaws, and the hind edge of the corslet is straight, or a little dilated behind.

CNODALON, Lat.—In which from the fifth articulation the antennæ are much compressed, and much toothed like a saw. The head is obviously narrower than the corslet.

CAMPSIA, Lepel. and Serv. *Camaria* of the same.—The antennæ slightly serrated from the sixth articulation. The head is as wide as the hinder edge of the corslet. The body is in proportion larger, less convex, with the corslet wider behind. Said to have but ten articulations in the antennæ, but we have always counted eleven.

In all other helops the mesosternum has no observable emargination, and the hind end of the presternum is not elongated into a point.

Here the body is sometimes ovoid or oval, sometimes more oblong, but narrowed at both ends. It is never cylindrical or linear, or very flat. Some sub-genera have been formed of certain helopes approximating to the first by their swollen body, as if gibbous behind.

Those whose body is nearly ovoid or short, with the corslet transversal, flat, or merely bent, compose the sub-genera following :—

SPHENISCUS, Kirb.—Which might be taken at first for erotyles, and which have, like the last, the last articulation of the antennæ dilated or serrated, and the corslet flat. (*S. erotyloides*, Kirb. Lin. iv. xii.)

ACANTHOPUS, Meg. Dej.—Shorter and more round than speniscus, with the antennæ simple, terminating in a larger and ovoid articulation. The anterior thighs enlarged and indented, at least in one sex; the legs nearly linear, with small spurs, or none; the anterior arched.

AMARYGMUS, Dalm. *Cnodalon*, *Helops*, *Chrysomela*, Fab —Near the last, but the antennæ simple, filiform; anterior thighs not enlarged or indented. All the legs straight, and terminated in a spur.

Those in which the corslet is enlarged underneath, ovoid and truncated at both ends, narrower in all its length than the abdomen, with the antennæ simple, thickened toward the end; all the legs straight, long, and bent or arched, are SPHENISCUS, of Kirby. Lin. Trans. xxi.

The same naturalist includes under the general denomination, ADELIUM, *Calosoma*, Fab. Lin. Trans. xxi.—Certain helopes of an oval form, with the corslet wider than it is long, nearly obicular, emarginated in front, truncated at the other end, dilated and arched laterally; the antennæ sub-filiform, with most of the articulations nearly in the form of a reversed cone.

The species with an ovalo-oblong body, insensibly arched and convex, or nearly straight above, with the antennæ simple, whether filiform or a little thicker toward the end, especially in the females, and the corslet nearly square or heart-shaped, elongated and truncated behind, form two other sub-genera.

HELOPS, Fab., properly so called, with most of the articulations of the antennæ formed nearly like a reversed cone, or cylindrical and narrow at their base. The corslet is transversal, scarcely so long as wide, either square or trapezoid, or heart-shaped, narrowed suddenly behind, terminated by pointed angles, and always applied exactly against the base of the elytra.

LÆNA, Mej. Dej. *Helops*, Fab. *Scaurus*, Sterm.—With the antennæ composed generally, at least in the females, of short top-shaped articulations; the last is thicker than the preceding, and ovoid. The corslet is formed like a truncated heart, elevated or convex above, separated from the abdomen by a piece with obtuse or rounded angles. The thighs, especially the anterior, are swollen.

The last of the helopes have the body elongated, narrow, nearly of the same size throughout, either thick and sub-cylindrical, or very depressed. The corslet is nearly square, or in form of a truncated heart.

Those with the body thick, sub-cylindrical or linear, with the corslet nearly square, not narrowed behind, form two sub-genera, STENOTRACHELUS, *Dryops*, Payk., and STRONGYLIUM, Kirb. *Helops*, Fab.

The former of these have the head elongated, narrowed behind, nearly like a neck, the antennæ terminated suddenly by three shorter and thicker articulations; the third is much larger than the following.

In the latter the head is not so elongated or narrowed behind, and the last articulation of the antennæ, a little more dilated, does not differ suddenly from the preceding; the third is only a little longer than the following.

Those whose body is flat, with the corslet narrowed behind, nearly in the form of a truncated heart, compose the last sub-genus, that of PYTHO, Lat. Fab.

The antennæ scarcely thicken at all as they proceed, but are nearly filiform, with the last articulation sub-conical; the third is scarcely longer than the preceding and the following.

Some Brazilian species approximate to pytho, but the second articulation is evidently shorter than the third, and the angles of the corslet are acute, as in this sub-genus.

The second tribe, CISTELIDES, is very like the last, but the insertion of the antennæ is not covered; the mandibles terminate in an entire point, or without emargination. The hooks of the tarsi are indented below, like a comb. Many of these live on flowers. The digestive canal is shorter than the helops, and the chylific ventricle has no papilla. This tribe forms the genus

CISTELA, Fab.

Some have all the articulations of the tarsi entire, the last of the maxillary palpi is merely a little larger, like a reversed cone, or triangular, and in them the corslet is thick, narrower than the abdomen, sub-orbicular, or nearly heart-shaped. The antennæ are thicker toward the end. The thighs are like a knob.

LYSTRONICHUS, Lat.—In these the corslet is depressed, trapezoid, as wide as the abdomen at the posterior edge, or scarcely more narrow. The antennæ are filiform, or slightly thicker at the end.

CISTELA, Fab.—These have the head advanced like a muzzle, the labrum scarcely so wide as long, most of the articulations of the antennæ either like a reversed cone, or triangular, dilated even and serrated; the last is always oblong. The body is ovoid or oval.

MYCETOCHARES, Lat. *Mycetophila*, Gyll. Dej. *Cistela*, Fab.—In these the head is not advanced, the labrum is short, transversal and linear; most of the articulations of the antennæ are short, nearly top-shaped, but the last is ovoid. The body, especially in the males, is narrow and elongated. The jaws and the labrum are soft.

Others have the punultimate articulation of the tarsi bilobed, and the last of the maxillary palpi much dilated and hatchet-formed. The body is generally more oblong.

The third tribe, SERROPALPIDES, is remarkable, as its name imports, by the large and inclined maxillary palpi, which are often serrated. The antennæ are inserted in an emargination of the eyes, naked, as in the previous tribe, and in general short and filiform. The mandibles are emarginated or bifid at the end, and the hooks of the tarsi are simple. The body is sub-cylindrical, in some oval, in others with the head inclined, and the corslet trapezoid; the anterior extremity of the head is not advanced, and the posterior thighs are not swollen, characters which distinguish these from many heteromera of the following tribe. The penultimate articulation of the tarsi, or the first four of them at least, is in general bilobed, and in those in which it is entire the hind legs at least are fitted for leaping; they are long, compressed, with the tarsi slender, sub-setacious, and with the first articulation elongated; the fore legs are often short and dilated. This tribe has its type in the genus

DIRCÆA, of Fab.

A few have the antennæ club-shaped. Such as ORCHESIA, Lat. *Dircæa*, Fab.

The last articulation of the maxillary palpi is hatchet-formed; the feet are formed for leaping; the penultimate articulation of the four anterior tarsi is bifid.

The others have the antennæ filiformed. These have the feet fitted for leaping, the body oval or ovoid, the antennæ always short, sub-cylindrical, the maxillary palpi simply a little thicker at the end, but by no means terminated by a hatchet-formed articulation, and all those of the tarsi are entire.

Eustrophus, Illig. *Mycetophagus*, Fab.—Their body is ovoid, with the corslet broad, emarginated in front, and the posterior angles elongated. The antennæ short, the four hind legs moderately elongated, and terminated by two long spurs.

Hallomenus, Payk. *Dircæa*, Fab.—Have the body more elongated, oval, the antennæ larger than the corslet, and the hind legs long and slender, with two short spines at the end.

These in general have the body straight and elongated, the maxillary palpi terminated by a hatchet-formed articulation, and the penultimate articulation of the tarsi, of the first four of them at least, bilobed.

Sometimes the antennæ are thick and composed of short articulations, like a reversed cone or a top.

In some, such as the two following sub-genera, the body is oval, with the corslet transversal or isometrical, spreading from front to rear.

Dircœa, Fab. *Xylita*, Payk.—The maxillary palpi are not serrated, and the last articulation is more advanced on the inner side than in the foregoing. The corslet is depressed insensibly on the sides. The scutellum is very small.

Melandrya, Fab.—The maxillary palpi are evidently serrated, the internal extremity of the second or third articulation being elongated to a point, and even with the fourth or last. The corslet is suddenly laterally depressed toward the outer angles, with the hind edge sinuous. The scutellum is of the ordinary size.

In the following sub-genus the body is narrow, almost linear. The corslet forms a long square, narrowed behind.

HYPULUS, Payk. *Dircæa*, Fab.—The antennæ are longer than in the last, slightly perfoliated, with the articulations more separated; the last three of the maxillary palpi form, united, an oval club.

Sometimes the antennæ are slender, composed of elongated sub-cylindrical articulations. The body is long and straight, with the abdomen elongated.

SERROPALPUS, Hellw. Payk. *Dircæa*, Fab.—The body is firm in substance, with the maxillary palpi strongly serrated, the corslet as long at least as it is wide, the four posterior tarsi long; all the articulations of the last two are entire, or without apparent divisions.

CONOPALPUS, Gyll.—The body is soft, with the maxillary palpi slightly serrated, the corslet transversal, and the tarsi moderately elongated; the penultimate articulation is bilobed to all.

The fourth tribe, ŒDEMERITES, though connected with the foregoing in some particulars, differ from them in the aggregate. The body is elongated, narrow, almost linear, with the head and corslet rather narrower than the abdomen; the antennæ are longer than those parts, serrated in some (*calopes*), filiform or setaceous, and composed of sub-cylindrical long articulations in others; the anterior extremity of the head is more or less elongated, like a little muzzle, slightly narrowed behind, with the eyes in proportion more elevated than in the preceding; the corslet is at least as long as it is wide, nearly square or sub-cylindrical, or slightly narrowed behind. The elytra are linear, or narrowed behind like an oval, and often flexible. These insects have some relations to *telephorus* and *zonitis*.

They will be included in one genus,

ŒDEMERA, Oliv.

Some, whose antennæ are always short, simple and inserted in an emargination of the eyes, whose posterior thighs are enlarged, at least in one sex, have the corslet as wide as the base of the abdomen, wider than the head, and the hooks of the tarsi bifid.

NOTHUS, Ziegl. Oliv. *Osphya*, Illig. *Dryops*, Schœnh.—The maxillary palpi are terminated by a large articulation in form of an elongated hatchet. The hind legs are in one sex very thick, with a stray tooth, and two small spurs underneath, near the end of their legs. The head is not elongated in front.

Perhaps we should place here, in a natural succession, *Rhæbus* of Fisch. (See the family Rhyncophorus.)

Others, whose antennæ are always longer than the head and corslet, whose feet are in general of the same thickness, have the corslet narrower than the base of the abdomen, a little more slender behind, and the hooks of the tarsi entire.

CALOPUS, Fab. *Cerambyx*, De G. Oliv. Col. iv.—Whose fore feet, in both sexes, are of the same thickness as the rest, or very little different, with the antennæ inserted in an emargination of the eyes, serrated, with the second articulation much shorter than the following, like a knot, and transversal.

SPAREDRUS, Megerl. Dej. *Pedilus?* Fisch.—Like calopus as to the feet and insertion of the antennæ, but with these last organs simple, with the second articulation like a reversed cone, as in the following, and one half at least of its length.

DYTILUS, Fisch., *Helops*, *Dryops*, *Necydalis*, Fab., *Œdemera*, Oliv. Feet equal in both sexes, but the antennæ filiform, and inserted before the eyes. The elytra not narrowed behind.

Œdemera, Oliv., *Necydalis, Dryops*, Fab. The anterior thighs much enlarged in one sex, the antennæ commonly long, and more downy toward the end, the elytra suddenly narrowed toward the end.

The fifth and last tribe of Stenelytra, Rhynchostoma, is allied in some particulars to the last, and others to the family Rhynchophorus. The head is evidently elongated in front, like a muzzle or flat trunk, having the antennæ, which are always entire, or without emargination, at its base, and in front of the eyes. These form a single genus,

Mycterus.

Sometimes the antennæ are filiform, and the muzzle is not enlarged at the end; the corslet is narrowed in front, like a truncated cone or trapezium, the ligula is emarginated, and the penultimate articulation of the tarsi bilobed.

Stenostoma, Lat. Charpent., *Leptura*, Fab., have the body narrow, the corslet like a truncated, elongated cone, the elytra flexible, narrow, elongated, and pointed; the antennæ composed of long, cylindrical articulations.

Mycterus, Clairv. Oliv., *Bruchus, Rhinomacer*, Fab., *Mylabris*, Schœff. The body ovoid, solid, and covered with a silky down, the corslet trapeziform; the abdomen squared, long, and rounded behind. The articulations of the antennæ for the most part like a reversed cone, twelve apparently in number, the last running suddenly to a point; last articulation of the maxillary palpi like a reversed triangle.

Sometimes the antennæ terminate in an elongated club, the muzzle is very flat, with an angle on each side, the corslet like a truncated heart, narrowed behind, the ligula entire, as are also all the articulations of the tarsi. They form the sub-genus Rhinosimus, Lat. Oliv., *Curculio*, Lin. De G., *Anthribus*, Fab.

SUPPLEMENT

ON THE

STENELYTR

Of the genus which is the type of the first tribe of this family, little can be said, and nothing at all of its subdivisions. The Helopes are found in spring and summer under the barks of dead trees, or in the clefts and fissures of living ones. Their mode of living is absolutely unknown. The larva of one species is frequently found in the tan formed by insects at the foot of trees. Its body is greatly elongated, cylindrical, composed of twelve articulations; the last of which is terminated in two small and raised points, between which is placed the arms. The first three articulations have each one pair of very short feet, formed of many pieces, and terminated by a very short hook. The head is as broad as the body, and provided above with a clypeaceous piece, which covers the mouth. On each side is seen a small antenna directed forwards. The mouth is furnished with strong jaws; the eyes are not apparent; the body of these larvæ is absolutely smooth, and often of a brilliant polish. They serve for food to the nightingales and warblers.

We are enabled, however, through the kindness of Mr. Children, to present figures of several new species proper to the subdivisions of Helops, viz.

A new species of *Cnodalon* of a copper-green colour, the thorax with two punctures, the elytra with a large tubercle, the anterior summit with a spine, the legs greenish black. It inhabits Brazil. We have given it the specific name of *nodosum.*

A new Adelium, with the specific addition *cupreum,* which is of a bright bronze copper-colour, with the margins of the thorax yellowish, the antennæ and legs testaceous. It inhabits New Holland.

A new species of *Helops,* which may be called *aterrimus.* It is hirsute greenish black, with the head and thorax and elytra punctated, the latter irregularly but strongly; the antennæ and legs are black. This is also from New Holland.

We would insert a new sub-genus under the name of STILPONOTUS, to follow that of Helops. Its characters are, the antennæ nearly as long as the head, and thorax much serrated; the palpi with the last joints securiform; thorax square, and the body in the form of an elongated cone.

The species *euripiformis* is fulvous red, with the head rather darker and the elytra slightly striated. It is from Brazil.

Geoffroy has designated, under the name of CISTELA, a genus of insects of this order, which almost all other entomologists have termed *Byrrhus,* with Linnæus ; but Fabricius has transferred this denomination to another genus, namely, the one of which we are now treating. This genus has considerable affinity with helops. The antennæ and palpi vary a little according to the species. There do not appear to be any very precise limits between this genus and that of *allecula,* which has been dismembered from it.

The larvæ of these insects are as yet unknown.

We might here enter into a lengthened examination of the mode in which authors have arranged those insects comprehended in the text under the designation of MELANDRYA; but as these observations would apply exclusively to the physical character of the insects which we would confine as much

Pl. 80

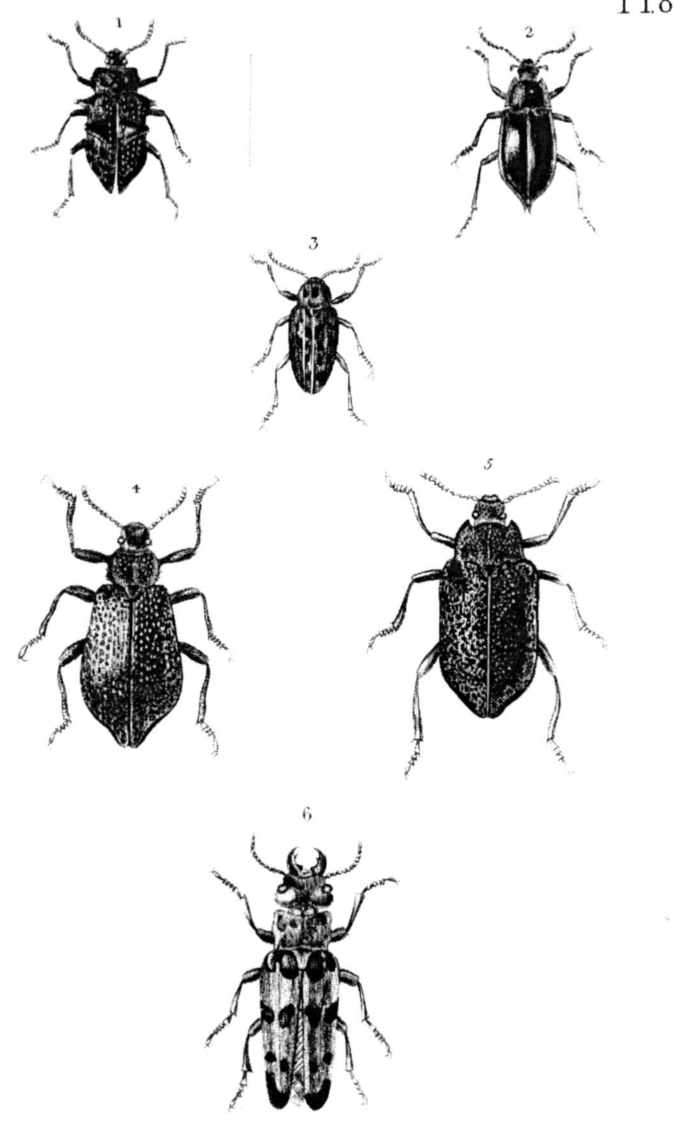

1. Cnodaltum *nodosum*. *G. R. Gray*. 2. Adelium *cupreum*. *G. R. Gray*. 3. Helops? *nigromaculatus*. *G. R. Gray*. 4. Metallonotus *denticollis*. *G. R. Gray*. 5. Helops? *rugosus*. *G. R. Gray*. 6. Horia *maculata*. *Fabr.*

London, Published by Whittaker & C.° Ave Maria Lane. 1831.

Pl 74

1. Cnodalon *nodosum.* 2. Adelium *cupreum.* 3. Diaperis *nigro punctatus.*
4. Lagria *gibbosus.* 5. Helops *aterrimus.* 6. Horia *maculata.*

London, Published by Whittaker & C°. Ave Maria Lane 1831.

as possible to the text of the author, we shall, in this instance, as in very many others, under all circumstances, however reluctantly, prefer brevity to detail.

Some species of the Serrapalpi of Illiger have been referred successively, by subsequent systematists, to the genera, *Elater*, *ptilinus*, and *tenebrio*. To put an end, if possible, to these vaccillations, it was necessary to fix, in a precise and rigorous manner, the character of the genera presented by this division of heteromera: this has been done by M. Latreille.

The Serropalpus, which Illiger named *Fusculus*, forms a particular genus, (*Seraptia*,) which is placed by M. Latreille in the family of the Trachelides. Its head is disengaged, the corslet is almost semicircular, and the terminal articulation of the labial palpi is very large, and semi-lunar. The *Dircœa ruficollis* of Fabricius presents the same distinctive characters; but yet, it may be isolated generically, in consequence of its antennæ, which have but ten distinct articulations, instead of eleven; and for the most part elongated and cylindrical. Another species of the same genus is to be found in the neighbourhood of Paris. The *hallomeni* of Helwigg, from which those must be separated, whose antennæ are thicker at their extremity, (*Orchesia*,) have all the articulations of the tarsi entire; while in all the following insects the penultimate articulation of the tarsi is perfect, at least in the four anterior tarsi. M. Latreille reduces the genus Serropalpus of Hillenius to a single species (*Striatus*). Here the four anterior tarsi alone have their penultimate articulation bilobate. The mandibles, which are shorter and thicker than in the other species placed by Illiger in the same section, have no very sensible denticulations on their internal side. The articulations of the antennæ are for the most part long and cylindrical; the middle of the internal side of the second articulation of the maxillary palpi, is advanced in form of a tooth. The inferior angle of the following articulation, and always on the same side, is also prolonged in the same manner; but the tooth is longer, more slender, very sharp, and a little curved. The last articulation, or the fourth, is very large, in the form of a narrow and elongated hatchet, with the internal side hollow in the middle of its length, and membranes upon its edges; so that one would imagine that this articulation was divided longitudinally in two. Its inferior and internal extremity is also prolonged into a point, so that instead of being inserted, the preceding articulation, by the side which forms the base of its triangular section, it holds these by a point, situated almost towards the middle of the exterior side. The superior extremity of the preceding articulation represents a sort of pedicle.

The Melandryæ inhabit wood. The majority remain concealed under the bark of trees.

Of the SERROPALPI we shall give a brief description, which is not to be found in the text of the only species known. This is the *Striatus* of which Fabricius makes a Dircæa (barbata). Its size varies considerably. The largest individuals are nearly eight lines in length. The body is narrow—almost cylindrical, narrowed behind, of a deep brown, silky, very finely rugose, with the elytra terminating in a point, and slightly striated. The antennæ are almost of the length of one half the body, of a clearer or reddish brown, as well as the palpi and the tarsi. Its larva lives in old dry wood, particularly that of the fir-tree. It there excavates cylindrical holes, which penetrate almost even to the sap. There also, or under the bark, is found, in the month of June, the perfect insect. It is also to be met with, at times, in houses.

Authors have also varied much in their methods with respect to the genus ŒDEMERA. These insects have been classed with *Necydales*, *Dryops*, and *Cantharis*. Olivier has, with propriety, formed them into a separate genus, which he has thus named from two Greek words, one signifying *swelled*, and the other *thigh*.

The Œdemeræ are found on flowers, in meadows. They fly with facility. Their larva, and the history of their metamorphoses, are as yet unknown. They constitute a very numerous genus.

The Rhinosimi are very near the Anthribi of Fabricius, or the Macrocephali of Olivier. These insects are very small, and live under the barks of trees, or in wood.

OUR SECOND GENERAL DIVISION

AND

FOURTH FAMILY

OF

HETEROMEROUS COLEOPTERA,

THAT OF

TRACHELIDES,

HAS the head triangular or heart-shaped, attached to a sort of neck or pedicle, and on the side of which it cannot, being as broad or broader in this point than the corslet, re-enter into the interior cavity; the body is in general soft, with the elytra flexible, without striæ, sometimes very short, in others they are slightly inclined. The jaws are never unguiculated. The articulations of the tarsi are often entire, and the hooks bifid.

We shall divide this family into six tribes, forming as many genera.

The first, LAGRIARIÆ, have the body elongated, narrowed in front, with the corslet subcylindrical, or square, or ovoid, and truncated; the antennæ, inserted near an emargination of the eyes, are simple, filiform, or gradually enlarged toward the end; in general, and at least in part, grained, with the last articulation larger than the preceding in the males; the palpi thicker at the end, and the last articulation of the maxillary the largest, like a reversed triangle; the thighs oval and club-formed, the legs elongated, straight, with the two anterior at least arched; the penultimate articulations of the tarsi bilobed, and the hooks of the last without fissure or indentation. This tribe is formed of the genus

LAGRIA, Fab. *Chrysomela*, Lin. *Cantharis*, Geoff. The species whose antennæ, thicker as they advance, are altogether or partially grained, with the last articulation ovoid or ovaliform, whose head is but little advanced in front, elongated and rounded imperceptibly behind, whose corslet is subcylindrical or squared, compose our genus Lagria, properly so called.

That which I have named *Statyra*, is formed of species similar at first sight to Agra, of the family of carnivorous pentamerous coleoptera. Here the antennæ are filiform, composed of subcylindrical articulations, with the last long and pointed. The head is elongated in front, decidedly and suddenly narrowed behind the eyes. The corslet is longitudinal, ovaliform, and truncated at both ends. The end of the elytra forms a tooth or spine.

We refer with doubt to the same tribe our genus *Hemipeplus*. (Famill. Nat. du Regne Anim. p. 398.) This genus does not belong to the tetramera, as I at first thought, but to the heteromera. The penultimate articulation of the tarsi is bilobed.

The second tribe PYROCHROIDES, is allied to the last as in the tarsi, the elongation and anterior narrowing of the body, but it is flat, with the corslet nearly orbicular or trapezoid. The antennæ, in the males at least, are like a comb or a feather; the maxillary palpi are slightly serrated, and terminated by an elongated hatched-formed articulation; the labial are filiform; the abdomen is elongated, entirely covered by the elytra, and rounded at the end. They form the genus PYROCHROA, Geoff. Fab. Dej. *Lampyris*, Lin.

The species whose antennæ are nearly as long as the body in the males, having long barbed threads; whose eyes are large and near each other behind, with the corslet in form of a truncated cone or trapezoid, with the body finally narrower, more elongated, as well as the tarsi, compose the genus Dendroides, Lat. *Pogonocerus*, Fisch.

Those whose antennæ are simply pectinated and shorter, whose eyes are wide apart, and whose corslet is suborbicular and transversal, are the Pyrochroa, properly so called.

The third tribe, Mordellonæ, have no constant character in their details, but it is easy to distinguish them from the other heteromera of the same family by the general formation of the body, which is elevated and arched, with the head low. In their antennæ, &c. they differ among themselves, and approximate other sub-genera, but they differ from all others in their great agility, and in the firmness and solidity of their ligaments. Linnæus made of them his genus

Mordella.

Some have the palpi of nearly the same thickness throughout, the antennæ of the males very much pectinated. The end of the mandibles without emargination. The articulation of the tarsi always entire, and their hooks indented or bifid. The middle of the hinder edge of the corslet is always much elongated behind, and is like the scutellum. The eyes are not emarginated. The larvæ of some of these (*Ripiphori*) live in the nests of certain wasps.

These Ripiphori (Bosc. Fab.) have the wings extended beyond the elytra, which are as long as the abdomen; the hooks of the tarsi are bifid, the antennæ inserted near the inner edge of the eyes, pectinated on both sides in the males, serrated or with a single row of short teeth in the other sex. The lobe terminating the jaws is very long, linear, and projecting, and the ligula, as much elongated, is deeply bifid.

Myodites, Lat. *Ripidius*, Thunb. *Ripiphorus*, Oliv. Fab. &c. have also the wings elongated, but the elytra are very short. The hooks of the tarsi are indented underneath.

The antennæ are inserted on the top of the head, much pectinated in both sexes. The jaws are but little elongated, the ligula is long and entire.

Pelocotoma, Fisch. *Ripiphorus*, Payk. Gyll., like the last, have serrated hooks to the tarsi, but the wings are covered by the elytra. The antennæ, inserted above the eyes, have but a single row of teeth or threads in both sexes. The scutellum is apparent. The jaws do not project. The ligula is emarginated.

In the others the wings are always covered by the elytra, elongated to near the end of the abdomen, and pointed. The posterior edge of the corslet is not at all, or but little lobed. The abdomen of the females terminates like a pointed tail. The eyes are sometimes emarginated. The maxillary palpi terminate in a large hatched-form articulation. The end of the mandibles is emarginated or bifid. The antennæ, even in the males, are at most serrated.

Mordella, Lin. Fab. have the antennæ of the same size throughout, slightly serrated in the males; all the articulations of the tarsi are entire, and their hooks have underneath one or more indentations. The eyes are not emarginated.

Anaspis, Geoff. *Mordella*, Lin. Fab. These differ from the last in the antennæ, which are simple and enlarge toward the end, the emargination of their eyes, and by having the penultimate articulation of the four anterior tarsi bilobed. The hooks are entire, and without apparent dentition.

The fourth tribe, Anthicides, have the antennæ simple or slightly serrated, filiform, or a little thickened toward the end, most of the articulations of which are like a reversed cone, with the exception of the last, and sometimes the penultimate also, which is rather large or oval; the maxillary palpi terminate in a hatchet-formed knob; and the articulations of the tarsi, to the penultimate articulation, bilobed. The body is narrower in front, with the eyes entire, or very slightly emarginated; the corslet is sometimes a reversed ovoid, narrowed and truncated behind, and sometimes divided into two knots, sometimes semicircular. They compose the genus

NOTOXUS of Geoff.

SCRAPTIA, Lat. *Serropalpus*, Illig. These are easily distinguishable from all other insects of this tribe, by having the corslet almost semicircular, the antennæ inserted in a slight emargination of the eyes, filiform, and with subcylindrical articulations. In appearance they resemble Mordella Cistela, &c.

STEROPES, Stev. *Blastanus*, Hoff. The antennæ terminate in three articulations, much larger than the last, and cylindrical.

NOTOXUS, Geoff. Oliv. *Anthicus*, Payk. Fab. The antennæ thicken imperceptibly, and are composed almost altogether of articulations formed like reversed cones. The corslet is in the form of a reversed ovoid, narrowed and truncated behind, or divided into two globular parts.

The last two tribes of the family, and of the section of Heteromera, have some characters in common, such as having the mandibles terminating in a simple point, the palpi filiform, or a little thicker at the end, but never clubbed or hatchet-formed, the abdomen soft, the elytra flexible, vesicant in most of them, all the articulations of the tarsi, except in a few, entire, with the hooks generally bifid.

The HORIALES compose the fifth tribe, and differ from the succeeding in the hooks, which are indented, and each with a serrated appendix. They have the antennæ filiform, at most of the length of the corslet; the labrum is small, the mandibles strong and sailliant, the palpi filiform, the corslet squared. The two hind legs very strong, at least in one sex. This tribe is composed of the genus

HORIA of Fab.

One species of these, the type of my genus *Cissites*, differs from the rest in having the head narrower than the corslet, and the hinder thighs much swollen.

The sixth and last tribe, CANTHARIDIA, is distinguished from the last by the hooks of the tarsi, which are deeply divided, and appear as if double. The head is in general thick, larger, and rounded behind. The corslet is generally narrowed behind, and approaches the shape of a truncated heart. The elytra are often a little inclined laterally, forming a compressed rounded ridge. This tribe is formed of the genus

MELOE of Lin.,

Which has been divided into many other genera, which may be subdivided by the varieties of form of the antennæ.

In one of these subdivisions there are only nine articulations in the antennæ in both sexes; the last of these articulations is very large, in form of an ovoid head; those of the males, as well as the maxillary palpi, are very irregular. The body is depressed. Such are the insects of the subgenus *Cerocoma*, Geoff. Schœff. Fab.

In all the others the palpi are the same, and are regular in both sexes. The antennæ have commonly eleven articulations, and when there are one or two less, they terminate regularly in a knob. The body is thick, with the wing-cases a little inclined.

In these, the antennæ, always regular and grained in both sexes, appear sometimes to be composed of only nine or ten articulations, and never longer than one half the body;

sometimes they terminate in an arched knob, or are evidently thicker toward the end, and sometimes they form, starting from the second articulation, a short cylindrical, subspindle-shaped stalk. These compose the genus *Mylabrum* of Fab.

Those in which the last two or three articulations of the antennæ unite at least in the female, and form rather a sudden, thick, and ovoid knob, the end of which does not pass beyond the corslet, and in which the total number of the distinct articulations of these organs is not more than nine or ten, form the sub-genus. HYCLEUS. Lat. *Dices*, Dej. *Mylabris*, *Oliv*.

Those in which the same organs are in proportion longer, have in both sexes eleven very distinct and separate articulations, progressively thickening or terminating gradually in an elongated knob, with the eleventh, or last, articulation separated from the others of a larger and ovoid form. MYLABRIS. Fab. Oliv. Lat.

The length of the antennæ respectively vary a little, and their modifications influence the form of the articulations, especially of the intermediate.

ŒNAS, Oliv. *Meloe*, Lin. *Lytta*, Fab. These seem to constitute the passage from Mylabris to the following heteromera. Their antennæ, scarcely larger than the corslet, are nearly of the same thickness throughout. The first articulation forms a knob, and is in the shape of a reversed cone; immediately after the following, which is very short, the stem forms an elbow in form of a cylindrical body or spindle, composed of short articulations, serrated and transversal, except the last, which is conoid.

The other heteromera of the same tribe have the antennæ always composed of eleven distinct articulations, nearly of the same thickness throughout, or more slender toward the end, and often much larger than the head and corslet. They are irregular in most males.

Meloe, properly so called, Lin. Fab., have the antennæ composed of short round articulations, the intermediate of which are thicker, and sometimes so disposed that these organs present in this point, in many males, an emargination or swelling. The wings are wanting, and the cases are oval or triangular, and crossing each other in a portion of their inner side, cover the abdomen partially, especially in the female, where it is very voluminous.

All the Heteromera, of the following sub-genera, have wings, and the elytra found as usual, cover longitudinally the upper side of the abdomen.

Among these sub-genera we shall first distinguish those in which these cases are not narrowed abruptly, like an awl, towards their posterior extremity, and entirely cover the wings.

Tetraonyx, Lat. *Apalus*, Fab. *Lytta*, Klüg. These have not, like the cantharides and zonites, the jaws elongated, and terminated by a silken thread bent beneath. The penultimate articulation of their tarsi is emarginated, and nearly bilobed, and the corslet forms a transvere square; these insects are, moreover, nearly allied to the cantharides, and belong to the new continent.

Cantharis, Geoff. Oliv. *Meloe*, Lin. *Lytta*, Fab.—These have all the articulations of the tarsi entire, and the corslet nearly ovoid, a little elongated, narrowed before, and truncated behind, which distinguishes them from the preceding sub-genus. The second articulation of the antennæ is much shorter than the following, and the last of the maxillaries is obviously thicker than the preceding.

The head is a little wider than the corslet. These characters remove cantharis from zonitis. The antennæ of the males are sometimes irregular, and even semipectinated.

Zonitis, Fab. *Apalus*, Oliv.—These have the antennæ more slender than in cantharis, especially in the males; the length of their second articulation equals at least the half that of the following. The maxillary palpi are filiform, with the last articulation subcylindrical. The head is a little elongated before, and is as wide as the corslet.

The males of the two following sub-genera possess one character quite peculiar: the lobe terminating the jaws is elongated into a sort of thread, more or less long, silky, and bent. Such are

NEMOGNATHUS, Lat. *Zonitis*, Fab. which have the antennæ filiform, with the second articulation shorter than the fourth; the corslet is square, rounded at the sides.

GNATHIUM, Kirb.—The antennæ are rather thicker at the end, and the second articulation is nearly as long as the fourth. The corslet is in the form of a ball, and narrowed behind.

Finally, the last sub-genus of this tribe, SITARIS, Lat. *Apalus*, Fab. is remarkable for the sudden narrowing of the posterior extremity of the cases, which leaves a portion of the wings exposed. These insects are greatly assimilated to Zonitis, and, like them, live in the larva state, in the nests of certain solitary mason bees. In the Apali, properly so called of Fabricius, the elytra are rather less narrowed, and the inner ends of the articulation of the antennæ are a little advanced or dilated, like small teeth.

SUPPLEMENT

ON THE

TRACHELIDES.

In the first genus of this family, Lagria, the species most known in Europe, *Lagria hirta*, had been placed with the *Chrysomelæ*, by Linnæus, and in the genus *Cantharis*, by Geoffroy. Degeer made it a tenebrio. Fabricius formed it into a peculiar genus, composed at present of fewer species, the *Dasytes*, which this author at first had referred to it, being now separated from it. The body and the elytra of these insects are generally soft or flexible.

The *lagria hirta* more particularly inhabits the woods, and lives upon the leaves of different vegetables. When it is seized, it folds back its feet and antennæ, and feigns death. This insect is about four lines in length, black, hairy, with the elytra yellowish and semi-transparent, and on each of which are discernible, from four to five raised, but not very marked lines. The corslet is almost cylindrical. The male is distinguished from the female by its eyes, being more approximate behind, and by the antennæ, the last articulation of which is as long as the four preceding united. A species much approximating to this is to be found in Spain and Barbary.

Our figure of *Lagria metallonotus*, a new species, has the head, thorax, and legs shining green, the elytra shining golden green, with irregular punctures. It inhabits Sierra Leone.

The insects of the genus Pyrochroa had, at first, been arranged by Linnæus with Lampyris, with which they have some relations of form. It was Geoffroy who separated them, to form a peculiar genus, under the name *Pyrochroa*. Fabricius had placed in this genus Pyrochroa, many insects, since separated from it, and classed under *Lycus*.

The larva is elongated, depressed, terminated by two points, with the head strong, and having a form analogous to that of the perfect insect. It lives under the barks of old trees. The perfect insect is found in highways, at the foot of hedges, in timber-yards, or in woods. The genus is not numerous.

The ancient naturalists designated, under the name of Mordella, the insects which proceeded from larvæ, or little worms, which feed upon the cabbage-stalk.

Linnæus, in the first editions of his *Systema Naturæ*, bestowed this denomination on a genus of coleoptera, in which the antennæ are filiform, with the last articulation globular, and in which the feet are generally proper for the purpose of leaping. The Mordellæ, properly so called, were thus united to the Altisæ, and some other insects of a very different kind. But it was not long before this great naturalist recognized the defects of this arrangement, and his genus Mordella, such as he has presented it, in the second edition of his *Fauna Suecica*, printed in 1761, is perfectly natural, and embraces the entire tribe of the Mordellonæ.

He had even corrected the same error in the tenth edition of his *Systema Naturæ*. Some species of this genus have the antennæ flabellate, others have them serrated; others, in fine, have them simple, and even a little thicker towards the end.

The Ripiphori, which M. Bosc had established in his fine collection, before Fabricius, the myodites, and pelecotoma, form in the tribe of the Mordellonæ, a very natural division, and remarkable for the disposition of the articulations of their antennæ, which, beginning from the second or third each, throw out on the interior side one or two branches, more or less long, the assemblage of which composes a plume, or fan. The manner in which the antennæ are inserted, the consideration of the form of the claws of the tarsi, and other parts, furnish characters which mark these three generic sections extremely well.

The domains of the rhipiphori extend from America to the East Indies, but these insects are generally rare, the males particularly. It appears, from many observations, that the *ripiphorus paradoxus*, which is found in autumn, or towards the end of the summer, lives, to the moment of its final metamorphosis, in the nests of the common, or other wasps. We may presume that the larvæ of other species are also parasitical.

The Myodites were at first termed *Myodes* by M. Latreille, who changed the latter appellation in consequence of its having been applied by M. de Lamarck, to a genus of diptera.

Olivier and Fabricius place with the ripiphori, an insect which, by the form of its antennæ, that of the palpi, and the general physiognomy, presents, in fact, the characters of this genus, but which is nevertheless removed from the other species by the shortness of its elytra, so that the wings are almost entirely uncovered. From this came the origin of its specific denomination *Subdipterus*. Dorthes, who first discovered this species, and gave its figure and description, was of opinion that it ought to form a new genus alongside of the *necydalis* of Linnæus. Our English naturalists named this genus *Dorthesia*, a denomination afterwards given by M. Bosc to a new genus of hemiptera, the type of which is the *coccus caraccias*, discovered and described also by Dorthes. It appears that the first of these genera corresponded to that which is now called *ripiphorus*. But the species described by Dorthes, as the type of this division, presents many essential differences, which induced M. Latreille to detach it from ripiphorus.

The Notoxi are very small, and very agile coleoptera, which are to be met with either on plants, or on the earth, and whose larvæ are unknown.

The genus notoxus of Fabricius, comprehended at first the same insects, and the coleoptera of the genus *Opilus* of the text, tribe *Clerii.* M. Paykull separated the first under the generic denomination of *Anthicus,* a change of which Fabricius has approved in his System of the *Eleutherata*. But he has united to *anthicus* some very different coleoptera, such as *pselaphus* and *Scydmenus.* Many species of notoxi are apterous.

The genus Horia, formed by Fabricius in his *Mantissa Insectorum,* is composed of two species, placed by this author in his earlier works with *Lymexylon.*

We know nothing of the manner of living of the horiæ, which are all foreign to Europe; but in the absence of positive knowledge, analogy, especially in natural history, is a tolerably sure means to guide us in our conjectures, and, according to the relations which exist in *Horia Cantharis, Mylabris,* and *Meloe,* we may well believe that the mode of life of all these insects should be the same, or, at least, not very different.

The genus named Meloe by Linnæus, comprehended the soft heteromera, which are known to possess vesicating properties, such as the *cantharides,* the *mylabres,* and some other insects of the same section forming the genus *notoxus* of Olivier, or that of *anthicus* of Paykull and Fabricius. He divides Meloe into two sections, the *apterous* and the *winged.* The first is that which since the time of Geoffroy, has exclusively formed the genus Meloe, or the *proscarabœi* of that author. It is, in fact, under the last-mentioned generic name, that these insects are designated by some ancient naturalists, and principally by Mauffet, who has figured numbers of them. According to him, they are the meloe of Paracelsus. The Germans name them *May-worms*, because the most common species usually appears at that season of the year. Some authors have called them *unctuous scarabæi*, because they send forth from the articulations of each knee of their legs, when they are seized, a yellowish, viscous fluid, similar to oil, and which, according to Frisch, has the colour of violet, but which Degeer says is devoid of smell.

Of all the genera of the section of heteromera, Meloe is one of the most distinct, as may be seen by the characters in the text, which we will not repeat here. Their body is most frequently of a deep black, bordering in many on bluish, and violet, and very much punctuated. These animals are heavy, and are found upon the ground, in the fields, tilled lands, or on the edges of highways. They appear to prefer sandy or calcareous places, exposed to the sun, and feed on various plants, which are raised a little above their natal soil. Some appear in the spring, and others in the autumn. In some parts of Spain, they are mingled with the cantharides, for the same purposes as these latter insects. They were formerly regarded as a specific against madness, and farriers employed, in some cases, oil in which these insects had been macerated. The earlier naturalists have enlarged very much on their pretended medical properties, as may be seen in Mauffet. M. Latreille thinks it probable that these animals were those which the ancients named *buprestis*, and which they regarded as a mortal poison for oxen, and even for man himself. He has largely explained the reasons of this opinion in a memoir, forming part of the Annals of the Museum of Natural History. Many males have the fifth, sixth, and seventh articulations of their antennæ broader than the rest, which gives to these organs an irregular figure. Seen in profile, their middle even presents, in consequence of the disposition of these articulations, a strong emargination or species of crescent. According to an observation of Mr. Sowerby, the males, in coupling, seize by means of their antennæ those of their females.

The individuals of this last sex are remarkable for the prodigious bulk, and tension of the abdomen. The number of eggs with which it is filled is very considerable, since one female, that of the *proscarabæus*, which Goëdart preserved and fed with the leaves of the anemone, or with those of the ranunculus, laid, from the twelfth of May to the twelfth of June following, two thousand nine hundred and twelve eggs, and he estimated at almost as many those proceeding from the same laying, but which he had been unable to count. These eggs were laid at two separate times, the insect burying each time in the ground, in a hole which it had made with the posterior extremity of its belly, and depositing its eggs there in a packet. These eggs are yellow, and resemble, according to this writer, small grains of sand pressed together. The larvæ which he obtained from them, and which he vainly endeavoured to bring up, though he presented them with a great variety of substances, both animal and vegetable, have the body long, cylindrical, sprinkled with hairs, composed of eleven rings almost equal, and with an oval head, provided with two eyes, and two antennæ of tolerable length. They have six feet, which appear large in comparison with the extent of the body. Its posterior extremity is terminated by two long appendages in the form of silky hairs. Those are not expressed in the figure which Frisch has given of these same larvæ. Geoffroy tells us that the larvæ very much resemble the perfect animal, that they are of the same colour, thick, heavy, soft, and no where scaly but on the head. According to him, they are found buried in the earth, where they undergo their metamorphosis.

Degeer, who comprehends under the generic name of *Cantharis* the insects thus designated in the text, together with *Mylabris* and *Meloe*, is of all naturalists, the one who has described most in detail the eggs and the larva of *Meloe proscarabæus*. A female, which he had shut up in a sand-box, half filled with earth, laid very deep in the earth, a large heap of oblong eggs, of a fine clear orange colour, applied one over the other, without being glued together, and forming a packet about the bigness of a nut. They are very small, and their number, consequently, very considerable. Seen with the microscope, they have the figure of a cylinder, rounded at the two ends, and their pellicle is coriaceous, flexible, but very tense. The larvæ were born a month after the laying, which took place the 18th of May.

These larvæ have six feet. They are of an ochreous yellow, with two black eyes. The head is oval, or little flatted, provided with two antennæ, composed of three articulations, and terminated by a hair ; with two very long mandibles, curved, and very much pointed, and four antennulæ. The body is composed of twelve rings. The feet are attached to the first three, which are much larger than the following. The last is terminated by four very fine threads, two of which are longer than the others. The feet are divided into three parts, and terminated by two very pointed hooks, between which is a broad, flatted piece, in the form of a spear-head. It is by means of these hooks that the larvæ fasten themselves to the objects upon which they walk. They also assist themselves in walking, and attaching themselves, by means of a nipple, which they have at the posterior part of the body.

We find, sometimes, in the body of a dipterous insect, the

eristalis intricarius of Fabricius, and on those of some species of *Apiariæ*, especially on the humble bees, a small insect almost similar to those larvæ, figured, and well described by Mr. Kirby, in his History of the Bees of England—(*pediculus melittæ*). Degeer having remarked this analogy, put along with the larvæ of this meloë two common flies, and another species of the same family. In less than half an hour, a very great number of these larvæ had found the means of getting in the body of one of the flies, and fixing themselves to its breast, and a part of the belly. The fly made some vain efforts to get rid of them. It perished on the second or third day, and the larvæ abandoned its carcass; but Degeer having furnished them for many days with living flies, they fastened themselves to them. As soon as a fly passed near them, there were always some which seized it immediately by the foot or wing, and from that moment they never let go, and very soon began to attack the body. Degeer, seeing, nevertheless, that they did not attain any visible growth, grew tired of supplying them with flies, and they all died, one after the other. "Who would have supposed," exclaims this writer, "that the young or larvæ of cantharides of this species would have been found on the body of flies? No one would have ever thought of looking for them in such a situation."

This fact is, in truth, very marvellous; and when we reflect on the consequences which result from it, we might doubt of its reality, and suspect that this observer had been mistaken. M. Latreille, in his Natural History of Crustacea, has combatted this opinion: and Mr. Kirby, before him, had testified the same incredulity. In this respect, nature, always wise, must have provided for the wants of the posterity of the meloë; and if their larvæ be destined to live parasitically on some species of diptera and hymenoptera, how shall they find, without trouble, at the moment in which they are disclosed, and in the place where they are assembled in such numbers, a sufficient quantity of these insects to provide them with food and dwelling? How could they be able, contenting themselves with the small degree of subsistence which they can obtain by a sort of suction, to grow in such a manner as to acquire a volume at least twenty times greater than that which they possess at their birth? How could they undergo the different moultings, which require a perfect state of inaction? Is it possible, moreover, that they can acquire their full growth in the short space of time for which these same insects exist? Their carcases do not serve them for aliment, as Degeer himself has observed; they quit them, as do the *lice*, the *ricini*, the *mites*, and other little parasite animals, when those which they gnaw, or suck, have ceased to exist.

M. Latreille has often found those pretended parasite animals which Mr. Kirby has named *pediculus melittæ*: none of them ever appeared to him to exceed the others in size, at least in any sensible degree. He carefully studied their organization. They presented all the characters of a perfect insect and one forming a new genus, approximating to that of *ricinus*. The mouth composed of mandibles, of jaws, and very small, but projecting palpi, is situated under the muzzle, or advancement of the head. This character is exclusively peculiar to the insects of the parasite order. The manner in which the feet of *pediculus melittæ* terminate presents us with a new analogy to the ricini, and indicates that these animals have identical habits. The figure which Goedart has given of the meloë, and that of the larva of *Cantharis officinalis*, which may be seen in the Journal of Natural History, entitled in German *Naturforscher*, are remote in some points, from the figure which Degeer has published of the insect, which he takes for the larva of *meloë proscarabæus*. If it be true, as he has said, that the larva of *cantharis officinalis* lives on roots, it is to be presumed that the larvæ of meloë, mylabris, and cerocoma, feed in the same manner. M. Latreille has sometimes seen the flowers in a very small field all covered with cerocomæ. There are likewise found in limited spaces, great quantities of cantharides. These facts indicate that these insects had passed the first age of their lives in these localities, and that vegetables much more abundant than animal substances, had supported them in this state, as they still do, when they are developed. This manner of life is common in general to all the heteromera. In truth, the larvæ of the *ripiphori* inhabit the nest of wasps, and those of *Zonitis* have for their domicile the nests of some solitary bees. But, besides that the first may have as aliment the matter which composes the nest, and that the second can appropriate to themselves the provisions which the apiariæ have collected for their young ; these heteromerous insects are not very common in their perfect state, and spread abroad here and there. We may well conceive that the mothers of the parasite insects have less facility of supplying the wants of their family than those of the herbivorous insects. Many species of *mites* living on insects, or other animals, are also found isolatedly, either on the ground or on plants. They afterwards fix themselves, when they have the means of so doing, on the bodies of these animals.

The insects, such as *eristalis intricarius*, the *andrenetæ*, the humble bees, on which the *pediculus melittæ* subsist, make their nests, and have passed their infancy in the earth. It is probable that in these circumstances this parasite animal, inhabiting in all probability the same localities, attaches itself to these insects, or that the female agglutinates upon them some of its eggs. We may presume then that the little animal which Degeer regards as the larva of the meloë proscarabæus, may be a species of the same genus as the *pediculus melittæ*, or perhaps a variety of this insect.

Might not its eggs be found mingled with the earth, which Degeer had put into the box in which he had enclosed the meloë proscarabæus? The female of this larva might herself have deposited the germs of her posterity in those of this very same insect. It is possible that many individuals of this larva might have lived upon the captive meloë, and having abandoned it after its death, might have deceived the inspection of this naturalist; for the eggs of this insect, for want of fecundation, might have been sterile. Such are the solutions which M. Latreille originally proposed concerning this curious subject, but none of which were founded on positive observation, and were consequently unsatisfactory. We have thought it right thus to lay before our readers the original opinions and arguments of M. Latreille on this subject, though, as it will be seen by the text, he has since found reason to change his sentiments. We must confess, however, that to ourselves the subject still appears to be full of doubt and obscurity.

Dr. Leach, in the eleventh volume of the Linnæan Transactions, has given a monograph of the genus meloë. Megerle had already published another; but the Doctor's is far more complete, and accompanied moreover with excellent figures. Dr. Leach has rectified the synonimy of some species, and his labours are so much the more useful, as their distinction required the most attentive and scrupulous examination. Many of these species, though essentially different, appearing on the first glance to constitute but one, it was especially necessary to become acquainted with the individuals of both sexes, a point to which the Doctor has peculiarly attended.

We now come to a genus of insects long celebrated for their extensive use in medicine—we mean the CANTHARIDES. Their mandibles terminating in an entire knob distinguishes them from *œdemera*, in which genus, moreover, the penultimate articulation of the tarsi is bilobate. They are removed from *meloë*, *mylabris* and *cerocoma*, by the form of their antennæ. They have more affinity with *zonitis*, *nemognathus* and *sitaris*, but the four palpi in these three genera are filiform. The *zonitis* have their antennæ long, almost setaceous, with the second articulation more elongated. The elytra are very much narrowed at their extremity in *sitaris*, and the jaws very much elongated in *nemognathus*.

The larvæ of the cantharides have the body soft, of a yellowish white, composed of thirteen rings; the head rounded, a little flatted, and furnished with two short and filiform antennæ; the mouth furnished with two tolerably solid jaws, and four antennulæ. There are six short and scaly feet.

These larvæ live in the earth, and feed upon various roots. When arrived at their full growth they change into the nymph state in the earth, and do not come out of the ground until they have assumed the form of the perfect insect.

The cantharis is one of those insects which have been most anciently and most universally known. Physicians, who were the first natural philosophers, and the first observers of nature, have made mention of the cantharides in the remotest times. But they have only considered them under that relation which was most suitable to their own profession, and as furnishing to medicine one of its most powerful agents. The naturalist, who is less anxious about becoming acquainted with the medicinal virtues of the dead, than with the peculiar habits of the living cantharides, is yet very far from having acquired in this respect certain, extensive, and satisfactory information. The only species which has been deemed to be endowed with useful properties has caused a forgetfulness of all the others which compose the entire genus; and all that we know, in general, respecting these insects, is, that in our European climates they live on plants, devour the leaves of certain trees, shun the cold, appear at the commencement of spring, and disappear at the beginning of autumn. We are therefore unable to do any more than present some general ideas respecting the cantharis which is peculiarly consecrated to the purposes of medicine.

The opinion of Baglivi that the usage of cantharides was first introduced by the Arabs, appears to be without foundation, for it is sufficiently proved that their employment was not unknown to Hippocrates himself. But it must be observed that the cantharides of the ancients, and those of the Chinese, are not the same as those of the Europeans. The Chinese employ a species of *mylabris*, and it appears from what Dioscorides has written on this subject, that the cantharides of the ancients were another species of the same genus, namely, the *M. chicorii.*

"The most efficacious cantharides," says Dioscorides, "are those of many colours, which have yellow transverse bands, with the body elongated, bulky and fat—those of a single colour have no virtue." This description does not at all accord with the character of our cantharis, which is of a fine green colour. But it is perfectly suitable, on the contrary, to the *mylabris chicorii*, very common in the country where Dioscorides resided, and throughout all the East.

It is more than probable that experiments on insects relatively to their utility in medicine and the arts, have been too much neglected in general Their diminutive size has doubtless caused them to be too much despised. It cannot, however, be doubted that there must be a great number of them whose virtues are at least equal to those of the cantharides, and many others which are less acrid and less caustic, might in many cases be taken internally, with less danger and a greater chance of success. We may rest assured that all the species which belong to the genus *cantharis* possess pretty nearly the same virtues as the species which is most generally known, and consequently in all the countries in which they are found the same usage might be made of them. Among the insects taken from other genera, which might furnish caustic and irritating particles, and which might be substituted for the cantharides, to a certain extent, we may range *meloë*, *mylabris*, *carabus*, *tenebrio*, *cicindela*, *scarites*, *coccinella*, &c. The cast skin of the majority of the caterpillars produces a dust, which scattered by the winds, raises pustules on the face with which it comes in contact. The same effect is occasioned by the hair and wool of certain phalenæ when they are touched. Merian found at Surinam some species of the larvæ of lepidoptera, which one could not touch without being suddenly attacked with inflammation.

It is in the course of the month of June that the cantharides assemble together for the purposes of pairing. This is the time which must be seized for collecting them, particularly in the evening, at the setting of the sun, or in the morning, at his rising. These insects are found almost throughout all Europe, but more commonly in the warmer climates. They vary prodigiously in point of size. The cantharides used in medicine are about nine lines in length, and two or three in breadth. They throw themselves upon ash trees, honey-suckles, lilacs, rose trees, poplars, elms, &c. of which they devour the leaves, and when this sort of pasture is wanting they attack the corn fields and meadows, and cause very serious damage. As they appear in troops or swarms, and are preceded by a fetid odour, somewhat resembling that of a *mouse*, it is easy to discover and collect them, using, however, certain precautions, which it is prudent never to neglect. There are two modes of proceeding in the collection of cantharides; the most simple consists in disposing under the tree which is loaded with these insects, one or more cloths, on which they are made to fall by shaking the branches. They are afterwards gathered on a hair sieve, and held over the vapour of vinegar, which causes them to die, or they are collected in a clear linen cloth, which is steeped several times in a vessel filled with vinegar, mixed with water. This is the method which is most generally adopted.

The second method of collecting the cantharides is more embarrassing and expensive than the first. Cloths are spread under the trees, and all around; vinegar is put into a state of evaporation, by causing it to be boiled in earthen pans, placed in chafing dishes. The trees are then shaken to cause the cantharides to fall, the latter are picked up immediately and promptly enclosed for twenty-four hours, in vessels of wood, earth, or glass, made for this express purpose.

When the cantharides are dead the next consideration is the proper mode of drying them. For this purpose they are exposed to the sun, or what is still better, placed in a well-aired garret, on hurdles covered with linen cloth or paper. They are stirred about with a little stick, or with the hands, provided, however, with gloves, for without this precaution the workmen would be exposed to heat of urine, and experience sharp pains round the neck of the bladder, opthalmia, and severe itching. When the cantharides have received a suitable degree of dessiccation, they become so light that fifty of them hardly weigh a dram.

The preservation of the cantharides is easy. They are kept in boxes or barrels, invested internally with paper, and closed. But before this it is necessary that they should be perfectly dry, for otherwise they would contract a most detestable odour, which would render them totally unfit for use.

The insects of the genus Cerocoma exhibit very brilliant colours, and proper to distinguish them. They frequent flowers, on which they are found during a great portion of the summer. They fly with very great agility, but are easily caught when they bury their heads in the calices of flowers to extract the honeyed juice from them. The habits of their larvæ are altogether unknown, but we may presume that they live in the earth, like the larvæ of the cantharides, and feed upon the roots of plants. They often appear in great abundance about the summer solstice.

The insects of the genus MYLABRIS were united to the meloës by Linnæus, and to the cantharides by Degeer. Fabricius separated them, and formed them into a peculiar genus, under the denomination of mylabris, which had already been applied by Geoffroy to a very different set of coleoptera.

The insects of this genus are peculiar to the hot and sandy countries of the Old World; they especially abound in Africa and the Levant. They are found on the flowers and leaves of divers vegetables, and particularly on those whose flowers are composite. When they are seized they fold back their antennæ and feet against the body, like dermestes, lycus, and many other insects of little agility, which seek to deceive their enemies by counterfeiting death. Their larvæ are unknown.

It appears from some passages of Pliny and Dioscorides, that the ancients designated these insects under the name of *cantharides*, for they mention that the best cantharides are those whose elytra are marked with yellow transverse bands; a character perfectly answering to some of these insects which abound in the south of Europe, and in the East.

They are still employed at the present day in Italy for pharmaceutical purposes, and particularly in Naples, in the place of our cantharides, or at all events mixed up with them. The Chinese also make the same use of the *mylabris pustulata* of Olivier, which is found in their country. It is very difficult to establish precise limits between the species of mylabris, because the spots of the elytra vary very considerably.

The *M. Chicorii* has been sometimes found by M. Latreille in the sandy plains, exposed to the sun, in the neighbourhood of Paris, on thistles, but it remains more peculiarly on the plants of endive or succory.

The Zonites are very much approximating to the cantharides. They are found on flowers, in the southern countries of Europe, in Africa, in Asia Minor, in Syria, and in Persia. It appears that their larvæ live like those of sitaris, in the nest of certain *apiariæ*.

The insects of the genus Sitaris, in the larvæ state, live in the nests of some solitary bees. Their habits are otherwise unknown.

THE

THIRD GENERAL SECTION

OF

COLEOPTEROUS INSECTS,

THAT CALLED

TETRAMERA,

INCLUDES exclusively those which have four articulations to all the tarsi.

All these insects feed on vegetable substances. Their larvæ have commonly short feet, but in many the feet are wanting, or are represented by small mammæ. The perfect insects frequent flowers, or the leaves of plants.

I shall divide this section into seven families. The larvæ of the first four or five live most commonly hidden in the interior of vegetables, and are generally without tarsi, or have them very small; many of them gnaw the hard ligneous parts of vegetables. These coleoptera are the largest of the section. The first family,

RHYNCHOPHORA,

Is distinguished by the anterior elongation of the head, which forms a sort of muzzle or trunk. Most of them have the abdomen thick, and the antennæ often bent into a club. The penultimate articulation of their tarsi is almost always bilobed. The posterior thighs are indented in many.

The larvæ have the body oblong, like a little worm, very

soft, white, with a scaly head, and are deprived of feet, or have in their stead mere little nipples. Their nymphæ are inclosed in a cocoon.

Some have the labrum apparent, the anterior elongation of the head short, wide, depressed, muzzle-shaped ; the palpi very visible, filiform, or thicker at the end. They compose the genus

BRUCHUS, of Lin.,

which is sub-divided as follows.

The species whose antennæ are club-shaped, or obviously thicker at the end, whose eyes are without emargination, and who appear to have five articulations to the four anterior tarsi, form the sub-genus RHINOSIMUS, which we have placed from this character with the heteromera, but which in many other particulars approximates to the sub-genus following.

Those with similar antennæ and eyes, but which have only four articulations to all the tarsi, with the penultimate bilobed, enter into the sub-genus ANTHRIBUS, of Geoff. and Fab., to which may be joined the rhinomacera of Oliv.

BRUCHUS, properly so called, or the *mylabris*, Geoff., have their antennæ filiform, often serrated, or like a comb, and the eyes emarginated ; the anus is uncovered, and the posterior feet are commonly very large.

RHÆBUS of Fischer is distinguished from the last by their elytra being flexible, and the hooks of the tarsi bifid.

XYLOPHILUS, of Bonnelli, differs from these in having the palpi terminating in a club.

The others have no apparent labrum, the palpi are very small, scarcely perceptible to the naked eye, and of conical form ; the anterior elongation of the head represents a bill or trunk.

Sometimes the antennæ are straight, inserted on the trunk, and composed of nine or twelve articulations.

Those in which the last three or four articulations are united into a knob, form the genus

ATTELABUS, of Lin., and more particularly that of Fabricius, or that of the *Becmares*, of Geoff.

The proportions of the trunk, the manner in which it terminates, as well as the legs, and the form of the abdomen, have given rise to the establishment of the four following sub-genera, viz., APODA, ATTELABUS, RHYNCHITE, and APION. The first is the most distinct. The head of these insects is narrowed behind, forming a sort of neck, and united with the corslet by a kind of patella. Their muzzle is short, thick, enlarged at the end, a character common to Attelabus, properly so called, but in which the head, as well as in the other two sub-genera, enters the corslet as far as the eyes. Here the muzzle is elongated in form of a proboscis. In the rhynchites it is a little enlarged at the end, and the abdomen is sub-quadrate.

Certain rhynchophora, very analogous to attelabus, but whose body is narrower and more elongated, have been formed into the following genera.

RHINOTIA, Kirby. *Belus*, Schœnh.—Whose antennæ thicken without forming a knob, and whose body is nearly linear.

EURHINUS, Kirby, in which they terminate in an elongated knob, the last articulation in the males being very long.

TUBICENUS, Dej. *Auletes*, Schœnh.—In which they terminate also in a knob, but perfoliated, and with articulations differing but little in length. The abdomen is moreover in a long square, not oval, as that of eurhinus.

Those in which the antennæ are filiform, or in which the last articulation only forms the knob or proboscis, often larger in the males than in the females, and often moreover otherwise terminated, is always carried forward, in which all the parts of the body are commonly much elongated, and in which the penultimate articulation of the tarsi is bilobed, form the genus

BRENTUS, Fab. *Curculio*, Lin.

These insects belong to the hot climates. Some have the body linear, and the antennæ filiform, or a little enlarged toward the end, and composed of eleven articulations. These include BRENTUS, properly so called.

Mr. Stevens has separated from these, under the collective and generic name of *arrhenodes*, certain species, with the head as if cut behind the eyes, with the muzzle short, and terminated by two narrow and advanced mandibles in the males.

Others, similar as to the form of the body, have only nine articulations in the antennæ, the last of which forms a little knob. Such are

ULOCERUS, Schœnh.

The last CYLAS, Lat., have ten articulations in the antennæ, the last forming an oval knob. The corslet is as if divided into two lobes; the posterior, that which forms the pedicle, is the smallest. The abdomen is oval.

Sometimes the antennæ are distinctly bent, the first articulation being much longer than the following. These form the genus CURCULIO of Lin.

We shall divide them into short and long-billed, according as the antennæ are inserted near the end of the proboscis, and even with the origin of the mandibles, or more behind, whether toward its middle or near the base.

The short-billed curculiones of this naturalist are divided by Fabricius into two genera.

Brachycerus, Fab.

Have all the articulations of the tarsi entire, and without brushes or cushions underneath. Their short slightly-bent antennæ have only at the end nine articulations, the last of which forms the knob. They are wingless. Their body is very rugged or unequal.

Curculio, Fab.

Have almost the entire under-part of the tarsi furnished with short serrated hair, forming cushions, and the penultimate articulation deeply divided into two lobes. Their antennæ are composed of eleven articulations, or even twelve, counting the false articulation which sometimes terminates them, the last of which form the knob.

This genus, although much more limited than in the system of Linnæus, includes still a vast number of species discovered since his time. Many naturalists, and more especially Germar and Schœnherr, have divided the genus into many. We may, in conformity with our observations, make two principal divisions of them.

Those whose mentum, more or less widened above, and more or less orbicular, occupies the whole width of the buccal cavity, hides entirely, or nearly so, the jaws, and whose mandibles have no very apparent indentations, or which present only below the point a slight sinus.

We may include in the first sub-genus that of Cyclomus, of the short-billed rhynchophora, which, like the foregoing, have the tarsi deprived of a brush, and the penultimate articulation entire, or slightly emarginated, without very distinct lobes. To these may be referred *Crylops*, *Deracanthus*, *Amycterus* and *Cyclomus*, of Schœnh.

All the rest have their tarsi furnished underneath with

brushes, and the penultimate articulation deeply bilobed. Some of these have wings.

Here the lateral furrows of the proboscis are oblique and directed underneath. The fore-feet differ little from the following in proportions. They form a first sub-genus, that of CURCULIO, properly so called, which comprehends a great number of genera of Germar and Schœnh., whose characters are of little importance, and often very equivocal. We may at most separate those whose antennæ are larger in proportion.

Among those whose antennæ are short, with the corslet longitudinal, like a truncated cone, the shoulders projecting, and of which have been formed the genera *Entimus, Chlorima*, &c., are arranged some South American species, remarkable for their beauty and often also for their size.

The genus LEPTOSOMUS, of Schœnherr, although formed of a single species (*C. acuminatus*), has nevertheless some characters so peculiar that it may be kept as a sub-genus. The head is elongated behind, with the proboscis very short. The corslet is sub-cylindrical. The elytra terminate like divergent spines. The antennæ are short.

We will pass to a third sub-genus, LEPTOCERUS, which differs from the first by having the two anterior feet larger than the others, with the thighs thick, the legs arched, and the tarsi often dilated and ciliated. The antennæ are commonly long and slender. The corslet is sub-globular or triangular, the abdomen being scarcely larger than it.

A fourth sub-genus, PHYLLOBIUS, will include other brevirostræ of the same division, and also winged, but in which the furrows of the sides of the proboscis are straight, short, and consist only of a simple fossa. This includes several genera, *Phyllobius, Macrorynus, Myllocerus, Cyphicerus, Amblirhinus*, and *Phytoscapus*, of M. Schœnherr.

The Brevirostræ, with the penultimate articulation of the

tarsi bilobed but apterous, and almost always without a cushion, all form other sub-genera, viz. OTHIORYNCHUS and OMIAS, including those with the antennary furrows straight, and PACHYRHYNCHUS, PSALIDIUM, THYLACITES, and SYZYGOPS, comprehending those with the like furrows bent. Othiorynchus is distinguished from Omias by the dilation, in form of a little ear, of the lateral and lower part of the proboscis, serving as insertions for the antennæ;—from Syzygops, a cyclops of M. Dejean, by having the eyes nearly united above;—from Psalidium, by having projecting mandibles arched or crescented. Thylacites differs from Pachyrhynchus by their slender antennæ, as long, or nearly as long, as the corslet. The abdomen is, moreover, very much swollen. To Omias and Thylacites should be united many genera of M. Schœnherr. That of HYPANTUS may be preserved very near to Othiorynchus, but distinguished from it by a corslet larger when compared with the abdomen, and nearly globular.

Our second general division of the genus Curculio of Fab. differs from the first by the narrowing of the mentum, which not occupying the whole extent of the buccal cavity, leaves the jaws visible on either side; the mandibles, moreover, are evidently indented The knob of the antennæ is often formed by the last five or six articulations.

Some have not more than two teeth to the mandibles. The labial palpi are distinct. The knob of the antennæ, sufficiently sudden, commences only at the eighth or ninth articulation, and has not at all the figure of an elongated spindle, nor is the body, though oblong, formed the same.

Some are apterous with the tarsi deprived of cushions. Their penultimate articulation is slightly bilobed. Such is the genus MYNIOPS of Schœnherr, to which may be united his *Rhytirrhinus*.

Others, also, apterous, have, like most rhynchophori, the

under part of the tarsi furnished with cushions, and the penultimate deeply bilobed. They will compose the sub-genus Liparus, which will also include several genera of the same naturalist.

These which have wings may form two other sub-genera, viz.—Hypera, Germ., *Phytonomus*, *Coniatus*, Schœn., in which the legs have no hooks at their extremity, or but one very small; and that of Hylobius, which have a very strong one at the inner extremity.

The others, whose mandibles have three or four teeth, have a mentum narrowed near the upper end, truncated, and with the palpi little observable or nullified. Their antennæ terminate almost gradually in a knob, formed like an elongated spindle. The figure of the body is often analogous to this. Olivier has confounded them with Lixus, from which, in truth, they differ very little. They will compose the sub-genus Cleonus.

The long-billed curculiones, or those whose antennæ are inserted on the side of the origin of the mandibles, often near the middle of the proboscis, which is generally long, include nearly all the species of the genera *Lixus Rhynchænus* and *Calandra* of Fab.

In the first two, the antennæ have at least ten articulations, but more generally eleven or twelve, and the last three, at least, form the knob.

Lixus, Fab.

These are very like cleonus, as well by the organs of manducation as by the knob like an elongated spindle, of the antennæ, the straight and elongated form of the body, and the arms or the legs.

Rhynchænus, Fab.

Has not such a general assembly of characters.

Sometimes the feet are contiguous to each other at their base, and without sternal fossæ, to receive the proboscis.

Some do not leap, and their antennæ are composed of eleven or twelve articulations. These have wings.

TAMNOPHILUS has the antennæ but little bent, short, with twelve articulations, terminated in an oval knob, and carried in a short proboscis, advanced, and but little arched; the eyes are near above; the end of the abdomen is visible; the legs are armed at their extremity with a strong hook. These will form the first sub-genus.

Other Rhyncheni are remarkable for their arched legs, furnished with a strong hook at the end; their tarsi are long, filiform, with but few hairs underneath, with the last articulation little dilated, simply heart-shaped. They will compose the sub-genus BAGOUS. They are small insects, which frequent marshes.

Some others, with the same habits, differ from their congenera in their tarsi, of which the penultimate articulation entirely shuts up the last between its lobes. These are, sometimes, without hook. They form the sub-genus BRACHYPUS.

BALANINUS displays some Rhynchophori remarkable by the length of the proboscis, which at least equals, and often much exceeds, the length of the body.

RHYNCHŒNUS, properly so called, differs from the last by negative characters, and from the following by having the antennæ composed of twelve articulations.

SYBINES: this has but eleven, seven of which are before the knob.

Those are deprived of wings. Such is the sub-genus MYORHINUS, Schœnh. *Apsis*, Germ., to which we shall unite the genera *Tanyrhynchus*, *Solenorhinus*, *Styphlus*, *Trachodes* (*Comasinus*, Deg.) of M. Schœnherr.

We may then pass to those which have only nine articulations to the antennæ, and which have the faculty of leaping.

Cionus, Clair., do not leap at all, and there are nine or ten articulations to the antennæ. Their body is commonly very short and sub-globular.

Then will follow those whose posterior thighs are very thick, enabling them to leap. The antennæ have eleven articulations. The body is short, and ovoido-conical.

Those with the antennæ inserted on the proboscis form the sub-genus Orchestes, Illig. *Salius*, Germ.

Those in which the antennæ spring from between the eyes form that of Ramphus, Clair.

Among the last Rhynchœni remaining to be described, the feet are widened at their insertion; the sternum, moreover, presents a cavity larger or smaller, which receives the proboscis, and even sometimes the antennæ.

Those in which this cavity does not exist may form the two sub-genera, Amerhinus and Baridius.

The first has the body oval or sub-cylindrical, convex above. In the second, it is depressed and rhomboidal.

The Rhynchæni of Fab., whose sternum has a fosset for containing the proboscis, have been distributed by M. Schœnherr into a great many genera, which we shall reduce as follows:—

They have wings, or are apterous.

Among the former, some have a sub-rhomboidal form, with the corslet narrowed suddenly like a tube near its anterior extremity; the abdomen is nearly triangular. They are connected with Baridius, but here the antennæ have twelve articulations.

Camptorhynchus, *Eurhinus*, Schœn. These are distinguished by the antennæ forming a thick perfoliated knob.

Centrinus have a distinct scutellum; the abdomen entirely covered by the elytra, the eyes wide, and the knob of the antennæ elongated. The chest often has on each side of its cavity a tooth, or little horn.

Zygops is very remarkable for its large eyes, approaching or united above, and by their feet generally long, and the posterior at least very wide.

Ceutorhynchus, whose scutellum is scarcely perceptible; the elytra are rounded at their extremity, and do not entirely cover the abdomen. The eyes are wide. The knob of the antennæ is oval, and the extremity of the legs is without spines. In these, the antennæ have only eleven articulations. Hydaticus.

Others have the body ovoid, short, much swollen above, with the abdomen embraced in its circle by the elytra. The thighs are channelled, and receive the legs therein. Their eyes are large. The antennæ have always twelve articulations. Orobitis.

Others having the body oblong convex, with the fore feet commonly larger, especially in the males; the antennæ with twelve articulations, the eyes wide, and the elytra covering the abdomen compose the sub-genus Cryptorhynchus.

Those which are apterous, or in which the wings are very imperfect, and are without scutellum, will form another sub-genus, that of Tylode, *Ulosomus*, *Seleropterus*, Schœn.

The remaining longirostra have in general nine articulations at most to the antennæ, and the last, or two last at most, form a knob with a coriaceous epidermis, and of which the extremity is spongy.

They may be divided in one genus,

Calandra,

Which may be divided into six sub-genera.

The first two are apterous, and have, as well as the preceding and the following, with the exception of the last, four articulations on all the tarsi, whose penultimate is bilobed.

The antennæ are inserted at a little distance from the middle of the proboscis, and are bent.

In the first ANCHONUS, Schœn. these organs have nine articulations before the knob. The tenth, and, perhaps, two others intimately united with the preceding, and indistinct, form a short ovoid knob.

In the second, ORTHOCHÆTES, Germar, it is the eighth which forms the knob.

The other four sub-genera have wings.

In the first three, the tarsi have only four articulations, with the penultimate bilobed.

RHINA, Lat. *Lixus*, Fab.—The antennæ are much bent, inserted near the middle of a straight proboscis advanced, and the eighth articulation of which forms a very elongated sub-cylindrical knob. The fore feet, at least of the males, are larger than the others.

CALANDRA, properly so called, have also the antennæ very much bent, but inserted near the base of the proboscis ; the eighth articulation forms a triangular or ovoid knob.

The fifth sub-genus, COSSONUS, Clairv., have the antennæ scarcely larger than the proboscis and the head, and eight articulations before the knob ; they are thick, and inserted toward the middle of the proboscis.

The last sub-genus, DRYOPTHORUS, *Bulbifer*, Deg., so far as the tarsi are concerned, is anomolous; they have five articulations, neither of which is bilobed Their antennæ have only six articulations, the last of which forms the knob.

SUPPLEMENT

ON THE

RHYNCHOPHORA.

THE larvæ of BRUCHUS, the first genus of this family, have the body tolerably thick, swelled, arched, and very short. It is composed of several rings, not very distinct one from the other. The head is small and scaly, and furnished with very hard and trenchant mandibles. There are nine stigmata on each side, through which is introduced the air which is necessary to the support of life.

It is in this larva-state that the Bruchi carry on such ravages on the different grains of the majority of leguminous plants, and of some kernel fruits, and particularly on beans, lentils, vetches, and pease; also in the grains of the *gleditsia*, the *theobroma*, the *mimosa*, and many species of palms. The larva passes the winter in the grain, a part of the internal substance of which it consumes. It then becomes changed into a nymph at the commencement of spring, or even before the end of the winter, and the perfect insect issues forth in spring. Before it undergoes its metamorphosis, it takes care to manage an opening for itself, by rendering, at

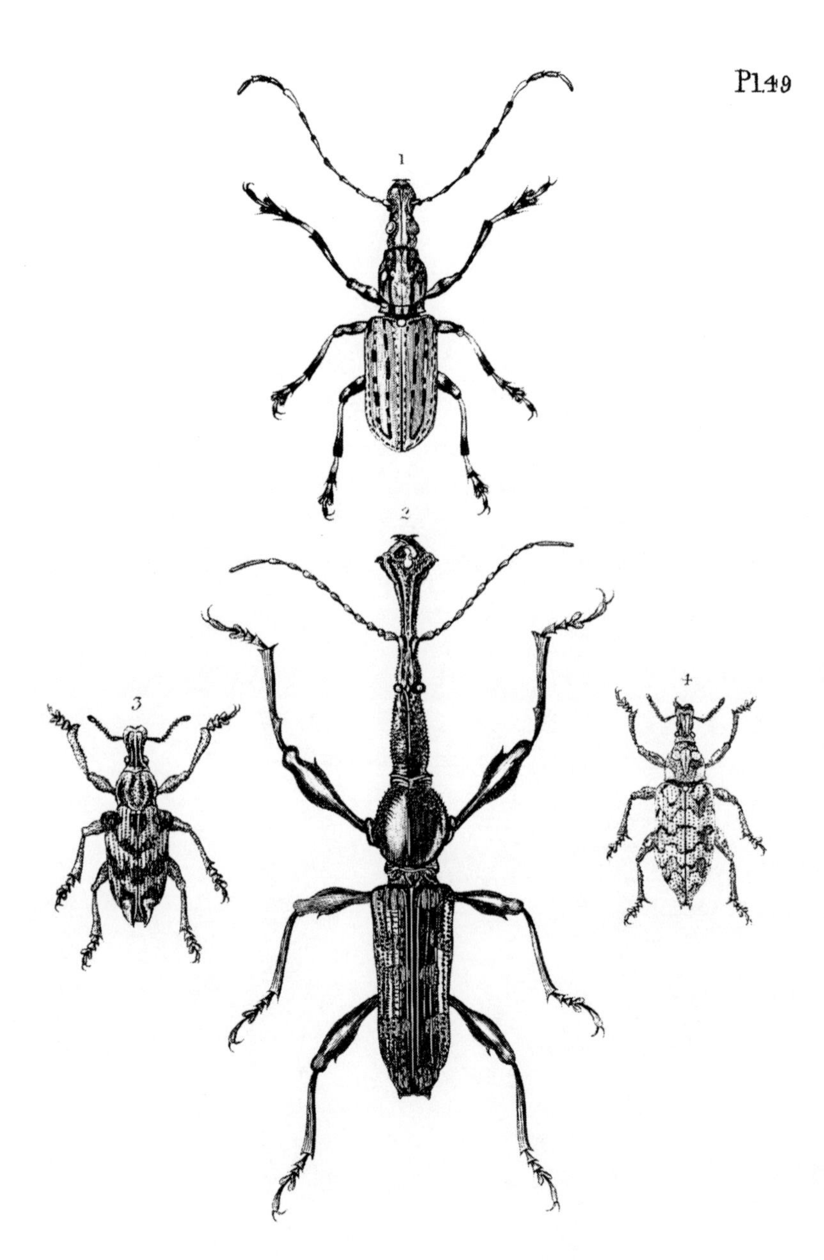

1. Dolicbocera *Childrenii.* *G.R. Gray.* 2. Brenthus *Temminchii.* *Kuhl.*

3. Chlorima *cyaneipes.* *G.R. Gray.* 4. Curculio *spectabilis.* *G.R. Gray.*

London, Published by Whittaker & C.º Ave Maria Lane 1837.

a certain part of the grain, the bark or exterior skin, so thin, that the slightest effort is sufficient to pierce it.

In its final state the bruchus does no sort of mischief to grains or fruits. These insects then frequent flowers, or different plants, and begin to pair. After pairing, the female returns on the young silicæ, and on the cods which are ready to be formed, for the purpose of depositing her eggs. She deposits usually but one egg in each grain. Nevertheless we sometimes find two of these larvæ in the beans of marshes.

These insects are not common in Europe. Some species are very much extended in the southern parts of France, in Spain, and in Italy. They are always to be met with more rarely as we advance towards the north.

In Europe it is more particularly beans, peas, and all kinds of vetches, that are the most exposed to the ravages of the larvæ of bruchus. The external envelope of these leguminous plants, does not manifest in any way the sojourn of the larva, and sometimes in opening a pea or a bean, we are surprised to find, in the middle of a considerable vacancy, the perfect insect dead, having, without doubt, been unable to manage for itself an opening.

The insect named *Dolichocera Childreni*, of plate 49, which appeared to us at first to be generically distinct from Bruchus, seems, on further examination, to belong to a sub-genus *Macrocephalus* of Oliv., which name, therefore, must be substituted for Dolichocera The species is fulvous, varied with lines of black and white, the antennæ are black, with the ninth joint white. The species is from Java.

The larva of the *Bruchus acaciæ*, an insect which inhabits North America, lives in the substance of the grains of the *robinia pseudo-acacia*. The *Bruchus palmarum* is found in all South America. Its larva feeds on the almond of a species of palm, named at Cayenne, *Counana*. It is the *cocos guineensis* of Linnæus.

The *bruchus pisi* is found on flowers in France, Germany, Italy, Spain, Greece, and North America. Its larva lives in the interior of peas, lentiles, chicklings, beans, and vetches, of every description.

Geoffroy, who established the genus ANTHRIBUS, has mentioned seven species. The first three, two of which he has figured, belong to the genus *anthribus* of Fabricius, or the *macrocephalus* of Olivier. The fourth is a *nitidula*. The three others are ranged with the *phalacri* of M. Paykull, and form the genus *anthribus* of Olivier.

But this celebrated entomologist has taken as a type of the genus, some insects which differ essentially from those which Geoffroy had in view, as the figures of the two species of anthribi, which he has given, sufficiently prove. M. Latreille, therefore, particularly designates, under the name of anthribus, those coleoptera of which Olivier composes his genus *macrocephalus*, agreeing in this with Fabricius, and all other naturalists, who followed, in this respect, the same nomenclature.

Some species are found on woods and under the bark of trees. The others live on flowers.

The genus ATTELABUS, in the method of Linnæus, is composed of coleoptera very different as to their organization and habits, and comprehends but a single species of the genus which bears that name at the present day. Geoffroy designates, in the same manner, the *hister* of this naturalist, and with one of his *attelabi*, and some other insects very analogous, forms a very natural genus, that of *rhinomacer*. Fabricius, in adopting it, has thought proper to preserve the Linnæan denomination of *attelabus*. This genus, through the labours of Herbst, Clairville, and Olivier, has since undergone many changes, so that the genus attelabus, properly so called, is now confined to the species which present the characters detailed in the text. But in general the coleoptera comprehended in the primitive genus of *rhinomacer* or *attelabus*, live pretty nearly in the same manner, and resemble each other considerably in their first state of existence. Their larvæ are soft, whitish worms, without feet, whose body is tolerably thick, and composed of twelve not very distinct rings. The head is hard, scaly, and armed with two rather solid jaws. They all live on vegetable substances. They attack the leaves, the flowers, the fruits, and the stalks of plants. They feed upon their substance, or roll the leaves and gnaw away the parenchyma. They change skin several times, and when arrived at their full bulk, they spin a cocoon of silk, or construct it of a sort of resinous matter tolerably solid, and then become transformed into nymphs, from which state they issue, after some time, in the form of the perfect insect.

When these larvæ are numerous, they do considerable damage to vegetables, either by depriving them of their leaves, or attacking the young shoots, or in fine, by gnawing the flowers and fruits. It is so much the more difficult to guard against them, as they only make their appearance in consequence of the ravages which they commit. They do not work openly, but enclosed in the centre of a stalk or fruit, which they gnaw insensibly. Thus we are not advertised of their presence until the evil is without a remedy.

It is usually on the plants which have nourished the larvæ, that the perfect insects are to be found. They are sometimes in different flowers, from which they extract the melliferous juice. Some of them also feed on the parenchyma of the leaves. But less dangerous and less voracious than their larvæ, the injury they cause to vegetables is much less considerable. Their length is from one to four or six lines.

The insects of the genus Brentus are distinguished from *Curculio* by straight and moniliform antennæ, independently of some slight differences in the conformation of the parts of the mouth.

These insects, remarkable for their extremely elongated form, are found only in the warm climates ; a single species only has been discovered in Europe. They live on flowers, or on the bark of trees. We insert a new figure, from the cabinet of the Reverend Mr. Hope, of *B. Temminckii*, described by Mr. Klüg. It is black, the elytra varied, with red spots, and strong striæ. This species is from Java.

The genus Brachycerus has been confounded with *Curculio*, which it much resembles ; but differs from it as stated in the text.

The Brachyceri do not frequent flowers, nor are they ever found on trees and plants, like the *Curculiones*. Having no wings, they cannot quit the surface of the earth, although, in general, the insects deprived of wings have received, in compensation, a greater agility in the legs, such as *Carabus*, *Tenebrio*, &c. This is not the case with the brachyceri. Although their limbs are tolerably long and thick, they cannot walk with any degree of rapidity ; in fact, they creep along very slowly indeed.

These insects are found only in the south of Europe, and in foreign countries. We have not been able, as yet, to acquire any positive knowledge concerning their larvæ. The body of some species is covered more or less, in certain places, with a scaly imbricated dust, which is easily detached, and which the insect loses in growing older.

The genus *Curculio*, though still very numerous, is much more restrained than formerly, and the study of it more facilitated. It comprehends, besides the species whose general characters are given in the text, those whose antennæ, equally composed of eleven articulations, and terminated in a knob, are inserted towards the middle of an elongated proboscis. These species have been designated by the epithet of *Longirostres*. Fabricius has separated them, and formed of them two distinct genera, *Lixus* and *Rhynchænus*. Olivier, from this last, has withdrawn some species, those of his genus *Liparus*; but it must be observed, that M. de Clairville refers to his genus *Rhynchænus* only, the *leaping curculiones*.

Fabricius, by giving more extent to this last division, has departed from the principles of M. Clairville, and thrown confusion into nomenclature; for his genus Rhynchænus no longer corresponds with that of the last-mentioned naturalist. To obviate this embarrassment, M. Latreille, in dividing the genus *Curculio* after the manner of Fabricius, had preserved this generic name to the species with long proboscis, and termed those in which this organ is short *Brachyrhinus*. But the nomenclature of Fabricius has, nevertheless, prevailed. Some German naturalists, however, have preserved the term *Brachyrhinus*, and apply it to a new group, dismembered from the Curculio of Fabricius.

The genus, as it stands at present, is composed of the largest species, especially of those preferred by amateurs, in consequence of their agreeable forms, very various, and often exceedingly brilliant colours, produced by imbricated scales, analogous in their arrangement to those of the wings of the lepidoptera. The majority of those handsome species, such as *Regalis*, *Imperialis*, *Chrysis*, *Fastuosus*, &c., are peculiar to Brazil and Peru. Those of the ancient Continent are generally smaller, and less ornamented. Some, notwithstanding, such as *taniarisci*, *viridis*, *argentatus*, &c. are still remarkable for the splendour of their costume.

The Curculiones are of a slow and timid character. They feed on leaves. There are even some of them, as *Curculio ligustici*, which sometimes ravage the fields, which are sown with ferns. Their larvæ would also appear to subsist, in like manner, in vegetable substances.

Our species, *C. cyanipes*, is of a brilliant, light-green colour, varied with irregular bands of black. It is from Brazil. *C. spectabilis* is of a bright emerald green, with spots and bands of golden copper, varied with black.

Degeer has given the history of the *lixus paraplecticus*, which Linnæus has thus named specifically, from his opinion that the larvæ of these insects being eaten by horses along with the plant on which they feed, gave them the malady called *paraplegia*, (in Swedish *staikra*,) as well as the plant itself, which is the *phellandrium aquaticum*—a sort of umbelliferous plant, very common in many marshes. The interior of the submerged portion of its thick stalks serves as a retreat to the larvæ of this lixus, which live there in a solitary manner, and with the head always placed towards the top. These larvæ feed upon the sap. They are about seven lines in length, and a little more than a line in diameter, entirely white, or the colour of milk, a little yellowish, with the head scaly, and of a yellowish brown. The body is almost of equal thickness in its whole extent, except towards its posterior extremity, where it is terminated in a cone. It is divided into twelve rings, the first three of which have each underneath, towards the sides, two nipples, which represent feet, but without assisting the progression of the animal. According to the observations of this great naturalist, it advances, holding itself reversed, or with belly upwards, elongating and shortening its wings, which have transverse wrinkles, cut rather deeply, and which form fleshy irregular eminences. The belly is smooth. All along the sides, the skin of which is smooth and bare, a sort of fold is visible, and a range of oval points, a little raised and ridged, of a clear brown; these are the stigmata. There are nine on each side. The hinder part of the body is a little curved, and a little forked at the end, with a small incision, in which the anus is found. The head is oval, with a scaly skin, divided into two semi-hoods, and an anterior and triangular piece. A whitish suture separates these parts from each other. The mouth is furnished with very small hairs, and composed of two coneous, strong, and very pointed mandibles, of two small lips, of two jaws, and four conical articulated palpi, the maxillary of which are larger, and forked at the end. The lower lip has three small conical parts, the middle one of which resembles the spinneret of caterpillars. Each side of the head presents a black point, which seems to be an eye.

The larva is transformed, in the commencement of July, into a nymph in the interior of the stalks, in which it has lived. This nymph is naked, or without shell or cocoon, of the same length as the larva, and thick in proportion; white, with the abdomen bordering on yellow, rounded, and armed with two scaly crooks at each posterior extremity; each of its rings has above a transverse range of scales, points short and brown. The proboscis is curved underneath the breast. The elytra and feet are applied upon its sides. The nymph holds itself straight, or with the head upwards, in its habitation. It has much vivacity, which it announces by the movements of the rings of its body, and by changing place, by means of the spines and crooks of its abdomen. When it is about to pass into the perfect state, it raises itself to a height, the level of which is above the surface of the water, knaws with its teeth a portion of the stalk, and makes a large and oval aperture, which serves it for a passage out. This last transformation takes place towards the end of the same month, and of July. Linnæus has erroneously advanced that it passes the winter in the stalk.

The *lixus odontalgicus* lives on the flowers of thistles, and of *cercium*. It has been vaunted as a specific for the tooth-ache.

The species in which the proboscis is lodged, in a hollow in the breast, compose the genus *Crypto rhyncus* of Illiger. The habits of the Rhynchæni are the same as those of the other curculionites. Olivier, who does not adopt the distinction of Illiger, has figured and described one hundred and sixty-nine species.

Passing over the other sub-genera, we shall proceed to the last genus of the rhyncophora, the Calandræ. These have been separated from the genus *curculio* of Linnæus, and are distinguished from all the others of the same family by character designated in the text.

Of all the insects of this genus, the *Calandra granaria* (Weevil) is the most common, and the most formidable to us, since it attacks the principal basis of our food. These insects are sometimes so numerous in a heap of corn that they destroy it altogether, leaving nothing behind but the chaff or envelope of the grain. A larva is always alone in a grain of corn. It is in this lodge that it acquires its growth, at the expence of the farina in which it feeds. In proportion as it eats, it enlarges its lodging, so that it may be large enough to contain it under the form of a nymph. This little larva, very white, has the form of an elongated and soft worm, and the body is composed of nine prominent and rounded rings. It is nearly a line in length, with a rounded, yellow, scaly head, provided with organs proper for the purpose of gnawing the grain.

When the larva has eaten all the flower, and is arrived at its full growth, it remains in the envelope of the grain, where it is metamorphosed into a nymph, of a clear white, and transparent. Under the envelope may be distinguished the proboscis, the antennæ, and the rest of the insect. In this state it takes no nutriment, gives no sign of life but in the lower part of the nymph, which is capable of some movements when it is disturbed. Eight or ten days after this first metamorphosis, the insect breaks the envelope which held it swaddled up, it pierces the skin of the grain to produce an opening, and let itself out of its prison. The weevil then appears under its final form.

In general, that which serves as nutriment to insects in their larva or caterpillar state, no longer suits them in their perfect form. Such, however, is not the case with the Calandra, if we are to believe some naturalists. Scarcely has it issued from its nymph-state than it proceeds to pierce the envelope of grains, to establish itself there anew, and again to feed on their farina. We might suppose, however, that the Calandra in its image state, or that of perfect insect, does not feed on the farina of corn except when it can find nothing better, and that if it appears to seek out the heaps of corn, it is for the purpose of depositing its eggs there. But, in fact, the first of these suppositions cannot be well entertained, for in visiting heaps of corn attacked by Calandræ, we often find the insect lodged in the interior of the grain; its black colour does not announce that it has recently issued from its state of nymph, since it is of a straw-colour at the time when it has just quitted its sheath. Nevertheless, we must doubtless believe that it occasions much less injury in this last state than in that of larva.

For a long time it was believed that a heap of corn heated, or the grains beginning to germinate through humidity, would engender weevils. Some naturalists, who, without doubt, had applied themselves but little to the observation of this species of insects, have assured us that the Calandra deposited its eggs in the ears of corn while the grain was yet in milk, and that it was transported with the corn into the granaries. More exact observations have destroyed these errors. The Calandra has no sooner issued from its envelope of nymph, than it is in a state for pairing, like the majority of insects, for the purpose of reproducing its species. Its pairing always bears a strict relation to a certain degree of heat. When the heat is under eight or nine degrees, these insects have not sufficient vigour to feel the desire of copulation. They live in a state of repose, and even lethargy, if the weather be cold, and are then incapable of doing injury. According to the season and the climate, the deposition of eggs commences sooner or later. In the southern parts of France it takes place in the month of April, and it often extends, these insects continuing to propagate, even towards the middle of September. Thus the destruction of grain in those countries must be much more considerable than in the more northern climates. As long as the weather continues warm, these insects pair very frequently. Their union lasts a considerable time. They may be swept or carried away without a separation taking place between them. The female, consequently, deposits her eggs in all months, when the temperature is up to a suitable degree. As soon as it begins to be cold in the mornings she ceases to lay.

From the moment of pairing to that in which the insect appears under the form of the calandra, there elapse about forty or forty-five days. By this we may see, that in a year there must be many generations of these insects, which multiply still more in the very hot climates. According to a table formed respecting the multiplication of the calandra, it results, that by adding together the numbers of each generation, we have the sum-total of six thousand and forty-five calandra proceeding from a single pair during five months, dating from the end of April, until towards the middle of September, while the mercury continues in the thermometer above fifteen degrees, and it scarcely ever falls lower in the southern parts of France. We can, therefore, no longer be astonished if enormous heaps of grain are destroyed by these insects.

As soon as the female of the calandra has been fecundated, she buries herself in heaps of corn for the purpose of depositing her eggs there, and concealing them immediately under the skin of the grains. She makes an insertion which keeps her a little raised in this part, and forms there a small elevation not very sensible to view. These holes are not perpendicular to the surface of the grains, but oblique or even parallel, and are stopped with a species of gluten, of the same colour as the corn. It appears that these insects commence by sinking between the skin and the substance of the grain, the little dart, which is concealed under the lower part of the proboscis. The female never puts more than one egg in each grain of corn. This egg very speedily excludes the young. At the end of some days there issues forth from it a small larva, which, lodged in the grain, is perfectly sheltered from the injuries of the air, because its excrements serve to close the aperture through which it has entered, so that there is no use in stirring the grain, as all the shaking in the world will not incommode it.

It is in heaps of corn that we usually find the calandræ, at some inches in depth, and not at the surface, unless they should happen to be disturbed in their retreat, and attempt to make their escape. There it is that they live, that they usually pair, and that the females deposit their eggs. It is hardly possible to know, by looking at the grains, which of them have been attacked by these insects, since they have the same form and the same appearance of those which are untouched. They may be known by the weight, and the most unequivocal criterion is when several handfuls of the grain are thrown into water. Those which, though they look well, swim upon the surface of the water, have lost a portion of their farinaceous substance by the devastations of the calandræ.

As long as the weather continues warm, the calandræ do not quit the heap of corn, of which they have taken possession, unless they are obliged to dislodge and abandon it, by shaking it with shovels, or passing it through a sieve. As soon as the mornings begin to become fresh, all the calandræ, young and old, abandon the heaps of corn, which no longer supply them with a sufficiently warm retreat. They retire into the clefts of walls, into the flaws in woods and floors. They are sometimes found behind tapestry, under chimneys, and in fine, every where, where they can find a retreat to protect them against the cold. It has, however, been seriously supposed, that the *calandræ* remain in a state of lethargy during the whole winter, and regain, at the return of spring, the heaps of corn which they have abandoned, and there recommence the deposition of their eggs. A general and constant rule in the class of insects is, that those which have paired perish shortly after, and that they do not pass the winter except in the egg, or in the larva-state. It is doubtless rare, that even those which have not been exhausted in fulfilling the intentions of nature, can brave the rigour of the season, and do not perish before the spring arrives.

It is natural to suppose, that some means of exterminating these most destructive insects have been anxiously sought after in all times. But all such means, even to the present moment, have experienced so little success, that they may be regarded as very nearly useless. They consist, for the most part, in fumigations made with decoctions composed of herbs, of a strong and disagreeable odour. The result of all such processes has been to communicate to the corn a fetid and disgusting odour, without injuring the calandræ, which being sunk in the heaps of corn, could not be incommoded by them.

Experience has, moreover, proved, that these odours which appear to us the most disagreeable, are by no means hurtful to the calandræ, and even if they could prove so, it would be very difficult to bring them to bear upon these insects. Those which might happen to be at the surface of the heap of corn, would sink down directly, or abandon the granary, and return when the disagreeable smell was dissipated. The odour of the essential oil of terebenthine does not appear to occasion to them the least annoyance. The smoke of sulphur, such an active agent in breaking the elasticity of the air is without success in suffocating and killing the calandræ, which have no need, for the purposes of respiration, of the same quantity of air which is necessary to the larger animals. All these fumigations are equally unsuccessful in the destruction of their larvæ, which, however, cause by far the greatest degree of devastation.

Some economists have supposed that to protect the corn against the ravages of the calandræ, it would be sufficient to put it into wainscotted cellars, or to pass it through the sieve in the winter season. But in putting the corn into the cellars, it would be difficult to preserve it from humidity, which would cause it to germinate and rot. Besides, the calandræ would only find themselves there in a more undisturbed state for the commission of their ravages with a greater degree of certainty. As for the use of the sieve, it is perfectly ineffective in winter, because the moment the weather begins to grow cold, these insects quit the heaps of corn. This means too, is very insufficient to detach the eggs, which are so well glued, and so adherent to the grain, that it is impossible to separate them by sifting it, or stirring it about with a shovel.

Experiments have proved that a sudden heat of nineteen degrees is sufficient to destroy the calandræ without burning them ; but this sudden rarefaction of the air would not suffocate these insects when they are sunk in a heap of corn. It has been observed, that a heat of sixty degrees was necessary to destroy the calandræ in a stove. But this excessive heat, which has also the advantage of destroying the eggs and the larvæ enclosed in the grain, is likely to dry up the corn too much, and even to calcine it, while at the same time it does not preserve it from the insects which remain in the granaries, and which will still proceed to attack it if they can find no other.

As the calandræ are incapable of doing mischief during the cold weather, since they then cease to eat and to multi ply, the idea of substituting cold instead of heat as a remedy against them has struck some persons. It has been proposed to have a ventilator, the effect of which would be to keep up in a granary a degree of air, sufficiently cold to reduce these insects to an incapacity of exercising any of the functions necessary to the preservation of their existence, and the multiplication of their species. By continuing the action of this ventilator during the whole summer, the calandræ might be obliged to dislodge, or by reducing them to a state of lethargic inaction they would become incapable of doing mischief. This method appears to be so much the more efficacious, as it bears a relation to the natural mode of life of these pernicious insects.

We shall not point out many other methods founded on false and gratuitous suppositions; but we shall make mention of a process equally cheap and simple, and which merits the attention of those who are interested in the preservation of corn. When it is observed at the return of spring, that the calandræ are beginning to spread in the heaps of corn that have passed the winter in granaries, it will be necessary to form a small heap of five or six measures, which are to be placed at a suitable distance from the larger heap. They then stir with a shovel the corn of the principal heap in which these insects have established themselves. The calandræ, who are singularly fond of tranquillity, being disturbed by this movement, seek to fly and to escape, and seeing another heap of corn along side of that from which they have been forced to remove they run thither to take refuge. If they try to gain the walls for the purpose of saving themselves, which is rare, the persons who watch their flight take care to collect them with a broom, which they should have in their hands, towards the heap to which the others have retired, or to crush them with their feet. This is so much the more easily to be done as the insect never stirs. It remains motionless, as if it were dead, from the moment in which it is touched. If it is brought back towards the little heap of corn, placed in reserve, it will endeavour immediately to enter and sink in there the moment it is no longer disturbed by the broom. When all the calandræ are collected together, boiling water is brought in a cauldron, it is poured upon the corn, which is stirred from time to time with a shovel, so that the water shall penetrate throughout before it grows cold. All these insects then immediately die, burned or suffocated at the moment. The corn is then spread out for the purpose of drying it, after which it is easy, by sifting it, to separate from it the dead calandræ. It must be observed that it is essential to perform this operation in the commencement of spring, so as to prevent the laying of these insects. If performed too late it will prove utterly inefficacious, because the eggs deposited in, and glued to the grain, do not separate from it though it be agitated with ever so much violence, and would produce a generation of calandræ which would destroy all the corn which we had been endeavouring to preserve. The generation which exists is only dangerous in giving birth to that which succeeds. We must therefore prevent the existence of the latter by destroying the former. This method may be performed on a large scale as well as on a small one, without occasioning any considerable expense, a thing which often prevents the execution of the very best projects.

"Sometimes," says Mr. Kirby, "this pest becomes so infinitely numerous, that a sensible man, engaged in the brewing trade once told me, speaking perhaps, rather hyperbolically, that they collected and destroyed them by bushels, and no wonder, for a single pair may produce, in one year, above six thousand descendants."

Among the largest of the exotic species is the *calandra palmarum*, or palm-weevil, which is entirely of a black colour, and measures more than two inches in length. Its larva, which is very large, white, and of an oval shape, resides in the tenderest part of the smaller palm-trees, and is considered, fried or broiled, as one of the greatest dainties in the West Indies. "The tree," says Madame Merian, "grows to the height of a man, and is cut off when it begins to be tender, is cooked like a cauliflower, and tastes better than an artichoke. In the middle of these trees live innumerable quantities of worms, which at first are as small as a maggot in a nut, but afterwards grow to a very large size, and feed on the marrow of the tree. These worms are laid on the coals to roast, and are considered as a highly agreeable food."

THE SECOND FAMILY

OF

TETRAMEROUS COLEOPTERA.

XYLOPHAGI

HAS the head terminated in the ordinary manner, without observable projection, like a proboscis or muzzle; the antennæ are thicker toward the extremity, or perfoliated from their base, always short, with less than eleven articulations in many of them, and generally with entire articulations to the tarsi, or with the penultimate enlarged and heart-shaped in others; in the latter case the antennæ always terminate in a knob, either solid and ovoid, or divided into three leaflets, and the palpi small and conical.

We shall divide this family into three sections.

First. Those whose antennæ, having ten articulations at most, sometimes terminate in a strong knob, in general solid, but of three elongated leaflets; in others, sometimes form from their base a cylindrical and perfoliated knob, and whose palpi are conical. The fore legs of most of them are indented and armed with a strong hook, and their tarsi, whose penultimate articulation is often heart-shaped or bilobed, can fold down upon them.

Some have the palpi very small, the body convex above, the head globular, partly buried in the corslet, and the antennæ terminated in a solid or three-leafed knob, and preceded by five articulations at least. These Xylophagi compose the genus

SCOLYTUS, Geoff.

Which Linnæus did not distinguish from Dermestes.

Sometimes the last articulation of the tarsi is bilobed. The antennæ have seven or eight articulations before the knob.

HYLURGUS, Lat. *Hylesinus*, Fab., have the club of the antennæ solid, sub-globular, obtuse, little or not at all compressed, annulated transversely, and the body sub-cylindrical.

HYLESINUS, Fab., have also the antennæ terminated in a solid knob, little or not at all compressed, and annulated transversely, but proceeding to a point. Their body is sub-ovoid.

In the two following sub-genera, this knob is still solid, but strongly compressed, and the lower articulations form concentric bends.

SCOLYTUS proper, Geoff. *Hylesinus*, Fab. *Eccoptogaster*, Herbst. Gyllenh., have their antennæ straight, barbless, inserted very little on the inner edges of the eye, which are very narrow, elongated, and vertical.

CAMPTOCERUS, Dej. *Hylesinus*, Fab. The males have the antennæ decidedly bent, but furnished without with long hair; they are inserted at a perceptible distance from the eyes, which are elliptical and oblique.

PLOIOTRIBUS, Lat. *Hylesinus*, Fab., recede from all other insects of this family by the knob of the antennæ being composed of three elongated leaflets.

Sometimes all the articulations of the tarsi are entire, and the knob of the antennæ, always solid and compressed, commences at the sixth or seventh articulation.

TOMICUS, Lat. *Ips*, De G. *Bostricus*, Fab. Their antennæ cannot be folded under the eyes, and the knob is distinctly annulated. The head is sub-globular. The corslet is without emargination on the sides. The legs are not striated. The body is cylindrical, with the eyes elongated, and a little emarginated.

PLATYPUS, Herbst. *Bostrichus*, Fab. The antennæ are shorter than the head, can be folded, and terminate in a large knob without distinct rings. The body is linear, with the head cut vertically in front, the eyes nearly round and entire, the corslet emarginated on each side to receive a portion of the anterior thighs, the two fore legs are divided on their posterior face by transverse crests, and the tarsi are long and thin, with the first articulation much elongated. The two hind feet are very much bent behind.

Others have the palpi large, apparent, and of unequal length. Their body is depressed, narrowed before. Their antennæ are sometimes of two articulations, the last being very large, flat, sub-triangular, or nearly ovoid, sometimes of ten and entirely perfoliated. The labrum is large. The elytra are truncated, and the tarsi short, with all the articulations entire. They compose the genus

PAUSSUS, Lin. Fab.

Those with antennæ of two articulations, include PAUSSUS proper.

Those with ten entirely perfoliated articulations, compose the sub-genus CERAPTERUS, Swed.

The second section will include the Xylophagi, whose antennæ have ten articulations, and whose maxillary palpi at least do not run to a point, but are of equal thickness, or dilated at the end. The articulations of their tarsi are always entire.

This is divisible into two principal genera, according to the terminations of the antennæ. In the first, the last three articulations form a perfoliated knob, that of

Bostrichus.

Bostrichus Proper, Geoff. *Apate,—Synodendron*, Fab. —*Dermestes*, Lin., have the body more or less cylindrical, the head round, sub-globular, capable of immersion in the corslet to the eyes; the corslet more or less convex before, and forming a sort of hood, and the first two, and the last, articulations of the tarsi elongated.

Psoa, Fab., differ from them only in having the body straighter and more elongated, with the corslet depressed, and nearly square. The jaws have but a single lobe instead of two.

Cis, Lat.—*Anobium*, Fab., have the body nearly oval, depressed, or but little elevated, with the corslet transversal, rounded, and bordered laterally, a little dilated or advanced to the middle of the anterior edge, and the last articulation of the tarsi much larger than the preceding. The head of the males is often horned or tuberculated.

Nemosoma, Desmar. *Ips*, Oliv. *Colydium*, Hellw., have the body long and linear, the antennæ scarcely longer than the head, the mandibles strong, projecting, indented at their extremity; the fore legs triangular, indented on the outside, and the tarsi slender and elongated.

The second genus of this second division,

Monotoma,

Is distinguished from the last by having the knob solid, or in the shape of a bud (the tenth articulation) of the antennæ. The body is long, depressed, and often parallelipiped, with the fore part of the head narrowed and slightly advanced, like a triangular and obtuse muzzle. The palpi are very small, and not projecting, as well as the mandibles.

In some the head is not at all separated from the corslet by any sort of neck, and can be withdrawn under it behind.

Synchita, Helw. Dej. *Lyctus*, *Elophorus*, Fab. The anterior extremity of the head is transverse, and without projection; the first two articulations of the antennæ are almost identical, and the corslet, obviously wider than long is separated from the base of the elytra by an observable interval.

Cerylon, Lat., *Synchita*, Helv., *Lyctus*, Fab., have the anterior extremity of the head advanced like an obtuse triangle. The first articulation of the antennæ much thicker than the following; the corslet applied behind against the base of the elytra, wider than long, or nearly isometrical, without borders; the body sub-oval, and the elytra without posterior truncature, and covering all the upper part of the abdomen.

Rhyzophagus, Herbst, Gyll., *Lyctus*, Fab., resemble the last in most particulars, but the body is narrow and elongated, with the corslet wider than long, bordered, and the elytra truncated.

Monotoma, Herbst., *Cerylon*, Gyll., have the head as big as the corslet, and separated therefrom by a sort of neck. The first two articulations of the antennæ are thicker than the following, and nearly equal (the first a little the largest). The upper end of the knob seems to shew vestiges of one or two articulations. The head is triangular, a little advanced into an obtuse muzzle. The body is elongated, with the corslet longer than it is wide.

The Xylophagi of the third division have eleven very distinct articulations in the antennæ; the palpi are filiform, or thicker at the end in some than in others, and all the articulations of the tarsi are entire.

We shall begin with those in which the knob of the antennæ has but two articulations. They will form the genus

Lyctus.

Some have the mandibles, and the first articulation of the antennæ, entirely covered. The body is narrow and elongated, nearly linear, with the eyes large, and the corslet elongated.

Lyctus, Fab.—In these the edges of the head entirely, or for the most part, cover the first articulation of the antennæ. The mandibles are not projecting.

Diodlesma, Mej. Dej., have the antennæ as long as the corslet; the body ovalo-oblong, convex, with the corslet nearly semi-orbicular, and the abdomen nearly oval.

Bitoma, Herbst. Gyll. *Lyctus*, Fab., have the antennæ shorter than the corslet; the body is long, straight, sub-parallelipiped and depressed, with the corslet square.

In the other Xylophagi, having antennæ with eleven articulations, the three or four last form the knob, or the last only is larger than the preceding; they may be sub divided as follows: —

Sometimes the mandibles are covered, or project but little; such are

Mycetophagus, Fab.

In these the antennæ, scarcely larger than the head, are inserted under the advanced edge of the head, and terminate abruptly in a perfoliated knob of three articulations.

Colydium, Fab., have the body linear, with the head very obtuse in part, the corslet of the width of the abdomen in a square more or less long, and the abdomen elongated.

The first two articulations of the antennæ are larger than the following. These to the eighth inclusive are very short, and transversal. There the antennæ are at least as long as the corslet.

These have the body oval, with the corslet transversal,

wider behind; the first and last articulations of the tarsi are elongated, and the antennæ terminate in a perfoliated knob, either elongated, and commencing towards the sixth or seventh articulation, or sudden oval, and formed only of the last three.

MYCETOPHAGUS proper, Fab. *Tritoma*, Geoff—The knob of the antenna begins at the sixth or seventh articulation. The last is sub-ovoid.

TRYPHYLLUS, Mej. Dej. *Mycetophagus*, Gyll.—In these the knob of the antennæ is shorter, formed only of the last three articulations, the last nearly globular.

Those have the body oblong, with the corslet narrower than the abdomen, at least behind; the first articulation of the tarsi is as long as the following, or scarcely longer, and the antennæ terminated by a straight knob, elongated, little or not at all perfoliated, formed of the last three articulations.

MERYX, Lat., distinguishable from the following by their maxillary palpi, (always projecting,) terminated by a larger articulation, like a reversed triangle.

DASYCERUS, Brong., which has only three articulations to the tarsi, but which is, nevertheless, allied to this family by other characters. The first two articulations of the antennæ are globular, the following slender, capillary, and downy, and the last three globular and downy. The head is triangular, and distinct from the corslet. The maxillary palpi are projecting, slender, and terminated like an awl. The corslet and elytra are furrowed. The abdomen is subglobular.

LATRIDIUS, Herbst. *Tenebrio*, Lin. *Dermestes*, Fab., have very short palpi, terminating like an awl, the head and corslet narrower than the abdomen. The first articulation of the antennæ very thick and globular, the following to the tenth inclusive, nearly like a reversed cone, smooth or simply pubescent; the last larger than the preceding, and ovoid; the corslet wider than long, or sub-isometrical ; the abdomen nearly square, or nearly oval.

SILVANUS, Lat. Gyll. *Dermestes*, Fab., have the body nearly linear or parallelipiped ; the corslet larger than wide, of the width of the abdomen before ; the first articulation of the antennæ nearly equal, nearly in the shape of a top, and the last nearly globular ; the palpi are almost filiform, and the anterior extremity of the head is a little advanced, and narrowed like a triangular and obtuse muzzle.

Sometimes the mandibles are entirely exposed or projecting, and strong. The body is in general narrow, elongated, and depressed. These insects form the genus

TROGOSITA, Oliv. Fab. *Platycerus*, Geoff.

Some have the antennæ shorter than, or as long as the corslet, terminated in a compressed knob, a little serrated, and formed of the three or four last articulations. The lingua is entire.

TROGOSITA, Fab.—The mandibles are shorter than the head, crossed ; the lingua, nearly square, is not elongated between the palpi, and the jaws have only a single lobe.

PROSTOMIS, Lat. *Megagnathus*, Mej. *Trogosita*, Fab., have the mandibles larger than the head, parallelly advanced; the lingua is narrow, elongated, advanced between the palpi, and there are two lobes to the jaws. The body is long, straight, sub-linear.

The others have the antennæ nearly as long as the body, of the same thickness to the tenth articulation inclusive ; the following and last is larger, in form of a reversed triangle, and truncated obliquely at the end. The lingua is bifid, PASSANDRA, Dalm. Schœnh.

SUPPLEMENT

ON THE

XYLOPHAGI.

THE first division of this family is composed of the genus SCOLYTUS, established by Geoffroy upon an insect which lives under the bark of trees, and in which he finds some relations with dermestes. In fact, it is in this last mentioned genus that Linnæus has placed some other coleoptera, very analogous to the scolyti of Geoffroy, but which Degeer has separated from it to form a peculiar genus, namely, his *Ips*.

Of all the Xylophagi, the insects of this tribe are the most destructive. Their larvæ gnaw and furrow in various directions, and often in the manner of rays, the first layers or strata of wood, and sometimes even penetrate deeper. When they are greatly multiplied in certain forests, particularly those of the pine and fir, they destroy in a few years a great quantity of trees, or render them perfectly useless for all the purposes of art. Some of them do considerable mischief to the olive tree. The *Scolytus destructor* is very injurious to the oak and elm.

As for the *Scolyti* proper, the type of this division, they live in carious and worm-eaten wood, not only when in the larvæ state, but also in that of the perfect insect. It is they which, in conjunction with the *Anobia*, pierce it in all directions, and destroy it by little and little, by converting it into very fine dust. This dust is nothing more than the substance of the wood, on which they feed, and which they have returned in the form of excrement. But the *Anobia* attack only the dead wood, while the Scolyti most generally feed upon the living tree. The latter sometimes cause the shoots, branches, and even the trees themselves to perish, by destroying their fibres, and extending or changing the character of the sap which gives them life. The larva is short and soft, provided with six feet, and a hard scaly head. It is armed with two very strong jaws, by means of which it gnaws and destroys the hardest wood. It undergoes its metamorphosis in the wood itself which has supported it, and does not leave it except when pressed by the desire of reproduction.

The larvæ of the Bostrichi resemble a soft short worm, a little inflated. Their body, usually curved into an arch, is composed of twelve distinct rings. It is provided with six scaly feet and a scaly head, tolerably hard, armed with two very hard, solid, and scaly jaws. These larvæ, similar to those of *Anobium*, live in dead wood, gnaw it, pierce it in all parts, and reduce it to dust. They do not arrive at their full growth but in the space of one or two years. They cast their skin, and undergo their metamorphoses, in the wood which they have gnawed, and they do not issue from it but under the form of the perfect insect. These larvæ may be brought up in the farina of rye or wheat. They live there very well, and almost always change there into the chrysalis state; but the perfect insect is very rarely obtained by this method.

It is around trees which are half dead, on the dead branches, under the half rotten bark of old trees, and especially of oaks, and in fine, on wood which has been cut down for some time, that we meet with the Bostrichi, either when they are coming out of such places, or returning thither to deposit their eggs.

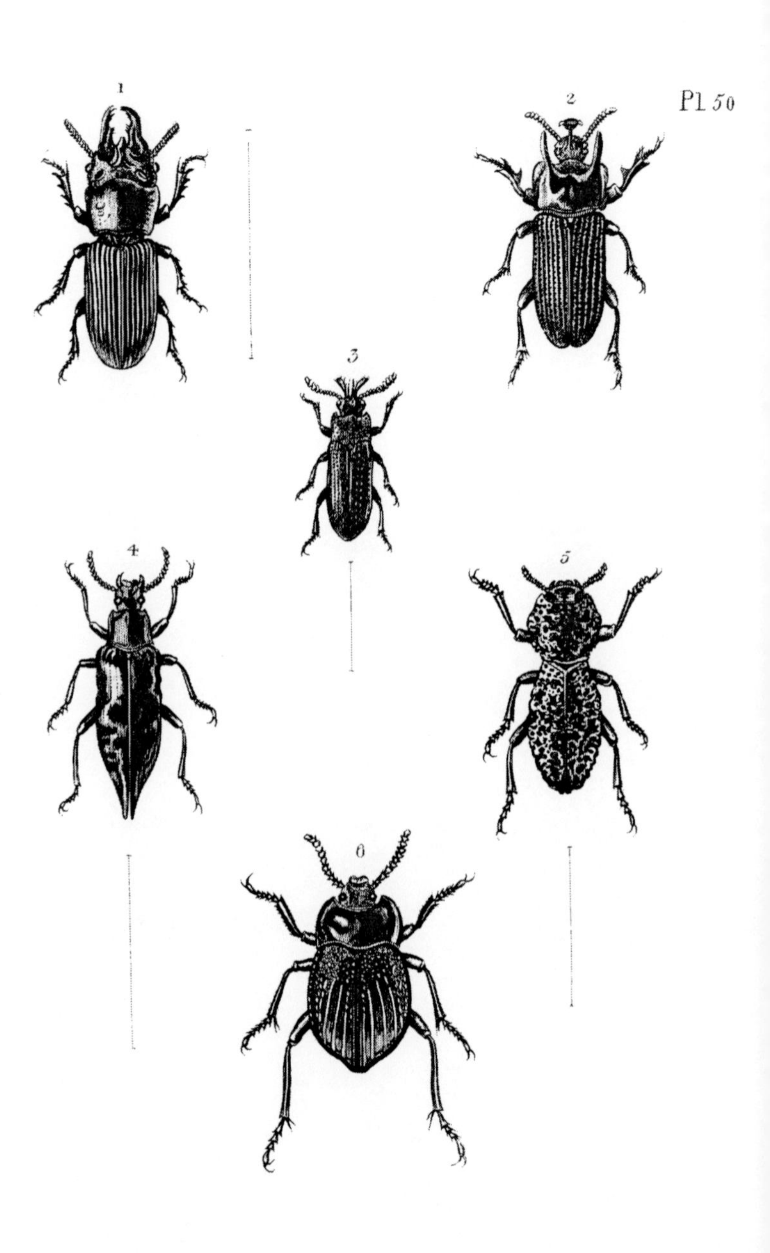

1. Phrenapates *Bennettii.* *Kirby.* 2. Antimachus *cornutus.* *Gebl.* 3. Toxicum *Richesiactum.* *Latr.* 4. Ryſsocheton *politus* *G.R.Gray.* 5. Zopherus *Mexicana* *Hope* 6. Nyctelia *Hopei.* *G.R.Gray.*

London. Published by Whittaker & Co. Ave Maria Lane. 1831.

1. Phrenapates *Bennettii.* 2. Rhyssochiton *politus.* 3. Antimachus *cornutus.* 4. Zopherus *Mexicana.*
5. Toxicum *Richesiachum.* 6. Nictelia *Hopei.*

London, Published by Whittaker & C.° Ave Maria Lane. 1831.

The late Mr. Eschcholtz, in his Zoological Atlas, has formed a new genus, under the name of *Trypananeus*, of the Bostrichus thoracicus of Fabricius, with the following characters; viz. The antennæ with a solid club, large, compressed; clypeus rostriform, the thorax much longer than broad, dentated anteriorly in the males. The species (*Denticollis*) we have figured. It is of uniformly shining black, very slightly punctuated, and is from Brazil.

The Rev. W. Kirby has described a new genus in his Fauna Boreali-Americana, under the name PHRENAPATES, with the upper lip transverse, hairy at the apex; the maxillary palpi filiform, with the last joint cylindrical; the labial palpi filiform, two first joints subclavated; the third straight internally, curving externally; the antennæ moniliform, robust, the three terminal joints rather larger than the rest; the last subovate. The species the above gentleman named *Bennettii*. It is black glossy; head with three horns intermediate, one incurved subemarginate at the apex, the lateral oblique, truncated. The length is fourteen lines. It is from Choco in Columbia.

M. Latreille has composed the genus CIS of several insects, which had been placed with *Anobium* and *Dermestes*, and which, nevertheless, are very remote from them in their characters, and even in their habits.

These insects are very small, and of sombre colours. They are found at the end of winter, and the commencement of spring, on the coriaceous excrescences which grow on willows, and the old trunks of oaks. They are sometimes there in very great quantities, and remain at the lower part of the excrescence. As soon as one approaches to seize them, they fold back their feet and antennæ against their body, and suffer themselves to fall down.

The species of this genus which is most common is the *Cis boleti*, and is to be found in the spring season. It is of an obscure brown. Its corslet and elytra present slight silky reflections. Its feet are testaceous. To the same genus must be referred the *Anobium reticulatum* of Fabricius, a species which is found in the forest of Fontainbleau.

The larva of the *Trogossita carraboides*, *Tenebrio Mauritanicus*, Lin., which is named *Cadelle* in French, has been observed by Dorthes, and forms the subject of a very interesting memoir. Dorthes shut up some of these larvæ in a bottle, with some corn. They lived there until the winter, but none of them changed into the nymph-state. On the first approach of cold weather, they all died. Having observed several of these larvæ climbing along granaries, and removing from the heaps of corn, he presumed that they did so for the purpose of taking refuge in holes, and there undergoing the transformation into the nymph-state. He then shut some of them up in boxes, containing corn on one side and earth on the other. These larvæ sunk themselves into the earth, and at the end of their metamorphosis, he beheld the perfect insect issue forth. Dorthes frequently saw these insects pair upon the corn, but he was unable to trace them to the oviposition, and it still remains uncertain whether the insect deposits its eggs on the grain, or whether the larva introduces itself there. These insects are never known in their perfect state to attack the corn, for when enclosed in a bottle with some grain they have never touched it, but rather endeavoured to devour one another. They have been found on the following day without antennæ and feet. They are often observed on corn, trying to devour the *tinæ*, or engaged in perpetuating their species. The *trogossita caraboides*, however, and the *blue trogossita*, have been found in old bread, the crumbs of which were gnawed away by them.

The larva, when arrived at its full growth, is about eight lines in length, and scarcely more than a line in breadth. Its body is whitish, composed of twelve tolerably distinct rings,

Pl. 60

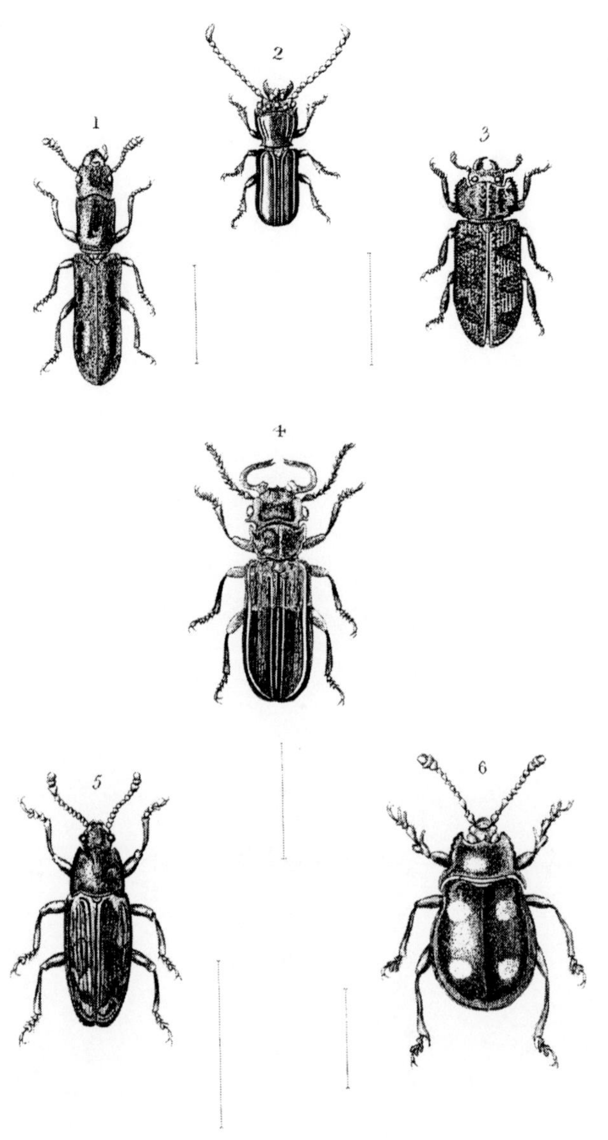

1. Temnoscheila *splendens.* *G.R. Gray.* 2. Parandra *fasciata.* *G.R. Gray.* 3. Trogosita *squamosa.* *G.R. Gray.* 4. Cucujus *Dejeani.* *G.R. Gray.* 5. Triplatoma *variegata.* *Westwood.* 6. Eumorphus *tetraspilotus.* *Hope.*

London. Published by Whittaker & C.° Ave Maria Lane. 1831.

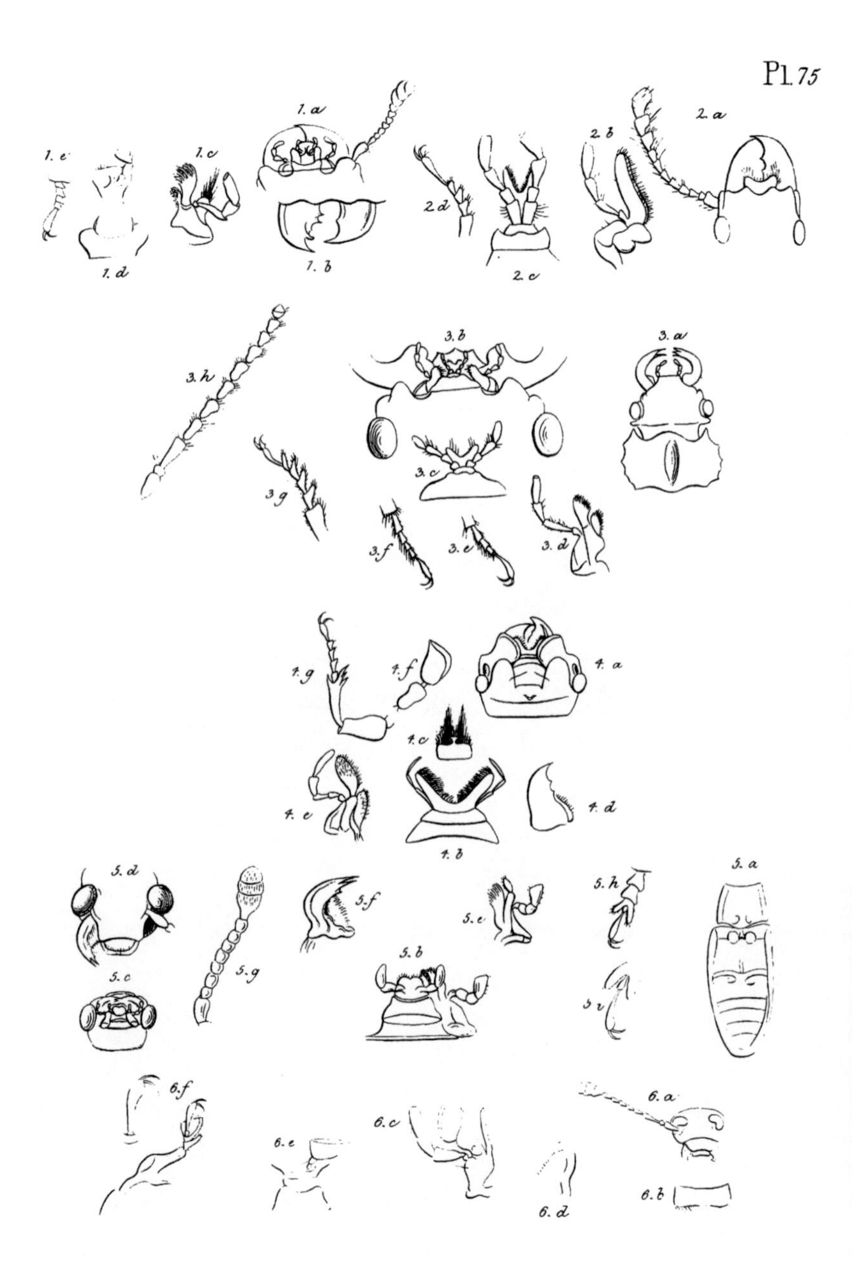

1. Temnoscheila *splendens*. 2. Parandra *fasciata*. 3. Cucujus *Dejeani*. 4. Trogosita *squamosa*. 5. Triplatoma *variegata*. 6. Eumorphus *tetraspilotus*

London. Published by Whittaker & Co Ave Maria Lane. 1837

and bristling, with some scattered, short, and rather stiff hairs. The head is black, hard and scaly, armed with two arched, trenchant, corneous, and very hard mandibles. Some obscure spots are observable, placed in the first three rings. The last is terminated by two corneous, and very hard crooks. It has six short, scaly feet, which proceed from the first three rings of the body. In the southern provinces it is very abundant, and does the greatest mischief imaginable to the grain. It is much more hurtful than the larvæ of *Curculio* and *tinea*, which remain in the interior of the grains which they inhabit, and which suffice for their whole support. But the *cadelle*, whose body is much larger, requires much more nutriment. Accordingly, it does not enter into the bottom of the grain. It attacks it from without, passes from one grain to another, and a single larva can destroy a considerable quantity of corn. It is principally towards the end of winter, when it has acquired its full growth, that it commits the greatest damage. At the commencement of spring it quits the heaps of corn, and proceeds to the holes, clifts, and crevices, in barns, and sinks into the earth or dust, for the purpose of undergoing its metamorphosis. The perfect insect shews itself in the spring, and during the whole summer.

We have figured two new species of Trogosita. Of the first, *Temnoscheila splendens*, the head and thorax are copperish green, the back reddish copper, with two large violet spots on each side. This sub-genus is described by Mr. Westwood in the Zoological Journal. The second species we have named *Trogosita squamosa.* It is black, covered with varied coloured scales, the thorax trilobed, and serrated on the sides. The length is seven lines and a half. This species is from Melville's Island, and may be formed into a separate sub-genus.

We have also figured a new species of *Passandra*, and named it *fasciata.* It is fulvous red, with the antennæ, the sides of the thorax, the suture and margin of the elytra, black; the length is nine lines. It is from Carthaginia.

We have passed over in silence most of the sub-divisions of this family, of which, however, nothing could, in the present state of knowledge, be said touching their instincts, manners, and habits; we might, indeed, in this and other places, enlarge on the history of these artificial divisions themselves, and on the modes in which different entomologists have admitted, combined, modified, or changed them; but this is a branch of entomological knowledge, if it may be so called, not perhaps altogether necessary in our supplements, which are intended rather to refer to Nature's works, than to the labours of systematists.

THE THIRD FAMILY

OF

TETRAMEROUS COLEOPTERA.

PLATYSOMA

APPROXIMATES to the last as well in their internal anatomy, and their tarsi, whose articulations are all entire, as in their habits; but their antennæ are equally thick throughout, or are thinner toward the end. Their mandibles are always projecting; the lingua is bifid or emarginated; their palpi are short, and the body is depressed, elongated, with the corslet sub-quadrate. They may be reduced to a single genus,

CUCUJUS, Fab.

Of these we may distinguish CUCUJUS, proper, whose antennæ, much shorter than the body in many, are composed of articulations in the form of a reversed cone or top, and nearly grained, the first of which is shorter than the head.

DENDROPHAGUS, Gyll. *Cucujus*, Fab. Pyk., have these organs in general formed of cylindrical articulations, elongated, the first of which is larger than the head, and the second and third shorter than the following. The labial palpi terminate in a knob.

ULEOIOTA, Lat. *Brontes*, Fab., have the antennæ analogous to the last, but the third articulation is as long as the following, and all the palpi are more slender at the end. The mandibles of the most common species (*flavipes*), and on which M. Dufour has given some anatomical observations, have, in the males, an elongation in the shape of a long sharp horn.

As we have nothing to add by way of supplement to this family, unless it be to observe that this sub-genus, CUCUJUS, differs from that so named by Geoffroy, which are the *Buprestes* of Linnæus, we proceed at once to the next family of the text, premising that we have inserted a figure of a new species of *Cucujus*, named *Dejeani*, which is fulvous, with the antennæ, tibiæ, tarsi, and the posterior half of the elytra, black; the mandibles are long, curved, and bifid at the tip. The females are like the other sex, except that the mandibles are very short. This species is from Brazil, and is seven lines long.

THE FOURTH FAMILY

OF

TETRAMEROUS COLEOPTERA.

Longicornes

Has the under part of the first three articulations of the tarsi furnished with brushes, the second and third heart-shaped, and a little swelling or knob, like an articulation, at the origin of the last. The lingua, supported on a short and transverse mentum, is commonly membraneous, heart-shaped, emarginated or bifid, corneous, and in very short and transverse segments of circles in others (*Parandra*). The antennæ are filiform or setaceous, in general, as long as the body or less, sometimes simple in both sexes, sometimes serrated, pectinated, or fan-shaped in the males. The eyes, in many, are kidney-shaped. The corslet is in form of a trapezium, or narrowed in front, in those whose eyes are round, entire, or a little emarginated; in this case, moreover, the feet are long and slender, with the tarsi elongated.

M. Leon Dufour, in opposition to the opinion of M. Marcel de Serres, denies the existence of a gizzard in these insects. The alimentary tube is in general beset with papillæ, is preceded by a crop, but less or little decided in *Lamia* and *Leptura*, which, in our method, terminate the family. The testicles are formed by distinct capsules, or sperm bags, which are pediculated, large, and vary in number in the different genera.

Their larvæ live almost entirely in the interior, or under the bark of trees; they are either without feet, or have them very small; their body is soft, whitish, thicker in front, with the head scaly, provided with strong mandibles, and without any other projecting parts. They do great damage to trees, especially those full grown, often bore very deeply, or cut channels in them. Some gnaw the roots of plants. The females have the abdomen terminated by a tubular and corneous oviduct. These insects produce a slight sharp sound by the friction of the pedicle of the base of the abdomen against the interior parietes of the corslet, when they cause it to enter there, and withdraw it alternately.

We shall first divide the longicornes into two sections. Those of the first have the eyes strongly emarginated or crossed, or elongated and narrow; their head is buried as far as these organs, in the corslet, without being distinguished from it by a sudden narrowing, forming a kind of neck. In many it is vertical.

Some have the last articulation of the palpi sometimes nearly in the form of a cone or reversed triangle, sometimes sub-cylindrical and truncated at the end; the lobe terminating the jaws, straight (not bent inward at the end); the head is commonly advanced, or simply bent, and in those which, by a very rare exception (*Dorcacerus*) it is vertical, its width is nearly equal to that of the body, and the antennæ are very wide at their insertion and have spines; the corslet is often very unequal or squared, and rarely cylindrical.

The longicornes are divisible into two principal sections, or small tribes.

Prionii, whose characters are, labrum none, or very small and indistinct; mandibles strong, or very large, especially in many males; the interior lobe of the jaws none, or very

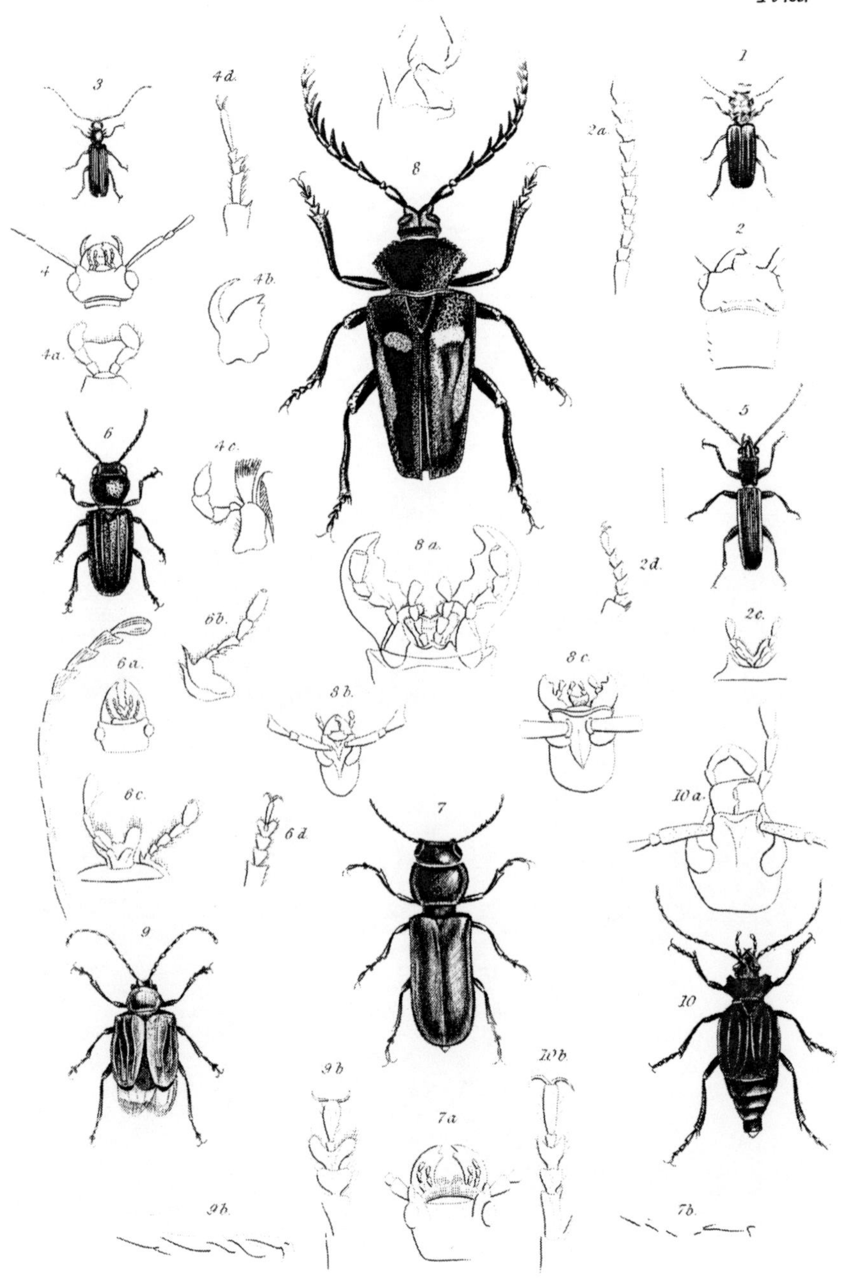

1. Cucujus mandibularis. Gory.
2. Details of Cucujus depressus. Fab.
3. Brontes spinicollis Gory.
4. Details of Brontes flavipes. Fab.
5. Dendrophagus crenatus. Payk.
6. Spondylus buprestoides. Fab.
7. Paranda lineola. Gory.
8. Prionus Desmarestii. Guer.
9. Anacolus sanguineus. Serv.
10. Prionapterus staphilinus. Guer.

London, Published by Whittaker & C.º Ave Maria Lane, 1832.

small; the antennæ inserted near the base of the mandibles, or the emargination of the eyes, but not surrounded by them at their insertion; the corslet, in general, trapezoid or squared, channelled, or indented laterally. The first genus,

PARANDRA, Lat. *Attelabus*, De G. *Tenebrio*, Fab.,

Have, like the following, the antennæ simple, nearly grained, compressed, of the same thickness throughout, as long or longer than the corslet, scarcely reaching to the end of the first articulation of the palpi, and distinguished, as well as some other genera of this family, by the lingua being corneous, formed like the segment of a circle, very short, transverse, without emargination or lobes; and by the tarsi, whose penultimate articulation is slightly bilobed; and the last, evidently longer than the preceding taken together, has between its hooks a small appendix with two threads at the end. The body is parallellipiped, depressed, with the corslet square, rounded at the hinder angles, without spine or teeth. They belong to America.

SPONDYLIS, Fab. *Attelabus*, Lin. *Cerambyx*, Deg.—These approximate to parandra in their antennæ, and the exiguity of the maxillary lobes, but differ from them as to the lingua, as in all the following longicornes the lingua is membranous and heart-shaped. They also differ with reference to the tarsi; the penultimate articulation is deeply bilobate, and the last is not longer than the preceding together, and is without appendix. These are also distinguishable from the following genera, by their sub-globular corslet, without edging, and deprived of teeth or spines. Their larvæ live in the firs and pines of Europe.

The third and last genus of this tribe,

PRIONUS, Geoff. Fab. Oliv.

Has the antennæ larger than the head and corslet, serrated

or pectinated, in some, simple, slender toward the end, and with the articulations elongated in others. The terminal lobe of the jaws is as long or less than the first two articulations of the palpi. The body is in general depressed, with the corslet square or trapezoid, indented or spiny, or angular at the sides.

This genus includes many species of various forms.

Some with the body nearly parallelipiped or elongated, straight, with the corslet much shorter than the abdomen, square or trapezoid, and much bent laterally, the antennæ simple, or slightly serrated (*P. scabricornis*, Fab.).

Others, in general, less oblong, a little bent in front, the corslet much indented, the elytra pectenated and serrated in the males (*Cerambyx coriaceus*, Lin.).

Others, belonging to Brazil, with the elytra small, triangular, not entirely covering the abdomen. They appear to form a new sub-genus, ANACOLUS *A.* (*Sanguineus lugubris*, Enc. Method.).

Lastly, other prioni, many with varied and metallic colours, have the body shorter and larger, sub-oval, with the head elongated posteriorly behind the eyes, &c. (*P. nitidus*, and others of Fab.)

The second principal section or tribe,

CERAMBYCINI,

Have the labrum very apparent, reaching the whole width of the anterior extremity of the head; the two maxillary lobes are very distinct and projecting; the mandibles of common size, and but little different in the two sexes; the eyes always emarginated, and surrounding, at least, in part, the base of the antennæ, which are commonly as long or longer than the body; the thighs, of at least the first four legs, are commonly knobbed, ovoid or oval, narrowed into a pedicle at the base.

First will come those whose last articulation of the palpi

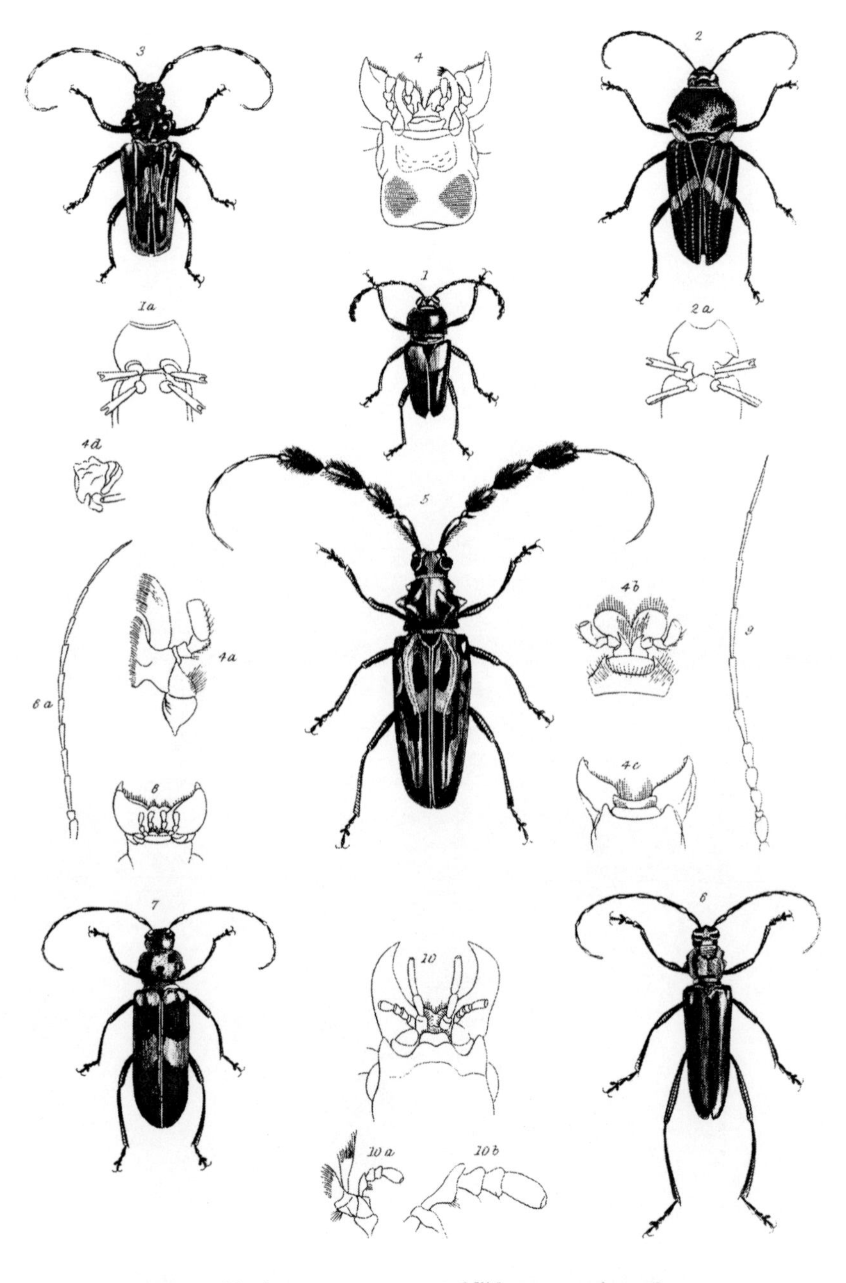

1. *Lissonotus unifasciatus.* – Gory.
2. *Megaderus stigma.* – Fab.
3. *Trachyderes nigrofasciatus.* – Gory.
4. Details of *Trachyderes. succinctus.* – Fab.
5. *Lophonocerus barbicornis.* – Oliv.
6. *Callichroma speciosa.* – Gory.
7. *Acanthopterus tripunctatus.* – Gory.
8. Details of *Acanthopterus. budensis.* – Goeze.
9. Ant of *Cerambyx. heros.* – Fab.
10. Details of Cer.[x] (*callichroma*) *hemipterus.* – Oliv.

London, Published by Whittaker & C.º Ave Maria Lane 1837.

is always manifestly thicker than the preceding, in form of a triangle or reversed cone ; the head is not perceptibly narrowed and elongated behind like a muzzle ; the corslet is not enlarged from front to rear, and has not a trapezoid figure, and the elytra are either very short or scale-formed, or narrowed abruptly a short distance from their base, and terminated like an awl. We may designate this division by the epithet *regular* cerambyx, in opposition to those which follow, which are in many respects anomalous, and the last of which appear to be connected with those of the tribe immediately succeeding this. They compose the genera *Cerambyx, Clytus, Callidium,* of Fabricius, a part of *Stenocores,* a different genus made before him by Geoffroy.

CERAMBYX.

Several South American species, shorter and larger than the following, with the antennæ often pectenated, serrated, or spiny, are remarkable for the spreading of the corslet, whose length nearly equals half that of the elytra; sometimes smooth, it is nearly semi-orbicular, unindented only at the posterior angles, sometimes very unequal and tubercular, their presternum is either squared or terminated in a point, either even truncated, entire or emarginated at its posterior extremity, which is applied to an anterior projection of the mesosternum. The fore feet at least are wide at their insertion. The scutellum is large in many ; the tarsi are short and dilated.

Those of this division in which the corslet, nearly semi-orbicular, and always very large, is united or simply chagrined, with a single tooth on each side to the posterior angles, whose posterior extremity of the presternum is flat, truncated, either without emargination or with, and applied on the mesosternum, whose scutellum is always large, and which have the legs very wide, form two sub-genera.

LISSONOTUS, Dalm. *Cerambyx,* Fab., with the antennæ

strongly compressed, serrated, or semi-pectenated, long, and with the posterior extremity of the presternum without emargination.

Megaderus, Dej. *Callidium*, Fab., with simple antennæ, shorter than the body, the posterior extremity of the presternum emarginated, receives the opposite end of the mesosternum, so that when joined they appear to form but a single plane.

Those whose corslet is very unequal, tubercular, or much indented, with the presternum squared or terminated behind in a point, have been dispersed into many sub-genera.

Here the antennæ are long, setaceous, simple, or at most a little spiny, or furnished with bundles of hair.

The corslet is always large, unequal, scarcely broader than it is long.

Dorcacerus, Dej. *Cerambyx*, Oliv., distinguished from all the others by their head being vertical, large, nearly as wide as the corslet, measured in its greater transverse diameter, plain, and downy in front. The antennæ are very wide, the presternum is not elevated into a keel, and terminates in a simple point. The scutellum is small.

Trachyderes, Dalm. *Cerambyx*, Fab.—The corslet is much wider than the head, with the hind extremity of the presternum, and often also its opposite carenated; the scutellum is elongated, the elytra are larger at their base and diminish, the antennæ are without bundles of hair.

Lophonocerus, Lat., have also the head narrower than the corslet, the presternum carenated, but the corslet and scutellum are smaller, and the elytra enlarge, or at least do not become narrow; the third and three following articulations of the antennæ have bundles of hair.

There the antennæ are shorter than the body, pectenated or serrated. The corslet is transversal, indented laterally. The elytra enlarge behind.

CTENODES, Oliv. Klüg.—Now the corslet, sometimes nearly square or cylindrical, sometimes orbicular, or nearly globular, is much shorter than the elytra, at least in those in which it increases in width; the presternum has no keel or elongated point at its posterior extremity. The scutellum is always small. The legs are near together at their insertion.

One sub-genus only, that of PHŒNICOCERUS, Lat., differs from the following in the antennæ of the male, the articulations from the third being long narrow laminæ, and forming a large fan. Only one species is known, which is from Brazil (*P. Dejeanii*).

In the other the antennæ are at most spiny, or a little serrated.

Many remarkable for their colours, and pleasant smell, are anamalous in reference to the relative proportions of the palpi. The maxillaries are smaller than the labial, and even shorter than the terminal lobe of the jaws, which often projects. The body is depressed, with the fore part of the head narrowed and pointed. The hind legs are often much compressed. These compose the sub-genus CALLICHROMA, Lat. *Cerambyx*, Fab., including *C. Moschatus*, Lin., and *C. Ambrosiacus*, Stev., belonging to Europe.

Others of the same division, with ordinary maxillary palpi, are distinguishable by the antennæ, which, in the males, have twelve articulations; they are long, setaceous, spiny, or barbed. The corslet is indented or spined on the sides. We shall unite them in the sub-genus ACANTHROPTERA, Lat. *Callichroma*, *Purpuricenus*, *Stenocorus*, Dej. Dal.

Some American species, with the corslet nearly square, or sub-cylindrical, and the elytra in general terminated with one or two spines, are the Stenocori of Dalman.

Others, with the body rather elevated, the corslet nearly globular, and the antennæ simple, are the *Purpuricenus* of Ziegler and Dej.

The following cerambyces have only eleven articulations to the antennæ.

Some of the males, at least, have the antennæ long, setaceous, the last articulation of the palpi formed like a reversed cone, the corslet either nearly square or a little dilated in the middle, oblong or sub-cylindrical; it is often rugose and tuberculated laterally. They compose the sub-genus CERAMBYX, proper, Lin. Fab.

HAMATICERUS has been distinguished generically, having the corslet unequal or rough, in general, spined or tuberculated, and dilated in the middle of its side, with the third, fourth and fifth articulation of the antennæ evidently thicker than the following, thickened and rounded at the end; these suddenly longer and slenderer, and sub-cylindrical, form with the preceding a sudden transition. These organs are much longer in the males, *C. heros*, Fab. Oliv. *C. cerdo*, Lin.

We refer to the same sub-genus several species of callichroma, of Dej., with the corslet united, but little unequal, longer in proportion, oval, and truncated at both ends, or sub-cylindrical.

We shall moreover unite to the sub-genus of capricorns, the *Gnoma*, of Dejean. Their corslet is much larger and cylindrical. The inner angle of the upper end of the articulations of the antennæ is a little dilated. The palpi are nearly filiform, and the mandibles have a tooth on the inner side. (*G. ruficollis*, Fab., and *G. sanguinea*, Dej.)

Those whose antennæ are not commonly longer than the body, and rather filiform than setaceous, the corslet always imperfect is sometimes nearly globular or orbicular, and sometimes narrower, sub-cylindrical, and simply dilated and rounded in the middle, with the palpi always very short, and terminated by an articulation rather thicker and wider than in the preceding, in form of a reversed triangle, compose in the first works of Fabricius, and in the Entomology of Oli-

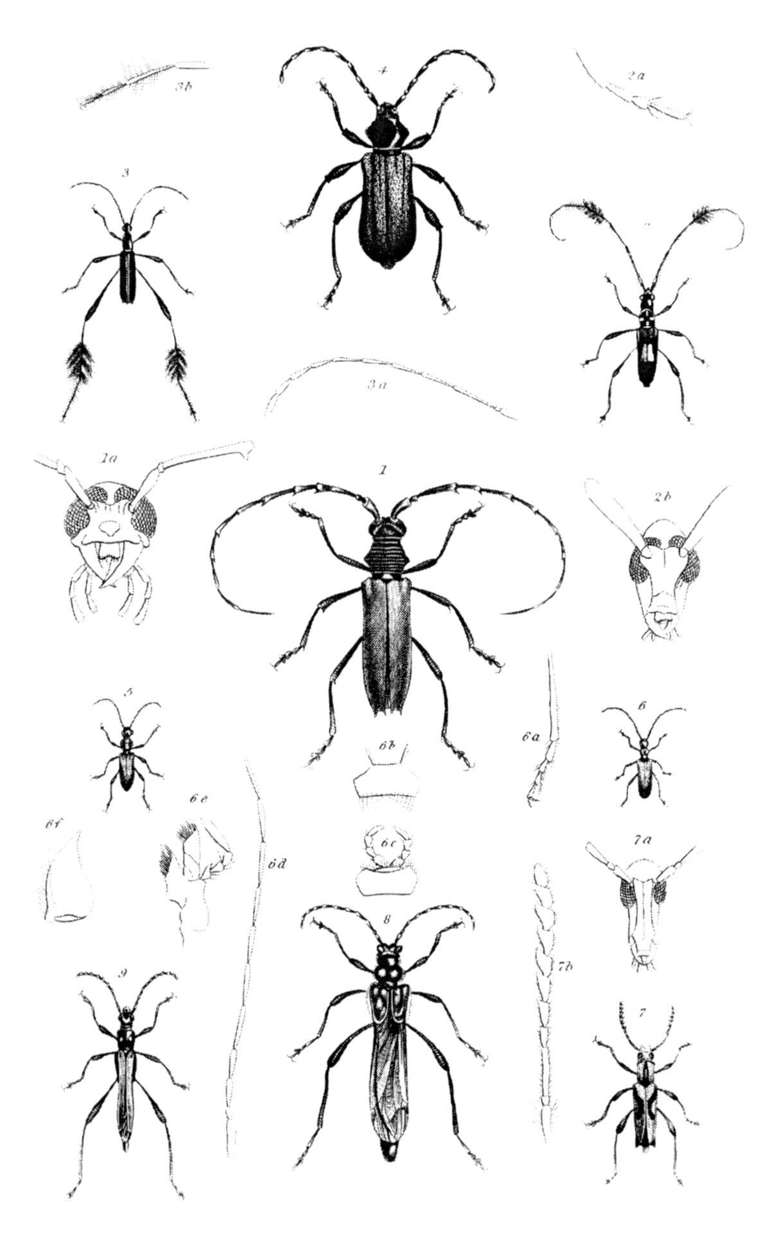

1. *Cerambyx rufipennis*._Gory.
2. *Cerambyx speculifer*._Gory.
3. *Cerambyx hirtipes*._Gory.
4. *Callidium insubrecum* Ziegler.
5. *Certallum ruficolle*._Fab.
6. *Obrium ferrugineum*._Fab.
7. *Rhinotragus coccineus*._Gory.
8. *Necydalis major*._Lin.
9. *Stenopterus elegans*. Klug.

London; Published by Whittaker & C° Ave Maria Lane 1832.

vier, the genus Callidium, which may nevertheless be divided into three.

Those with the head at least as wide as the corslet, which is sub-cylindrical, and simply dilated and rounded in the middle, compose the genus *certallum*, of Megerle and Dejean.

Those with the head narrower than the corslet, elevated nearly globular, compose that of *Clitus*, of Fab.

Lastly, those with the corslet also wider than the head, flatted and orbicular, have still the generic name of callidium, *Cer. sanguineus*, Lin. *Leptura arcuata*, Lin.

We shall terminate this tribe by insects, which, in reference to the palpi, the form of the head, the corslet, and that of the elytra, as well as in their proportions, have exceptions or remarkable anomalies.

We shall begin with those the form of whose corslet is very analogous to that of the preceding, and especially the certallum. It is as wide as the head and the base of the elytra, or very little narrower, either sub-cylindrical or rounded, or nearly orbicular, and in both larger towards the middle. The last articulation of the palpi is sometimes slender at the end, and terminating in a point, sometimes thicker and truncated, and formed like a reversed cone. All the thighs are like knobs, supported on a sudden, slender and elongated pedicle. The elytra of most of them are either very short, or contracted suddenly at a short distance from their base, and finally subulated.

Then follow these whose elytra are like those of the preceding insects. The first genus,

Obrium, Mej. Dej. *Callidium*, *Saperda*, Fab.,

Have the head rounded, and not elongated before, like a muzzle, the palpi filiform, with the last articulation terminating in a point; the antennæ long and setaceous, the corslet long, narrow, sub-cylindrical, or like a truncated oval.

The second genus,

RHINOTRAGUS, Dalm.,

Differs from the last by having the head elongated, and narrowed in front like a muzzle; the last articulation of the palpi is a little thicker than the preceding, and truncated at the end; the antennæ are shorter than the body, a little dilated and serrated at the end, and the corslet is sub-orbicular. These insects are allied to the following genus,

NECYDALIS, of Lin.

The only one of this tribe, with the elytra very short, or scale-shaped, or elongated, as in common, to the end of the abdomen, but contracted suddenly a little from their insertion, then very narrow, and proceeding to a point, or terminated like an awl. The last insects resemble *œdermera*, with which Fabricius united them in this respect only. The last articulation of the palpi is a little larger, and nearly in the form of a reversed cone, and compressed. The abdomen is long, narrow, contracted, and as if pedicled at the base. The wings are folded only at their extremity.

Those with subulated elytra will form the first sub-genus, STENOPTERUS, Illig., from which may be separated several exotic species, with the antennæ shorter, thicker, and more serrated toward the end.

Those with short, scale-formed elytra, compose the sub-genus, NECYDALIS, proper, which corresponds with the genus, *Molorchus*, of Fab.

Some Polynesian insects, which in a natural order might be placed between *Lamia* and *Leptura*, will terminate the division of Cerambyx.

The palpi are nearly filiform, with the last articulation

Pl. 95.

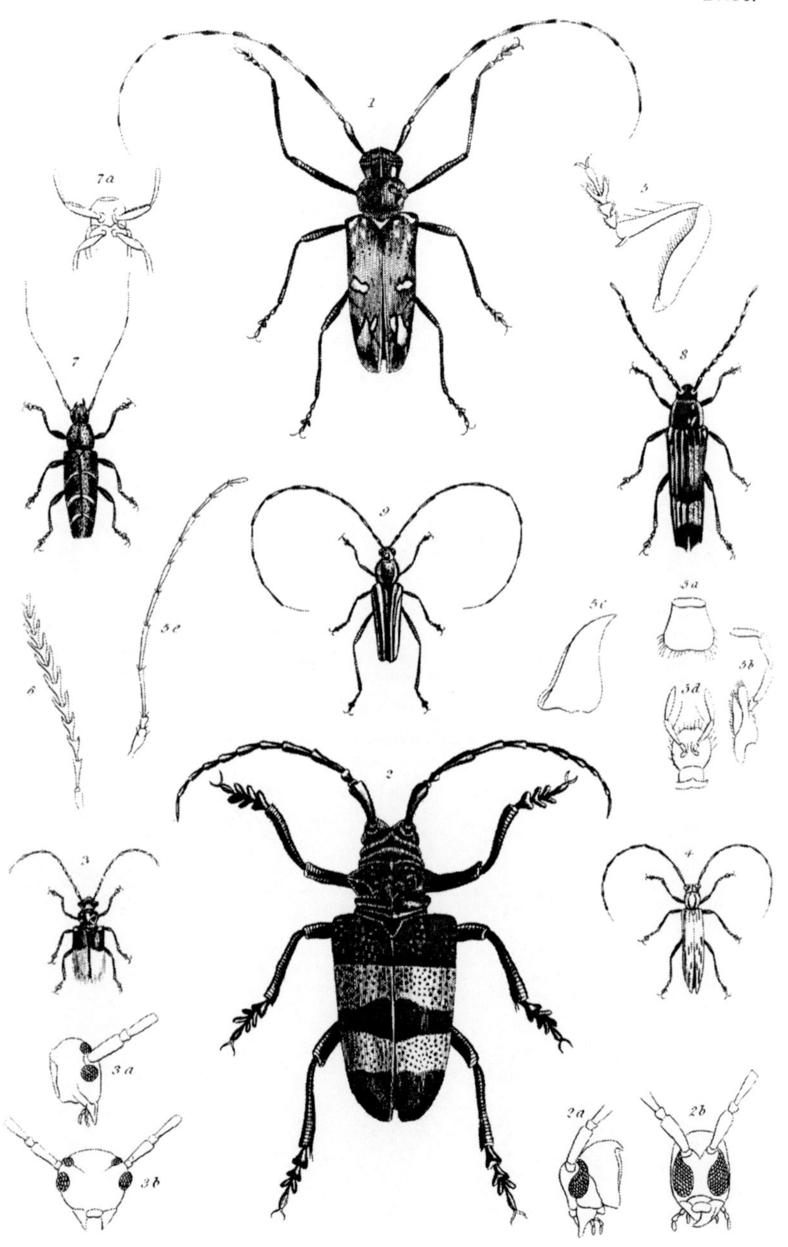

1. *Acrocinus trochlearis*.. Gory.
2. *Lamia aurocinta*.. Gory.
3. *Tetraopes dimidiata*.. Gory.
4. *Saperda albicans*.. Guer.
5. Details of *Saperda Atkinsoni*.. Curtis.
6. Ant of *Distichocera maculicollis*.. Kirby.
7. *Tmesisternus bizonulatus*.. Guer.
8. *Tragocerus bidentatus*.. Donov.
9. *Leptocera bilineata*.. Gory.

London, Published by Whittaker & C° Ave Maria Lane, 1832.

sub-cylindrical, rather more slender at the base; the corslet in general smooth, or but little uneven, without pointed tubercles, widened from front to rear, in form of a trapezium, or a truncated cone, as in the last tribe of this family; the abdomen is nearly in the shape of a reversed triangle in most, and the elytra are truncated at the end. These insects will form four genera.

DISTICHOCERA, Kirby.

In which the antennæ of the male dilate toward the end, with the articulations beginning at the third, furcated at the end.

TMESISTERNUS, Latr.,

With simple setaceous antennæ, larger than the body, the corslet lobed behind, with the back part of the presternum elongated, truncated, and received into an emargination of a projection of the mesosternum.

TRAGOCERUS, Dej.,

Without presternal projection; the antennæ filiform, a little shorter than the body, and slightly serrated; the corslet is unequal, rather sinuous latterly, and the elytra form a long square.

LEPTOCERA, Dej.,

Also without any projection on the presternum; but the antennæ are setaceous, much larger than the body especially in the male; the corslet is united, and forms a truncated cone; the abdomen and elytra are sub-triangular.

The Longicornes of our third tribe (*Lamiariæ*,) are distinguished by a vertical head, nearly filiform palpi, but terminated by a pointed ovoid articulation; the outer lobe of the jaws is narrowed at the end, and bends towards the inner division. The antennæ are often setaceous and simple, and the corslet, except the tubercles or spine on the sides, is nearly of the same width throughout. Some species are apterous, which is peculiar to this division.

This tribe is composed of the genera *Lamia*, *Saperda* of Fab., some of his Stenocores, and some of Coloboth and Cerambyx of Dej.; but I have not as yet discovered characters which may sufficiently separate the first of these genera from the following.

C. longimanus of Lin. does not belong to this genus, but is *sui generis* of the tribe of Lamiariæ.

ACROCINUS, Illig. *Macropus*, Thunb.,

Distinguished from all this family by having a moveable tubercle terminated in a point, but not a spine, on each side the corslet. The body is flat, with the corslet transversal, the antennæ long and slender, and the fore feet longer than the others; the elytra are truncated at the end and terminated by two teeth, the outer one being the strongest.

All the others compose but one genus,

LAMIA,

Which we shall divide into two sections, 1st, with the corslet tubercular, ridged, or spiny, and 2nd, with it smooth and cylindrical. The first is divisible into winged and apterous.

Those with the body short, wide, depressed, the corslet transversal, the abdomen nearly square, the legs strong, and tarsi dilated, include the genus *Acanthocinus* of Meg. and Dej. (*Lamia ædilis*, Fab. of Europe.)

Others analogous to these in form, but with the antennæ barbed, or with tufts of hair, form the genus *Pogonocherus* of the same naturalists.

Others, *Tetraopes*, with the body a little cylindrical and elongated, have each eye entirely divided into two by the tubercle, from which the antenna springs.

Other Lamia of Fab., with the body long and narrow, the antennæ very long, a strong spine on each side of the corslet, the fore legs a little bent, and the intermediate with a tooth on the inner side, compose the genus *Monachammus* of Dej., but as he has not given the character, I refer to them by presumption.

In the catalogue of Coleoptera of Dej., if we except the apterous species, the other *Lamiæ* of Fab. preserve that generic name, but it appears by another catalogue of M. Dahl. that two species of France, *Curculionoides*, *nebulosa*, have been formed into the genus *Mesosa* of M. Mégerle. If we suppose that saperda differs from Lamia by the want of lateral points to the corslet, these species approach saperda in this respect, but the body is shorter and wider, and in this respect approaches to Lamia. (*L. curculionoides*, Fab. *Cer. textor*, Lin.) These species compose the genus *Dorcadion* of Dalman adopted by most entomologists.

M. Mégerle has formed some small species into the genus *Parmena*, but they do not appear to me to differ from the others, except in having the antennæ longer than the body, and the articulations more elongated, and rather cylindrical than conical. Other larger species (*tristis*, *lugubris*, *funesta*,) should be joined to them. *C. fuliginator*, Lin. is among these.

Other Lamiariæ have the corslet cylindrical, and without lateral tubercles or spines. Their body is always elongated, and nearly linear in many. They compose the genus

SAPERDA of Fab.

That which he names *Gnoma*, as to the direction of the head and the parts of the mouth, resembles Lamia, but the corslet is as long as the abdomen, cylindrical, a little narrower in the middle, without spines or tubercles. The antennæ are longer than the body, sometimes furnished with tufts of hair; the fore feet are elongated.

The Count Dejean has separated from Saperda the genera *Adesmus, Apomecyna, Colobothea.*

Adesmus has the first and third articulation of the antennæ elongated.

Apomecyna has the body cylindrical, the antennæ filiform, short, ending in a sharp point, with the third and fourth articulation long, and the next short.

Colobothea have the antennæ very near together at their insertion, the body compressed, and as if squared laterally, the cases emarginated or truncated at the end, with its outer angle elongated like a tooth or spine.

Other saperdæ, of Brazil, whose corslet is as wide as the elytra, or nearly so, whose antennæ have the third and fourth articulation, or at least the preceding, very elongated or dilated, furnished with hair, and the last suddenly shorter, and whose elytra are enlarged and rounded at the end, form another division.

Many others, whose body is always long and narrow, with twelve instead of eleven articulations to their antennæ, form another sub-genus.

Cer. Carcharias, Lin., and *Cer. linearis,* Lin. are among the Saperdæ proper.

Other species, with the body still narrower, and the antennæ excessively long, nearly as slender as a hair, have been described.

The fourth and last tribe, *Lepturetæ*, have the eyes round, entire, or slightly emarginated, whose antennæ are inserted in part, or, at most, at the anterior extremity of this slight emargination; the head is always bent, elongated behind the eyes in many, or narrowed suddenly, like a neck, at its junction with the corslet; this last part is conical or trapezoid, and narrowed before. The elytra become narrow as they

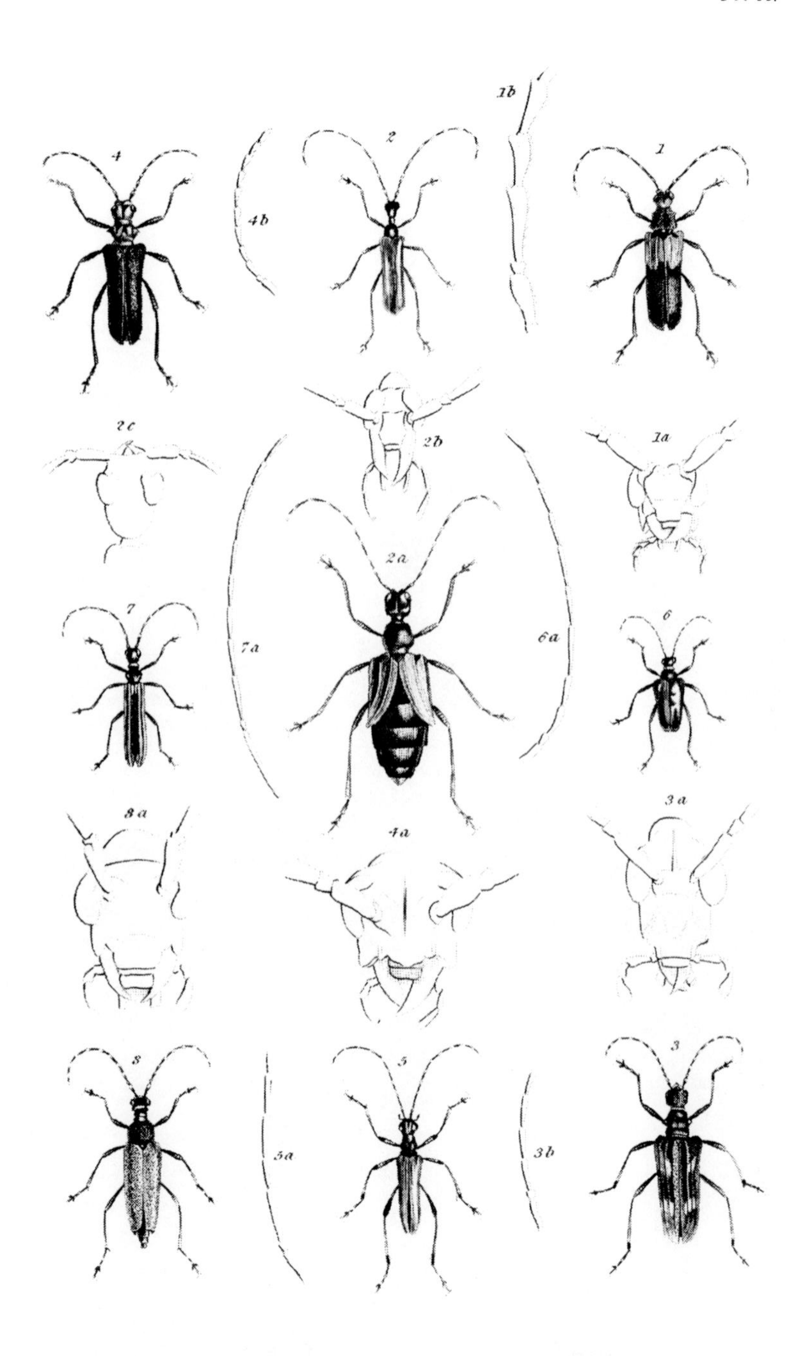

1. *Desmocerus cyaneus.* Fab.
2. *Vesperus Græcus.* Guer.
3. *Rhagium bifasciatum.* Fab.
4. *Rhamnusium salicis.* Fab.
5. *Toxotus meridianus.* Fab.
6. *Pachyta Laportii.* Guer.
7. *Stenoderus ceramboides.* Kirby.
8. *Leptura annulata.* Gory.

London, Published by Whittaker & C° Ave Maria Lane 1832.

advance. This tribe, after deducting certain species belonging to the preceding, compose the genus,

LEPTURA of Lin. *Stenocorus*, Geoff. *Rhagium*, *Leptura*, Fab.

Sometimes the head is elongated behind the eyes. The antennæ, often shorter than the body, are near each other at their base, inserted out of the eyes on two little tubercles, and separated by a deep line. The corslet is commonly tubercular, or spiny on the sides.

Here the palpi are filiform; the last articulation of the maxillary is sub-cylindrical, and the same of the labial ovoid; the third of the antennæ and the two following are dilated at their outer angle, bent, and silky, particularly in the male. These are DESMOCERUS, of Dej. The corslet is in form of a trapezium, without tubercles or points on the side, with the posterior angles very pointed. The jaws and labrum appear to me to resemble lamia. (*Stenocorus Cyaneus*, Fab.)

There the palpi are enlarged at the end, and terminated by a cone-shaped articulation. The antennæ are regular, smooth, and merely pubescent.

Some differ from the rest by the male only being winged. Their corslet is conical, united without spine or tubercles.

They compose the genus, VESPERUS, Dej. *Stenocorus*, Fab. Their head is large, supported on a sort of patella. The antennæ are long, slightly serrated, with the first articulation shorter than the third. The last of the palpi is subtriangular. The eyes are oval, slightly emarginated. The elytra of the females are short, soft, and open.

In the following both sexes are winged, the corslet is spiny or tubercular, laterally unequal, and as if bent at both ends (*Rhagium*, Fab. *Stenocorus*, Oliv., with some of the lepturæ of the former). Subsequent entomologists have divided them into five genera, which may be reduced to four.

Rhagium, Dahl. Antennæ simple, half as long as the body; the last articulation of the palpi forms a triangular knob; head large, and eyes entire; a spine-formed tubercle on each side the corslet.

Rhamnusium, Meg. Antennæ a little shorter than the body, serrated; the third and fourth articulations shortest; eyes emarginated.

Toxotus, Pachyta, Meg. Dej. Antennæ simple, as long, at least, as the body; first articulation much shorter than the head; eyes entire, or nearly so; abdomen triangular, narrowed behind.

Stenoderus, Dej. *Cerambyx*, Fab. *Leptura*, Kirb. *Stenocorus*, Liv. Antennæ long; first articulation as long, at least, as the head; body long, narrow, sub-linear; eyes entire.

Sometimes the head is narrowed suddenly behind, the antennæ wide apart at their insertion, with the two eminences where they spring nearly confounded; corslet almost always united, without lateral tubercles. These are Leptura, Dej. Dahl.

Some have the corslet nearly even above, trapezoid, or conical. *L. armata*, Gyll. *L. calcarata*, Fab.

Others have the corslet much more elevated, and round, or nearly globular (*L. tomentosa*, Fab.).

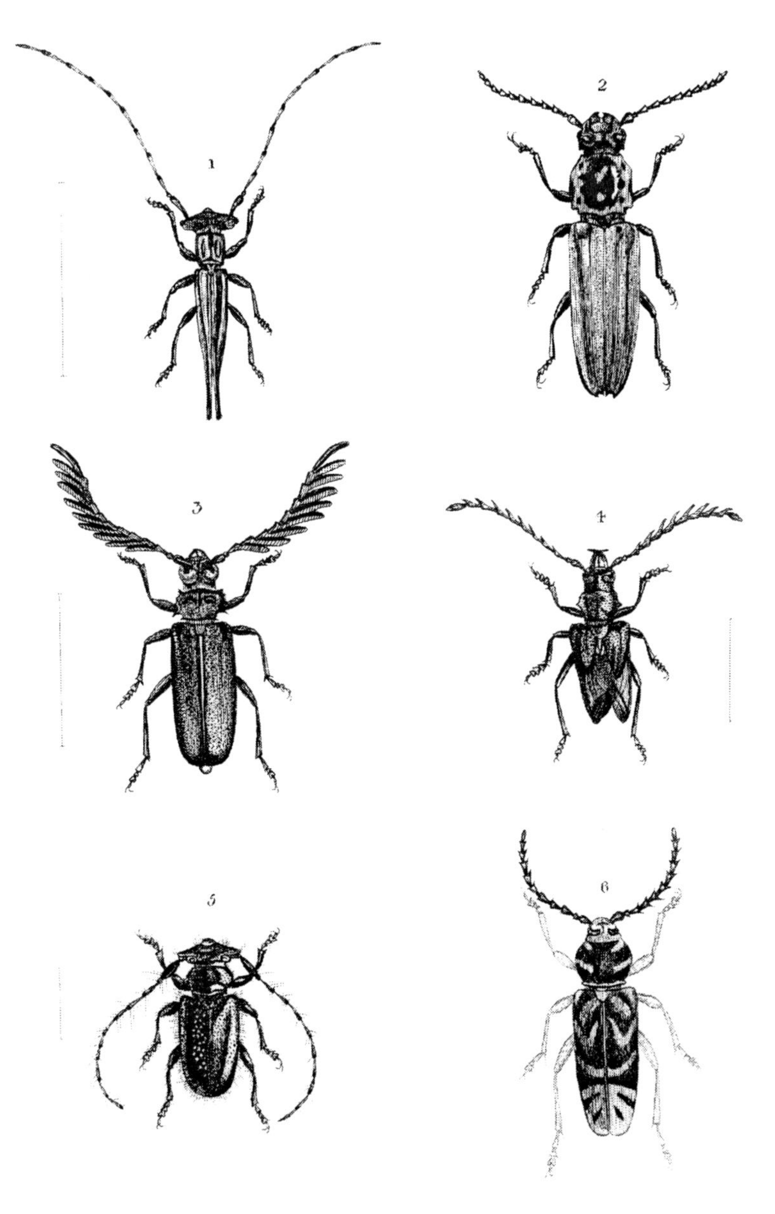

1. Emicodes *Fichteli.* *G.R.Gray.* 2. Coptocephalus *Brasiliensis.* *G.R.Gray.* 3. Phænicocerus *Dejeani* *Lat.* 4. Anacolus *Lugubris.* *Serv.* 5. Eurycephalus *niger.* *G.R.Gray.* 6. Clytus *Hayii.* *G.R.Gray.*

London, Published by Whittaker & C.º Ave Maria Lane, 1831.

I. O. Westwood. del.

1. Coptocephalus *Brasiliensis*. 2. Enicodes *Fichteli*. 3. Anacolus *lugubris*. 4. Phœnicocerus *Dejeani*. 5. Eurycephalus *niger*. 6. Clytus *Hayii*.

London, Published by Whittaker & C.º Ave Maria Lane, 1831.

SUPPLEMENT

ON THE

LONGICORNES.

Of the first genus of this family, (Parandra,) we shall merely observe that, considered in all their natural relations, they appear to conduct us from the cucuji to the spondyli, and the rest of the Longicornes. They were originally left by M. Latreille in the last family.

With the metamorphoses of these insects we are not acquainted. But it is highly probable that they live in wood, or under the bark of trees, after the manner of the *cucuji* and *spondylis*, with which they possess so close an affinity; but the known species are of a brown-maron colour, shining, smooth, and even, and inhabit for the most part the continent of America. The males are a little larger than the females, and have longer mandibles.

The insect of which Fabricius has composed his genus, Spondylis, had been successively placed with *buprestis*, *attelabus*, and the *capricorns*.

The larva probably lives in the interior of wood. Its manners, as well as those of the perfect insect, are unknown.

The insect figured under the name *Coptocephalus Brasiliensis*, appears to form a new sub-genus, but nearly allied to Spondylis. The antennæ and legs are black, the thorax fulvous, with the middle black, the elytra fulvous, with a reddish tint.

It might seem proper that the Prioni should be placed at the head of this family, from their gigantic size, and many strongly marked characters, and very near the Capricorns, with which they have great relations. It is even difficult to establish the certain and precise limits of these two genera, which approximate so closely to each other, both in forms and habits.

These insects, as we have just observed, are very large, and the females are in general more bulky than the males. They are found in large woods and forests. During the day they remain concealed in the holes which their larvæ have made in the trunks of old trees. They issue forth in the evening to fly, and seek out an individual of their own species with which to pair. Their flight is heavy, and the least shock throws them down.

The larvæ of these insects inhabit the trunks of the largest trees, and such as are in a rapid progress to decay. They hasten, too, this decay, by the quantity of holes which they make in them. They differ little from those of the other Coleoptera, which live in wood. They resemble a thick white worm, the body of which is divided into twelve rings. Their head is a little broader than the rest of the body, and of a consistence a little more solid. It is armed with two short and strong mandibles, which serve to cut the wood on which they feed. They have three pair of scaly feet, so small that they are of no kind of use. But, however, their organs of locomotion are formed upon another model, and perfectly appropriate to the places inhabited by these larvæ. Nature has provided them with a multitude of small nipples, which cover the last nine rings of their body; they rest them against the sides of the holes which they inhabit, when they are desirous of traversing them. Afterwards they contract and elongate their rings successively, and push themselves forward with facility.

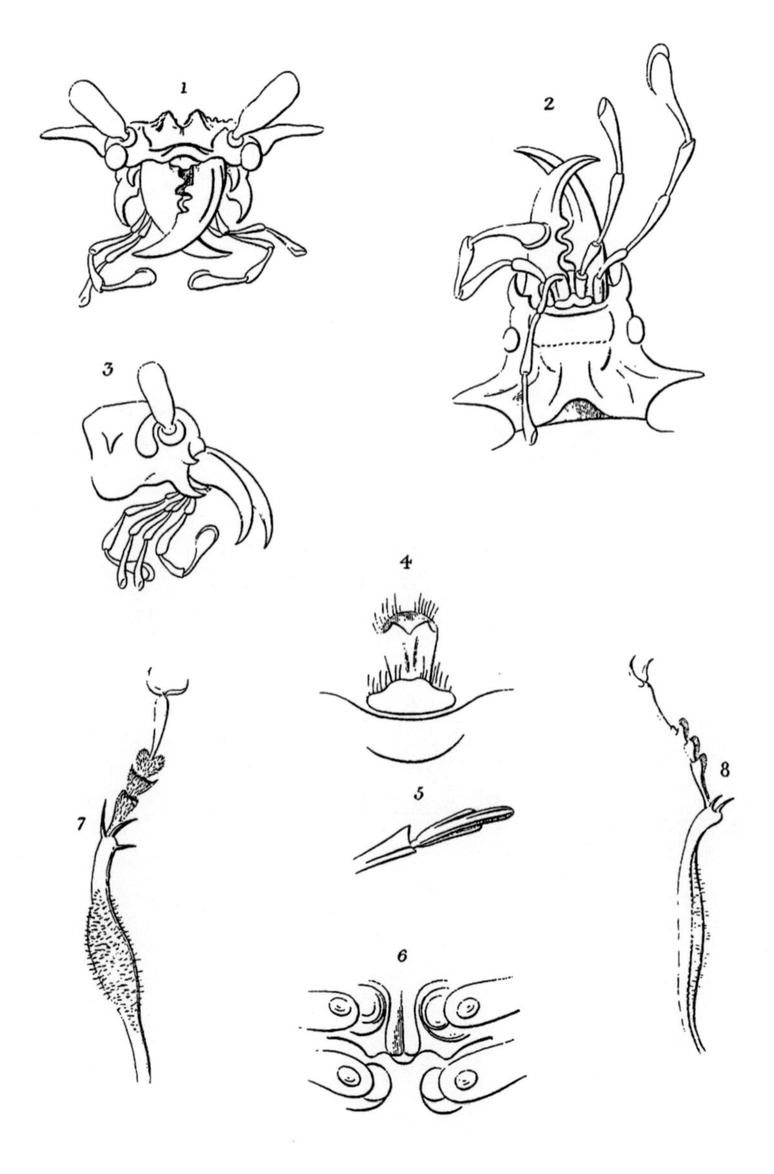

I. O. Westwood. delt.

Details of Psalidognathus *Friendii.*

London. Published by Whittaker & Co. Ave Maria Lane. 1837.

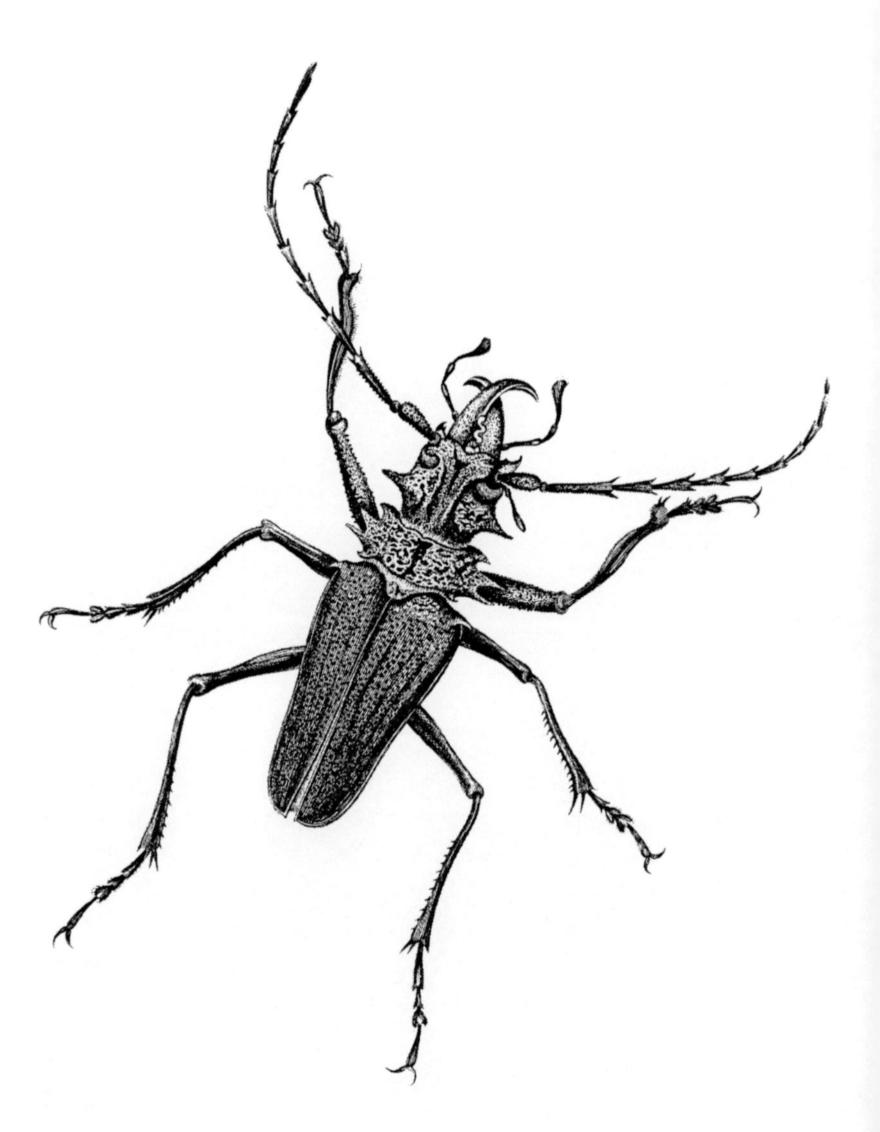

Coll. J. G. Children, Esqr *I. O. Westwood. del*[t]

Psalidognathus *Friendii.* *G. R. Gray.*

London. Published by Whittaker & C° Ave Maria Lane. 1831.

When these larvæ have acquired their full growth, they spin a thick cocoon, composed in a great measure of saw-dust. They then change into the chrysalis. But before they undergo their metamorphosis they approach the surface of the tree, that they may issue more easily from their hole, when they assume the form of the perfect insect.

The female Prioni lay a considerable number of yellowish oblong eggs, which they deposit in the clifts and fissures of the wood, by the assistance of a sort of corneous tube, which is inclosed in their abdomen, and which, at this period, they protrude forth.

The *prionus arvicornis* is a native of America; its larva inhabits the wood of the *Bombax*, Linn. The people of the country use it as food, and esteem it as no small delicacy. The *coriarius* is a native of Europe, and has been described by Geoffroy. It inhabits the trunks of old oak trees. It never flies but by night, and in the evening.

We have formed a new sub-genus under the name of PSALIDOGNATHUS, which belongs to this division. The antennæ long, with spines at the apex of each joint; basal joint long, oblong; the second joint short, globose; third long as the fifth and sixth together; seventh to eleventh moderate, and grooved on the outer side. The labrum coriaceous, very small, rounded. The mandibles long, curved under, the exterior ridge rounded; the apex with a sharp edge interiorly, the base with three small teeth; the maxillary palpi as long as the head and thorax; the first joint as long as the third, and the second as long as the fourth; the latter has an enlarging, flattened, and rounded club; the head square, with a strong spine on each side, also a spine at the base of the mandibles; the thorax much broader than long, with three spines on each side; the scutellum triangular, rounded at the tips; the body long, broader at the base, narrowing towards the tip; the legs long; the anterior tibia flattened, grooved, and hairy beneath; the tarsi rather slender. The species we have named *Friendii*, after Lieutenant Friend, of the Royal Navy, who discovered this fine insect in Columbia. It is rugose, of a metallic green, tinted with purple; the antennæ and legs of a metallic purple; the length of the body is two inches and a half. It is in the cabinet of J. G. Children, Esq.

We also insert figures of *Anocolus lugubris* (see p. 99), which is black and punctated; the elytra do not cover the abdomen, and *A. quadripunctatus*, which is fulvous, with the antennæ, tibiæ, and elytral spots, black. Both are from Brazil.

The insects of the genus CERAMBYX, or the *Capricorns*, must have been long ago distinguished by the fine proportions and the varied colours which the majority of the species exhibit, and especially by the length of the antennæ, which characterizes the genus. Their body is elongated. The antennæ differ in their length, even in the same species. The males, in general, have them longer than the females. Their walk is neither slow nor precipitate, and they frequently make use of their wings. When they find themselves seized, they seek to defend themselves, and send forth a sharp and tolerably strong sound, by rubbing their corslet against the base of the scutellum. The capricorns are usually met with in woods, and on the trunks of trees. They are seldom to be seen upon flowers. They feed upon the wood, or on the juices which flow from the trees. The female employs a sort of tail, or auger, which she has at the end of the abdomen, to pierce the wood, and introduce and deposit her eggs therein.

The body of the larvæ is elongated, tolerably soft, and composed of thirteen very distinct rings. Their head is scaly, and rather hard. The mouth is furnished with two very strong jaws, by means of which these larvæ gnaw the sub-

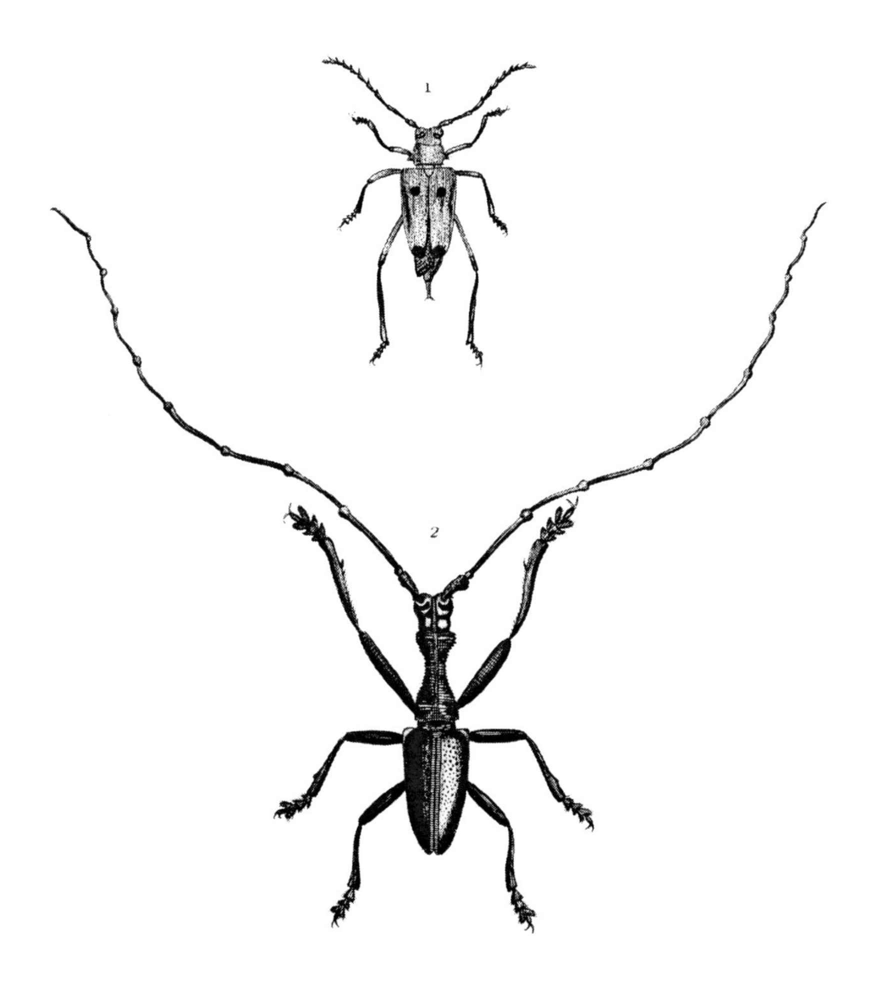

J. O. Westwood del.

1. Anacolus *quadri punctatus*. *G.R.Gray.* 2. Gnoma *suturalis*. *Westw.*

London, Published by Whittaker & Co Ave Maria Lane 1837.

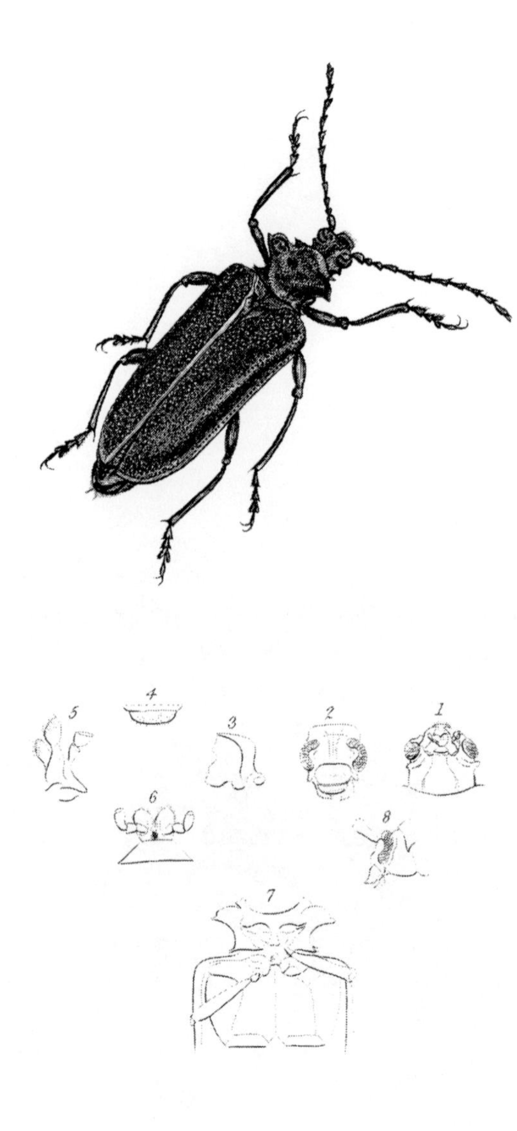

J. O. Westwood, del

Cheloderus Childreni. G. R. Gray.

London, Published by Whittaker & C.º Ave Maria Lane, 1832.

stance of the wood, which supplies them with their nutriment. They change their skin several times, remain two or three years in their first state, and finally change into a nymph, from which the perfect insect issues forth in the course of some little time.

One of the largest of these insects in Europe is the *Cerambyx heros.* Its larva does material injury to oak trees, in the trunks of which it makes very deep holes.

We have re-figured the *Cerambyx Fichteli*, of *Schreiber*, It is of a bronze colour, the antennæ brown ; the head broad, pointed on each side; the elytra much longer than the abdomen, rather pointed, divergent, and rather hairy at the tip. This insect we have formed into a new sub-genus, under the name of *Enicodes ;* it is from New Holland.

The insects of the genus Callidium are found for the most part in forests, on the half-rotten trunks of trees, and in timber-yards, where they are oftentimes caught, at the moment in which they are issuing forth from the wood in which the larva was nourished. They sometimes are seen to enter into apartments. Many species frequent flowers, and live upon their nectar.

These insects make a noise occasionally, by the rubbing of the corslet against the base of the scutellum, which is chagreened. This noise increases in proportion as they are more disturbed, and as the movements of the flexion and elevation of the head are more precipitate.

The *Callidia* often make use of their wings. They rise with facility to their flight, which is tolerably sustained.

Their larvæ resemble soft and elongated worms. Their body is composed of thirteen rings, and six very small scaly feet, which are distinguished with difficulty. Their mouth is armed with two strong jaws, which answers the purpose of gnawing and reducing to powder the wood upon which they feed. Accordingly, it is only in the furrows which they trace in the wood that they are to be found ; and as they advance in gnawing the wood, they fill the vacancies which are left with their excrement, which is nothing more than the dust of the wood which has served them for aliment. It is somewhat compact, but very friable, and it preserves the colour of the wood.

These larvæ remain in their first state for two years. During this time they change their skin several times, until, being arrived at their full growth, they quit it altogether, and appear under the form of a nymph. This last differs from the larva. Its body is shorter, and more compact. Its rings are less apparent, and the elytra may be distinguished through the envelope which covers them. They are short and folded back, much in the same manner as the wing of the butterfly is in the chrysalis of that insect.

To the sub-genus *Clytus*, of this division, we have added a fine species, which we named *Haiji*. It is black, with yellow various formed markings; the antennæ black, and the legs yellow. The length is one inch and a line, the specimen is from the Tannaserim Coast.

No observations have been made on the metamorphoses of the NECYDALES, but we may presume, however, that they take place in the interior of wood. The conical tube which Degeer observed at the anus of one species, renders the induction which we may derive from analogy the more probable. These insects in summer, are found upon flowers.

Linnæus, and other naturalists, have placed the insects of the genus LAMIA with the *Capricorns* (*Cerambyx*). Fabricius has separated them, and formed of them a peculiar genus.

The lamiæ, like all the longicornes, send forth a sharp sound, produced by the friction of the anterior parietes of the corslet against the base of the abdomen. This genus is composed of a great number of species, but more abundant, and of a greater size in the wooded countries, situated between the tropics. South America furnishes a considerable number, especially of those which have the body flatted. The majority of the species of the ancient continent, and those of Africa, among others, belong to the last section of the genus.

The larvæ of the majority of the lamiæ, live in wood, after the manner of those of other insects of the same family. It is there also, and particularly in timber-yards, that the perfect insect is to be found. Some species, composing the last division, remain constantly on the ground. It is presumable that their larvæ make their habitation there, and feed on the interior of the roots of divers vegetables.

We have formed here a new sub-genus, and named it *Eurycephalus*; the head flat, rather triangular, with the sides projecting and pointed, from which the antennæ spring, of which the first joint is very large, the second very small, the rest linear. We have named this species *Niger*. It is hirsute, black, with the antennæ reddish. Inhabits Brazil. The length is five lines.

The genus SAPERDA is formed from the fourth family of the capricorns of Linnæus, and the first of the *lepturæ* of Geoffroy. Fabricius, in establishing it, gave it the name of *Saperda*, which had been applied by some Greek authors to a fish, which is unknown to us.

The Saperdæ derive their nutriment from the substance of vegetables, and many of them frequent flowers. But the greater number attach themselves, in preference, to the stalks and branches of different shrubs and trees, and remain there almost motionless. They seldom fly off, unless they are heated by the rays of the sun, or excited by the desire of reproduction.

Rœsel has described the metamorphoses of one species (*Cylindricollis*). The larva feeds on the sap of the pear and plum-tree. Its form is elongated, pointed posteriorly, narrowed towards the first rings, and afterwards widening abruptly. The head is scaly as well as the upper part of the first ring, and it is furnished with very strong mandibles. The feet either do not exist, or are but little apparent. It is in the cavities which it has hollowed, in taking its food, that it changes into an elongated nymph, provided in miniature with all the organs which the perfect insect is to possess. According to Goedart, the larva of the *Saperda carcharias*, lives in the oak. It is apodal, elongated, a little depressed, soft, broader anteriorly, and armed with very strong mandibles. Its body grows narrow insensibly towards the extremity, and is terminated by an abrupt and rounded swelling. One of the industrious means which it employs to advance more and more, and find the wood, is to form a resting point by contracting and reducing itself almost into a ball. The effect of gravitation being thus diminished, the anterior part of the body is more at liberty, and the action of the mandibles becomes more powerful. Its lodging being once enlarged and completed, it resumes its natural position. This larva is transformed into a nymph at the end of October, and the perfect insect issues forth in the month of June, in the following year.

In the genus *Gnoma*, we have figured a new species, which Mr. Westwood has named *Suturalis*. It is black, with two lines down the thorax, the suture and summit of the elytra dusky yellow. This species is from New South Wales.

The LEPTURÆ are found in wood, on the trunks of trees, and often also on flowers. Their larvæ feed on rotten wood, and resemble in all essential characters those of the other *longicornes*.

THE FIFTH FAMILY

OF

TETRAMEROUS COLEOPTERA.

EUPODA

Is composed of insects, the first of which (*Donuciæ*) are so nearly allied to the last, Longicornes, that Linnæus and Geoffroy have confounded them together, and the last of which are so nearly related to chrysomela, the type of the following family, that the first of these naturalists placed them in this genus. The organs of manducation have the same affinities; thus in the former the lingua is membranaceous, bifid, or bilobed, the same as that of the longicornes; their jaws also greatly resemble theirs, but in the last, Eupodæ is nearly square or rounded, and analogous to that of Cyclica; nevertheless the maxillary lobes are membranous or but little coriaceous, whitish or yellowish; the exterior enlarges toward the extremity, and has not the figure of a palpus, characters which give to these parts more resemblance to the corresponding parts of the longicornes than to those of cyclica. The body is more or less oblong, with the head and corslet narrower than the abdomen; the antennæ are filiform or thicker as they advance, and are inserted before the eyes, which in some are entire, round, and rather prominent, and in others are slightly emarginated; the head enters behind into the corslet, which is cylindrical or a transverse square, the abdomen is large compared with the rest of the body, forming an elongated square or triangle; the articulations of the tarsi, except the last, are furnished underneath with cushions, and the penultimate is bifid or bilobed; the posterior thighs are much enlarged in most of them, whence the name of the family is taken.

We shall divide this family into two tribes. First, *Sagrides*, including the genus

SAGRA,

Whose mandibles terminate in a sharp point, the lingua is deeply emarginate or bilobed.

Some have the palpi filiform, the eyes emarginate, and the posterior thighs very thick, with the legs arched.

MEGALOPUS, Fab., have the anterior extremity of the head advanced like a muzzle, the mandibles strong and bent, the palpi terminated by an elongated and very pointed articulation, the lingua divided deeply into two elongated lobes, the body short, with the corslet square or trapezoid and transversal, the antennæ thicker as they advance, and termiminated in an elongated knob, the third articulation being longer than the second and fourth; the hind legs are long.

SAGRA, Fab., have the palpi terminated by an ovoid articulation, the divisions of the lingua short, the corslet cylindrical, antennæ filiform, with the lower articulations shortest, fore-legs rather thick, not long, and angular.

Others have the palpi thicker at the end, the eyes entire, and the thighs nearly of the same thickness. The body is always elongated, narrow, rather depressed, or very slightly elevated, with the corslet narrowed behind and nearly heart-shaped.

ORSODACNA, Lat. Oliv.—*Crioceris*, Fab. The antennæ are filiform, composed of articulations like reversed cones, the last of the palpi only is a little longer than the preceding, and nearly like a truncated ovoid; the corslet is at least as long as it is wide.

PSAMMŒCUS, Boudier.—*Anthicus*, Fab.—*Latridius*, Deg., have the antennæ composed of short and serrated articulations which thicken progressively; the maxillary palpi terminate abruptly in a strong triangular knob. The corslet is wider than it is long. The body is more depressed than in the last, with the antennæ shorter, and the eyes more prominent.

The second tribe, CRIOCERIDES, has the end of the mandibles truncated, or with two or three teeth. The tongue is entire, or but slightly emarginate. It is composed of the genus

CRIOCERIS, Geoff. *Chrysomela*, Lin.

Which we shall divide thus:—

Sometimes the mandibles go to a point, and have at the end two or three teeth.

DONACIA, Fab. *Leptura*, Lin., have the posterior thighs large, the antennæ equally thick throughout, with elongated articulations, the eyes entire, and the last articulation of the tarsi enclosed for the most part of its length by the lobes of the preceding.

HÆMONIA, Meg. Dej. The penultimate articulation of the tarsi is very small in these, in form of a knot, nearly entire, and the last is very strong.

PETAURISTES, Lat., united by Fab. with *Lema*, or our *Crioceris* proper, have also the posterior thighs thick, but the eyes are emarginated; the antennæ, as well as in those, are generally composed of shorter articulations, and the lobes of the penultimate articulation of the tarsi is much less elongated, and encloses only the root of the following.

CRIOCERIS, Geoff. Oliv.—*Lema*, Fab.—*Chrysomela*, Lin., differ from the last in having the hind feet like the rest; the antennæ thicken a little as they advance, and are nearly grained, the articulations not being much larger than they are wide. The eyes are elevated and emarginated. The hind extremity of the head forms behind the eyes a sort of neck.

AUCHENIA, Thunb., differs from the last, from which they were not heretofore distinguished, by having the eyes entire; their palpi narrowed, and ending in a point; the last seven articulations of the antennæ are longer, and their corslet is dilated toward the middle, on each side, like an angle or tooth.

Sometimes the mandibles are truncated, the palpi terminated by an enlarged truncated articulation, with a small elongation, like a ring, presenting the appearance of another articulation. The antennæ are slender, composed of very elongated sub-cylindrical articulations.

MEGASCELIS, Dej. Lat. The eyes are a little emarginated. The mandibles are thick. The outer maxillary lobe is narrow, cylindrical, and bent inward. The labial palpi are nearly as large as the maxillary. These insects, proper to South America, appear to be allied in some respects to *Colapsis*; but, by their general form, they are arranged with eupodes.

SUPPLEMENT

ON THE

EUPODA.

On the first genus of this family, or, more properly speaking, the sub-genus, which forms the type of the tribe, namely, Sagra, we have not much to say. These insects seem to form a passage from *orsodacna* to *donacia* and *crioceris*. They approach to the second of these genera, in the brilliancy of their colours, the size of their posterior feet, and especially that of their thighs. Like *Crioceris*, their eyes are elongated and narrow, with an impression between them in the manner of a cross, thus—X. They are all winged and large insects, peculiar to the southern countries of Africa and Asia.

The species named *tristis* by Fabricius is not, as he has asserted, a native of Cayenne, but of the coast of Angola. Some other species come from China, and often constitute a part of the boxes of insects which the merchants of that country sell to Europeans. But it does not appear that the habitation of these insects extends further to the east. It is not known if any have been received from the Moluccas or New Holland. Their manners are unknown. Fabricius presumes that the differences hitherto regarded as specific, are only sexual : a point not very easy of verification, in consequence of the paucity of specimens.

We have added specimens of two new sub-genera of Mr. Macleay, near to Sagra.

1st, CARPOPHAGUS. The labrum transverse, sub-bilobed, with the margin ciliated; the mandibles strong, arched; the tip sharp, with a small tooth near the middle, and the base ciliated; the maxillary palpi short, with the last joint obtuse, oval; labial palpi with the last joint ovate, the tip truncate, the posterior femora enlarged, with one large tooth. The species *Banksiæ*, is black brown, sprinkled with white hairs; the head and thorax punctated, divided by a smooth line; the scutellum cinereous; the elytra rugose, with four sub-elevated lines; body beneath, and feet cinereous.

2nd, *Mergamerus*, with the labrum transverse, square; the mandibles formed like the last; the maxillæ, large, oval, and ciliated; the maxillary palpi with the last joint securiform, compressed; the labial palpi with the last joint triangular, compressed, or securiform; the labrum bilobed. Our species, *Kingii*, is blackish brown; the labrum and palpi, piceous; the thorax scarcely punctated; posterior, rugose; the elytra with punctated stria, with a channel on the shoulder; the body beneath, and legs cinereous. Both these insects are from New Holland.

The insects called ORSODACNÆ are among the number of those which appear to connect the family of the *longicornes* with that of *Cyclica*. These insects inhabit the leaves of trees, but their metamorphoses are unknown. The most common species, which is found in spring on cherry-trees, plum-trees, &c., is that which M. Latreille names *Chlowtica*.

The CRIOCERES are insects which, though rather small, are remarkable for a handsome form, a little elongated, and decorated in some species with brilliant colours. They sometimes make their appearance at a very early period of the spring. It is on the flowers of gardens, fields, and mea-

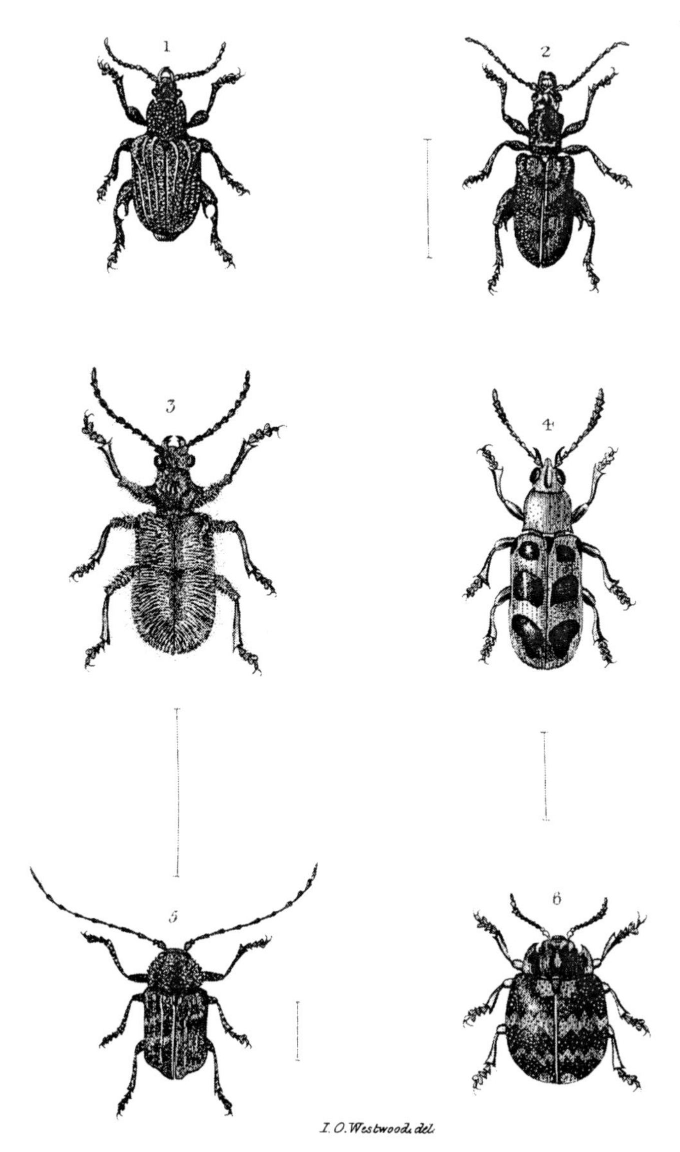

I. O. Westwood. del.

1. Megamerus *Kingii.* *Mac Leay.* 2. Carpophagus *Banksii.* *Mac Leay.* 3. Goniopleura *auricoma.* *Westw.* 3. Chelobasis *bicolor.* *G. R. Gray.* 5. Crytocephalus *rugicollis.* *G. R. Gray.* 6. Doryphora *princeps.* *G. R. Gray.*

London, Published by Whittaker & C.o Ave Maria Lane 1837.

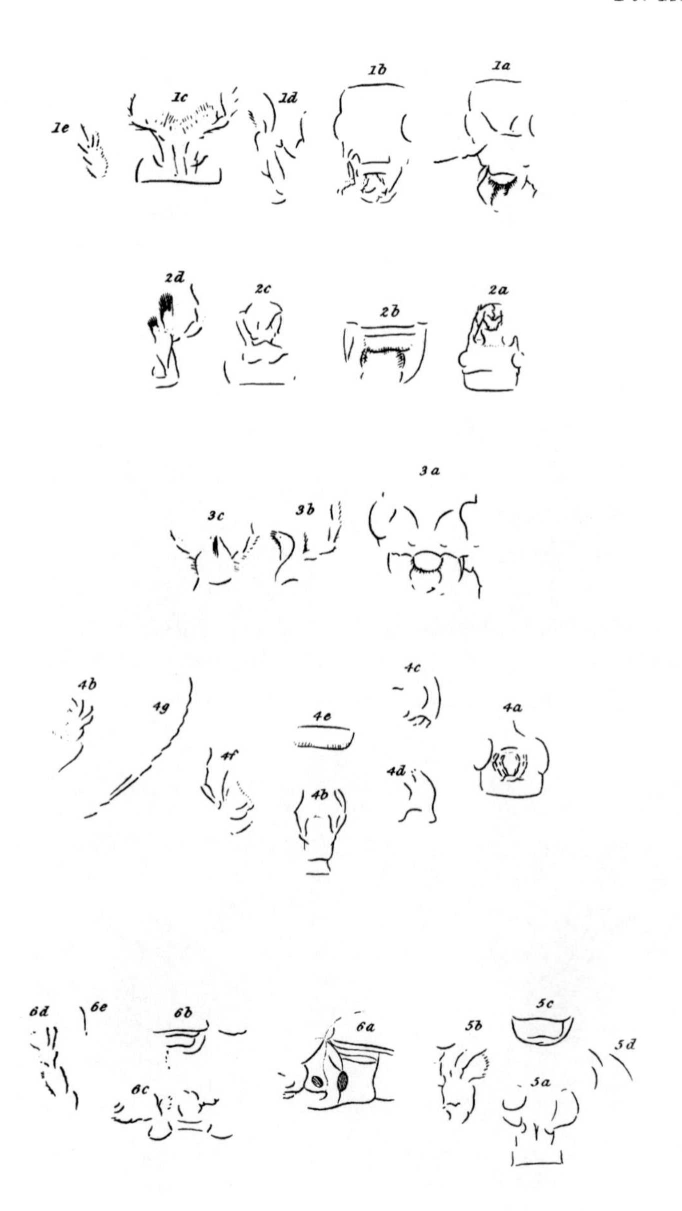

1. *Mergamerus Kingii.*
2. *Carpophagus Banksiæ.*
3. *Goniopteura auricoma.*
4. *Chelobasis bicolor.*
5. *Crystocephalus rugicollis.*
6. *Doryphora princeps.*

London, Published by Whittaker & C. Ave Maria Lane 1832.

dows, that they seek for repose and sustenance. When they are caught, they send forth a sort of little cry, produced by the rubbing of the superior extremity of the base of the abdomen against the interior parietes of the corslet. They pair very speedily in the flowers on which they live. Their intercourse lasts at least for a day, and sometimes longer. After pairing, the female walks over the flower, in search of a spot which may be suitable for the deposition of her eggs, which is usually the under part of some leaf. She there arranges them, one near the other, but with but little art and regularity. Each egg issues from the body invested with a fluid which is proper to glue it on the leaf, against which it is afterwards applied. The female deposits eight or ten together, and, doubtless, one of these little parcels does not constitute the entire of the ovi position. These eggs, in certain species, are oblong. Those recently laid are reddish, and even tolerably red. They grow brown, when the viscous fluid which covers them begins to dry. At the end of fifteen days, the little larvæ make their appearance, without, however, leaving behind them any vestige of a shell, or any remains of the envelope which connects them. But, perhaps, these shells are difficult to be found, because they are so very much attenuated, or because the movements which the insect makes to complete its emancipation from them, detach them from the leaf, and cause them to fall. Be this as it may, as soon as the little larvæ of the same nest are in a state to walk, they arrange themselves side by side, pretty nearly in the same manner, as do the common caterpillars. Their heads are on the same line. They live together, and eat only the substance of the leaf, on the side of which they are placed. In proportion as they grow, they separate from one another, and finally disperse over different parts of the leaf, and over different leaves. Then the larva attacks, sometimes the end of the leaf, sometimes one of its edges. Pretty often it pierces it in the middle, and eats throughout its entire thickness. This larva gives itself but little exercise. It walks but seldom, or at least does not proceed forward but when the leaf which it has attacked totally fails, or when there remain in the neighbourhood of the place which it has gnawed only some parts which are too much dried up. While it is eating, it makes, from time to time, a step backwards ; for its mode of eating is not to proceed, and take what is before it, but that which is towards the under part of its body.

The larvæ of the crioceres are thick, short, compact, and heavy. The body is soft, and covered with a fine and delicate skin. They have a scaly head, and six feet equally scaly. As much as the perfect insect agreeably attracts our regards by the beauty of its form, so does the larva repel us by a very different kind of aspect. It is not that it is more ill-conformed than many other larvæ, but it is the appearance of its clothing which renders it deformed and hideous. After having drawn its nutriment from the leaves, the substance of these same leaves still answers a useful purpose to the larva: it serves to clothe it. On the leaves which have been gnawed are to be seen little heaps of humid matter, of the colour and consistence of these same leaves, a little macerated and bruised. Each of these little heaps has a tolerably irregular figure, but, however, rounded, and a little oblong. All that is then perceivable is the matter which serves for a covering to each larva, and which covers it almost entirely. If we examine a little closer, we can soon distinguish the black head of the insect operating on the leaf, with the two teeth with which it is provided. We may also perceive on each side, and pretty near the head, the three pair of black and scaly feet, terminated by two small hooks, which the insect fastens into the surface of the leaf. This foreign matter is but little adhesive ; it is easily removed by a slight friction, and when the larva is stripped of it, it is found to be pretty similar to most other larvæ ; but its skin appears to be very delicate. Its transparency would lead us at once to conclude so, as it allows us to see the movements of most of the interior parts. Accordingly, nature has taught this little insect a most singular method of sheltering itself against the impression of the external air, or that of the rays of the sun. It has instructed it to cover itself with its own proper excrements, and has disposed every thing to enable it to do so with facility. The aperture of the anus of other insects is placed at the end, or near the end of the last ring, and usually in the side of the belly. The anus of this larva is a little removed from the posterior end. It is placed at the junction of the penultimate ring with the last, and, what is remarkable, on the side of the back. The disposition of the rectum, or of the intestine, which conducts the excrements to the anus, and that of the muscles which protrude them, must correspond to the end which Nature has intended, in thus disposing this aperture. The excrements which issue from the body of insects in general are expelled backwards in the line of the body. Those of the larva of the crioceris are raised above the body, and directed towards the region of the head. They are not, however, sent very far. When they are entirely out of the anus, they fall upon the part of the back which is nearest. They are retained there by their viscosity; but, however, they are retained but feebly. Without even changing place, the insect gives certain movements to its rings which, by little and little, conduct the excrements from the place in which they have fallen as far as the head. We may easily imagine the mode in which it prepares inclined planes for them, one after another, by swelling the part of its body on which they are, and contracting the part which follows towards the region of the head. The larva does more: it folds and raises the part of the rings which precede that on which the excrements are situated. From which it is clear, that when it extends the folded portion without lowering it, this part, in being developed, pushes the excrements into the depression which has been prepared for them. The form of the back itself is such, that when a portion of the excrements has been conducted to a certain distance from the anus, it finds a declination from these to the head.

To see distinctly how all this takes place, it is necessary to strip the insect naked, place it on a young and fresh leaf, and then observe it with a microscope. It will very soon begin to eat, and a little time after, the anus is observed to swell. It exhibits edges which were not perceptible before. Finally, the anus half opens, and the end of a little mass of excrements issues forth. What the insect casts forth is a sort of cylinder, the two ends of which are rounded. We have already mentioned that when this grain of excrement issues forth, it is directed towards the head. Nevertheless, a little after it has come out, it is found to be placed transversely, or at least a little inclined towards the length of the body. The friction which it meets with, and the little regularity with which it is pushed on, give it this direction. There are times when these grains are arranged with tolerable order, placed parallely one to the other, and perpendicularly to the length of the body. But this seldom happens except on the posterior part, and when the anus has furnished a considerable number of them, in a short space of time. The insect which has been stripped, has need to eat for about two hours, before the anus can furnish, at different times, the quantity of matter which is reauisite to cover all the upper part of the body; at the end of two hours this covering is complete, but it is very thin, not being more in thickness than a single grain of the excrement. By slow degrees it grows thicker. The same mechanism which has conducted the grains to the neighbourhood of the head, forcing them to press against each other, to give place to the excrements which issue forth, the excrements which are in the neighbourhood of the hinder part of the body, are pushed on and carried forward. They are soft, yield to the pressure, and flatten in one direction and rise in another, namely, in that which renders the layer thicker. This layer, which covers the body, thickens then by little and little, and at last to such a degree that if we remove it at a certain time, from above the body of the larvæ, we shall see that this covering is of a volume three times larger than that of the insect itself, and of a weight which would seem to overburthen the animal. The thicker the covering is, the more irregular is its figure, and the browner its colour.

We have said that the excrements of which it is formed have the colour and consistence of bruised and macerated leaves. Accordingly, they are nothing else. They are, at first, of a greenish yellow, but their upper surface dries by little and little, and assumes a browner shade, gradually approaching towards black. When this clothing becomes too stiff, or too heavy, the insect apparently divests itself of it, for we sometimes see these larvæ naked, or nearly so. But they do not remain long in this state. They easily disembarrass themselves of a too heavy covering, either entirely, or in part. They have only to place themselves so that it shall touch or rub against some part of the plant, and then draw themselves forward. When the insect preserves its covering for a long time, it sometimes overflows the head, and that which covers the first rings is often black and dry, while the rest is humid and greenish. This dry part, which goes beyond the head, sometimes falls in shreds. The *Cassidæ*, as we shall see hereafter, adopt a similar mode of protecting their sensible and tender skin from the danger of external impressions. But they do it in somewhat a different manner, which we shall explain in the proper place.

In fourteen or fifteen days the larva of *crioceris* have acquired all their growth. Then they are no longer covered with their excrements. They are seen, either entirely, or in part naked. Their body assumes a more deeply coloured tint. They walk, and do not appear as tranquil as they were before. They are now near the time of their metamorphosis. It is in the earth that this change is to take place, and it is to proceed thither to conceal themselves, that they are in motion.

A little time after these larvæ have entered the ground, they set to work to make themselves a shell, the exterior of which is covered with some grains of the earth which surrounds them. These shells are so well covered that they might be taken for little masses of the ordinary and uneven earth. They are in general about the bulk of a small bean, or a large pea. When they are pressed between the two fingers, and slightly enough to be recognised, they send forth a little sound, similar, in miniature, to that of a bladder, prest to bursting. From this it is plain, that the shells within which these larvæ are transformed, are bladders, very close, and filled with a very elastic air, since a trifling compression causes this air to break the shell, with noise. When these shells are opened, their interior is found to have the polish of satin. It is of a shining and silvery white. In a word, these shells, or cocoons rather, resemble that which some caterpillars make of fine and lustrous silk, and cover with earth. Nevertheless, this sort of stuff is very differently, and much more simply fabricated. The caterpillars spin, for the purpose of making their cocoons, while these larvæ employ a sort of froth, or slaver, which is less thick than the fluid of which the silk is composed, but is nevertheless analogous to it. This froth being dry, forms shining and flexible leaves, such as they would be if they were actually of silk.

When one of these larvæ is preparing for its transformation, it lodges itself in a sort of box, made of grains of earth, cemented, apparently by the fluid. But this fluid serves more especially to invest the parietes of the cavity. The larva can furnish a sufficient quantity of it, so that that which is dried shall form a silky lining of a sensible thickness. When earth is wanting to the larva, when it has been unable to form a cavity, the solid parietes of which shall be proper to receive and sustain its frothy fluid, it is difficult for it to employ this fluid usefully. The slender layer which commences to assume consistence is often broken, or, at least, rumpled by the motions of the insect.

Two or three days after the larva has been shut up in its shell, it is metamorphosed into a nymph, similar in the disposition of its parts to the other nymphs of the coleoptera of the same family. In fine, in about fifteen days after the insect has entered into the earth in its larva form, if it be in the summer season, it is in a state to appear in its final shape. It pierces the shell, issues from the earth, and proceeds in search of the plants, whose leaves and flowers are most suitable for its nutriment.

The DONACIÆ form a sub-genus, composed of a small number of species, which may be ranged among insects of the middle size. They are endowed with an agreeable form, enhanced by a striking brilliancy of colour. They live among aquatic plants, such as the *reed*, the *iris*, &c.; we may also suspect that their larvæ live in the stalks or roots of these same plants. The nymph of the *Donacia crassipes*, according to Linnæus, is found in the form of a brown shell, on the root of the *Phellandria*.

THE SIXTH FAMILY

OF

TETRAMEROUS COLEOPTERA.

CYCLICA

HAVE also the first three articulations of the tarsi spongy, or furnished with cushions underneath, with the penultimate divided into two lobes, and the antennæ filiform, or a little thicker toward the end. The body is generally round, with the base of the corslet as wide as the elytra in the small number of those, in which the body is oblong. They have jaws, whose exterior division by its narrow, sub-cylindrical form, and deeper colour, has the appearance of a palpus; the interior division is larger, and without scaly crook. The tongue is nearly square or oval, entire, or slightly emarginated.

All the larvæ with which we are acquainted are provided with six feet, have the body soft, coloured, and feed as well as the perfect insect on leaves, to which they commonly fix them by means of a viscous humour. It is there also that many of them are changed into nymphs, to the posterior end of which is engaged and folded into a ball, the last skin of the larva. These nymphs are often of various colours. Other larvæ enter into the earth.

These insects in general are small, often ornamented with metallic and brilliant colours, and have the body without hair.

With reference to the different habits of the larvæ, the Cyclica are divided into four principal sections:—

1. Those, whose larvæ cover themselves with their own excrement.

2. Those which live in tubes, which they carry about with them.

3. Naked larvæ.

4. Those whose larvæ live in the interior of vegetables, and feed on their parenchyma. *The leapers.*

We shall divide them into three tribes, according to the mode of insertion of the antennæ.

The CASSIDARIÆ, which form the first tribe, have the antennæ inserted in the upper part of the head, near each other, straight, short, filiform, and sub-cylindrical, or thickening gradually toward the end; the mouth is altogether underneath, the palpi are short, sub-filiform, and sometimes arched, sometimes received in part into the cavity of the presternum; the eyes are ovoid or round, the feet contractile, short, with the tarsi flat, the lobes of the penultimate articulation totally enclose the last. The body being flat underneath, these insects have the faculty, by means of their tarsi, of sticking immoveably on the surface of leaves; otherwise the body is in general orbicular or oval, and overedged all round by the corslet and elytra. The head is hidden under the corslet, or received into an anterior emargination. Their colours are very varied, and distributed under the form of spots, points, and rays. Those of their larvæ which are known, cover themselves with their own excrement. They compose two genera.

HISPA, Lin.,

Whose body is oblong, with the head entirely uncovered and disengaged, and the corslet in the form of a trapezium. The mandibles have only two or three teeth, the outer maxillary lobe is shorter than the inner; the antennæ are filiform, and carried forward.

ALURNUS, of Fab., differ only in the form of the mandibles, the upper end of which forms a strong-pointed tooth, with a small one on the inner side. The ligula is corneous. This includes the largest species.

HISPA proper, Lin. Fab.,[1] have the mandibles short, ending in two or three small teeth, nearly equal. Some have the upper part of the body, and even a portion of the antennæ very spiny, as *Hispa atra*, Lin.

CHALEPUS, Thunb. Taking *H. spinipes*, of Fab., for type of this sub-genus, it differs from the last, by having the legs long, slender, and arched, with the anterior armed internally with a long spine. The third articulation of the antennæ is also longer in proportion.

Some others, *Monoceros*, Oliv. *Porrecta*, Schœn. *Rostratus*, Kirby, are remarkable for a horn-like projection above the head, and probably form another sub-genus, CASSIDA, Lin. Fab., distinguished from Hispa by the following characters. The body is orbicular or sub-ovoid, and nearly square in a few. The corslet more or less semi-circular, entirely hides the head, or encases it in its anterior emargination. The elytra often elevated in the region of the scutellum, overedge the body. The mandibles have four teeth, at least, and the outer maxillary lobe is as long, or less so, than the inner.

IMATIDIUM, of Fab., differs from Cassida only in having the head uncovered, and engaged in an emargination of the corslet. Both have the body depressed, nearly round, formed like a buckler or little tortoise, often slightly elevated, pyramidically, on the middle of the back, and overedged all round by the sides of the corslet and elytra; underneath it is flat, so that these insects are as if stuck to the object they rest on.

In the second tribe, CHRYSOMELINÆ, the antennæ are inserted above the eyes, and are wide apart. These insects do not leap. They compose with those of the following tribe, and some of the preceding family, the genus chrysomela, of Lin. They form two genera.

CRYPTOCEPHALUS,

With the head buried in a vaulted or convex corslet, like a hood, so that the body, which forms in general a short cylinder or sub-ovoid, and narrowed behind, appears, when seen in front, as if truncated, and deprived of its head. The antennæ of some are more or less serrated or pectinated, of others, are long and filiform. The last articulation of the palpi is always ovoid.

Sometimes the antennæ are short, pectinated, or serrated, from the fourth or fifth articulation. Here the outer edge of the elytra is straight, or has but a slight emargination; the posterior angles of the corslet are rounded, and not vaulted; the anterior are not feathered underneath. The body forms always a short cylinder, with the antennæ always free, the eyes entire, or but little emarginated. The males often have the head wider, with the mandibles stronger and more advanced, and the fore-feet longer. CLYTHRA, Leach, Fab.—*Melolontha*, Geoff. (*C. quadripunctata*, Lin.)

At other times the elytra, much dilated outwardly at their insertion, are suddenly narrowed behind, and have a deep emargination. The posterior angles of the corslet are sharp, vaulted, and form a roof; the anterior are much bent underneath. The eyes are obviously emarginated in many. The upper part of the body in these—and they are the most numerous in which it is less short and less convex—is commonly very unequal.

CHLAMYS, Knoch. Formed like a short cylinder or cube, with the corslet suddenly elevated like a hump in the middle, and elongated to the middle of the outer edge, or unilobed. The body is generally very rough. The labial palpi are forked in some.

LAMPROSOMA, Kirb., have the corslet nearly globular, very convex, smooth, with the corslet short, wide, and gradually elevated, and slightly lobed in the middle of the outer edge. The last five articulations of the antennæ are serrated, but less dilated than in the last.

Sometimes the antennæ obviously longer than the head and corslet, are simple and filiform, or thicker toward the end, or even terminated in a knob, and, moreover, in general serrated, but beginning only at the seventh articulation. The body of many of them is ovoid, and narrowed behind. The last articulation of the antennæ is appended, so that their number appears to be twelve. These have the body cylindrical, with the corslet as wide as the abdomen, all its length.

CRYPTOCEPHALUS proper. These have the antennæ and palpi of the same thickness throughout.

CHORAGUS, Kirb., have the antennæ terminated by three thicker articulations, forming a knob, and the palpi slender at the end. These have the body narrowed in front, and sub-ovoid.

The last five articulations of the antennæ are often longer, more or less compressed, and more or less dilated or serrated. The maxillary palpi are thicker at the end, or nearly terminated in an ovoid knob, formed either by the last articulations, or by that and the preceding united.

EURYOPE, Dalm., have the mandibles very strong, and the second articulation of the antennæ manifestly larger than the third.

EUMOLPUS, Fab., have the mandibles of ordinary size, but the second articulation of the antennæ is shorter than the following: *E. vitis*, Fab. This sub-genus is allied to the genus

CHRYSOMELA,

Whose body is in general ovoid or oval, with the head projecting, advanced, or simply bent; the antennæ are simple, about half as long as the body, and in general grained and thickened insensibly.

Some, whose body is always ovoid or oval, winged, and whose palpi terminate in a point, approximate to Eumolpus, and differ from the following by their filiformed antennæ, more than half the length of the body, composed of elongated sub-cylindrical articulations, with the eleventh or last terminated with an appendix or false articulation, whose length nearly equals the half that of the preceding portion of this articulation. Such are COLASPIS, Fab., which have no projection to the mesosternum.

PODONTIA, Dalm. The mesosternum advances into a short and conical point, received at the end into a posterior emargination of the presternum.

The first and last articulation but one of the tarsi are very large and dilated, the second small. The last of the maxillary is conical. The body is oblong, depressed, or slightly elevated.

In the following of the same tribe the antennæ are shorter, composed of reversed conical articulations, more or less grained, and thickening as they advance; the false articulation is very short and indistinct.

Some have the maxillary palpi thicker, and truncated at the end.

Among these, some have the last two articulations of these palpi united, and forming together a truncated knob; the last is shorter than the preceding, whether transverse, or in the form of a short truncated cone.

PHYLLOCHARIS, Dalm., has no mesosternal projection.

DORYPHORA, Illig., has on the contrary the mesosternum

advanced to a point, like a horn. These are American; the last are from New Holland and Java.

PAROPSIS, *Notoclea* Marsh, is another Australian sub-genus, distinct from all others of this family by the maxillary palpi, whose last articulation is much longer, and hatchet-shaped.

In the two following sub-genera the corresponding articulation, well detached from the preceding, and as large or larger than it, is more or less semi-ovoid.

TIMARCHA, Meg. Dej, which have been arranged with the chrysomelæ, includes those which are apterous. Their body is gibbous, with the antennæ grained, especially underneath; the elytra are united, and the tarsi in general very much dilated, at least in the males.

Among those whose corslet is narrowed behind, approaching a crescent shape, is found, *Tenebrio lævigatus*, of Lin.

CHRYSOMELA proper, includes those of Olivier which have wings, and whose maxillary palpi, after the sub-divisions above established, have the last articulations of the palpi as large or larger than the preceding, in form of a truncated ovoid or reversed cone. Such are *C. sanguinolenta*, Lin. *C. cerealis*, Lin. *C. populi*, Lin.

We shall terminate this tribe by those whose maxillary palpi are slender at the end, and terminated in a point. They will compose two sub-genera.

PHÆDON, Meg. *Colaphus*. Ejusd., whose body is ovoid or orbicular.

PRASOCURIS, Lat., *Helodes*, Fab., whose body is narrower, more elongated, nearly parallelipiped with the diameters of the corslet, nearly equal. The four or five last articulations of the antennæ are dilated, and nearly form a knob.

The third and last tribe of Cyclica, GALERUCITÆ, has the antennæ always as long, or less than half the length of the body, of the same thickness throughout, or insensibly thicker towards the end, inserted between the eyes at a little distance from the mouth, and commonly near each other at the base, and near a little longitudinal keel. The maxillary palpi are thicker toward the middle and terminate in two articulations in form of a cone, but opposed or united at the base, the last being short, either truncated, obtuse, or pointed. The body is sometimes ovoid or oval, sometimes nearly hemispherical; many, especially the smallest species, have the posterior thighs very thick, which enables them to leap. This tribe is composed of the genus

GALERUCA,

Which we shall divide into two principal sections, viz., leapers and non-leapers.

Some foreign species with the penultimate articulation of the maxillary palpi dilated, and the last much shorter and truncated, form the genus ADORIUM of Fab., *Oides* of Weber.

Those, in which the last two articulations of the maxillary palpi do not differ in size, and whose antennæ, composed of cylindrical articulations, are at least as long as the body, have been distinguished by the generic name of LUPERUS by Geoff.

Others, with the palpi terminating in like manner, have the antennæ shorter, and composed of articulations like reversed cones, are the GALERUCA proper of Geoff. Such as *C. Calmariensis*, Lin., *C. Tanaceti*, Lin.

The leaping Galerucitæ, or those whose posterior thighs are enlarged, dispersed by Fabricius among the genera *Chrysomela*, *Galeruca*, *Crioceris*, are united into one (*Altica* or *Haltica*) in the systems of Geoff., Olivier, and Illiger.

OCTOGONOTES (Drapiez Annal. des Science, Phys., III.) differ from the rest in the form of the maxillary palpi. As in Adoria the penultimate articulation is thick, top-shaped, and the last is short and truncated, the labial palpi terminate in an awl-shaped point, the same as in all the following sub-genera, but here the maxillaries have the same conformation, or are equally subulated at the end. The last articulation of the posterior tarsi is suddenly enlarged, and rounded above with two small hooks.

ŒDIONYCHIS, Latr., distinguished by this last character from the following sub-genera. We refer to this the first two families of Illiger's marograph (*A. Marginella*, Oliv.).

In the other sub-genera the corresponding articulation of the tarsi is elongated and gradually thickened, and the two hooks, of ordinary size, are situated, as usual, at the extremity, and in a longitudinal direction.

PSYLLIODES, Latr., have the first articulation of their posteria tarsi very long, inserted above the posterior end of the leg; this extremity is elongated like a conical appendix, compressed, hollow, slightly indented on the edge, and terminated by a little tooth.

DIBOLIA, Latr., formerly *Altitarsus*, whose head is for the greater part withdrawn into the corslet, and whose hind legs terminate in a forked spine.

ALTISA proper, Latr. The head is projecting, the hind legs are truncated at the end, without particular elongation or forked spine; the tarsi spring from this extremity, and their length does not equal half that of the leg. *C. oleracea*, Lin., *C. nitidula*, Lin.

LONGITARSUS, Latr, have all the character of Altica proper, but the posterior tarsi are also at least as long as the legs to which they are attached.

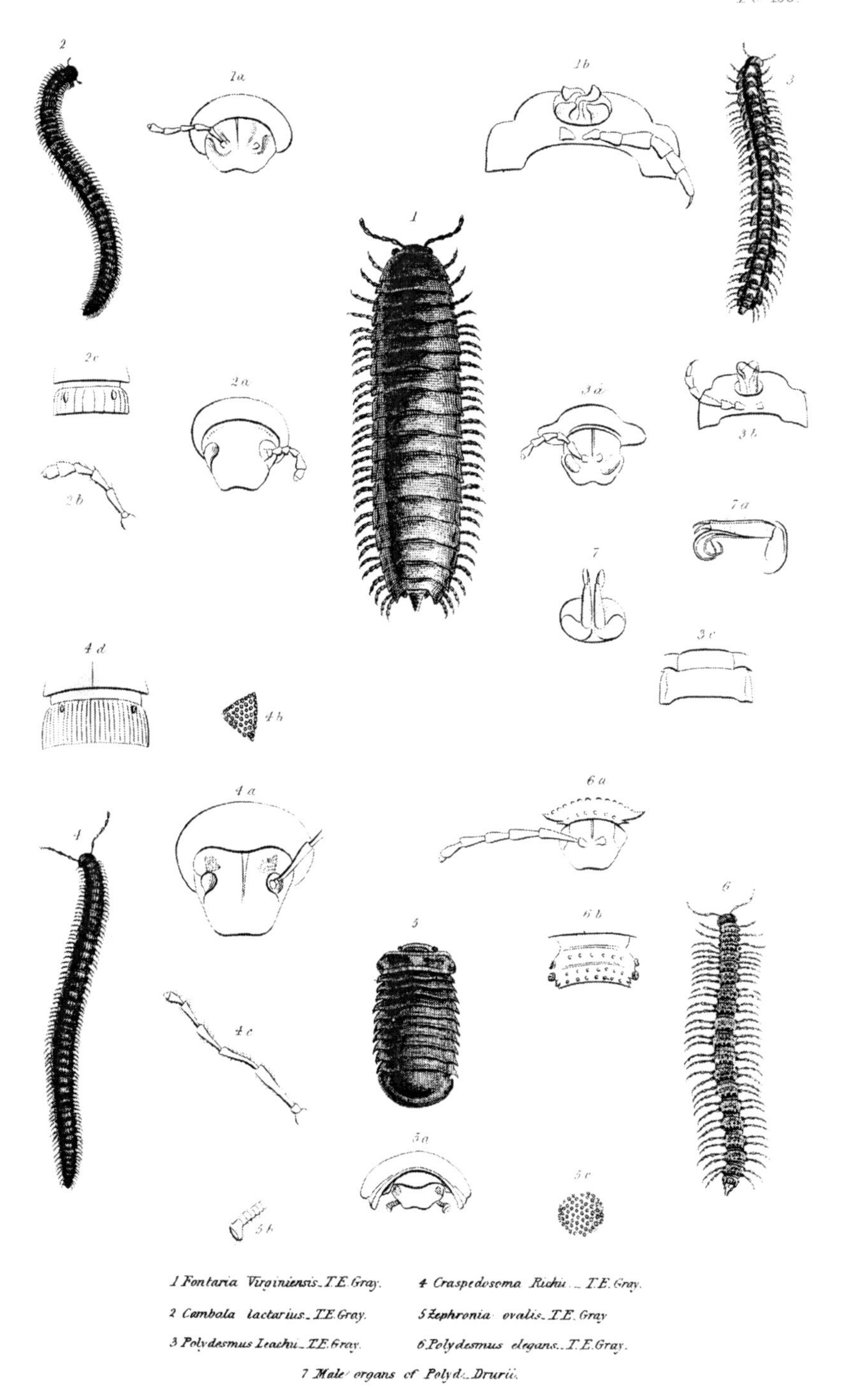

1 Fontaria Virginiensis._T.E. Gray.

2 Cambala lactarius._T.E. Gray.

3 Polydesmus Leachii._T.E. Gray.

4 Craspedosoma Richii._T.E. Gray.

5 Zephronia ovalis._T.E. Gray.

6 Polydesmus elegans._T.E. Gray.

7 Male organs of Polyd: Drurii.

London. Published by Whittaker & Co. Ave Maria Lane. 1832

SUPPLEMENT

ON

THE CYCLICA.

The genus Hispa was established by Linnæus. Their metamorphoses have not yet been observed. These insects fix themselves on different vegetables on which they feed, and they suffer themselves to fall to the ground as soon as one attempts to seize them. They contract their feet and shew no sign of life.

We have formed a new sub-genus to follow Hispa, *Chelobasis*. The antennæ with the first joint large, claw-shape; the third long, cylindrical; the other enlarged towards the tip, the labrum transverse, the sides rounded, the maxillary palpi, with the last joint, oval, acute; the labial palpi, with the last joint, long, club-shaped; the head pointed in front, and all the tibiæ compressed; the species was named *Vicolor*. It is yellowish-white, with six black spots on the elytra, the posterior ones the largest. It is from South America.

The insects of the genus Cassida received this name from the peculiar contour presented by their body. The corslet, and the elytra in which it is, as it were, encased, pass it considerably on the sides, which gives it the appearance of a *helmet*. These insects are therefore very easily recognized. They live on plants, which constitute their nutriment. They are very rarely seen to run, and still more rarely to make use of their wings. Most part of the species are enriched with fine golden or silvery colours, which disappear, however, when the insect is dead and preserved in cabinets, but which may be made to reappear by means of hot water, in which the cassidæ are put to soften for about a quarter of an hour. Beside the perfect insect, the larva is often found on the same plant, and well merits to fix the attention of naturalists.

The larvæ of the cassidæ have the body smooth, broad, short, flatted, edged on the sides, with ramous and spiny appendages, and provided with six scaly feet. The head is small and scaly, furnished with teeth, with three small tubercles and four black points on each side.

What particularly should engage our attention is the singular form of the tail, which is curved above the body, terminates in a sort of fork, and which is about the length of half the body. The two branches or prongs of which it is composed are like conical threads, which terminate in a tolerably fine point. They have sorts of short spines from their origin to a certain distance of their extent, but only on the external side. Between the two prongs, at the extremity of a nipple more or less curved, and raised at the will of the insect, the anus is seen, which has the form of a cylindrical tube, and which is placed so that the excrements which issue from it slide along a fork, which is inclined, and disposed so as to receive them. When they are heaped too near the origin of these little prongs, the nipple in which the anus is can push them on, and make them go farther. Perhaps the rings, and the spines which border them, still further assist to make the excrements go forward. By little and little they accumulate, are glued one against the other, and then pushed insensibly beyond the points of the prongs they form a mass or a roof capable of covering the whole insect. Such are the means, as simple as worthy of remark, which nature has contrived to protect the soft bodies of these larvæ from the impressions of external objects which might otherwise prove injurious to them. This roof is most frequently immediately above the body. It touches it without burdening it, and sometimes it is almost perpendicular to the plane of the body. It is often placed a little above, and almost parallel. All the different positions of this species of parasol are varied, as are those of the forked tail which sustains it. This covering, though pretty well cemented of itself, is still further fortified by the cast skin of the insect, which sometimes serves it as a basis.

Before its metamorphosis the larva must change skin several times. The spoil which it abandons is incomplete, the prongs themselves must fall, and this is the longest, and, perhaps, the most difficult part in the whole operation of moulting. It is upon the same leaf where the larva has lived that it is to undergo its metamorphosis, without forming either shell or envelope of any kind. To prepare itself for this, it ceases to hold the tail raised up ; it carries it then extended behind, and in the same line with the body. By friction against the leaf, it contrives to quit the skin, with the prongs, and causes this covering to fall, for which it has no further use. It fixes itself afterwards against the leaf by the two rings of the body, which follow that to which the last pair of feet is attached. Thus fixed, it remäins tranquil for two or three days, and afterwards quits its skin, to appear under the form of a nymph, which is to remain engaged in the skin, then reduced to a clew of thread, the only support which the nymph can have, and which it is also to preserve. The nymph has accordingly a forked tail, but the threads are finer, and not so long as those of the larva, and they have neither hairs nor spines.

This nymph, not so long as the larva, is of an oval and flatted figure. It has an ample corslet, pretty nearly of a semi-lunar form, the contour of which is bordered by a rank of spines, short, and simple, or without hairs. The belly is bordered on the two sides, with flat appendages or laminæ, in the form of leaves, pointed at the end, and furnished with spines or sorts of hairs. On each side of the back are seen four small tubes, which are the stigmata. In observing the nymph underneath, we there perceive all the parts of the perfect insect, which issues forth at the end of fifteen days, through the rupture made at the anterior part of the skin above.

The perfect insect deposits its eggs on the leaves. They are ranged one beside the other, and often form packets covered with excrements.

The Cryptocephali are insects, the majority of which are remarkable, not for their size, which is below the middle standard, but for the brilliancy and beauty of their colours. They live on plants, and are capable of causing much injury to them, by gnawing the young shoots in proportion as they issue from the germ. They do not cut, but macerate them, and occasion them to dry up and fall.

Naturally timid, the cryptocephalus has accordingly recourse to the artifice, useless enough, at least as far as man is concerned, of counterfeiting death. It walks slowly, and in a dull and heavy style. On the slightest touch it suffers itself to fall, folding its antennæ and feet underneath its body. At the same time it withdraws its head under the corslet, and only this last part and the body are then apparent. It is on willows that the majority of the species are principally to be found. The larva of cryptocephalus proper is unknown; but probably it much resembles that of Clythra.

Our figure, *C. rugicollis*, appears to be of a new species. It is brown, rugose, with a black spot on the thorax. The antennæ of the males are twice the length of the body, while those of the other sex are very short. It is from New Holland.

The larva of *C. quadripunctata* lives in a sort of tube, of tolerably solid matter, wrinkled at its exterior surface, almost cylindrical, firm, and rounded behind, and open at the other end. This habitation the insect transports along with itself wherever it goes.

Among the species of the genus Eumolpus, there is one very remarkable for the ravages which it commits. This is the *Eumolpus vitis*. It is small, smooth, and altogether black, with the exception of the elytra, which are a brownish fawn colour. Its larva lives upon the vine-tree, and often causes great damage by devouring the leaves, the young shoots, and sometimes the grape itself. This larva has the body pretty nearly oval, and of an obscure colour. It has six scaly feet, and a scaly head, armed with two small jaws, strong enough to gnaw the leaves, the new stalks, and even the grapes. It appears in spring, and more especially attaches itself to the young shoots of the vine-tree. It gnaws the pedicle of the clusters at the moment when, tender, pulpy, and full of sap, it issues from the germ. It exhausts it, destroys its organization, and causes it to fall, entirely dried up and withered; or if it should survive the attack, it always feels the effects of the wounds which it has received at its developement. It transmits to the cluster juices, neither sufficiently abundant, nor sufficiently elaborated. The grapes languish, and those parts of the cluster which correspond to the wounded fibres, remain feeble or sterile, bearing only abortive fruits, or not producing at all, while the other parts are developed, and fructify. This insect is but too well known to cultivators, especially in the wine-countries, where it causes in certain years very considerable ravages, and spreads desolation throughout all the families which derive their subsistence from the cultivation and produce of the vine. To this plague remedies of no great efficacy have hitherto been opposed, and doubtless it is extremely difficult to imagine any which may prove capable, not of exterminating this race of devouring insects, but merely of diminishing and hurting their population, and of opposing with success the evils which it induces. Perhaps the best plan would be to watch the moment in which the insect deposits its eggs. The method of making women and children detach and carry off the larvæ, as is done in the wine-countries in France, is a tedious process, and one in which a risk is run of increasing the evil, by breaking the young shoots.

The CHRYSOMELA, in general, are rather small. The largest are not above five or six lines in length, and three or four in breadth. Their form, which is very agreeable, and usually enriched with the finest colours, such as scarlet, azure, blue, and golden green, must have caused them to be pursued with eagerness, by amateurs desirous of embellishing their collections. Naturalists also have too frequently encountered them in their walks, not to have marked them in their descriptions. No species is found with hair ; they are all very bare, smooth, and without any sensible hairs, in consequence of which the brilliancy of their colours shines forth in perfect purity. They live on trees and plants, feed upon their leaves, and there deposit their eggs. The female in some species is so fruitful, and has the belly so full of eggs, and consequently so swelled, that the elytra can scarcely cover it.

The larvæ have six scaly feet, articulated, and tolerably long. Their body is elongated, divided into rings, and terminated in a point, furnished at the end with a fleshy nipple, which constitutes a seventh foot. They place it on the plain on which they walk, and as it is usually covered with a gluey matter, they make use of this to keep themselves fixed upon the leaves. Their head is scaly and rounded, furnished with teeth, with small antennæ, and with small barbles. Many species of those larvæ are fond of living in society in a single leaf, which they gnaw in common. For their transformation, they make use of the same precautions as the larvæ of the coccinellæ. They attach themselves somewhere, usually on the leaves, with the nipple of the hinder part of the body. Then they cause the skin of the larvæ to slip as far as the end of the body, where it remains, reduced to a mere clew of thread. There are, however, some species which enter the earth, for the purpose of undergoing this transformation into the nymph state.

These nymphs are usually of an oval figure, more or less elongated, and resemble in general those of many other coleoptera. They remain engaged by the hinder part in the skin of the larva, as above described, and attach themselves only by this place to the leaf. The chrysomela usually remain under the nymph-form only some weeks, and in many instances only for a few days.

Mr. Westwood has formed a new genus, near to chrysomela, GONIOPLEURA. The maxillary palpi have the last joints oval, acute, and ciliated; the labium large, pointed, bifid at the tip; the labial palpi has the last joint smaller than the third, oval, acute, ciliated; the mandibles large, arched, bifid at the tip; the labrum rather semicircular. The species *auricoma*, has the antennæ, head, thorax, legs, and the anterior half of the elytra, reddish brown, the other half copperish green, and the insect is covered with golden hair.

We also insert the figure of a new species of *Doryphora*, under the epithet *princeps*, from Brazil. It is blueish copper, with lines and spots of golden yellow.

The COLAPSES are more particularly found in America, and in all probability have the habits of the other insects of the same family.

THE GALERUCÆ walk slowly, seldom use their wings, are timid, suffer themselves to fall when menaced by any danger, or otherwise under the influence of fear, remain without motion, and try to deceive their enemy by appearing to his eyes as if deprived of life. They love shady and fresh places, woods, the edges of rivers, and occasionally meadows. The larvæ of the galerucæ have six feet, a scaly head, the body soft and pulpy. They live on the substance of the leaves, which they gnaw and devour. They fix themselves on one of these leaves, and cease to eat when they are about to undergo their metamorphosis.

To the history of the galerucæ, as to that of most insects, consecutive and more extensive details are wanting. These species are a little more particularly known, namely, *Galeruca tanaceti*, *G. calmariensis* and *G. nympheæ*. The first species lives on the common yellow tansy, and it is also on the leaves of this plant that the larvæ feed. The females are sometimes filled with eggs, which swell them so much, that the elytra can no longer reach farther than one half of the length of the belly, so that the last three rings are entirely uncovered. The larvæ are found in considerable quantities towards the month of June. They are all black, and but little more than five lines in length. They have six scaly feet, furnished at the extremity with a single crook, and at the hinder part of the body a fleshy nipple, which serves them as a seventh foot, and from which issues a gluey matter, which fixes the larvæ on that on which it walks. On the body there are many tubercles, ranged transversely, and furnished with six or seven small hairs. They walk slowly, and suffer themselves to fall to the earth, rolling their body in a circle the moment the plant is touched to which they are fixed. They are transformed into nymphs, of a fine yellow, bordering a little upon orange, with many small black and stiff hairs, some of which are placed on tubercles. The belly is curved into an arch. On these nymphs may be seen all the external parts of the galeruca, as the eyes, the antennæ, the six feet, and the sheaths of the elytra and the wings. Towards the sides of the body are observable some small black points, which are the stigmata. They are not fond of giving themselves any movement, and will remain tranquil even when they are touched. In three weeks the perfect insect is ready to quit the envelope of nymph.

The elms are sometimes, especially towards the commencement of autumn, altogether covered with galerucæ, which live particularly on these trees, and have been named from this their habitat. The leaves are pierced full of holes by their bites. On the first sensible approach of cold the insect seeks to avoid it. It takes refuge in, and penetrates into the houses, in the neighbourhood of its former residence. Casements which face the south are sometimes seen covered with them.

The *Galeruca nymphecæ* remains and lives in the month of June, and the rest of the year, on the leaves of the *potamogeton*, the *water-lily*, or other aquatic plants which are out of the water, and seldom removes from them. The larva which is found in the month of July, lives in a state of society on the large leaves, more particularly on those of the water-lily, which are suspended at the surface of the water, and these young insects are often found traversing these leaves in large numbers. The larva gnaws the upper substance of the leaf, leaving the lower membrane entire, and in eating it always proceeds forward. The gnawed places appear on the leaves like so many brown spots. These larvæ which are black, and four lines in length, are in general like those of other galerucæ, and like those of chrysomela. The twelve rings of the body, covered with coriaceous plates, are very well marked by deep incisions, and along the two sides there are elevations, in the form of tubercles, and each ring has, besides, on the upper part, a transverse line, in the form of an incision. When the larva bends the body, or when it elongates it considerably, the membranous skin which unites them together appears between the rings. The excrements rejected by these larvæ are found under the leaf, in the form of long tortuous threads, of a greyish brown. To operate their metamorphoses, the larvæ attach themselves by the nipple, which is behind, to the leaves on which they have lived, and then assume the figure of a nymph, stripping themselves of their skin, which they cause to slip backwards, and which they do not abandon altogether. The extremity of the belly of the nymph remains engaged in the folded skin, which serves the larva as a support and resting point, that they may remain attached to the leaves, as is observed in other larvæ of the genera *chrysomela* and *coccinella*. The nymph exhibits nothing very particular. It is short and thick, having at first only a yellow colour, like that of the under part of the larva, but which finally changes into a shiny black. The rings of the belly have on the upper part some tubercles, in the form of short points. As these larvæ, as well in their first as second form, are often exposed to be submerged in the water, particularly when the large leaves which they inhabit are agitated by the wind, their natural constitution is neither to fear the water, nor to receive any injury from it. Nevertheless, they appear more at their ease on the surface of the leaf, which remains dry above the water. They know how to remain, after a fashion, or at least to creep upon the superficies of the stream, and thus transport themselves from one place to another. In less than eight days they are metamorphosed into galerucæ, which still delight to remain on the leaves of the same aquatic plant, which they gnaw and feed upon, as they did in their first state. It has been observed, that when these larvæ have been drawn even from under the water, they are not wet, and appear altogether dry. It is yet unknown whether it be an unctuous transpiration, or an aëriform envelope which protects them from the contact of the water; nor are we acquainted with the mechanism by which they repose underneath the surface of this element.

The ALTICÆ are in general very small. The largest species of Europe are little more than two lines in length, and those of the warmest climates no more than three. They are commonly found in spring in fresh, humid, and rather fertile places, often spread in considerable quantities over potage plants, the leaves of which they gnaw in holes. The majority are adorned with the finest colours. All are shiny and extremely smooth, that is to say, without hair or down. The most common species in kitchen-gardens, are sometimes termed *garden fleas.*

THE SEVENTH AND LAST FAMILY

OF

TETRAMEROUS COLEOPTERA.

Clavipalpi

Are distinguished from all others of this section, which, like them, have the under part of the first three articulations of the tarsi furnished with brushes, and the penultimate bifid, by having the antennæ terminating in a very distinct and perfoliated knob, and by having their jaws armed internally with a nail or corneous tooth; a few have the articulation of the tarsi entire, but they differ from other analogous tetramera, by having the body sub-globular, and capable of contracting into a ball.

Their body is in general rounded, often, indeed, very convex and hemispherical, with the antennæ shorter than the body, the mandibles emarginated or indented at the end, the palpi terminating in a thicker articulation, and the last of the maxillary very broad, transverse, compressed, nearly crescented. The form of the manducatory organs indicate that they are gnawing insects, accordingly we find them in the fungusses of trees, under bark, &c.

Some have the penultimate articulation of the tarsi bilobed, and do not roll themselves into a ball. They may be united into one genus,

EROTYLUS, of Fab.,

Which have the last articulation of the maxillary palpi transversal, nearly crescented, or hatchet-shaped.

ERYTOLUS proper, Fab., from which *Ægithes*, of Fab., do not appear to be essentially distinct, have the intermediate articulation of the antennæ sub-cylindrical, and the knob formed by the last, oblong; the inner and corneous division of the jaws terminate in two teeth.

TRIPLAX, TRITOMA, Fab., differ, by having the antennæ grained, terminating in a short ovoid knob; the inner division of the jaws is membranous, with a single small tooth.

These with an hemispherical shape form the genus Tritoma, of Fab., *T. bipustulatum*, Oliv. Those with an oval or oblong body form Triplax of the same.

The others have the last articulation of the maxillary palpi elongated, and more or less oval.

LANGURIA, Lat. Oliv. *Trogosita*, Fab., have the body linear, and the knob of the antennæ with five articulations.

PHALARCRUS, Payk. *Anisotoma*, Illig. Fab. *Anthribus*, Geoff. Oliv. have the body sub-hemispherical, and the knob with three articulations.

The other clavipalpi have all the articulations of the tarsi simple, and the body sub-globular. They form the genus AGATHIDIUM, Illig. *Anisotoma*, Fab.

Of the few insects belonging to this family, scarcely any thing is known beyond those physical characters, from which they have been separated from the other coleoptera, which characters are sufficiently pointed out in the text. We have, therefore, nothing to add by way of supplement to this family.

THE FOURTH GENERAL SECTION

OF

COLEOPTEROUS INSECTS.

TRIMERA,

HAS only three articulations in all the tarsi. They compose three families. Those of the first are nearly allied to the last Tetramera. Their antennæ, always composed of eleven articulations, terminate in a knob formed by the last three, compressed, and like a cone or reversed triangle. The first articulation of the tarsi is always very distinct, the penultimate is commonly bilobed, and the last, with a knob at its base, always terminates with two hooks. The elytra entirely cover the abdomen, and are not truncated. Those of the third family approximate in this and other respects to the brachilytrous pentamera, and other coleoptera of the same section. The first family,

FUNGICOLÆ,

Have the antennæ longer than the head and corslet, the body oval, with the corslet trapezoid, the maxillary palpi are filiform, or a little thicker at the end, but not terminated by a hatchet-formed articulation. The penultimate articulation of the tarsi is always deeply biloped. This family may be reduced to one principal genus,

EUMORPHUS.

Some have the third articulation of the antennæ much longer than the preceding and following. Such are EUMORPHUS proper, Web. Fab., in which the knob of the antennæ is sudden, serrated, much compressed, and triangular. The maxillary palpi are filiform.

DAPSA, Zieg. The same knob is narrow and elongated. In the others the length of the third articulation little exceeds its preceding and succeeding articulations.

ENDOMYCHUS, Web. Fab., have the four palpi thicker at the end, the last three articulations of the antennæ separated laterally, larger than the rest, and forming a triangular knob.

LYCOPERDINA. *Eudamychus*, Fab., have the maxillary palpi filiform; the last articulation of the labial palpi larger than the preceding, and sub-ovoid, from the fourth to the ninth, inclusive of the antennæ, nearly grained, and the last two large and triangular.

Of these insects we have only to observe that they are named from the habitat in which they are most commonly found, viz., in fungi, though some are met with in dead wood, and under the bark of trees. We are ignorant alike of their changes and of their habits.

THE SECOND FAMILY

OF

TRIMEROUS COLEOPTERA.

Aphidiphagi

Is composed principally of insects, nearly hemispherical; the corslet is very short, transverse, and nearly crescented; the antennæ terminate in a compressed knob, like a reversed cone, composed of the last three articulations, and are shorter than the corslet; the last articulation of the maxillary palpi is very large, hatchet-formed, and the penultimate articulation of the tarsi is deeply bilobed. In the other trimera of the same family the articulations of the tarsi are simple, or the penultimate at least is slightly bifid, which distinguishes these from the fungicolæ. The antennæ have eleven articulations, the last forming a knob, like a reversed cone. They compose the genus,

Coccinella.

Lithophilus, Frôhl. Body ovoid, corslet bordered laterally, and narrowed behind, the penultimate articulation of the tarsi slightly bifid.

Coccinella proper. Body sub-hemispherical, corslet short, nearly crescented, not at all, or slightly bordered, penultimate articulation of the tarsi deeply bilobed. *C. 7 punctata. C. 2 punctata. C. pustulata*, Lin.

Clypeaster, Andersche. *Cossyphus*, Gyll. have the body very flat, buckler-shaped, with the head hidden under the corslet, nearly semicircular. The antennæ have distinctly but nine articulations, and terminate in an elongated knob. The articulations of the tarsi are entire. The presternum forms in part a sort of chin-cloth.

SUPPLEMENT

ON

THE APHIDIPHAGI.

THE insects forming the type of the first genus of this family are well known. Their scientific name is COCCINELLA, and they are commonly designated in this country *Lady-birds*.

The coccinella, whose entire body forms a hemisphere, or a segment of a sphere, are very easy of recognition. They do not arrive to any great size. The majority of the largest are scarcely more in diameter than a large pea. These insects are very handsome. Their elytra, which are very brilliant, and well applied one against the other, appear to form a shining scaly arch, of a single piece. Their colours are not very varied; but they have almost all of them some spots which distinguish them. These spots are severally arranged in a regular and agreeable manner. Their hemispherical figure constitutes one of their most prominent characters. There are some, however, that have the body a little more elongated, and bordering on the oval, but their number is but small. It is more particularly when the insect lowers its head, which it usually does when touched, that it appears most spherical. The coccinella have, moreover, some other characters which are sufficiently remarkable. When they are in a state of repose, they bend the legs by the side of the thighs, and apply them both together against the body; so that in looking at them from underneath, one would imagine that they were without legs, for they are so short that they cannot be perceived. When the lady-bird is touched a little, it causes to issue from the end of the thighs a drop of yellow fluid, which is mucilaginous, of a penetrating odour, very strong, and stinking. Although one might well suppose an aperture to exist at the extremity of each thigh, it nevertheless remains as yet undiscovered. All that has been observed is, that the fluid appears to escape from the joint itself, which unites the thigh to the leg. It is doubtless there that this aperture must be found, perhaps within the articulation.

These little insects do not walk very fast, but they fly with facility. They appear to open the elytra which covers their wings with much ease, and this they never fail to do before taking ground, when they are thrown into the air. The coccinellæ feed on aphides, and are found on all sorts of plants and trees, peopled with these little animals. In this way they do much good. Mr. Kirby observes, "that the lady-bird, or lady-cow (coccinella), the favourite of our childhood, as well as most of its congenus in the larva state, feeds entirely on aphides; and the havock made among them may be conceived, from the myriads upon myriads of these interesting little animals which are often to be seen in years when the plant-louse abounds. In 1807, the shore at Brighton, and all the watering-places on the south coast was literally covered with them, to the great surprise and even alarm of the inhabitants, who were ignorant that their little visitors were emigrants from the neighbouring hop-grounds, where, in their larva state, each had slain his thousands and ten thousands of the aphis, which, under the name of the *fly*, so frequently blasts the hopes of the hop-grower.

Professor Reich has stated that the larvæ of some species of this genus feed entirely on leaves, and especially the leaves of the common heath; but Mr. Kirby seems to think that there is something erroneous in this assertion.

There can be no other reason than their pursuit of the aphides, to account for the prodigious number of coccinellæ which are to be found, especially in autumn, on the banks of large rivers, and on the sea-shore. Mr. Kirby tells us, that "many years ago, those of the Humber were so thickly strewed with the common lady-bird *(Coccinella septempunctata)*, that it was difficult to avoid treading upon them. Whether the latter and their devourers cross the sea has not been ascertained; that the coccinellæ attempt it, is evident from their alighting upon ships at sea, as I have witnessed myself."

The *C. bipunctata* ejects from its joints a yellow fluid, of a very powerful, but by no means pleasing, scent of opium.

The lady-birds in general live out during the winter, and are the first insects which re-appear in spring. They then couple, and lay their eggs on the plants on which they have lived.

The larvæ of the coccinellæ are hexapod. Their body is elongated, of a conical figure, diminishing towards the hinder part, and divided into twelve rings. The first ring less broad, but longer than the following, is oval, flatted above, and covered with a scaly, or at least, a coriaceous and hard skin, and has the appearance of a little corslet. The skin of the other rings is membranaceous; but the second and third have each two oval plates, of a deeper colour than the rest, which are also scaly. In some species all the rings are bristling with spines, above and towards the sides; in others, they have raised and conical tubercles, all bristling with small points, in the form of blunt spines; while others again have the skin altogether smooth, and without spines. The last ring is small, and the larva often pushes forth from it a fleshy and tolerably thick nipple, which it sometimes rests on, and which then answers the purpose of a seventh foot. All the under part of the body is furnished with plenty of hair.

The head is small, scaly, a little flatted, and with rounded corners. It has two small, short, conical antennæ, divided into articulations; two lips, and four barbles, the external ones of which are large and thick, but the other two short and conical. The jaws, which are placed between the lips, are of a marone colour, and furnished with denticulations at the end. Some hairs are seen here and there upon the head, and on other parts of the body.

The six scaly feet, tolerably long, and almost of equal thickness in their whole extent, are divided into three parts, but their conformation is rather different from that of the feet of many other hexapod larvæ. The first part, united to the body, is short and thick; the second is long and cylindrical; and the third is like the preceding in thickness, and nearly so in length. The end of the foot is as thick as the rest, and terminated by a single crook, formed like the claw of a bird. On the two long parts of the feet are many hairs, some long, and the others short; but what is most singular is, that the little hairs, which are very numerous towards the end of the foot, on the interior side, are thicker at the end than elsewhere. They are terminated, as it were, in a small elongated knob, and are transparent. A microscope is necessary to observe all this. As these larvæ adhere strongly to the objects on which they walk, we might be led to believe that these hairs terminating in a knob, may furnish some gluey matter, proper for the purpose of better fixing the feet, although the crooks serve principally for this use. These larvæ live on plants and on trees of every description, loaded with aphides, which constitute their only nutriment. They are very voracious, and consume a great number of those little insects, which they seize with their fore-feet, and carry them thus to their mouth to devour them. They hold them fixed there by means of the two large barbles. They do not even spare their own species, but occasionally devour one another. When they are collected in the same sand-box, the little and the most feeble become the prey of the strongest. When they are about to be transformed, they attach themselves on the leaves and branches with the fleshy nipple of the rump, from which they eject a glutinous fluid, which cements it against the plane of position. At the end of two or three days they get rid of their skin, and appear in the form of nymphs. They cause the skin to slip by little and little towards the hinder part of the body, where it is gathered into a ball, in which the nymph remains engaged by the end of the body.

The nymphs are usually spotted with black and other colours, and the only movement which they make is that from time to time, and particularly when they are touched, they raise and lower the body alternately. They often raise themselves up perpendicularly on the rump, and remain some moments in this position, the rump serving as a hinge to the body; but in a state of inaction the head reposes. The coccinellæ quit the envelope of nymph often at the end of six days, and at other times after ten or eleven. When they are nearly issued from this envelope, the elytra are usually of a dirty and yellowish white, without any spot, and they are then of a soft and flexible consistence; but in proportion as they grow harder by the action of the external air, the spots commence, by degrees, to make their appearance. The under part of the body is of the same yellowish colour in the beginning; but at the end of some hours this colour becomes black, red, or yellow, according to the different species.

THE THIRD AND LAST FAMILY

OF

TRIMEROUS COLEOPTERA.

PSELAPHII

HAVE a certain resemblance to the Brachelytræ in their short truncated wing-cases, which cover only a portion of the abdomen, which is short, wide, obtuse, and rounded behind. The antennæ ending in a knob, or thicker toward the end, have sometimes only six articulations; the maxillary palpi are commonly very large, all the articulations of the tarsi are entire, and the first much shorter than the following, is not at all, or but little apparent at first sight; the last is most commonly terminated with a single hook.

These, with eleven articulations to the antennæ, form the genus PSELAPHUS, Herbst. *Staphylinus*, Lin. *Anthicus*, Fab.

A few have two hooks to the tarsi.

CHERNNUM, Latr. The first ten articulations of the antennæ are nearly equal, lenticular; the last is longer and subglobular. The palpi do not project.

DIONIX, Dej. The third to the seventh articulation very small, transverse, and grained; the four last are thicker, cylindrical, as long as the first seven together; the two penultimate, conical, equal, and the last ovoid, elongated, pointed, the thickest of all. The maxillary palpi project, but are shorter than the head and corslet together; they have four cylindrical articulations. The labial palpi of three articulations are short, directed forward, with a point at the end.

The others have only a single hook at the end of the tarsi.

In these the maxillary palpi bent, or folded, are, at least, as long as the head and corslet; their second and fourth articulations are much elongated, bent at the base, and terminate in a knob.

Sometimes the antennæ, evidently longer than the head and corslet, terminate in a knob formed by the last three articulations, which are manifestly larger than the others, and the last of which is sub-ovoid, or ovoido-conical.

PSELAPHUS proper, Herbst. Sometimes the ninth and tenth articulations are scarcely larger than the preceding, and the last is only a little thicker, nearly spherical (with a point at the end).

BITHYNUS, Leach. In these the second articulation is thicker than the first, and dilated like a tooth on the inner side.

ARCOPAGUS, Leach. The second articulation is on the contrary thinner than the first, and is sometimes even dilated.

There the maxillary palpi are shorter than the head and corslet taken together; the fourth articulation, at least, is short, or but little elongated, ovoid, or triangular.

CTENISTES, Reich. Very distinct from all the insects of this family, by reason of the three last articulations of the maxillary palpi, whose outer edge has a point or tooth, with a terminal thread.

BRYAXIS, Leach. *Euplectus, Tychus*, Ejusd. The maxillary palpi are not like the last; the last articulation is elongated into a cone or hatchet. The corslet is short, or scarcely longer than it is wide, and rounded.

The last have this peculiarity, viz., that the antennæ are

composed of only six articulations, or even of a single one. They form the genus

CLAVIGER.

CLAVIGER proper, in which these organs have distinctly six articulation. These insects have no apparent eyes. The maxillary palpi are very short, without distinct articulations, with two small nails at the end; the first two articulations of the tarsi are very short; the third and last are very long, with a single hook at the end.

ARTICERUS, Dalm. The antennæ do not appear to be composed of more than a single articulation, forming a cylindrical knob, elongated, and truncated at the end. The eyes are distinct, and the tarsi are terminated by two hooks.

Nota.—The tarsi of *Dermestes atomarius*, of De Geer, appearing to M. Laval as composed of only a single articulation, we have formerly with this, and some other insects, formed a new section of Coleoptera (MONOMERA), which has been adopted by M. Fischer, who has made a new genus of that insect, under the name of *Clambus*. But it appears that M. Schuppell has made the same section under the denomination *Ptilium*. M. Gyllenhal had united its species to scaphides; and we think that it is, in fact, near these last that this new genus should be placed: the section Monomera should then be suppressed.

We have only to say, by way of supplement to this family, that the species in general are very small, and live on the ground, in fresh and humid places, among plants. They are sometimes also found under the bark of trees, stones, and moss.

THE SIXTH ORDER

OF

INSECTS.

ORTHOPTERA.

ULONATA, *Fab.*

MIXED for the greater part by Linnæus with the Hemiptera, united by Geoffroy to the Coleoptera, but forming a special division of them, have in general a body less hard than the last; the wing-cases are soft, and half membranaceous, beset with nervures, and not joining each other for the most part of them by a straight line at the suture; the wings fold lengthwise, and in general like a fan, and are divided, in like manner, by membranaceous nervures; the jaws always terminate in a corneous piece, indented and covered with a galea, a piece corresponding with the outer division of the jaws of coleoptera; finally a sort of tongue or epiglottis.

These insects are subject to a mere partial metamorphosis, all their mutations consisting only in the increase and development of the elytra and wings, which begin to display themselves under a rudimentary form, or as little stumps, in the nymph. In other respects the nymph and larva resemble the perfect insect, and walk and feed in the same manner.

The mouth of the orthoptera is composed of a labrum, two mandibles, two jaws, a lip, and four palpi; those of the jaws always have five articulations; the labial palpi, as in the coleoptera, have only three. The mandibles are always very strong and corneous, and the ligula is always divided into two or four straps. The form of the antennæ varies less than in the coleoptera, but they are generally composed of a greater number of articulations. Many, besides the reticulated eyes, have two or three small, smooth eyes. The under part of the first articulations of the tarsi is often fleshy or membranaceous. Many of the females have a true auger formed of two blades, for the deposition of their eggs, which eggs are often under one common covering. The posterior extremity of the body has in general an appendix.

All the orthoptera have first a membranaceous stomach, or crop, followed by a muscular gizzard, armed internally with scales or corneous teeth, according to the species; round the pylorus there are, except in the Forficulæ, two or more close intestines, furnished at the bottom with many small biliary vessels. Numerous other vessels of the same kind are inserted toward the middle of the intestines, which are the same in the larvæ as in the perfect insect.*

All the known orthoptera, without exception, are terrestrial even in both their first states. Some are carnivorous, or omnivorous, but most of them feed on living plants. Those of our climate lay their eggs but once a year, toward the end of the summer, which is also the period of their last transformation.

We shall divide the Orthoptera into two principal families.

Some have all the feet alike, and equally fitted for running, others have the thighs of the posterior pair much broader than the rest, which enables them to leap. The males moreover produce a sharp noise or sort of stridulation. These may be called musical leaping orthoptera.

* M. Marcel de Senes has studied the anatomy of these insects particularly, to whose labours we can merely refer.—Ed.

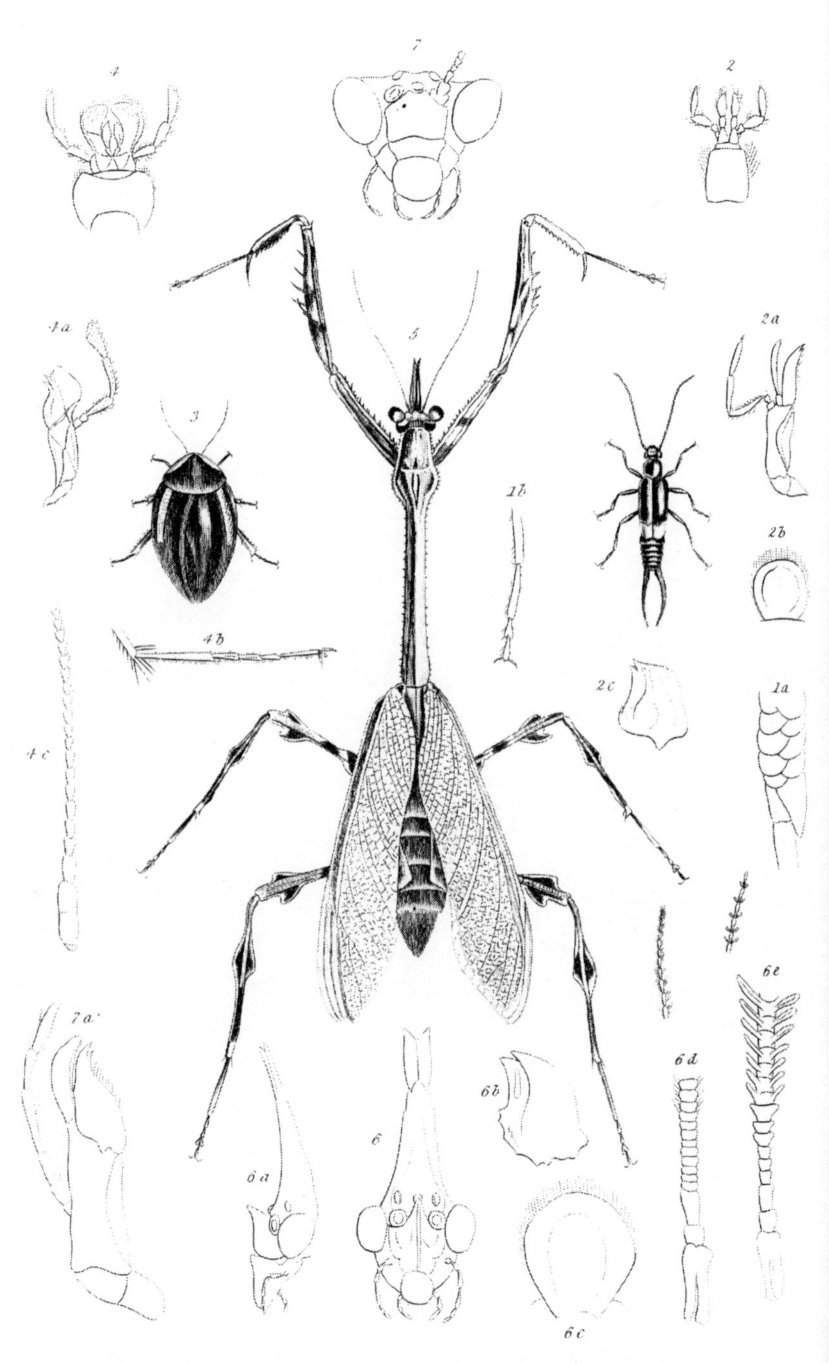

1. Forficula (Spongiphora Serv.) croceipennis. _ Serv.
2. Details of Forficula auricularia. _ Lin.
3. Blatta (Phorasphis Serv.) picta. _ Fab.
4. Details of Blatta Ægyptiaca. _ Fab.
5. Empusa lobipes. _ Olivier.
6. Details of Empusa pauperata. _ Fab.
7. Details of Mantis religiosa. _ Lin.

Published by Whittaker & C° Ave Maria Lane.

The first Family, or the Runners,

CURSORIA,

Have the hind legs, as well as the rest, fitted for running. Almost all of them have the cases and the wings hidden horizontally on the body. The females have no corneous auger. They form three genera. First,

Forsicula, Lin.,

With three articulations to the tarsi, the wings placed like a fan, and lying crosswise under the crustaceous cases, which are very short, and with a straight suture. The body is linear, with two large scaly moveable pieces forming pincers at its posterior end. The head is uncovered. The antennæ are filiform, inserted before the eyes, and composed of twelve or thirteen articulations, according to the species. The galea is slender, elongated, and sub-cylindrical. The ligula is forked. The corslet is in the shape of a plate.* The common *earwig*, *Forficula auricularia*, Lin. *F. minor*, Lin.

Blatta, Lin.,

Have five articulations to all the tarsi; the wings are folded only longitudinally; the head is concealed under the plate of the corslet, and the body is oval or orbicular, and flatted. The antennæ are like threads inserted in an internal emargination of the eyes, long, and composed of a great number of articulations. The palpi are long. The corslet has the form of a shield. The wing-cases are commonly as long as the abdomen, coriaceous, or half membranaceous, and cross each other a little at their suture. The posterior extremity of the

* In the 13th Vol. of the Annales des Sciences Naturelles, will be found the details of M. Dufour on his anatomy of this genus. Want of space obliges us to omit even an abridgment of these highly curious facts and observations.—Ed.

abdomen has two conical and articulated appendices. The legs have small spines. Their crop is longitudinal, and the gizzard has internally strong crooked teeth. Eight or ten cœcums may be counted round the pylorus. *B. orientalis*, Lin. *B. Lapponica*, Lin. *B. Americana*, De G. *B. Germanica*, Fab.

MANTIS, Fab.,

In which also five articulations are found in the tarsi, and the wings are simply folded longitudinally, but their head is exposed, and the body is narrow and elongated. They differ, moreover, from Blatta by their short palpi ending in a point, and by their ligula being divided into four.

Some have the two fore-feet larger than the others, long with a hip, the thighs strong, compressed, and armed with spines underneath, and the legs terminating in a strong hock. They have three smooth, distinct eyes, arranged in a triangle. The first segment of the body is very large, the four lobes of the ligula nearly of the same length, the antennæ inserted between the eyes, and the head triangular and vertical.

These orthoptera form the sub-genus MANTIS. Those whose front is elongated like a horn, and whose males have pectinated antennæ, form the genus *Empusa*, of Illig. They have at the end of the thighs a rounded and membranaceous appendix like a ruffle. The edges of the abdomen are festooned in many. Those without a horn on the head, and whose antennæ are simple in both sexes, compose alone the genus Mantis of the same naturalist, *M. religiosa*, Lin.

The others have the fore-feet like the following, the eyes are smooth, little distinct, or none ; the first segment of the body is shorter, or at most as long as the following ; the interior divisions of the ligula are shorter than the lateral ; the antennæ are inserted before the eyes, and the head nearly ovoid, and projecting, with the mandibles thick, and the

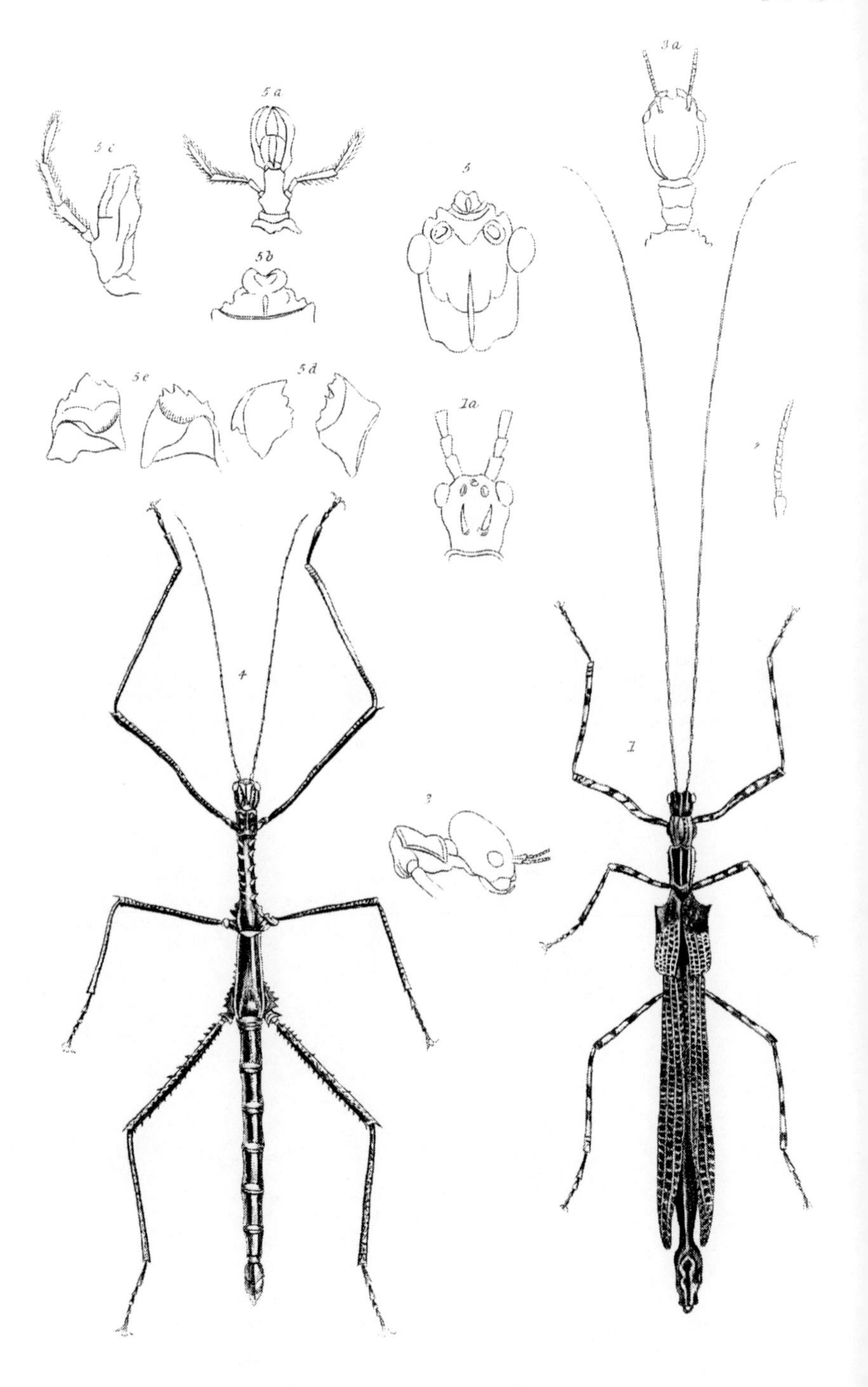

1 *Phasma annulata* _Serv 3 Head of *Cyphocrana.*

2 Ant of *Bacillus* 4 *Bacteria scabrosa* _Perch

5 Details of *Cladoxerus roseipennis* _Guer

London. Published by Whittaker & Co. Ave Maria Lane, 1832

palpi compressed. The two sexes frequently differ considerably. They form the sub-genus SPECTRUM, of Stoll.

They have been divided into two others.

The species whose body is filiform or linear, like a stick, form PHASMA of Fab. Many are altogether wingless, or with very short cases. Some very large are found in the Moluccas, and in South America. The south of France has *P. rossia*, Fab.

The species whose body is very flat and membranaceous as well as the feet, compose the genus *Phyllium*, of Illiger. *Mantis siccifolia*, Lin. The inhabitants of the Sechell Islands raise this species as an object of commerce and natural history.

The second family of Orthoptera, the Leapers,

SALTATORIA,

Whose two hind legs are remarkable for the size of their thighs, and their spiny legs are calculated for leaping.

The males call their females by a noise, commonly called a song. Sometimes this is produced by rubbing rapidly one against the other a part of each wing-case, which part is more membranaceous, and is like talk or glass. Sometimes it is produced by rubbing the posterior thighs, or the elytra and wings; these thighs doing the office of the bridge in a violin. Most of the females deposit their eggs in the earth. This family is composed of the genus

GRYLLUS, of Lin.,

Which we shall divide thus.

Some, whose males have the apparatus of song, formed by an inner portion of the cases, like a glass or drum-head, and whose females have very often a very projecting auger, formed like a dagger or sword, have antennæ either much thinner at their extremity, or of the same thickness throughout, but very short, and nearly in the shape of a chaplet. The wings and their cases are concealed horizontally on the body of those, a few in number, which have less than four articulations to all the tarsi. The ligula has always four divisions, the two intermediate being very small. The labrum is entire.

Sometimes the cases and wings are horizontal; the wings form, when at rest, a sort of straps or threads elongated beyond the cases, and the tarsi have but three articulations, as in the genus GRYLLUS, of Geoff. and Oliv.

These form four sub-genera,

1. GRYLLO-TALPA, Lat., whose legs and tarsi of both anterior feet are broad, flat, and indented like hands, or calculated for digging. The other tarsi are of ordinary shape, terminated by two hocks, and the antennæ are more slender at the end, elongated and composed of a great many articulations. (*Gryllus, gryllo-talpa*, Lin.)

2. TRIDACTYLUS, Oliv., *Xya*, Illig. These also dig into the earth, but with the fore-legs only. Instead of the posterior tarsi, they have moveable, narrow, bent appendices, formed like fingers. The antennæ are of equal thickness, very short, and the articulations are round. *Xya variegàta*, Illig., is found in the south of France on the river sides.

3. GRYLLUS, properly so called, which have no proper digging feet, and whose females have at the end of the body a projecting auger. Their antennæ are always elongated, more slender toward the end, and terminating in a point. Their smooth eyes are less distinct than in the tridactyle and mole-cricket.

There is found in Spain and Barbary a very singular Gryllus, (*G. umbraculatus*, Lin.) The male has on the front a membranous elongation which forms a sort of vail.

M. M. Lèfevre and Bibron have brought from Sicily a new and large species, which the former has described under the

Pl. II 3.

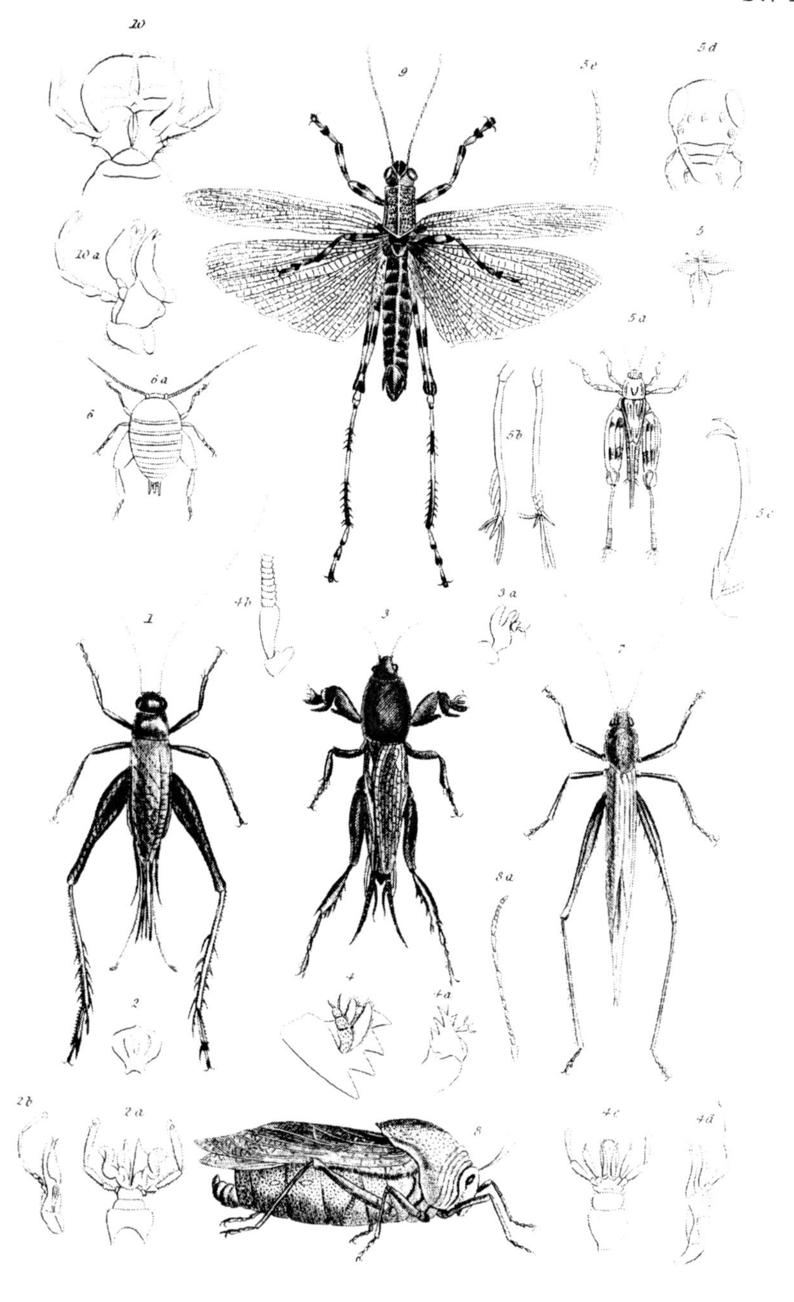

1. *Gryllus Servillei* _Guer.
2. Details of *Gryllus campestris* _Fab.
3. *Gryllotalpa didactylus* _Latr.
4. Details of *Gryllus vulgaris*.
5. *Tridactylus fasciatus* _Perch.
6. *Myrmecophila acervorum* _Panzer.
7. *Locusta erythrosoma* _Encycl.
8. *Pneumora inanis* _Fab.
9. *Acrydium tarsatum* _Serv.
10. Details of *Acrydium migratorium*

London, Published by Whittaker & Co. Ave Maria Lane, 1832.

name of *megacephalus ;* its stridulation continues for half a minute, and may be heard nearly a mile.

In (*G. monster,*) the wings fold in many spiral turns at their extremity.

4. MYRMECOPHILA—*Spærium* Charpeut. These have no wings, and their body is oval. In the antennæ, and in the absence of simple eyes, they resemble gryllus proper. Their posterior thighs are very thick. The only known species (*B. acervorum*, Panz.) lives in ant-hills.

Sometimes the cases and wings are ridged, and the tarsi have four articulations. The antennæ are always very long, and like a hair. The mandibles are less indented, and the galea is wider than in gryllus. The females always have an advanced auger, compressed and formed like a sabre or cutlass. There are only two cœcums, as in the last, but the biliary vessels surround the middle of the intestine, and are inserted there direct. These orthoptera are herbivorous, and form the genus

LOCUSTA, Geoff. Fab. *Gryllus tetigonia*, Lin.

L. viridissima, Fab. Rees, two inches long, green without spot, the auger straight. *L. verrucivora*, Fab.

Many species of this genus have no wings, or have only very short cases, as *L. ephippiger*, Fab.

Others, whose males produce their noise only by the friction of the thighs against the cases or wings, and whose females have no projecting auger, are distinguished moreover from the last by the antennæ, sometimes filiform and cylindrical, sometimes like a sword, or terminating in a knob, and always as long or less than the head and corslet. All have the cases and wings ridged or inclined, and three articulations to the tarsi. They have five or six cœcums, and their biliary vessels are inserted, as in most of this order, immediately at the inter-line. The ligula of most of them has only two divisions. All have three distinct simple eyes, the labrum emarginate, the mandibles much indented, the abdomen conical, and laterally compressed. They leap better than the last, have a stronger and more elevated flight, and feed on vegetables, of which they are very voracious. They may be included in a single genus,

ACRYDIUM, Geoff.,

Which may be sub-divided as follows.

Some have the mouth exposed, the ligula bifid, and a membranous pellet between the hooks at the end of the tarsi. These form PNEUMORA, of Thunb., part of *Gryllus bulla*, of Lin., which are distinct from the following by their hind legs being shorter than the body, and less calculated for leaping, and by the abdomen being vesicular, at least in one sex. The antennæ are filiform.

PROSCOPIA, Klüg. Apterous, with the body long and cylindrical, whose head, deprived of the simple eyes, is elongated forward like a cone or point : the hind legs are large, long, and near the middle feet, which are more elongated than common from the fore-feet.

TRUXALIS, Fab. *Gryllus acrida,* Lin., which, by their compressed prismatic and sword-shaped antennæ, and their pyramidical head, differ from all other orthoptera.

Some species of the sub-genus following, as *G. carinatus,* Lin. *G. gallinaceus,* Fab., are, with reference to the antennæ, intermediate between Truxalis and Gryllus proper, and form the genus *Xyphicera,* Lat. *Pamphagus, Thunb.*

GRYLLUS proper, *G. locusta,* Lin., and some *G. bulla.* Differ from Pneumera by having the hind legs longer than the body, and the abdomen solid and not vesicular, and from Truxalis by their ovoid head and antennæ filiform, or terminating in a knob.

Certain species, called by travellers Migratory Locusts,

emigrate in multitudes exceeding all calculation. A great part of Europe is often ravaged by *G. migratorius*, Lin.

The south of Europe, Barbary, and Egypt, suffers similar losses from *G. Ægypytius*, *tataricus*, Lin., which differs but little from *G. lineola* of Fab., found in the south of France. These two, and many other species, have a conical projection on the presternum, and compose my genus ACRYDIUM. Among those which have not this character, and whose antennæ are also filiform, some have the elytra and wings perfect in both sexes, and belong to the genus I have named ŒDIPODA. Of this number are *G. striaculus*, Lin., *G. cærulescens*, Lin.

Others without wings, and with the antennæ filiform, have the upper part of the corslet much elevated and compressed, forming a sharp crest rounded and pointed behind, *Acrydium armatum*, Fisch.

One of the sexes, at least in others (*G. pedester Giornæ*, of Charpent), have very short wings, and elytra not at all fitted for flight. I have formed these into a new generical section, PODISMA.

Those whose antennæ are enlarged at the end like a knob, whether in both sexes or in one only, form another genus, according to Thunberg, GOMPHOCERUS, as *G. Sibiricus*.

In the second division of the genus gryllus, the presternum receives into a cavity a portion of the underpart of the head; the ligula is quadrifid; the tarsi have no pellet between their hooks. The antennæ have only thirteen or fourteen articulations. The corslet is elongated behind in the shape of a large scutellum, sometimes longer than the body, and the cases are very small. These form the genus TETRIX, Lat. *Acrydium*, Fab. part of *Gryllus-bulla*, Lin. It includes only a few species.

SUPPLEMENT

ON THE

ORTHOPTERA.

The name of this order, borrowed from two Greek words, ορθος (straight), and πτερα (wings), indicates the peculiar disposition of these latter organs. But it is not on this peculiarity of conformation that this truly natural order is to be established. It is especially the mode of transformation or metamorphosis which characterises these insects. In this they approximate to the hemiptera, and are removed from the coleoptera, with which all authors had placed them before the time of Degeer.

Linnæus indeed had originally marked the distinction of this grand section of insects; for though he ranged them with the coleoptera, he nevertheless terminated that order with them.

Fabricius, in his System of Entomology, designated the same order under the denomination of *Ulonata*, because he observed that their jaws are engaged in a sort of membranaceous piece or gum, which he called *galea* (helmet). The term *Ulonata* comes from two Greek words, ουλον, *external gum*, and γναθος, *jaw*.

Olivier, who in the establishment of the first sections in his method, gave the priority to characters derived from the wings, observed that those of the insects of this order are folded longitudinally, and from this consideration he gave them the name of orthoptera. Degeer's denomination is better founded. In the method of Olivier the genus *forficula* terminates the coleoptera, and is thus very much removed from the orthoptera. It evidently, however, forms the passage from one order to the other, and Mr. Kirby, on the suggestion of Dr. Leach, formed it into a separate order, which he calls *dermaptera.* Their body is generally elongated, of a less firm consistence than that of the coleoptera, and often even soft or fleshy. Accordingly some African nations use them as an article of food, and from this circumstance were named by the ancients *acridophagi.* These insects are, of all others in the collections of natural history, the most exposed to be devoured by the *dermestes, anthreni, ptini,* and other destructive animals of the same class.

Independently of the peculiarities to be observed in the external configuration of the orthoptera, internally these insects exhibit a mode extremely singular. Most of the orthoptera, such as *locusta, gryllus, gryllo-talpa, acrydium,* &c. are remarkable for the real, or at least apparent, multiplicity of their stomachs. Four have been attributed to the *gryllo-talpa*, which is, of all insects, the one whose internal organization has been the most anciently studied. Its œsophagus is in the form of an elongated canal. It leads to a first stomach, rounded and membranous, and from which proceeds a very short canal, which conducts to a second stomach shorter than the preceding, but muscular, and with thicker parietes. It is furnished with parts which may be compared to the jaws, which are found in the stomach of the crustacea. There are small laminæ of a silken appearance, disposed on five longitudinal ranges, which are each composed of ten or twelve small laminæ, performing a sort of peristaltic motion, by the muscular action of this species of gizzard. Without doubt, the use of these pieces is to act upon the aliments. The other two stomachs, that is, the third and fourth, are similar one to the other, and placed opposite to each other, at the orifice of the intestine. These are wrinkled, thicker than the first, but less so than the second, and of a spongy contexture.

This insect, as well as other orthoptera, analogous in the same point, has been considered as ruminant, or possessed of the faculty of bringing back to the mouth the aliments contained in the digestive organ. But according to M. Marcel de Serres, these pouches, or cæcums, which have been taken for stomachs, are nothing of the sort, and contain only a salivary and biliary fluid, which the animal often disgorges or vomits when it is seized. The long and slender vessels which adhere to the intestinal canal, pour in their digestive fluids, which are secreted by those of the common mass of the humours.

According to M. Cuvier, the crop of the grylli often forms a lateral pouch; they have at the pylorus but two large cæcums, and the biliary vessels insinuate themselves into the intestine by a common canal.

The *locusta* have also but two cæcums; but the biliary vessels surround the middle of the intestine, and are inserted there directly. We see, by comparison, that the two parts of the digestive organ of the *gryllo-talpa*, which have been considered as the third and fourth stomachs, are only two cæcums analogous to the preceding. The *forficula* or *earwigs* are the only orthoptera whose pylorus presents no cæcum. From five to six are reckoned in *Acrydium*, and from eight to ten in *Blatta*. In these the crop is longitudinal, and the gizzard has internally some strong crooked teeth. The stomach of the mantis is nearly similar.

Most part of the orthoptera feed on vegetable substances only, and these matters being less proper for the purposes of animalization, than substances already animalized, it follows that these insects eat much more in proportion than those of the other orders, which feed on insects, carrion, or other corrupted substances. The herbivorous orthoptera are, then, of all insects, those which eat the most, and their voracity is excessive. Fields, entire countries, are despoiled of their verdure, by clouds of locusts, which fall there at once, and after having desolated the country speedily die of hunger. Their destruction even proves a no less terrible scourge than that which they have inflicted in their living state. In putrefying they fill the air with infectious miasmata, and pestilence follows hard upon the steps of famine.

The orthoptera are extremely fruitful. Their eggs, often very numerous, are usually large, and of an elongated form. But they must not be confounded with the Capsula which encloses them, as some observers have done with relation to the *Blatta*.

The larva of these insects differs from the perfect state only in the total absence of wings, and the nymph is distinguished from the larva only by the presence of the rudiments of these same organs. It is agile, and feeds upon the same substances as the larva. Towards the end of summer, and in autumn, these insects are most abundant in our climates. It is also seldom, but at this period, that they are found in a perfect state.

The insects of the genus Forficula (vulgò *Earwigs*) partake pretty equally of the character of the coleoptera, with which they have been ranged by some writers, and of those of the orthoptera.

The male forficulæ differ a little from the females in regard to the pair of pincers situated at the end of the abdomen. This is easy of observation in the *forficula auricularia*, or common earwig, the species which we most generally encounter. The hooks of the pincers are larger and more arched in the male than in the female. The forficulæ are very frequently to be met with either on the ground or on plants, and principally under the barks of trees, where they are often assembled in numerous societies. They feed on various matters either animal or vegetable. The form of their mandibles sufficiently announces them to be gnawers. They do considerable damage to fruits, and to the flowers of the pink more especially.

Frisch and Degeer have observed, that the female earwig watches with all possible care over the preservation of her eggs, which are found in the commencement of April, in moist places, under stones, and are collected in heaps. "This curious insect,' observes Mr. Kirby, "so unjustly traduced by vulgar prejudice—as if the Creator had willed that the insect world should combine within itself examples of all that is most remarkable in every other department of nature—still more nearly approaches the habits of the hen in her care of her family. She absolutely sits upon her eggs, as if to hatch them—a fact which Frisch appears first to have noticed—and guards them with the greatest care. Degeer, having found an earwig thus occupied, removed her into a box where there was some earth, and scattered the eggs in all directions. She soon, however, collected them, one by one, with her jaws, into a heap, and assiduously sat upon them as before. The young ones, which resemble the parent, except in wanting elytra and wings, and, strange to say, are, as soon as born, larger than the eggs which contained them, immediately upon being hatched, creep like a brood of chickens under the belly of the mother, who very quietly suffers them to push between her feet, and will often, as Degeer found, sit over them in this posture for some hours. This remarkable fact I have myself witnessed, having found an earwig under a stone, which I accidentally turned over, sitting upon a cluster of young ones, just as this celebrated naturalist has described."

These eggs are tolerably large, white, and smooth, and they disclose the young in the month of May. The little ones appear very large, relatively to the volume of the egg, which supposes that they must be very much compressed. The movement of the dorsal vessel is very sensible in the young larvæ, which have neither elytra nor wings, as is the case with all the other larvæ of the orthoptera. Their body is less thick at the two ends, and formed of thirteen rings. The first three have each one pair of feet, and correspond to the corslet and the breast. The two pieces of the pincers are conical, and a little divergent. The antennæ have as yet but eight articulations; the palpi and the feet are inflated. The corslet of the larvæ is distinct in the nymph. The sheaths of the elytra and the wings are flat, and pasted on the back. The two pieces of the pincers have their usual curve.

It is scarcely necessary to advert again to the vulgar prejudices to which these insects have given rise. It has been supposed that the earwig which flies the light, and seeks narrow cavities, introduced itself, during sleep, into the auditory conduit, pierced the tympanum, and even penetrated into the brain. Of this the populace of most countries are still persuaded, and it is wonderful to find even the great legislator of Natural History observing: *Aures dormientium interdum intrans, spiritu frumenti pellenda.* In consequence of this notion, a general proscription is extended against this race of insects, though we believe the attribution to be unsupported by a single fact. It is by no means probable that the instruments with which this insect is provided are capable of piercing such a membrane as the tympanum—not to mention that the wax of the ear is a secretion, calculated from its taste and odour to keep off all such intruders; nay, in all probability, intended for that very purpose.

We insert a figure of a new Brazilian species of *Forficula Brasiliensis*. It is obscure brown; the elytra paler, hinder margin of the thorax pale, transparent, with the last joint of the abdomen reddish-brown. The tail fulvous; it is from Brazil.

The BLATTÆ, or *Cock-roaches*, are easily distinguished from the other orthopterous insects.

It was Linnæus who gave to this genus the name of *Blatta*, derived from the Greek verb, *βλαπτω*, *to hurt*, a name perfectly appropriate to the insects which compose it, as we shall presently see.

Like the other orthoptera, the Blattæ do not undergo a complete metamorphosis. Their larvæ and nymphs resemble the perfect insect. They differ only in the circumstance of being destitute of wings, or merely possessing the rudiments of these organs.

The Blatta was known to the ancients: they named it *lucifuga*, or that which shuns the light; because in fact this insect loves obscurity, and is never seen except at night. It runs with considerable swiftness. Many species live in the woods. Some, as those of the east, and of America, establish themselves in our habitations, and there do a vast deal of injury, because they devour sugar, and all the animal and vegetable substances which have not been carefully shut up in closets, &c., where they cannot penetrate. They destroy clothing, leather, cotton, wool, eatables of all kinds, especially cheese, and the crumb of bread. They have a very disagreeable odour. They are seldom seen during the day, because they retire into the holes of walls, between floors, and under chests, &c; but in the evening they all issue from their retreat, as soon as the lights are withdrawn, and in the silence of the night. Then they cover tables in

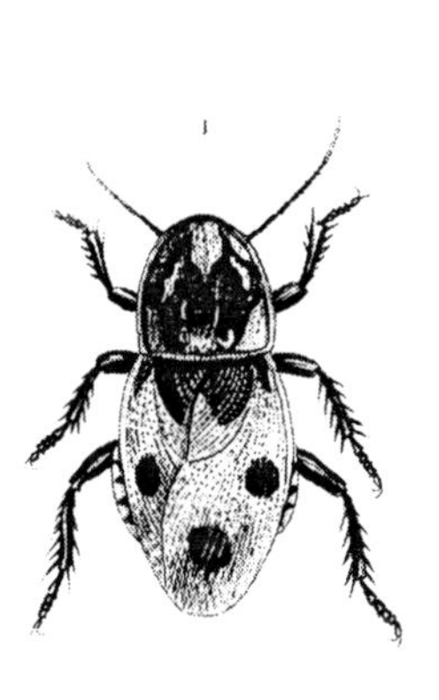

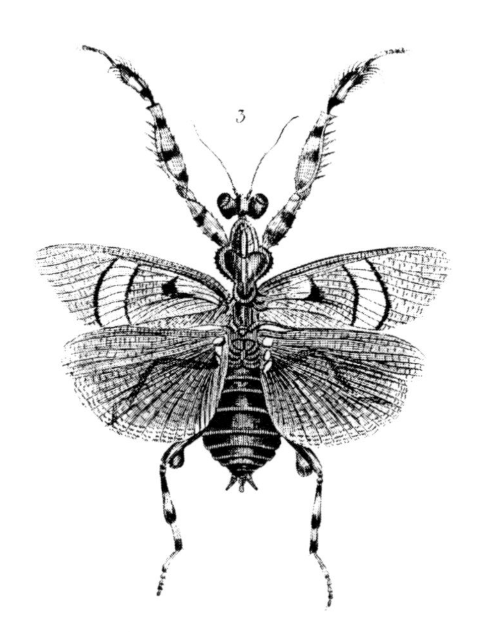

1. Blatta *maculata. Bilb.* 2. Forsicula *Brasiliensis. G. R. Gray.* 3. Blepharis *elegans. Westw.*

London. Published by Whittaker & Co. Ave Maria Lane. 1831.

kitchens, fall with voracity on the remains of eatables, of which they do not leave an atom remaining. They escape on the least appearance of danger, run very fast, and are very difficult to be caught. They communicate an unpleasant smell to the articles over which they have passed.

The following curious description of the ravages of this genus, or rather of one particular species of it, which seems to be that, termed *gigantea* by Linnæus, is extracted from Drury's work on exotic insects, an observer who had contemplated these animals in their native country :—

"The cockroaches are a race of pestiferous beings, equally noisome and mischievous to natives and strangers, but particularly to collectors. These nasty and voracious insects fly out in the evenings, and commit monstrous depredations; they plunder and erode all kinds of victuals, drest and undrest, and damage all sorts of clothing, especially those which are touched with powder, pomatum, and similar substances; every thing made of leather, books, paper, and various other articles, which, if they do not destroy, at least they soil, as they frequently deposit a drop of their excrement where they settle, and some way or other by that means damage what they cannot devour. They fly into the flame of candles, and sometimes into the dishes; are very fond of ink and of oil, into which they are apt to fall and perish. In this case they soon turn most offensively putrid, so that a man might as well sit over the cadaverous body of a large animal, as write with the ink in which they have died. They often fly into persons' faces and bosoms, and their legs being armed with sharp spines, the pricking excites a sudden horror, not easily described. In old houses they swarm by myriads, making every part filthy beyond description wherever they harbour, which, in the day-time, is in dark corners, behind all sorts of clothes, in trunks, boxes, and, in short, in every place where they can be concealed. In old timber, and deal houses, when the family is retired at night to sleep, this insect, among other disagreeable properties, has the power of making a noise, which very much resembles a pretty smart knocking with the knuckle upon the wainscotting. The Blatta gigantea, of Linnæus, in the West Indies, is therefore frequently known by the name of the Drummer. Three or four of these noisy creatures will sometimes be impelled to answer one another, and cause such a drumming noise, that none but those who are very good sleepers can rest for them. What is most disagreeable, those who have not gauze curtains are sometimes attacked by them in their sleep; the sick and dying have their extremities attacked, and the ends of the toes and fingers of the dead are frequently stripped both of the skin and flesh."

It does not appear that this destructive insect is at present known in Europe, though we have other species brought here in ships from warmer regions, which cause no small annoyance and mischief. Yet Mouffet gives the following account of the discovery of an insect in this country a long time ago, which scarcely seems referrible to any other than the species mentioned by Drury :—

"I have heard from persons of good credit, that one of these blattæ was found and taken in the top of the roof of the church in Peterborough, which was six times larger than the common blatta, and which not only pierced the skin of those who attempted to seize it, but bit so deep, as to draw blood in great quantity ; it was a thumb's length and breadth in size, and being confined in a cavity in the wall, after two or three days made its escape, no one knew how."

The female blattæ lay one or two capsular bodies, almost as thick as one half of their belly, and nearly of an oval form, each of which contains sixteen eggs. According to Frisch, that blatta which lives in kitchens, keeps during six or seven days at the orifice of the vulva, the body which she is about to lay. The nymphs have between the corslet and the abdomen, two broad and flat rings, which out-edge the breast considerably. It is from these parts that the wings issue forth. The most common species in this country is the *blatta orientalis*, or black cock-roach, most generally, but most erroneously, called the black-beetle. It is an importation from the east into Europe, through the commerce of the Levant, being brought over in ships' cargoes. These blattæ are fond of warmth ; accordingly they are continually found in kitchens of all kinds, especially in those of houses of entertainment, where much cooking, and where the warmth and steam of boilers, &c. are considerable. They are also found in bakehouses, where they inhabit the clefts of the walls, near the ovens. They are a complete pest in all such places. It is said that the *gryllus campestris* will destroy these insects. We have figured a Brazilian species under the name of *blatta rugicollis*. It is yellowish white, the thorax very rugose, disc black, with a white basal and apical spot ; the elytra with a large triangular spot at the base of the wings, also with two spots on the left elytron, and one on the right. The antennæ black, tipped with white, legs black.

We now come to that very extraordinary genus, the MANTIS.—" Imagination itself," as Dr. Shaw well observes, " can hardly conceive shapes more strange than those exhibited by some particular species."

Their two anterior feet are very large, and assist them in seizing and piercing the insects on which they feed. As they extend them very frequently, people have imagined that they could divine and indicate events. From this fantastical notion comes their Latin name of *mantis*, which signifies *diviner*. This name, however, is Greek in its origin, and was bestowed upon insects, which appear to be the same as those at present under consideration. We find, in fact, in one of the Idylls of Theocritus, this word employed to designate a thin, young girl, with slender and elongated arms. *Præmacram ac pertenuem puellam* μαντιν. *Corpore prælongo, pedibus etiam prælongis, locustæ genus.* Rondelet, Mouffet, Aldrovandus, and Linnæus, have adopted this denomination to indicate the same insects. The first of these authors tells us, that in Provence, these insects are indifferently named *devin*, and *prega-diou* or *preché-dieu*, in consequence of their having their fore-feet extended as if they were preaching or praying. He also adds, with naïveté—*Tam divina censetur bestiola, ut puero interroganti de viâ altero pede extento rectam monstret, atque rari vel nunquam fallat.*

The mantes may be called truly anomalous insects, and differ from the great majority of this class by the length of their corslet, which can be raised upright on the abdomen, and by the mode of articulation and conformation of their fore-feet, of which the insect makes use to carry its aliment to the mouth, like hands, the first articulation of these tarsi having the form of a crook, and making with the leg a sort of pincers.

Few of the mantes are found in the north, but they are very frequently observed in the south, in the three states of larva, motile nymph, and perfect insect. They feed on soft insects, which they devour all alive. The females lay their eggs in masses, disposed by beds, and enveloped with a glutinous or gelatinous matter, which dries in the air, and yet nevertheless remains flexible. These masses are found on the stalks of plants and shrubs; they resemble small wasp-nests, in which the eggs, enveloped in a sort of parchment, are disposed in two banks.

In the nymph-state, the mantes have on the back four flatted pieces, which are sheaths enclosing the elytra and wings. They walk and act like the living insect, live by rapine, and eat all the insects which they can catch by means of their anterior feet, which perform the office of pincers.

Roësel has preserved *Mantes*, feeding them with flies and other insects, which they seized with much address. "They are so cruel and carnivorous that they kill and eat one another, without being compelled to do so by hunger." The same author has seen little ones newly disclosed attack each other with fury, raising their corslet in the air, and holding their two anterior feet joined and ready for combat. Wishing to witness the sexual intercourse of these insects, the same writer shut up, in a sand-box, two of them, a male and female. They attacked each other fiercely, and the combat finished by the death of one of them. M. Poiret states that having also enclosed under a glass a male and female, the latter, with the sharp points of her feet, seized the male and cut off his head. As these insects are very tenacious of life, the male continued to live for some time after; the female then received his caresses, and ended by devouring him.

The *Mantis viatoria* is of a beautiful green colour, about three inches long, and of a slender shape. This is the species which, from its peculiar sitting posture, is held sacred by the vulgar. It is a stranger to our island, but is found in the warmer countries of Europe. "In its real disposition," says Dr. Shaw, "it is very far from sanctity; preying with great rapacity on all the smaller insects which fall in its way, and for which it lies in wait, with anxious assiduity, in the posture at first mentioned, seizing them with a sudden spring when within its reach, and devouring them. It is also of a very pugnacious nature, and when kept with others of its own species in a state of captivity, will attack its neighbours with the utmost violence, till one or the other is destroyed in the contest. Roësel, who kept some of these insects, observes, that in their mutual conflicts, their manœuvres very much resemble those of hussars fighting with sabres; and sometimes one cleaves the other through at a single stroke, or severs the head from its body. During these engagements the wings are generally expanded, and when the battle is over the conqueror devours his antagonist."

The Chinese exhibit these insects fighting, after the manner of cocks and quails; for it is probable that it is to this insect, or one very analogous to it, that Mr. Barrow alludes in his Travels in China.

"They have even extended their inquiries after fighting animals into the insect tribe, and have discovered a species of gryllus or locust, that will attack each other with such ferocity, as seldom to quit their hold without bringing away at the same time a limb of their antagonist. These little creatures are fed and kept apart, in bamboo cages, and the custom of making them devour each other is so common, that during the summer months, scarcely a boy is to be seen without his cage of grass-hoppers."

The mantis is as cowardly as it is cruel, for it will fly away from the ant, though it will destroy abundance of helpless flies. The *mantis causta* is a native of Africa, and has been supposed, but erroneously, to be an object of worship to the benighted Hottentots. It is, however, held by them in the highest veneration, and if it happen to alight on any person he is considered as the peculiar favourite of Heaven, and ever afterwards holds the rank of a saint.

We have figured a species which belongs to the sub-genus *Blepharis*, of M. Sevoille, named *elegaus* by Mr. Westwood. Its general colour is testaceous, the upper wings green, with the anterior and posterior margins, also a spot near the base, and the fascie pale testaceous. The two latter edged with black, the lower wings testaceous, the margin with several rows of small brown spots; the legs pale testaceous, spotted with brown, the abdomen brown, with the margins of each seg-

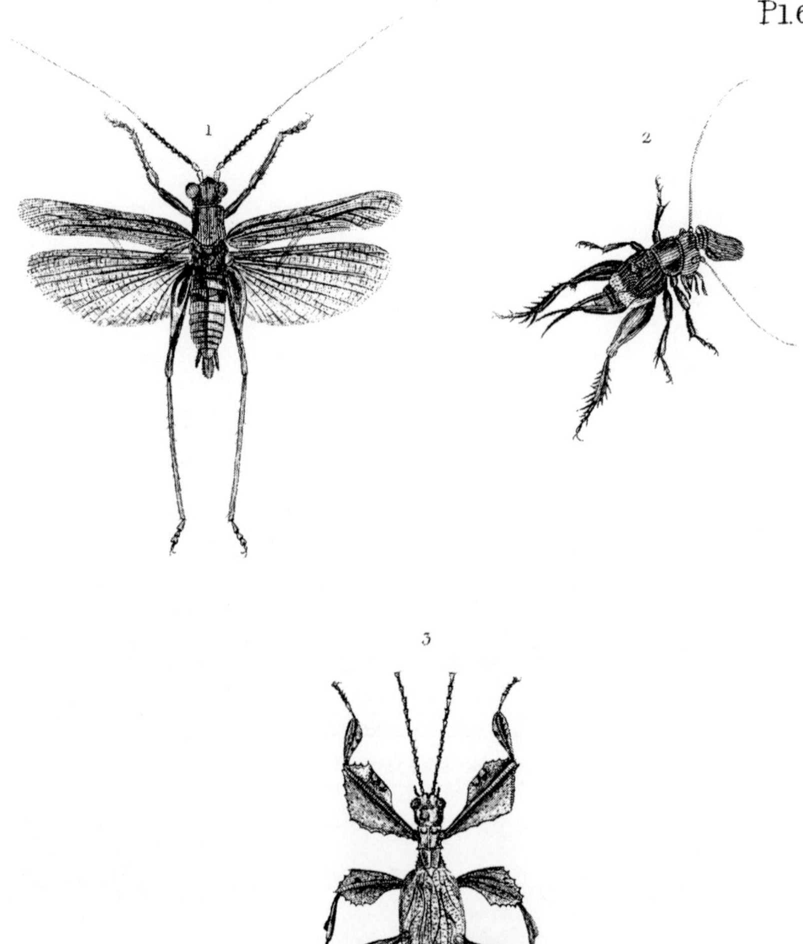

Westwood. del. Bull. sc.

1. Scaphura *Kirbii.* *Westw.* 2. Gryllus *umbraculatus. Lin.* 3. Phyllium *bioculatum.* *G.R.Gray.*

London, Published by Whittaker & C.º Ave Maria Lane 1831.

ment paler. It is from the Tanesserim coast, and is in Mr. Plimpton's cabinet.

The insects, Phasma, must of necessity be separated from mantis, and form with *phyllium* a peculiar division. If the latter resemble leaves, the former imitate the part which supports them: namely, the stalk or branch.

The East Indies produce species of great size, such as the *phasma gigas*, eight inches in length. This insect runs into many varieties, or perhaps, what is more probable, the species are more numerous than is generally supposed.

Few insects have so extraordinary a form as the *phyllia*. But it is only in large collections that we have any opportunity of seeing them, for they inhabit the most eastern parts of India. When they are placed on an orange-tree, or on a laurel, the man who is most accustomed to observe them, cannot distinguish them at the first glance. One is the more deceived, as their colour is green or yellowish; their elytra have truly all the appearance of a leaf, both in their figure and the disposition of the nervures, and their feet are folded under their body.

The *phyllia* have been named by some authors *walking leaves*. Like the other insects of the division *spectrum*, they live on vegetable substances. Some inhabitants of the Sechells Islands rear them, either from simple curiosity, or for the purpose of selling them to amateurs of natural history, or to such persons as make them an object of commercial speculation.

We have figured a new species of *phyllium*, under the name of *bioculatum*. The general colour is light green; the upper wings very short, and partly transparent; the lower wings nearly as long as the abdomen, narrow, white, and very transparent; the abdomen is very much dilated, and of a darker green colour, with two dark spots in the middle.

The habitat unknown. It is in the cabinet of the Rev. F. W. Hope.

We now come to the *leaping* division of the orthoptera, embracing the great genus Gryllus, of Linnæus. We shall treat the sub-genera in detail, according to the order in which they stand in the text.

The Gryllo-talpæ, or *mole-crickets*, have many relations with *gryllus* (*acheta*, Fab.), but are especially distinguished from them, by the form of their anterior feet, and the absence of any prominent auger, or ovipositor in the female.

We shall add a word or two here, to the description of their characters in the text. Their eyes are small, oval, and of a shining brown; near the internal side of each of them is remarked a small yellowish point, in relief. These are the simple eyes. The males have the internal part of their elytra more coriaceous, and elastic like parchment. These are their instruments of music. The sounds which they produce to call their females are less sharp and noisy than those emitted by the male grylli. This species of song is even tolerably sweet, when one is at a certain distance from the insect. It takes place after the setting of the sun, or before its rising; but unfortunately it announces the presence of an enemy to agriculture.

There is remarked at the extremity of the body, two setaceous, long, soft, and articulated appendages, and which are also proper to some other orthoptera, such as *Blatta*, *Mantis*, &c. We are ignorant of the use of these organs. Some authors have conjectured that these parts in the mole-crickets are two sorts of tentacula, which, when these animals are occupied in digging in the earth, advertise them on the slightest impression of the presence of an enemy, which would attack them behind. This explication is very ingenious, but *quere*, is it true? This is a point not so easily to be cleared up.

These insects have too singular a form, and are too hurtful not to be very well known: some authors have imagined them to be the *Staphylinos*, of Aristotle; Cordus takes them for his *Sphondylus*. Their history, however, goes back as far as the times of Mouffet, Goëdart, and Menzelius.

They have received the name of *mole-crickets*, in consequence of the form of their anterior feet, which are broad, and similar to the fore-feet of moles. They use them too in the same manner as the moles do, in the construction of subterraneous galleries. These insects are spread throughout the four quarters of the globe. The very northern parts of Europe are, perhaps, the only countries which have no reason to complain of their fatal devastations

The ravages of the gryllo-talpæ are but too well marked. Those large yellow spots where vegetation is extinct, which may be remarked in the country, are most generally their work. But their only object in this labour is to frame subterraneous galleries leading to their domicile; for they do not feed on vegetables. They destroy, on the contrary, many insects which are hurtful to us.

The gryllo-talpæ being nocturnal insects, it is difficult to ascertain whether they fly or not. The small extent of their wings, and the relative weight of the body of the insect, would lead us to believe that these organs of motion have no very powerful action. They must, nevertheless, be of some utility to the animal in running or leaping.

The extremity of the anterior feet of these insects has been compared to a hand; the four serrated talons of the legs, to fingers; and the tarsus, which moves like the branches of a scissors above the four fingers, and which serves to clean them, has been considered as the thumb. Without exactly adopting the details of this comparison, it seems more just to apply the name of *hands* to the extremities of the fore-feet in gryllo-talpæ, than to give this denomination to the same parts in some bees.

The force which these hands possess has been estimated. It can overcome an obstacle of that kind which the insect most usually meets with, namely, a load of earth of the weight of three pounds upon a smooth level. From this the conclusion is drawn, which is unfortunately but too true, that these insects can pierce and excavate a tolerably compact soil.

The females form in the commencement of summer, a burrow of almost half a foot in depth, and which, with its aperture, has the figure of a bottle with the neck curved, and the belly ovaliform. They smoothen its interior parietes, and lay there three or four hundred elongated eggs, shining, and of a yellowish brown. A progeny so numerous, must be to them a very painful burthen. The cradle which they intend for their posterity is of a very fine and tolerably compact earth. They close it accurately. Those who are fond of the marvellous, may read in Goëdart and other writers, that these tender mothers extend their solicitude, even to excavating a ditch around the nest of their young ones; place themselves there as sentinels, and during dry weather raise the lump of earth to which they have confided their dearest hopes, so that the heat of the sun may penetrate into it, and warm it more effectually. Another kind of foresight has also been attributed to these animals; some will have it, that equally wise with the *ants*, they transport into their asylum, like the latter insects, grains of corn, alimentary substances, &c. But for what purpose should they employ such useless care and pains?

The edge of the hands of the mole-crickets being vertical, it is easy to comprehend that to dig holes in the earth, or furrow it, these animals have only to bring their fore-feet together, to sink, and then separate them, holding them always in a perpendicular direction, and repeating the same manœuvre. The denticulations of these hands, are necessarily formed so as to cut the earth.

The little ones are disclosed at about the end of a month. These young ones are at first, when they leave their first skin, completely white, and continue so for about an hour. This change of colours takes place in all the moultings. It is unnecessary to say any thing respecting the form of these young insects. They resemble, with the exception of the wings, which are wanting, those from whom they received their existence. Their skin, however, is less velvety.

This first moulting is for them the period of an especial trial of their powers. They then disperse, and each takes his own way. In proportion as they grow, their tints become browner. The grey becomes converted into blackish.

Towards the third moulting, which arrives at the approaches of the winter, they think of taking up their winter quarters, and prepare for themselves a deep retreat in the earth. If the temperature of the atmosphere is not rigorous, they improve during the winter, and appear of a tolerably good size, at the return of the fine weather, or in the month of March. Their wings, however, are still wanting. They must moult a fourth, and even a fifth time, before they become completely adult. Then, being bolder, they quit the neighbourhood of the places which beheld their birth, transport themselves into fields and gardens, especially into the beds, when they become an atrocious scourge. Their line of march is often covered, and the traces of their passage are only indicated by a slight elevation of the earth above the surface of the soil. There can be no doubt that these animals, hurtful as they are, enter into the plans of that Supreme Wisdom which knows how to compensate every thing, and establishes between all forces a necessary equilibrium. They destroy a great number of hurtful plants, which pullulate too luxuriantly. Their devastations are calculated to draw man out of his carelessness, and awaken his activity. However, it is very certain that the gryllo-talpæ are extremely pernicious insects, and the entomologist should come to the assistance of the agriculturist, to supply him with weapons against one of his most mischievous enemies. The wisdom of the Creator presents us with the first means for the destruction of these ravagers. It has pleased Him that birds, ants, and other insects, should declare war against the young mole-crickets, or at least against the germs of this hurtful race. The plough of the labourer, in preparing the furrows where nutritious plants grow, crushes and destroys the insect which menaces them.

Those singular orthoptera, the Tridactyli, have the greatest relations with the *acheta* of Fabricius, and especially with the sub-genus which has been last under consideration. They are equally leaping and digging insects. But the Tridactyli cannot hollow the earth, but with their anterior legs.

These orthoptera resemble the gryllo-talpa. But they appear to possess the faculty of leaping in a more eminent degree. This is at least certain, as far as the species which inhabits the southern parts of France is concerned, which M. Illiger names *Xyla variegata*, and which, perhaps, ought to form a separate genus. It remains in the sand on the banks of rivers. M. Savigny brought the same species from Egypt.

The females of the Grylli proper have, at the extremity of the body, a scaly, cylindrical, elongated oviduct, a little swelled at the end, rising a little above, and formed of two concave pieces in the interior, the union of which composes a tube. The extremity of the swelling goes into a point. Each end is bifid in the *gryllus campestris*. The feet are strong, espe-

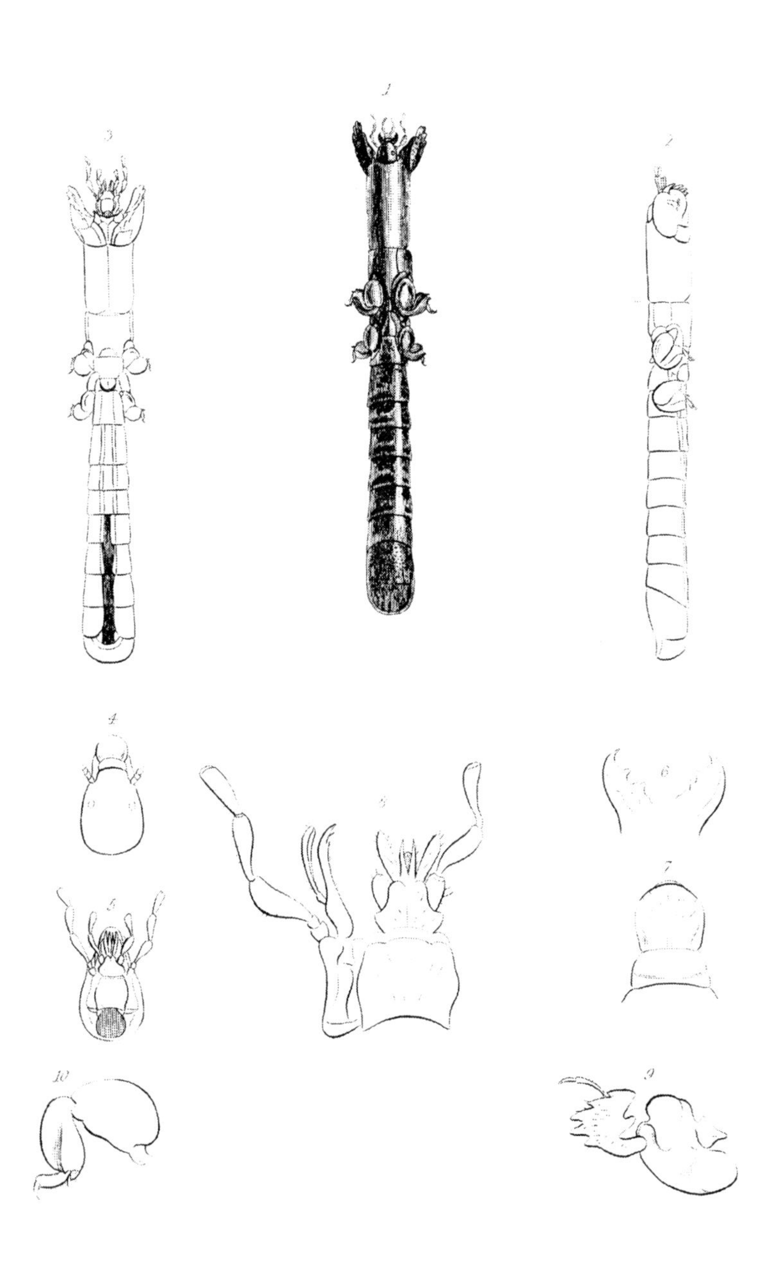

Cylindrodes Campbellii _ G.R.Gray.

London, Published by Whittaker & Co Ave Maria Lane 1832.

1 Gryllus migratorius (The Locust)

2 Gryllus Gryllo-talpa (The Mole Cricket)

3 Tetrix (Acridium Fab.) bipunctata

4 Gryllus domesticus (The House Cricket)

5 & 5a Tridactylus paradoxus

London; Published by Whittaker & C.º Ave Maria Lane; 1832.

cially the hinder ones, the thighs of which are very large, and the legs and tarsi are furnished with a double rank of spines.

These insects are generally known under the name of *crickets*. This name (*cri-cri*, in French) has been given them in consequence of the noise they are continually making. This noise is produced by the friction of their elytra one against the other. The two most common species are the *gryllus campestris* and the *gryllus domesticus*. The latter lives in houses. It delights, in preference, in kitchens, behind chimneys, in the holes and clefts of walls, and near the ovens of bakers. During the day it remains concealed ; but as soon as night approaches it quits its retreat, and proceeds in search of its aliment, which, according to some writers, consists in bread, flour, and other provisions of the same kind. But it is more likely that it feeds on insects, like the *gryllus campestris*. The male is very troublesome from its continual cry, which is proper only to itself, the female being mute. When he wishes to make himself heard, to advertise her of his presence, he raises his elytra so as to form an acute angle with his body. Then he rubs one against the other, with a horizontal and very quick motion. We have already remarked a difference between these parts in the two sexes. The elytra of the male are of a drier and more elastic nature, which renders them fit to excite by friction a sound similar to that produced by the rumpling of parchment. The chirruping of the cricket has long been formidable to the ears of superstition. Cold is very destructive to them. This is no doubt the reason why these animals invariably haunt the warmest parts of houses, and establish themselves there. It appears that the female, by means of its auger, places the eggs, which are oblong, in rubbish, or earth. The little ones are disclosed at the end of twelve days. It is only after three moultings that they acquire the appearance of wings, or are changed into nymphs. They require four months to undergo their final transformation. The females, notwithstanding, for some time before, are perfectly distinguishable by the presence of their auger. The insect immediately after its moulting is white.

The *field-crickets (G. campestris)* do not differ from the house-crickets as to form, but instead of being yellowish, they are almost black. Their size is somewhat larger; they are found during the whole summer in the fields. It is in the earth that they establish their dwelling and build their nest. It appears that when the winter is mild they pass it under ground, where they remain in a lethargic state; but when the cold is rigorous, they die. In the fine season, towards the setting of the sun, and during the whole night, the male is heard chirruping. The more remote we are from them, the louder and sharper appears their cry; in proportion as we approach them it is said that it grows milder, and even ceases altogether when we get very near them. They remain in pasture grounds, and meadows exposed to the sun, in preference to shady places. Children in the country amuse themselves by hunting them. They throw into their hole an ant attached to a hair. The field-crickets never fail to issue from their retreat to pursue their prey, and thus deliver themselves into the hands of their enemies. This mode of catching them was also in use among the ancients. It is even sufficient to introduce into the hole of this gryllus, the stalk of some plant, to oblige it to issue forth. Hence comes a proverb which is common in France, *il est sot comme un grillon*.

The larva of this gryllus is equally distinguished from the perfect insect by the want of wings and elytra. It walks, leaps, and takes nourishment. After some moultings, it passes into the nymph or pupa state; then there appears upon the back four flatted parts, which are the sheaths of the wings and elytra. These four sheaths are in the form of oval, thin laminæ. All the other parts of the body are similar to those of the perfect insect.

The two sexes have at the extremity of the abdomen, one on each side, two appendages, which are long, like conical threads, furnished with a great number of hairs, and which are often equal to the belly in length. These parts are not articulated like the antennæ, being of a single piece, and yet, nevertheless, very flexible. Besides those two pieces, the rump of the female is furnished with an auger of the length of the belly, which is straight, scaly, formed like a stiletto, and composed of two pieces, thicker at their extremity, and cut in this part something like the slit of a pen. Along their internal edge is a sort of groove, which serves as a conduct to the eggs, which the female deposits in the earth or other places. The female lays in July or August, nearly three hundred eggs. The little ones are disclosed fifteen days after, and feed, as is reported, on tender herbs, or on their roots. Their first moulting takes place before the winter season. As soon as the cold begins to be felt, they try to protect themselves from it by concealing themselves in the earth, and there they take no nutriment. The gentle warmth of spring causes them to re-appear. They then dig for themselves a sort of grotto, which serves them for an habitation, and where they remain in ambush. They are not in a condition to engender until June or July.

The *house-cricket* in general prefers houses that have been recently built, because the softness of the mortar enables them to form their retreats better.

Mr. White observes that, "though they are frequently heard by day, yet their natural time of motion is only in the night. As soon as it becomes dusk the chirping increases, and they come running forth, and are often to be seen in great numbers, from the size of a flea to that of their full stature. As one would suppose from the burning atmosphere which they inhabit, they are a thirsty race, and shew a great propensity for liquids, being frequently found dead in pans of water, milk, broth, or the like. Whatever is moist they are fond of, and therefore they often gnaw holes in wet woollen stockings and aprons that are hung to the fire. These crickets are not only very thirsty, but very voracious, for they will eat the scummings of pots, yeast, salt, and crumbs of bread. and kitchen offal, or sweepings of almost every description.

"In the summer they have been observed to fly, when it became dusk, out of the windows, and over the neighbouring roofs. This feat of activity accounts for the sudden manner in which they often leave their haunts, as it does also for the method by which they come to houses where they were not known before. It is remarkable that many sorts of insects seem never to use their wings, but when they wish to shift their quarters, and settle new colonies. When in the air, they move in waves or curves, like woodpeckers, opening and shutting their wings at every stroke, and thus are always rising or sinking. When their numbers increase to a great degree, they become pests, flying into the candles, and dashing into people's faces. In families, at such times, they are like Pharaoh's plague of frogs, 'in their bed-chambers, and upon their beds, and in their ovens, and in their kneading-troughs.'"

Cats catch heath-crickets, and, playing with them as they do with mice, devour them. Crickets may be destroyed like wasps, by phials half filled with beer, or any liquid, and set in their haunts; for being always eager to drink, they will crowd in till the bottles are full.

In certain countries of Africa, these insects are reported to constitute an article of commerce. Some persons rear them, feed them in a kind of iron oven, and sell them to the inhabitants, who are fond of their music, which they think induces sleep. Some have averred that this noise continues, and the insect remains alive for some considerable time after its head has been removed.

The *field-crickets*, like their brethren of the hearth, are nocturnal, and do not issue from their subterranean dwellings until the setting of the sun. They are difficult enough to be observed, for their timidity and caution are extreme. They return rapidly to their habitations, on hearing the slightest noise, or perceiving the most distant approach of danger. There they are certain to remain, until their suspicions are lulled asleep.

Though the wings of these insects are very curious, it is a singular fact that they are never observed to make use of them; and notwithstanding the length and muscularity of their hinder limbs, their movements, at least when driven from their holes, are neither vigorous nor active. Consequently, on such occasions, they are taken with facility.

The males, during the breeding season, are sometimes known to carry on a destructive warfare upon each other. Mr. White, desirous of preserving some of them, put them into the crevices of an old wall; but the first who obtained possession expelled with great ferocity the intrusion of any others of their fellows.

The cells which those insects construct are very curious and regular, and the instruments which they employ in perforating and rounding them, are their powerful jaws, which are denticulated like the claw of a lobster. These formidable arms, however, are of no use to them when caught, or at least they do not, upon such occasions, employ them as weapons of self-defence. They feed indiscriminately on such plants as grow in the neighbourhood of their dwellings, from which, in the day-time, they never remove to any distance. They are heard to chirp all night long, and in hot weather they make the hills and plains reverberate with their song, which is more especially louder during the dark and silent hours of midnight.

So much confusion has arisen from the change and substitution of names, that we feel some degree of difficulty here in the proper application of the name Locust. The general name of *gryllus* was given by Linnæus to the entire family of *Saltatoria*, or leapers. Subsequent naturalists have found it necessary to make several separations from this Linnæan genus; but in this, as in most other cases of the same kind, they have not invariably exercised a spirit of the soundest discretion, and have occasioned much confusion in nomenclature, by an injudicious transposition of synonimes. Without further criticism, however, we shall endeavour properly to discriminate here, the insects of which we are about to speak, under the general name of *locusts*.

The *locusts* proper of the text, *Locusta* of Geoffroy, and *Sauterelles* in French, are in fact, what we commonly call *grass-hoppers*. The true *locusts*, whose name is so formidable in the history of the earth, constitute the genus *Acridium* of Geoffroy, to which the same naturalist has given the French name *criquets*, and which we must carefully avoid confounding with our *crickets*, a genus already treated of. The true locusts also more particularly belong to that subdivision of *acrydium* named in the text, "*criquets* proprement dits," the *gryllus* of Fabricius, *gryllus-locusta* of Linnæus.

The first of these genera (*Locusta* of Geoffroy, and *Acrida* of Mr. Kirby) may be speedily dispatched. They have some resemblance to the true locusts, but the characters which distinguish them from these latter insects, are their tarsi of four articulations, and their long setaceous antennæ composed of a great number of articulations, not very distinct. The true locusts (*acridium*, G.) have but four articulations to the tarsi. The antennæ are short, filiform, or swelled at their extremity, and have from ten to twelve perceptible articulations.

The grass-hoppers are frequently found in meadows, leap a considerable distance, by the assistance of their hinder feet, which are much longer than the others. The males make a noise, more or less loud. It is always produced by the reciprocal friction of the elytra.

The females deposit their eggs in the earth; they lay a considerable quantity at once, assembled together in a slender membranous envelope. The larva which proceed from those eggs do not differ from the perfect insect, but in the total want of wings and elytra. Arrived at the nymph-state, they have these organs enclosed in a sort of germs, placed upon their back. But, like all other insects, they are unfit for reproduction until after the entire development of these parts, which development does not take place until they abandon the covering of nymph or pupa.

In their different forms these insects feed on herbs and plants, and eat a considerable quantity. An observation of Degeer proves that they can be carnivorous when occasion offers. This naturalist having shut up several individuals of the species *verrucivora*, one of them which died was devoured by the others. But they have not been observed to kill each other for the purposes of food.

These insects, which have the faculty of leaping to such a distance, also, at times, fly very high and very far.

The males are most generally distinguished from the females by an absence of an elongation of the abdomen of various forms, which is a true ovipositor, formed of separable laminæ, between which glide the eggs of the females. These laminæ are sometimes straight, sometimes curved with a convexity underneath, like a cutlass. The use of this instrument is to enable the insect to deposit its eggs in the ground, enveloped in a sort of mucous matter, which dries and becomes a true membrane, divided into a great number of lodges, from which issue forth the little larvæ. The manners of these insects are, otherwise, very imperfectly known.

We now come to the true Locusts, the *Acridium*, or *criquets*, of Geoffroy, and *Gryllus* of Fabricius. These pernicious animals leap very well, and shoot to a very considerable distance. Some species fly rapidly and very far; but in general they walk badly, and slowly. Like the last mentioned genus, they live on plants, and are accordingly found in great quantities in cultivated fields and meadows. The migratory locusts, of which we shall speak more at large by and by, are but too well known in the Levant, and in Africa. They multiply considerably, and appear in enormous troops, spreading devastation and famine far and wide.

The larvæ, like those of the preceding genera, differ from the perfect insect only by having no wings or elytra. After many moultings they pass to the nymph-state, and then have cases which enclose those parts. In both forms they walk and act like the perfect insect, and feed in the same manner. Their larvæ proceed from eggs; some females deposit theirs in the earth, where the heat causes them to disclose—others attach them to the stalks of gramineous plants, and enclose them in a frothy matter, which at first is soft, and subsequently hardens.

The locusts often send forth a sharp and interrupted sound. This is produced by rubbing their hinder thighs against their elytra and wings with considerable force. They never execute this movement with the two thighs at the same time, but employ one and the other alternately.

Olivier tells us that on each side of the belly a large and tolerably deep aperture is visible, the contour of which borders on the oval, and which is partly closed by an irregular piece in the form of a flat lamina. This lamina is scaly, but it is covered above with a flexible and wrinkled membrane, and its edges are furnished with some little hairs. The space of the tissue which the lamina leaves uncovered, is in some sort formed like a half-moon. At the bottom of this aperture is a white pellicle, very much stretched, and shining like a little mirror. On the side of the aperture, nearest to its head, is a small oval hole, into which it is easy to introduce the point of a stylet. On removing the pellicle, we discover a large cavity in this part of the body. This appears to be the organ of song in the same species.

If some travellers are to be believed, the locusts, which are such a plague to certain countries, serve as food for the people who inhabit the uncultivated lands towards the coasts of Barbary. As these insects are very abundant, and very large in this place, the inhabitants collect, and roast and eat them. They also preserve them in brine, after having removed the elytra and wings. This even becomes an object of commerce. In the southern parts of France, as M. Latreille informs us, the children are very fond of the fleshy thighs of these insects.

This is the insect whose ravages have been the theme of naturalists and historians in all ages, and upon a close examination we find it to be peculiarly fitted and furnished for the execution of its office. It is armed with two pair of very strong jaws, the upper terminating in short, and the lower in long teeth, by which it can both lacerate and grind its food—its stomach is of extraordinary capacity and powers—its hind legs enable it to leap to a considerable distance, and its ample vans are calculated to catch the winds as sails, and so as to carry it sometimes over the sea—and although a single individual can effect but little evil, yet when the entire surface of a country is covered by them, and every one makes bare the spot on which it stands, the mischief produced may be as infinite as their numbers.

The first record of the ravages of locusts, which we find in history, is the account in the Book of Exodus, of the visitation to the land of Egypt. Africa appears to have been most generally the quarter of the globe most severely subjected to the inroads of the locust tribe. A law was enacted and enforced in the territory of Cyrene, according to the account of Pliny, by which the people were obliged to destroy these insects in the egg, in the larva state, and in the image. A similar law prevailed in the island of Lemnos, where each person was forced to furnish annually a certain quantity of locusts. According to Orosius, A M. 3800, the north of Africa was so infested by them, that every vestige of vegetation vanished from the face of the earth. After this, he adds, that they flew off to sea and were drowned, but their carcases being cast upon shore, emitted a stench equal to what might have been produced by the dead bodies of one hundred thousand men. We are told by St. Augustine, that a pestilence arising from the same cause, destroyed no less than 800,000 people in the kingdom of Numidia, and many more in the countries along the sea-coast.

Blown from that quarter of the globe, the locusts have occasionally visited both Italy and Spain. The former country was severely ravaged by myriads of those desolating intruders, in 591, A.C. These were of a larger size than common, as we are informed by Mouffet, who quotes an ancient historian; and from their stench when cast into the sea, caused a plague, which carried off infinite numbers, both of men and cattle. A famine took place in the Venetian territory in 1487, occasioned by the ravages of these insects, in which 30,000 persons are reported to have perished. Mouffett mentions many other instances of the same kind, which have taken place in Europe at different periods. They entered Russia, in immense divisions, in three different places, in 1600, darkening the air with their numbers, and passed over from there into Poland and Lithuania.

In many parts they lay dead to the depth of four feet. Sometimes they covered the surface of the earth like a dark cloud, loaded the trees, and the destruction which they produced exceeded all calculation. They fall sometimes upon corn, and in three hours will consume an entire field, as happened once in the south of France. When they had finished the corn, they extended their devastations to vines, pulse, willows, and in short, to every thing else wearing the shape of vegetation, not excepting even hemp, which was not protected by its bitterness.

In 1748, considerable numbers of locusts visited this country, but, luckily, they did not propagate, and all soon perished. These were stragglers from the swarms which, a year previously, had desolated Wallachia, Moldavia, Transylvania, Hungary, and Poland. An account of this formidable invasion may be found in the forty-sixth volume of the Philosophical Transactions.

A friend of Mr. Kirby informed him, that at Poonah an immense cloud of locusts ravaged all the Mahratta territory, and was thought to have come from Arabia. This, indeed, was a most astonishing swarm, if Mr. Kirby's friend was correctly informed. The column extended five hundred miles, and was so dense, as thoroughly to hide the sun, and prevent any object from casting a shadow. This horde was not composed of the migratory locust, but of a red species, which imparted a sanguine colour to the trees on which they settled.

Dr. Clarke compares the progress of these animals to a snow-storm, when the flakes are carried obliquely by the wind. His carriage and horses were completely covered by them. This flight consisted of two species, *L. tatarica*, and *L. migratoria*. The former is considerably larger than the latter, and precedes it in its course. Mr. Barrow informs us, that in South Africa (in 1784 and 1797) two thousand square miles were literally covered by them. Being carried into the sea by a north-west wind, they formed, for fifty miles along shore, a bank three or four feet high; and when the wind was in the opposite point, the horrible odour which they exhaled was perceptible a hundred and fifty miles off.

In the empire of Morocco the ravages of the locust tribe have been dreadful beyond description. They caused a famine there in the year 1780. The poor wandered over the country, in search of a wretched existence from the roots of plants. They picked from the dung of camels, the undigested grains of barley, and devoured them with eagerness. Numbers perished, and the streets and roads were strewed with carcases. "When they visit a country," says Mr. Jackson, in his Travels in Morocco, "it behoves every one to lay in provision for a famine, for they stay from three to seven years. When they have devoured all other vegetables, they attack the trees, consuming first the leaves, and then the bark. From Mogador to Tangiers, before the plague in 1799, the face of the earth was covered by them. At that time a singular incident occurred at El Araiche. The whole region from the confines of the Sahara was ravaged by them; but on the other side of the river El Kos, not one of them was to be seen, although there was nothing to prevent their flying over it. Till then they had proceeded northward; but upon arriving at its banks, they turned to the east, so that all the country north of El Araiche was full of pulse, fruits, and grain, exhibiting a most striking contrast to the desolation of the adjoining district. At length they were all carried by a violent hurricane into the Western Ocean; the shore, as in former instances, was covered by their carcases, and a pestilence was caused by the horrid stench which they emitted; but when this evil ceased, their devastations were followed by a most abundant crop. The Arabs of the Desert, "whose hands are against every man," and who rejoice in the evil that befalls other nations, when they behold the clouds of locusts proceeding from the north, are filled with gladness, anticipating a general mortality, which they call *El Khere* (the benediction); for when a country is thus laid waste, they emerge from their arid deserts and pitch their tents in the desolated plains.

The peculiar noise produced by the locusts in their flight is well described by Mr. Southey:—

> "Onward they came, a dark continuous cloud
> Of congregated myriads numberless,
> The rushing of whose wings was as the sound
> Of a broad river, headlong in its course,
> Plunged from a mountain summit, or the roar
> Of a wild ocean in the autumn storm
> Shattering its billows on a shore of rocks."

It appears that in many of the instances which have been recorded, these devastating insects must have crossed the seas, though Haselquist asserts that they have no capability of so doing. He says, "The grasshopper or locust is not formed for travelling over the sea—it cannot fly far, but must alight as soon as it rises; for one that came on board us, a hundred certainly were drowned. We observe in the months of May and June a number of these insects coming from the south, and directing their course to the northern shore; they darken the sky like a thick cloud; but scarcely have they quitted the shore, when they, who a moment before ravaged and ruined the country, cover the surface of the sea with their dead bodies. By what instinct do these creatures undertake this dangerous flight? Is it not the wise institution of the Creator to destroy a dreadful plague to the country?"

Notwithstanding all this, we have seen that locusts do perform very considerable flights. Much in this way must depend upon their age and species—much also on the strength and point of the wind, for there is no doubt but that a swarm of these insects may be carried by a strong gale over a broad river, or across a narrow sea, nay, even to a considerable distance out to sea, as appears from an account quoted by Mr. Kirby, from an American newspaper, of what befel a vessel in 1811, which was becalmed at 200 miles distant from the Canary Islands, which was the nearest land; the account runs thus:

"A light air afterwards sprung up from the north-east, at which time there fell from the cloud an innumerable quantity of large grasshoppers, so as to cover the deck, the tops, and every part of the ship they could alight upon. They did not appear in the least exhausted; on the contrary, when an attempt was made to take hold of them, they instantly jumped, and endeavoured to elude being taken. The calm, or a very light air, lasted fully an hour, and during the whole of this time these insects continued to fall upon the ship and surround her: such as were within reach of the vessel alighted upon her; but immense numbers fell into the sea, and were seen floating in masses by the sides."

Some of these insects were preserved. They were of a reddish colour, with gray and speckled wings.

Fortunately for Europe, the countries of the east are the most frequently exposed to the ravages of the locust nation. They come in innumerable armies, so as literally to cause an eclipse of the sun. In this there is no exaggeration, for it is a point upon which all witnesses are unanimous. From time to time they abandon the deserts of Tartary and Arabia, which appear to be their cradle, and assembling in innumerable swarms, emigrate, and, like the hordes of barbarians that once issued from the same regions, carry misery and desolation into the bosom of the civilized world. An east-wind is generally most favourable to the flight of these predatory armies. Haselquist indeed observes, that they migrate in a direct meridian line from south to north, passing from Arabia to Palestine, Syria, Carmania, Natolia, Bithynia, Constantinople, Poland, &c.,—and that they never turn to the east or west. But this is utterly contradicted by many facts. They are known to emigrate not only from east to west, but also from north to south.

The island of Formosa, according to many accounts, is frequently exposed to the incursions of these desolating invaders. Charles the Twelfth, in Bessarabia, imagined himself assailed by a hurricane, mingled with tremendous hail, when a cloud of locusts suddenly falling, and covering both men and horses, arrested his entire army in its march. They have even crossed the Baltic, and visited Sweden in 1749. This was during their grand invasion of Europe, of which we have already spoken. Nothing could be heard from all quarters at that period but lamentations on lamentations. Their descents from the air were compared to a sudden storm, to a heavy fall of snow, to a hurricane, and even to a cloud of smoke, extending and spreading with rapidity. From that period we are told, in a German publication, that they have only made their appearance in Germany as isolated individuals here and there, the roughness of the climate being fortunately unfavourable to their reproduction. The last time in which they visited that country in numerous colonies, they began by devouring the finest and tenderest plants. After this, hunger compelled them to attack the leaves and bark of trees. They devour with the most incredible rapidity, but, like all voracious animals, they can fast for a very long time. Grundler observed their voracity with very great exactness. He put some of these locusts into a glass vessel with earth, in which there was some fresh barley, newly shot up. They first cut the stalk in two, devoured from the top the part which was left standing, as quickly as if it had entered their bodies of its own accord, and finally they demolished the part which their bite had caused to fall aside. All this was performed with an agility which it is impossible to describe. In their own country it is the hot summers, abundant in vegetation, which are most favourable to their reproduction—but serene and dry weather is the best for their aërial voyages. Their fecundity is so great, that in the places where they stop, entire sacks can be filled with their eggs, and even thirteen hogsheads full of them have been collected within a space of very moderate extent. A more striking idea still may be formed of their fecundity from a passage of locusts which took place in France in the year 1613, and another which occurred on the side of Bontzhida, in Tranyslvania, in 1780. The first entirely cut up, even to the very roots, more than fifteen thousand acres of corn in the neighbourhood of Arles, and had even penetrated into the barns and granaries, when, as it were sent by Providence, many hundreds of birds, especially starlings, came to diminish their numbers. Notwithstanding this, nothing could be more astonishing than their multiplication. Upon an order issued by government, for the collection of their eggs, more than three thousand measures were collected, from each of which would have issued nearly two millions of young ones. To prevent the fatal consequences which would have resulted from the other passage near Bontzhida, fifteen hundred persons were ordered each to gather a sack-full of locusts, part of which were crushed, part burned, and part interred. Notwithstanding this, very little diminution was remarked in their numbers, until very sharp and cold weather had come on. In the following spring there were millions of eggs disinterred and destroyed by the people, who were levied "en masse" for the operation; and notwithstanding all this, many places of tolerable extent, were still to be found, in which the soil was covered with young locusts, so that not a single spot was left naked. By dint of sweeping these were pushed into ditches newly excavated, the opposite edges of which were provided with cloths tightly stretched, and in which the locusts were crushed.

Providence usually opposes a great number of enemies to insects of this formidable character. A high wind, a cold rain, a tempest, may destroy millions of them in an instant. Foxes, hogs, birds, lizards, and frogs, devour an immense quantity of them. They sometimes wage cruel war upon each other. Some people of Arabia, and of some other countries of the east, take them in great quantities, have them dried, ground, and made into a sort of bread, when their crops have failed. At Bagdad, they are brought to market, and by this means, the price of other provisions is said to be considerably lowered. According to report, the locusts have something of the flavour of a pigeon. One man can easily despatch two hundred of them at a meal. The modes of cooking them are various. The Bedouins of Egypt roast them alive upon the coals, and eat them as a great delicacy, having first removed the wings and feet. They also remove, at least in some places, the intestines. The women and children of some parts of Arabia Felix, string them together, and thus sell them. The Arabs roast these insects, and steep them in butter, and when they wish to carry their luxury to an extreme, they give them but a single boil in water, and afterwards fry them in butter. The inhabitants of Morocco dry them on the roofs or terraces of their houses, and eat them either smoked, or broiled, or boiled. Other people of Barbary preserve them in pickle. According to Forskaël, there is no great relish in this aliment, and if used to too great a degree, it thickens the blood, and becomes injurious to melancholic temperaments. This traveller tells us, that he encountered infinite quantities of these insects at every step he travelled. They are hunted from one field to another, with a piece of cloth, attached to a long stick. The noise which they make in the air, says Forskaël, when in large swarms, resembles the rushing of a mighty cataract. They do not attack cereal plants when they are arrived at maturity, and a certain species of thrush belonging to this country, (*turdus gryllivora*) destroys about ten thousand of them per day. It has been pretended, that the *acridophagous* people, or such as feed on locusts, are subject at a certain age to an extraordinary malady, namely, that some winged gnats grow on the exterior of their body, which devour the flesh by little and little. The absurdity of this fable is sufficiently manifest.

Notwithstanding the opinion of Mr. Kirby respecting the species of locust which causes all these devastations which we have been describing, M. Latreille is somewhat doubtful whether that species be the *gryllus migratorius*, or migratory locust. Respecting the species which has several times ravaged the different countries of Europe, there can be no controversy, that being most undoubtedly the species last mentioned. But it is not so easy to say what that species may be, whose destructive swarms are so much dreaded by the people of the East, of Arabia, and of Barbary, and which these same people use as an article of food. Some authors have believed that it was the *gryllus cristatus* of Linné. This great naturalist, however, has fallen into two errors on this point. He unites under this name two species. The first is peculiar to Cayenne, and the second does not differ from the *gryllus dux* of Fabricius, figured by Drury. Both these species belong to America, and have been observed in no collections from the East, though Linnæus affirms that his *G. Cristatus* is also to be found in Asia. It seems not improbable that the *L. tatarica* which we have before mentioned, is the locust of the bible and of the eastern writers.

Pl 64

Westwood, del. Bull, sc.

Gryllus *monticollis*. *G.R. Gray*.

London, Published by Whittaker & C.º Ave Maria Lane. 1831.

We have selected from the extensive family of the locusts, a remarkable species from Poonah, which we have named *Gryllus monticollis.* It is of a light reddish brown colour; the thorax is covered with minute tubercles, and is very much elevated into a compressed ridge, which extends about one third the length of the abdomen; the hinder part of this ridge is expanded, beneath which is of a light greenish yellow, as are also the elytra; the latter and wings are shorter than the abdomen; the posterior tibiæ are light green.

THE SEVENTH ORDER OF INSECTS.

THE HEMIPTERA—*Ryngota*, Fab.,

Terminate the division of wing-cased insects, and are the only ones amongst them that have no proper mandibles or jaws. Their place is supplied by a tubular articulated piece, which has the appearance of a bill, which along its upper face has a gutter or canal, from which three stiff, scaly hairs can protrude, very firm and pointed, and covered at their base by a tongue. These, by their union, form a sucker, of which the tubular piece above mentioned is the sheath. The lower hair is composed of two threads, uniting a little beyond their origin. There is a tongue, properly so called, but bifid at the end. The palpi are wanting, yet there are some vestiges of them in *trips*.

The mouth of the hemiptera is then adapted only for suction. Like the other suckers, these insects have salivary vessels.

The wing-cases in the majority are coriaceous, or crustaceous, with the posterior extremity membranous. They almost always cross. In some, they are simply thicker and larger than the wings, and seem membranaceous.

The first segment of the trunk, which we have hitherto called corslet, is much less extended in many, and is incorporate with the second. Many have simple eyes, and often but two in number. The only change which the hemiptera undergo, consists in the development of the wings, and growth of the body.

This order is divided into two sections. In the first, HETEROPTERA, the bill originates from the forehead. The cases are membranous at their extremity, and the first segment of the trunk alone forms the corslet. The elytra and wings are horizontal, or but slightly inclined.

This section is composed of two families. The first, that of

GEOCORISÆ,

With naked antennæ, longer than the head, inserted between the eyes near their internal edge; three articulations to the tarsi, the first very short. They form the genus

CIMEX, of Linnæus.

Some *(Longilabres)* have the sheath of the sucker of four distinct and naked articulations; the labrum much prolonged, awl-formed, and striated above. Three distinct articulations always to the tarsi, the first equal to, or longer than the second. Antennæ, always filiform, are sometimes composed of five articulations. The body is short, oval, or rounded.

SCUTELLERA, Lam. *Tetyra*, Fab. Scutellum covers the whole abdomen.

PENTATOMA. Scutellum covers but a portion of the upper part of the abdomen.

Other pentatoma, whose metasternum, or mesosternum, rises into a keel, or exhibits a sort of spine, are distinguished by the name of EDESSA. Fabr.

A pentatoma of Cayenne, with cylindrical head, and anterior feet in a semi-ovaliform palette, forms a new generic group, HETEROSCELIS.

Sometimes the antennæ have but four articulations, and

the body is oblong; and here the antennæ are either filiform, or ovaliform.

Some exotic species approach the preceding in the ovoïd form of the body, but differ from the subsequent, by its being flat, membranaceous, and with dilated and angular edges; or because the corslet is prolonged hindwards, like a truncated lobe, and the sternum is horny. The latter form the sub-genus TESSERATOMA of M. Peletier and Serville. (*Edessa papillosa*, and *E. Amethystina*, Fabr.) Some others of his *Edessæ*, without thoracic elongation, but antennæ with four articulations, are the sub-genus DINIDOR.

A Brazilian species, analagous to *Aradus*, of Fab., is the type of the sub-genus PHLÆA, Lep. et Serv.

All the subsequent *geocorisæ* are generally oblong, and otherwise do not present the characters of the preceding. Some have the antennæ inserted near the lateral and upper edges of the head. Then come those whose body is more or less oblong, but without being filiform or linear.

COREUS, Fab. Body ovaliform, last articulation of the antennæ ovoïd or filiform.

The antennæ of the other geocorisæ of the same sub-division, terminate in an elongated articulation, cylindrical, or fusiform. They form a portion of LYGŒUS, and all ALYDUS, Fab. Most are exotic. To lygœus are referred the species whose simple eyes are separated from each other by an interval pretty nearly equal to that which separates each of them from the neighbouring complex eye.

HOLHYMENIA, Lepel. Serv. Second and third articulations of the antennæ palette-formed.

PACHYLIS, L. and S. Third alone has this form.

ANISOSCELIS. Antennæ filiform, without dilation.

Geocorisæ of the same division, with narrow and elongated body, projecting eyes, and corslet a little narrow in front, will form the sub-genus ALYDUS.

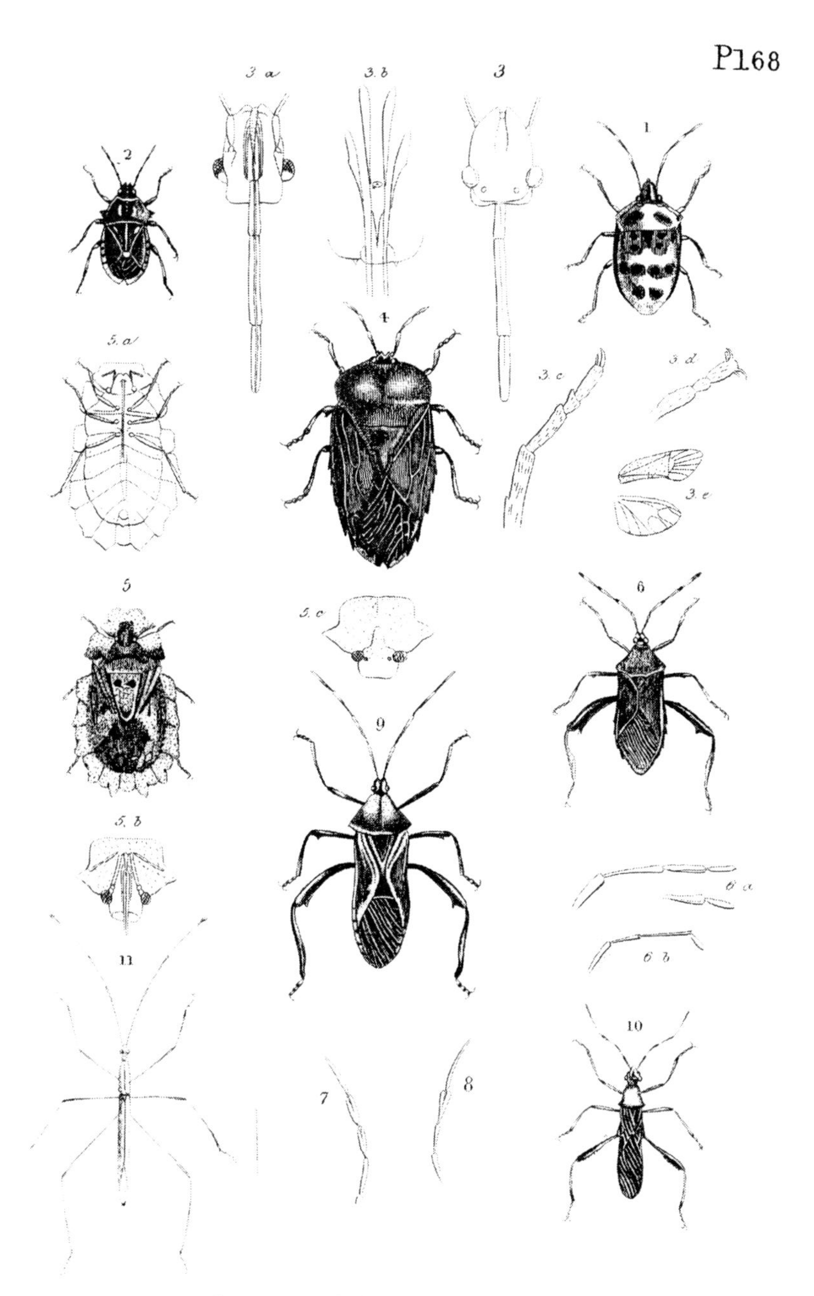

1. Scutellera *Dives. Guer.* 2. Pentatoma *yolofa. Guer.* 3. *Details of* Pent. *grisea. Latr.* 4. Tesseratoma *Sonneratii. Serv. & St Farg.* 5. Phlæa *cassidoides. Serv. & St Farg.* 6. Coreus *rubiginosus. Guer.* 6a. *Ant. of* Gonocerus. 6 b. *Ant. of* Syromastes. 7. *Ant. of* Pachlis. 8. *Ant. of* Holhymenia. 9. Anisoscelis *profanus. Fab.* 10. Alydus *annulicornis. Guer.* 11. Neides *tipularia. Latr.*

London, Published by Whittaker & Co. Ave Maria Lane 1837.

Now come geocorisæ, whose body is long, very narrow, filiform, or linear. LEPTOCORISA, Lat. Strait antennæ. NEIDES, Lat. Bent antennæ.

Geocorisæ, whose antennæ are filiform, or thicker towards the end, with four articulations, inserted lower than the preceding. Sometimes the head does not diminish into a neck. LYGŒUS, Fab. The head narrower than the corslet, and the latter trapezoïd.

Species with anterior thighs swelled, form the genus PACHYMERE of MM. Lepel. and Serv.

SALDA, Fab. Head broader, or as broad as the corslet. Prominent eyes; corslet equal and squared.

MYODOCHA, Latr. Head ovoïd, and diminishing into a neck.

We then come to the Geocorisæ longilabres, whose antennæ of four articulations diminish towards their extremity, or are setaceous.

ASTEMMA.

MIRIS, Fab. Like the last in antennæ, but corslet broader behind than before.

CAPSUS. Corslet trapezoïd, but second articulation of antennæ slender towards the base, and very setaceous.

HETEROTOMA. Quite distinct from the preceding in the size and breadth of the first two articulations of the antennæ.

The other hemiptera of this family have but two or three apparent articulations to the sheath of the sucker. Short labrum, not striate. First articulation of tarsi, and often the second, very short. Sometimes the feet are inserted at the middle of the breast, and end in two distinct hooks.

We afterwards separate the species whose bill is always straight, sheathed at base, or in its length. Eyes common size, and no neck. They compose most of the ACANTHIA, Fabr. He. has separated from it, SYRTES, Fab. Anterior feet, with monodactylous claw. TINGIS, Fab. Body flat, and antennæ terminating button-like; and ARADUS, antennæ cylindrical.

BUGS proper. CIMEX, Lat., *Acanthia*, Fab. Antennæ ending abruptly, like a hair.

In the other geocorisæ of this sub-division, the bill is naked, arched, or sometimes straight; labrum prominent, and a neck. Some have very large eyes. Among those which have not, is the primitive genus,

REDUVIUS, Fab.,

Short bill, but very sharp. Antennæ very slight towards the end.

This genus has been thus divided: HOLOPTILUS, Lep. et Serv. Three articulations to antennæ; last two furnished with long hairs, disposed in two ranks.

In the other species, the antennæ have four articulations, without hair, or scarcely any.

REDUVIUS (proper), Fab. Body oblong, feet of middle length. *Nabis*, Lat., and *petalocheiris*, Paliss., may be joined to them.

ZELUS, Fab. Linear body, very long and slender feet.

PLOIARIA, Scop. Like the last, but the two anterior feet with elongated haunches.

Next are geocorisæ with large eyes, no apparent neck, yet a separation between head and corslet. Some with short and arched bill, and silk-formed antennæ, are LEPTOPUS, Latr.

In others the bill is long and straight, and labrum projecting from its sheath; antennæ rather filiform. These are, SALDES of Fabricius, which Latreille thus divides: his ACANTHIA, with antennæ the length, or half the length of the body; and PELOGONUS, with antennæ much shorter, and folded under the eyes.

Sometimes the four hinder feet, long and slender, are inserted at the sides of the breast, and much separated at their origin. The hooks of the tarsi small, and feet adapted for aquatic purposes. They belong to the genus,

HYDROMETRA, of Fabr.,

Which Latreille divides into three sub-genera; HYDROMETRA (proper). Antennæ silk-formed, head muzzle-like, receiving the bill in an under channel. GERRIS. Antennæ filiform, and sheath of sucker with three articulations; and, VELIA. Sheath with but two articulations.

The second Family of HEMIPTERA,

HYDROCORISÆ.

Antennæ inserted and concealed under the eyes, shorter, or nearly so, than the head. These are equatic and carnivorous. Their tarsi have usually but one or two articulations, and the eyes are generally remarkably large.

Some *(Nepides)* have the two anterior feet like claws, the thigh very thick, or very long, with a canal underneath to receive the lower edge of the leg, and the tarsus very short, or confounded with the leg, and forming a large hook. In some the body is oval and depressed, in others linear. These insects form the genus

NEPA, of Linnæus.

Thus divided:

GALGULUS, Latr. Tarsi cylindrical, with two distinct articulations and two hooks at the end of the last. Antennae with three articulations, the last the largest.

Those of the following have four, and the anterior tarsi terminate in a point or hook.

NAUCORIS, Geoff. Fab. Labrum not sheathed, large, triangular, and covering the base of the bill. Their body is nearly ovoïd, depressed, the head round, and the eyes flat. Tooth-formed antennæ. Last four feet ciliate, and two articulations to the tarsi.

In the three following sub-genera, the labrum is sheathed, and there are two threads to the abdomen.

Belostoma. Two articulations to all the tarsi, and antennæ semi-pectinate.

Nepa, Latr. (proper). Anterior tarsi one articulation, posterior two. Antennæ forked. Thighs much broader on the anterior feet. Body elongated, and almost elliptical. Abdomen terminated by two silk threads, serving for respiration.

Ranatra, Fab. Linear body. Bill directed forward. Anterior thighs long and slender. In the others *(Notonectides)* the two anterior feet are curved underneath, and the tarsus pointed and ciliated. The body cylindrical or ovoid, and pretty thick. The hinder feet are oar-like, and have two not very distinct hooks. They compose the genus,

Notonecta, Lin.

Thus divided:

Corixa, Geoff. No scutellum; bill short, triangular; wing-cases horizontal. Anterior feet short, tarsi, single articulation; other feet long, and two hooks on the middle pair.

Notonecta. Distinct scutellum. Bill long and articulate. Wing-cases raised. All the tarsi with two articulations. Four anterior feet elbowed; tarsi with two hooks.

In the second section of Hemiptera (Homoptera) the bill originates from the lowest part of the head, near the breast, or even between the two anterior feet. The cases, throughout, are semi-membranaceous, and sometimes almost like the wings. The three segments of the trunk are united, and the first shorter than the next. The ovipositor is scaly, like a saw, and composed of three denticulated plates.

The first Family,

THE CICADARIÆ,

Have three articulations to the tarsi, small, conical antennæ, of from three to six pieces, comprehending a very fine hair with which they are terminated. Some which have these organs with six articulations, and three simple eyes, form the genus

CICADA, Oliv. *Tettigonia,* Fab.

The wing-cases are almost always transparent and veined. These insects do not jump, and the males produce a loud monotonous sound, occasioned by a certain apparatus.*

Other *Cicadariæ* have but three articulations to the antennæ, and two simple eyes. They jump. No sonorous organs. The cases are often coriaceous. Some *(Fulgorella)* have the antennæ immediately under the eyes, and the forehead often prolonged like a muzzle. Such is the genus

FULGORA, Lin, Oliv.

Those whose front is advanced, who have two simple eyes, and no appendage under the antennæ, are the *Fulgoræ* proper, of Fabr.

Others, without simple eyes, but with two small appendages under each antennæ, form the

OTIOCERUS, of Kirby, or *Cobax* of M. Germar.

Those which have little or no advancement in the forehead, compose divers genera in Fabricius, to which some others must also be associated.

Sometimes the antennæ are shorter than the head, inserted outside the eyes. Here two simple eyes are very discernible.

LYSTRA. Like small cicadæ proper; body and elytra

* We have described this in our supplement to this order.—ED.

elongated. Second articulation of antennæ, globular and granulated.

CIXIUS. Like the last, but second articulation of antennæ cylindrical and smooth.

Under the generic name of TETTIGOMETRA, are separated some insects analogous to the preceding, but whose antennæ are lodged between the lateral and posterior angles of the head, and those of the anterior extremity of the corslet. The eyes do not project. No simple eyes.

The species whose elytra are large, and prothorax sensibly shorter in the middle than the mesothorax, compose the sub-genus, PÆCILOPTERA, Latr. Germ., *Flata*, Fab.

Those in which it is as long as the mesothorax, and the elytra scarcely longer than the abdomen, are dilated at the base, and afterwards contracted, form that of ISSUS, Fabr.

Sometimes the antennæ are as long as the head, and usually inserted in a lower emargination of the eyes. ANOTIA, Kirby, approach the preceding as to the mode of insertion of the antennæ.

ASIRACA, Lat. *Delphax*, Fab. When they are inserted in a lower emargination of the eyes. They are of the length of the head and thorax, and the first articulation longer than the second. No simple eyes.

DELPHAX, Fab. Antennæ not longer than the head; first articulation shorter than the next. Simple eyes.

DERBE, Fab. Supposed to come after the preceding.

In the last cicadariæ, the antennæ are inserted between the eyes. They form the genus

CICADELLA.

Thus sub-divided:

We begin with the species (*Ledra* excepted) which originally composed the genus MEMBRACIS, of Fabricius. The head is inclined in front, prolonged into an obtuse point, and more or less semi-circular. Antennæ small, terminated by a hair, and inserted in a cavity under the edges of the head. The prothorax is sometimes dilated and horned on each side, prolonged behind into a point; sometimes longitudinally raised along the back, and sometimes advanced and pointed in front.

Some have no scutellum properly so called. The limbs, especially the anterior, are sometimes very compressed, and foliaceous. The upper part of the head always forms a semi-circular hood.

MEMBRACIS (proper), Fab.—Prothorax raised, compressed and foliaceous.

TRAGOPA, Lat.—A horn on each side of the prothorax, which prolonged posteriorly into a vaulted point, replaces the scutellum.

Sometimes the legs are not foliaceous.

DARNIS, Fab.—The posterior prolongation of the thorax covers the upper part of the abdomen and elytra, in the form of an elongated triangle.

BOCYDIUM, Latr.—Elytra naked or nearly so. Scutellary prolongation of prothorax narrow, and lanceolate.

In the others, the scutellum is partly uncovered, although there be a prolongation of prothorax. A transverse suture is at the extremity of the latter. CENTROTUS, Fabr.

Next come species in which the head is level with the prothorax and horizontal. No middle elevation, or posterior lengthening to prothorax, but lateral dilatations. Mesothorax formed like scutellum. Elytra entirely naked. Hinder legs spiny. In many the corslet is an irregular hexagon.

ÆTALION, Latr.—The head, from above, is only a transverse ridge. The forehead is abruptly inclined, and the two simple eyes are situated between the others. The antennæ very small, are inserted under an ideal line, drawn from one eye to the other. The legs are neither ciliate nor denticulate.

In the three succeeding sub-genera the vertex is triangular, with two simple eyes. Antennæ inserted in or above an ideal line, from one eye to the other.

Ledra.—Have the head very much flatted, in the form of a transverse hood, arched and terminated in the middle of the anterior edge by an obtuse angle. The sides of the prothorax are raised like rounded horns. The hinder legs compressed, and with an external denticulated membrane.

Ciccus, Latr.—The antennæ terminate after the second articulation, in a thread of five articulations. The anterior extremity of the head is usually advanced.

Cercopis.—Third articulation of antennæ conical, and terminated by an inarticulate thread.

In the concluding cicadariæ, the prothorax is scarcely prolonged, and terminates at the origin of the elytra in a straight line. The scutellum occupies a considerable portion of the breadth of the body.

Eulopa, Fallen.—Two very prominent eyes, head little advanced, but depressed in front; and feet without spines or teeth.

Eupelix, Germ.—Head triangular, elongate, and flat. Simple eyes.

Penthimia, Germ.—Antennæ inserted in a large fosset narrowing the space between the eyes. Head rounded, and its edges advanced above the fossettes. Simple eyes in the middle of the vertex. Body short. Near this sub-genus should be placed Gypona of M. Germal.

Jassus, Fab. Germ.—Vertex very short, either linear or arched, and very little advanced. Antennæ terminated by a long thread.

Cicadella (proper), or Tettigonia, Oliv., Germ.—*Cicada*, Lin. Fab.—Head triangular, but not much elongated or flatted.

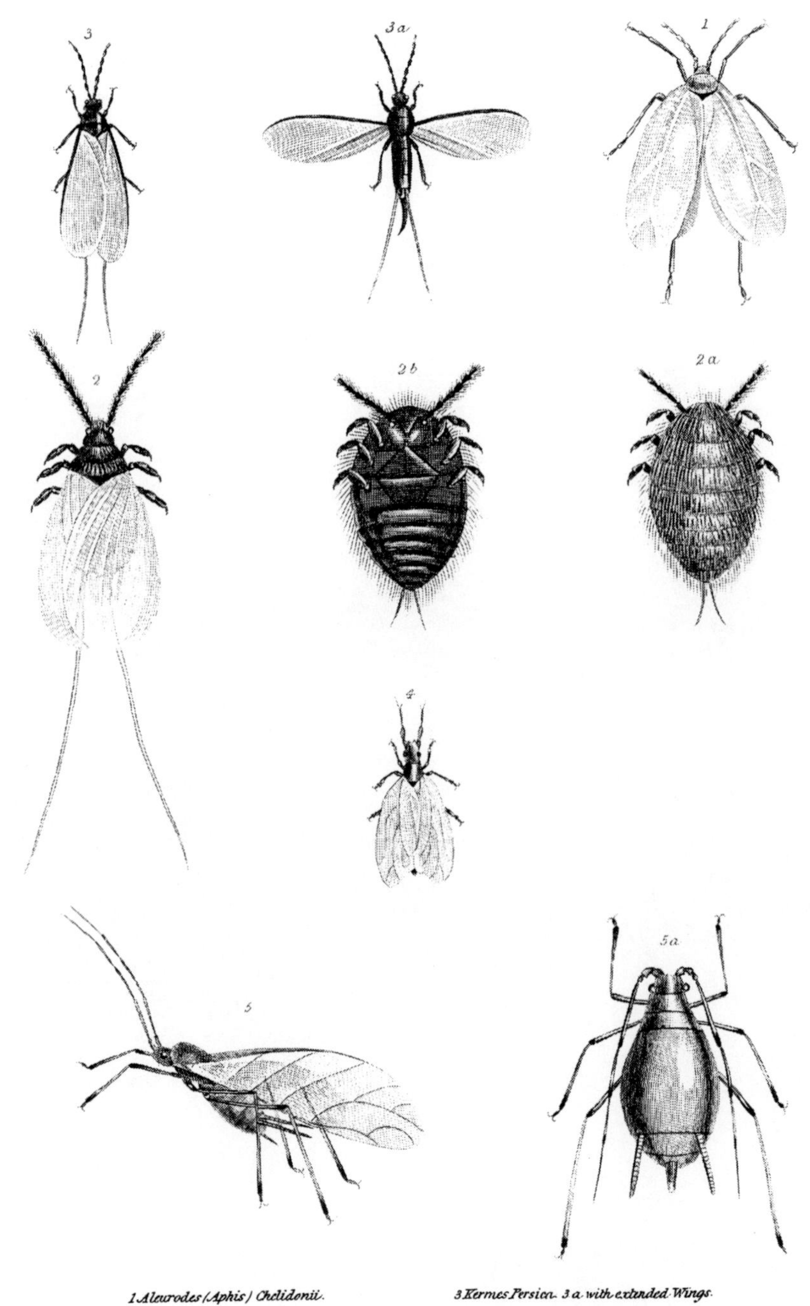

1 Aleurodes (Aphis) Chelidonii.
2 Coccus Cacti (The Cochineal) male.
2 a The female upper side. 2 b Female under side
3 Kermes Persica. 3 a with extended Wings.
4 Livia Lat. (Aphis) Juncorum.
5 Aphis Rosæ 5 a Larva of same.

London, Published by Whittaker & C.º Ave Maria Lane. 1832.

The second family of Homopterous Hemiptera, fourth of the order,

The APHIDII,

Have the tarsi with but two articulations, and antennæ filiform, longer than the head, and with from six to eleven articulations. The winged individuals have two elytra and two wings. They are small insects with a soft body, and the cases scarcely differ in consistence from the wings.

Some have ten or eleven articulations to the antennæ, the last of which is terminated by two threads. They jump; and compose the genus

Psylla, Geoff. *Chermes*, Linn.

Latreille has formed with the species which lives in the flowers of reeds, a genus Livia. Antennæ much thicker interiorly, than at their extremity.

The other aphidii have but six or eight articulations to their antennæ, and no threads.

Sometimes the cases and wings are linear, fringed with hairs, and the body almost cylindrical. The bill is very small, or scarcely distinct. Tarsi terminated by a vesicular articulation. Eight graniform articulations to the antennæ. Thrips, Lin.

Sometimes the wings and cases are oval or triangular, and assume the form of a roof. Bill distinct. Tarsi terminated by two hooks. Antennæ, six or seven articulations. Such are Aphis, Lin., thus divided:

Aphis (proper).—Antennæ longer than the corslet, seven articulations, third the longest. Eyes entire, and two nipples at the extremity of the abdomen.

Aleyrodes, Lat. *Tinea*, Lin.—Short antennæ, six articulations, and emarginated eyes.

The last family,

GALLINSECTA (a particular order with Degeer).

One articulation to the tarsi, with a single hook. Male without bill, and only two horizontal wings. Its abdomen terminated by two threads. Female without wings, but with a bill. Antennæ filiform, and, usually, with eleven articulations. They compose the genus COCCUS, of Lin.

Geoffroy divides the gallinsecta into two genera,—*Chermes* and Coccus.

A male coccus of Java, with twenty-two articulations to the antennæ, grained, and well furnished with hairs, and two wings almost coriaceous, is the type of the genus MONOPHLEBA of Dr. Leach.

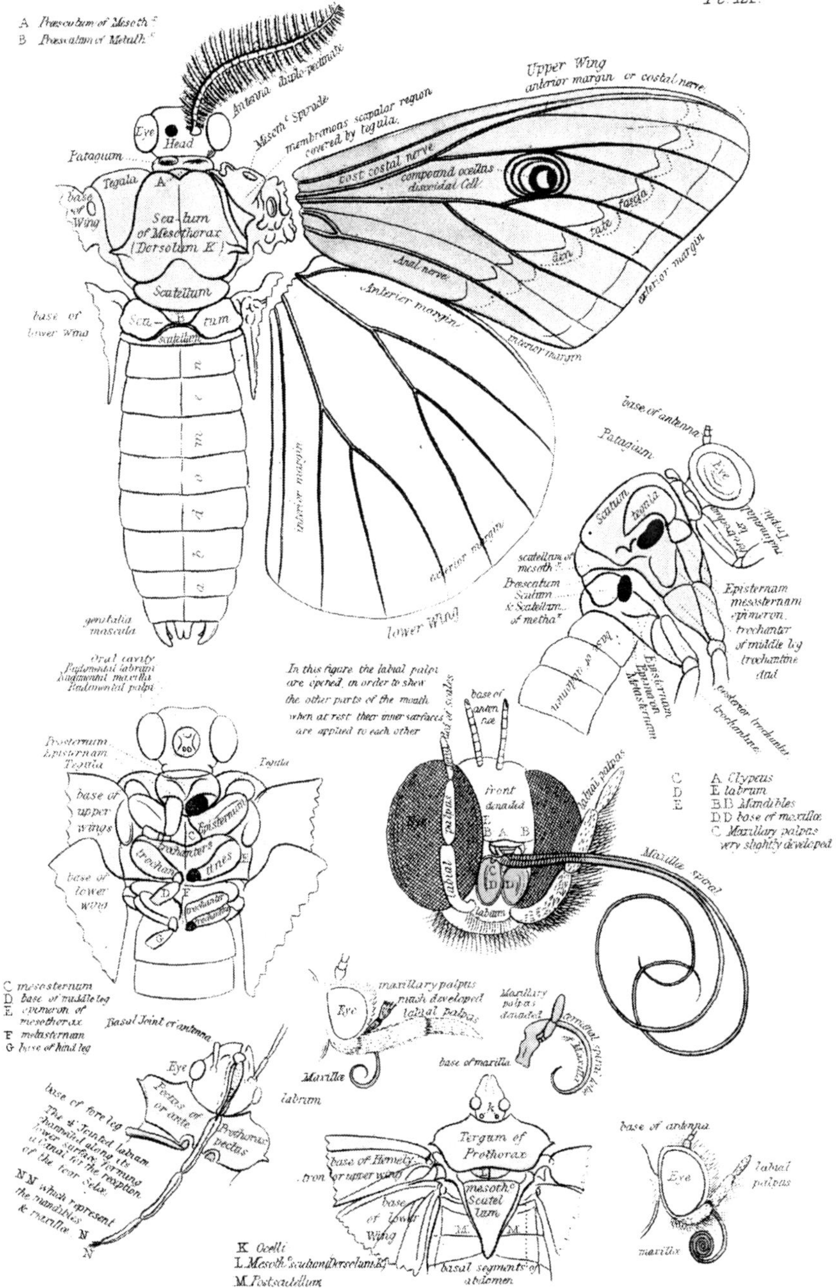

London Published by Whittaker & Co Ave Maria Lane. 1832.

SUPPLEMENT

ON THE

ORDER HEMIPTERA.

THE upper wings of a great number of insects of this order, especially some of the *cimex* genus, by reason of their consistence, participate at once, both of the nature of the elytra of the coleoptera, and of that of wings properly so called; for they are crustaceous, but terminated abruptly, and in the manner of an appendage, by a membranaceous part. From this circumstance is derived the name *hemiptera*, which is formed of two Greek words, one signifying *half*, and the other *wing*.

Linnæus, in the primitive institution of this order, founded it only on characters taken from the form and direction of the organs of manducation, *os (rostrum) subthorace inflexum;* but having afterwards adopted as the first basis of his method, relatively to insects provided with wings, the number and consistence of those parts, he injudiciously associated with the hemiptera, the blattæ, the mantes, the locusts, and other insects, composing at present the order orthoptera, and which he had originally placed at the end of the Coleoptera. Geoffroy, in this respect, followed the old plan of this great naturalist; and Degeer, who came after him, also adopted it; but he rendered it more perfect by establishing two new orders: one, that of *dermaptera* (*orthoptera* of Olivier), and the other exclusively formed of the genus *Coccus*, making part of the *hemiptera*. Since this period, all naturalists have approved of the change; but without admitting the last order instituted by Degeer. The order of *hemiptera* thus modified corresponds exactly to the *rhyngota* of Fabricius, having, as an essential character, *a beak (rostrum) for mouth; its sheath articulated.*

Of all the insects provided with elytra and wings, the hemiptera are the only ones which have neither mandibles nor jaws, properly so called, as appears by the text.

The rostrum of the hemiptera, which was named *sting* by the ancient naturalists, is only proper for extracting fluid matters. The attenuated stilets of which the sucker is formed, pierce the vessels of animals and plants, and the nutritive fluid, successively compressed, is forced to ascend along the sucker, and arrives at the œsophagus. The sheath of the sucker, when it is very much elongated, as in many *Geocorisæ*, is then often bent like a knee, or forms an angle with the sucker. The insect not being able to inspire the air, does not, properly speaking, feed itself by means of suction; and this expression, which has been employed for want of another more suitable, should not be taken rigorously either in this case, or in any other, in which we speak of insects under the improper appellation of *suctorial.*

The wings and elytra vary much in their forms. In the *bugs* a part of the elytra is hard, coriaceous, and resembles the elytra of the coleoptera, while the other part is membranaceous, and similar to the wing. In *cigala* and *aphis* they are membranaceous, and often clear and transparent. They have rather more consistence in *tettigonia*, *membracis*, *&c.* Those of the *aleyrodes* are flowery, and of a milky transparence. This occasioned Geoffroy to place these insects in the order of *tetraptera with farinaceous wings* under the name of *phalène de l'eclaire.*

Among the insects of this order, there are some which have no wings. Such are the *bed-bug*, some *lygæi;* among which may be remarked some aphides, and the female *cocci*. The males of these last have only two membranaceous wings. These anomalies, however, should not induce us to separate these insects from the hemiptera, for they are otherwise perfectly related to them, by the conformation of the mouth, and the manner in which they take their food.

The abdomen of the hemiptera has no remarkable peculiarity, if we except the mode in which its posterior extremity is conformed in some insects of this order. The female cigalæ have at the end of the abdomen a sort of point, or augur, concealed between some scales, which serves for the purpose of depositing their eggs. The aphides have at this same end sometimes two points or horns, sometimes two tubercles. The *cocci* have this part provided with threads more or less long.

All these insects undergo the metamorphoses of those of other orders; that is to say they pass successively through the different states of larva, nymph, and perfect insect. But the mode in which this change is executed and accomplished is different from that which we remark in the coleoptera. The larva does not resemble, like that of the majority of the last mentioned insects, a dull and heavy worm. It is a being almost similar to the one from which it received its birth, and differs from it only in the absence of wings and elytra, and a smaller size, susceptible of increase, which eminently distinguishes this state from that of the perfect insect.

To this first state succeeds that of nymph. The larvæ of the hemiptera arrive at this state by simply dropping their skin, which they change in their moulting. Arrived at this second state, they re-appear under the same form which they had before, with a very slight difference. They have then upon the back, precisely at the place where the elytra and wings should originate, two sorts of tubercles, or germs, which are concealed under the skin of the larvæ.

It is in this tubercle are also concealed the wings and the elytra, which only appear upon the body of the perfect insect. It is in the development of these parts, that the final metamorphosis of the hemiptera consists. Those, however, must be excepted, which are destitute of wings. All the change which they undergo, consists merely in the different moultings, and divers changes of the skin.

In our description of each genus of the hemiptera, we shall give all the details which the subject can furnish, relatively to the habits of these insects. Some are found in waters, as the *naucori*, the *corisæ*, the *nepæ*, the *ranatræ*, the *notonectes*. Others remain merely on the surface of the water, and seem to traverse it with their long feet. Such are the *gerris*, and the *hydrometra*. Others live on vegetable substances, remaining continually on trees and plants, to suck the sap therefrom. These are principally the *tettigoniæ*, *cigalæ*, *aphides*, and many *geocorisæ*. Others, in fine, attack animals. These are all the aquatic species, and moreover the *reduvii*, and other *geocorisæ*, *&c.*

The first family belonging to this order, is the Geocorisæ, or land-bugs. It is formed of the genus *cimex*, of Linnæus. The majority of the species spread a very fetid odour, and suck a variety of insects. There are some, however, which live on vegetables. The sub-genus Scutellera, was first established by M. Lamarck, and Fabricius afterwards named it *tetyra*. Linnæus first indicated it, by forming with these insects a particular division, that of *cimices scutellati*.

The scutelleræ, like the following genus *Pentatoma*, are found on plants. Their habits and metamorphosis are the same as the majority of the geocorisæ, which shall be explained hereafter. The differences of the general form of the body, or of its section, those which the head and the antennæ present, considered under the relation of the relative propor-

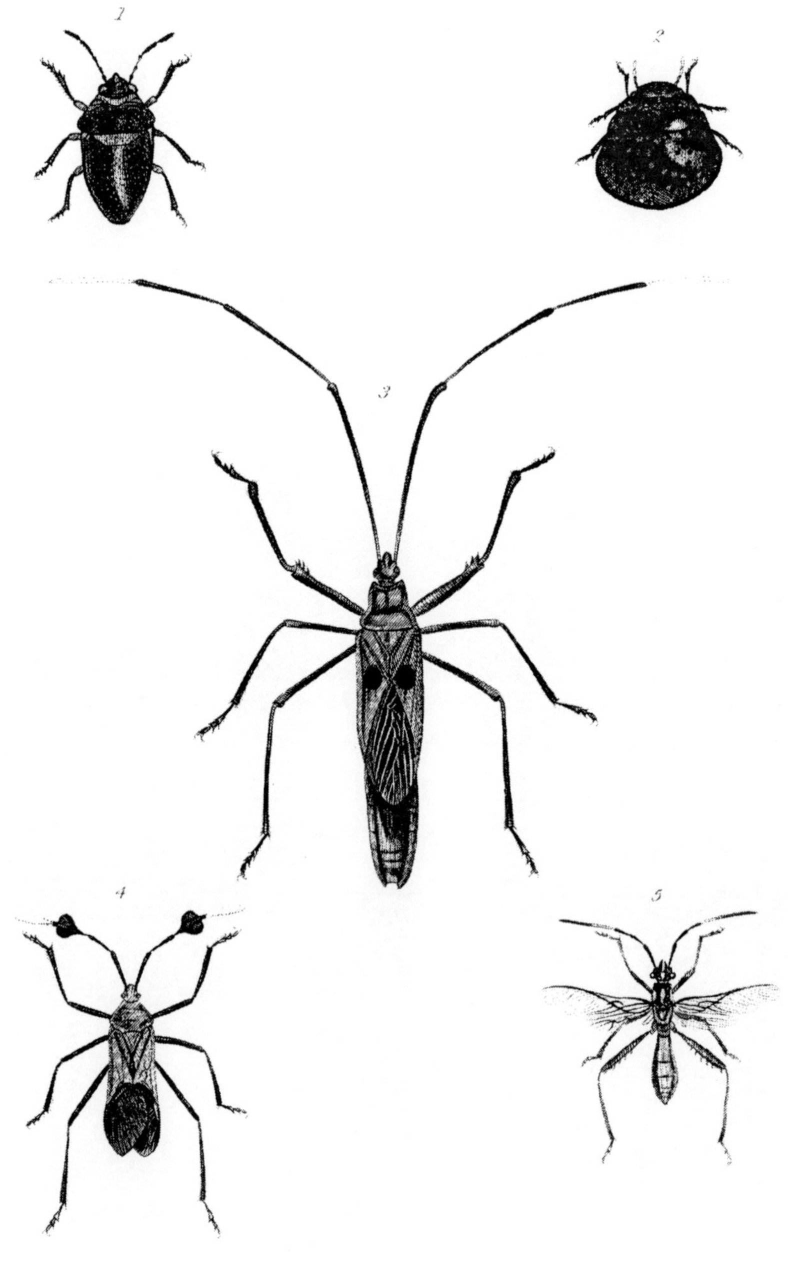

J. O. Westwood del.

1. *Scutellera basalis* _ G. R. Gray. 3. *Lygœus grandis* _ G. R. Gray.
2. *Canopus punctatus* _ Leach's MSS. 4. *Coreus bicolor* _ G. R. Gray.
5. *Alidus pellucidus* _ G. R. Gray.

London, Published by Whittaker & C.o Ave Maria Lane, 1832.

tions of their articulations, may be employed in the division of this genus, which is tolerably numerous.

The *scutellera nigro lineata*, is found in temperate and southern Europe. It is common in the south of France, on the leaves of the cherval, and on other plants. Another species (*S. hottentota*) is found on rye at the epoch of its maturity.

The scutellera, from the Melville Island, which we name *basalis*, is of a rich blue; and some specimens vary to a fine green colour, with the base of the thorax, scutellum, and the femora, fulvous.

We have figured a species of scutellera from the British Museum, which Dr. Leach considered to belong to the genus *Canopus*, of Fabricius, which Dalman says differs from scutellera, in having the body much more convex, concave underneath, with the edges of the scutellum overhanging the sides. The species is named by the above gentleman, *punctatus*. It is dark bronze-black, sprinkled with red spots; the legs reddish: this species is from Fantee.

We have next to speak of the genus PENTATOMA.

Geoffroy divided the *bug* genus, *cimex*, of Linnæus, into two families, having for characters the number of the articulations of the antennæ: to wit, four and five. Olivier has made of this second family, or of the bugs, whose antennæ have five articulations, a genus, which, from this consideration he has named *pentatoma*, from a Greek word, signifying five pieces.

These insects are found on plants, and feed upon their juices. They are often also found, and sometimes in a troop, having their bill advanced, and sunk to the end in the body of a caterpillar, or some other insect. They frequently shed a strong and disagreeable odour, which they communicate to the bodies over which they crawl. Their larvæ and nymphs do not differ from the perfect insect, but that the first have neither elytra or wings, and the second have only the rudiments of both.

We shall here notice a curious species, the *pentatoma betulæ*, or *cimex betulæ*. It is of a greenish or reddish grey. The antennæ are grey, with the extremity black. The scutellum is marked with a black spot. The upper part of the belly is black, with spots of a clear yellow, or flesh-colour, and black spots disposed alternately on the edges.

This species lives on the birch-tree, the leaves of which serve it as food. Degeer found, at the commencement of July, many females, accompanied by their little ones. Each of them had around her twenty, thirty, and sometimes even forty of them, and remained constantly beside them. As soon as one of these mothers quitted her place, and began to walk, all the little ones followed her, and halted whenever she stopped. She thus promenaded them from one place to another, conducting them as a hen leads her chickens, keeping guard over them to protect them. The same observer has seen one of these mothers beating her wings without ceasing, with a very rapid movement, without, however, changing place, as if for the purpose of keeping off some enemy which approached her. Modeer observes, that it is especially against the male that this unquiet mother is obliged to put herself in defence, because he endeavours to destroy his own posterity. The little ones proceed from the guardianship of their mothers, when they are strong enough to have no further need of their succours.

"It has happened to me," says Degeer, "to observe on one of these young *bugs*, placed under the microscope, that its proboscis was entirely disengaged out of the furrow of the sheath. It hung then at the end of the tongue, like a very long thread. I saw again, that at the end of the thread, the three pieces of which it was composed, were separated one from the other. The following day I observed on the same *bug* that every thing was restored to its proper place—that the proboscis was placed, as before, in the furrow of the sheath. It appears, then, that the bug can withdraw its proboscis out of the sheath, and put it back when it thinks proper. I drew the proboscis out of its sheath once again; I then saw how the intermediate part of the proboscis and of the point played—how the bug elongated and shortened it alternately. I saw some drops of fluid come out of, and re-enter, the proboscis. The two semi-sheaths which accompany it, played also alternately in front and rear. I was attentive to observe how the *bug* caused its proboscis to re-enter into the furrow of the sheath, and at last I achieved this, after having observed it, without interruption, for more than a quarter of an hour. It first of all puts its proboscis in a parallel line with the sheath, or, at least, it holds it extended, the entire length of the sheath. Afterwards it gives an inflexion to the sheath, about the middle of its extent. It folds it like a knee. It then applies this knee against the middle of the proboscis, or against that part of the proboscis which is opposite to the knee. The anterior feet then come to its assistance. The bug presses its proboscis with its feet against the sheath, so that this portion of its proboscis is then stopped in the groove or furrow. Finally, it presses the rest of its proboscis against the sheath with the same feet, and thus causes it to slide into the groove. As soon as the proboscis has once re-entered, it remains there.

Another species is called *Cimex baccarum*, or *P. baccarum*, from its frequenting berries and other fruits of a similar nature. Its antennæ are almost entirely variegated with black and white. The projection on which they are inserted terminates in a point. The hood is emarginated. The appendages of the wings are not punctated with brown, as in the species (*griseus*). The under part of the body has no advancement in the form of a spine. The body is hairy, and the edges of the abdomen are not denticulated. Degeer has found it on the white *petty-mullein.* It also frequents different sorts of fruit-trees with berries, especially currant and gooseberry bushes. It stinks very much, and pierces with its proboscis the elytra of the coleoptera, which it wants to suck.

The *Cimex ornatus, P. ornata,* is black, with four red spots on the corslet, and another of the same colour, and forked on the shield. The cases are red, with three black spots, and the edges of the belly alternately varied with those two colours.

This insect is found in abundance on the cabbage, and many other such-like plants. Its eggs are numerous, and ranged in many crowded lines. They have the form of a little grey barrel, and punctated with brown in the middle, and fasciated with brown at both ends. They are cemented by the lower extremity. The upper extremity is brown, with a narrow grey circle, and a point of the same colour in the middle. This extremity opens like a covercle, when the larva is disclosed.

The genus *bug, cimex,* of Linnæus, comprehends that insect to which the name is more constantly applied, and which is unfortunately but too well known by the disagreeable odour which it exhales, and by the torments which it occasions. We mean the *bed-bug* (*cimex lectularius*). In dismembering this genus, Fabricius has placed this species in that of *Acanthia,* while some insects, which differ from the preceding by many essential characters, are united in another generic section, termed *Cimex* by that naturalist. Some of these we have already considered under *pentatoma.* As it is incumbent upon naturalists as much as possible to make themselves understood by every one, and, wherever occasion will allow, to use the habitual and general language of mankind, this fantastic change in the nomenclature of these insects has been very properly rejected by M. Latreille.

His genus Bug or Cimex, therefore, has for its type the disgusting animal in question, namely — the *bed-bug*, or *cimex lectularius*. Any descriptive details of this insect are unfortunately quite superfluous, and it were a consummation devoutly to be wished that we were less learned upon this subject, or rather that we were in the most profound state of ignorance concerning it.

There is, we believe, scarcely a human being in these countries who has not had occasion to curse the insupportable odour of the bug, and who has not experienced its sanguinary disposition. It takes up its abode with our household goods, and conceals itself the more easily from our sight, that, its body being flat, it has the facility of lodging itself in the most narrow nooks which our apartments present to it, our furniture, and more especially our beds. It seldom or ever issues from its retreat but during the night. It is known but too well that it lives in a numerous society, that it germinates most prodigiously, and that its posterity, in despite of our most indefatigable pursuit, is certain to escape destruction. It comes to trouble our repose, and to torment us, at a season when sleep is positively most necessary for the refreshment of our bodies after the labours, cares, and fatigues of the day. Nature has endowed this insect with a singular industry to render useless all the precautions which we may take, to keep off the hateful nuisance. If it be unable to climb up our beds from below, it has the address to mount along the wall, to gain the ceiling, and to let itself drop, when it finds itself immediately above the bed. The greatest possible cleanliness, an extreme attention in frequently examining the places where the bugs have secured to themselves a favourable retreat, also in stopping all the holes and clefts of the walls, may protect us from these troublesome insects, or at all events must contribute to diminish their numbers. We may introduce into the places where they remain concealed, and it must be done as deeply as possible, some essence of venetian terebinthinus, oil of petrolium, &c. The gas produced by a strong solution of nitric acid and copper, the communications with the external air being closed, or the vapour of sulphur, will reach them in every direction, and with greater facility. But the greatest care must be taken to get out of the apartment directly, not to re-enter it until after the lapse of some days, and the utmost precaution. The most fatal accidents have occurred, for want of the necessary observance of these matters.

It is generally supposed that this bug was first introduced to this country in the fir-timber which was brought over for the purpose of rebuilding this metropolis, after the great fire of 1666. It is generally said that bugs were not known in England before that time; and many of these insects were found almost immediately afterwards, in the newly-erected edifices. This may probably be true, as in many of the remoter parts of the empire they are unknown to the present day, or at all events extremely rare, and it is pretty well known, that they generally attack newly-arrived visitants from the country with the greatest severity. They generally attack the young with more avidity than the old. The blood is in a purer state, and the skin more tender and easily pierced.

Independently of blood, for which they evince so voracious an appetite, they will feed on dried paste, size, deal, beech, osier, and some other sorts of timber, of which they suck the sap. They can subsist on any of these, and it is fortunate, perhaps, that unlike some of our other tormentors of the insect tribe, they can thus vary their food. It is said that they will not touch oak, walnut, cedar, or mahogany wood. Several pairs, confined within wood of these kinds, soon died,

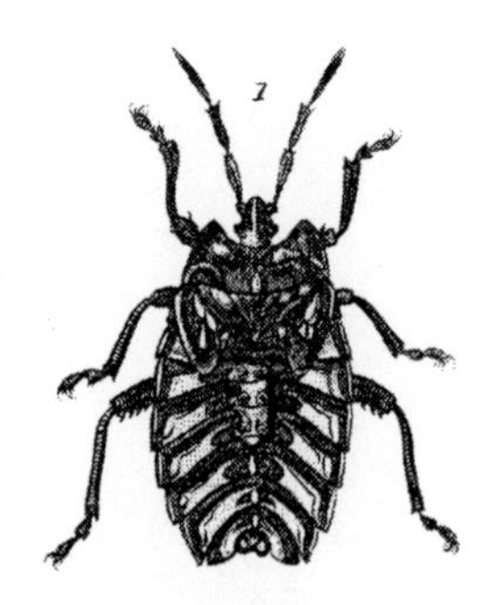

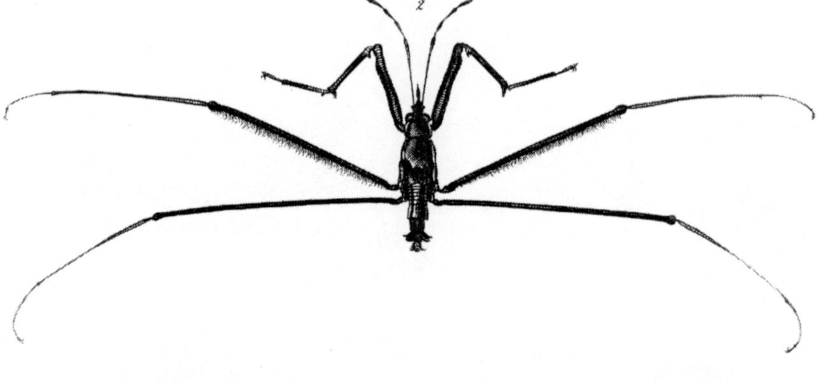

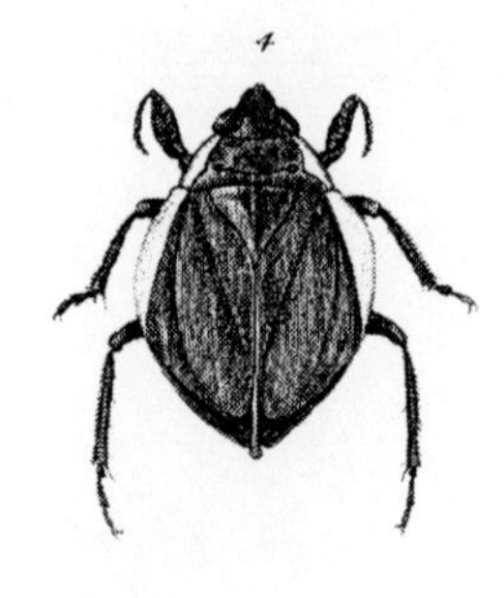

J.O. Westwood del.

1. *Tesseratoma ossa-cruenta. Hope MSS.*
2. *Gerris laticauda Hardw.*
3. *Belostoma marginata. G.R.Gray.*
4. *Syrtis fasciata. G.R.Gray.*

London. Published by Whittaker & C.º Ave Maria Lane 1832.

whilst others continued to live in the other woods during the whole year.

The female lays about fifty eggs at a time. These eggs are white, and at first are covered with a viscous matter, which afterwards hardens, and fixes them wherever they are laid. The young bugs come forth in about three weeks. The usual times of laying, are in the months of March, May, July, and September. Two hundred young ones may be produced from every female bug that lives through the season. Thus it may be seen, what a numerous increase there may be of those disgusting vermin, when proper care is not taken to destroy them.

The young bugs for some time remain perfectly white, but turn brown in about three weeks. They arrive to their full growth in about eleven weeks. They are watchful and cunning, and so very fierce that they will sometimes fight among themselves with the utmost fury. They seldom give over until one is killed, and sometimes both perish in consequence of the combat. Spiders are very great enemies to them, and often seize upon and devour them.

Bugs abound in all warm climates, and our merchant vessels are consequently in general swarming with them. This is also one cause of their being so very numerous in seaport towns, and particularly in London, being brought here in bales of goods, &c. Every package brought from vessels should be carefully examined before it be admitted into houses and warehouses.

It is generally supposed that bugs do not lie torpid during the winter season, but take less nutriment, and do not leave their retreat so frequently as in summer. This is most certainly the case with some individuals.

Mr. Hope has named a species of *Tesseratoma? ossa cruenta*, in his MSS. which we have figured. It is red, with an oblong yellow spot on each side of the segments of the abdomen. The middle near the base is yellow. This insect is from India.

The Corei belong to the numerous family of the bugs, with which they have been placed by Linnæus and Geoffroy. They are found during all the fine season on plants, often in company with their larvæ and nymphs. These last resemble them in their forms and colours; but the larvæ are destitute of wings and elytra, and the nymphs have but the rudiments of those organs.

Like other insects, the *Corei* are not in a state to couple until after they have acquired wings. The females lay a great number of eggs, which they place upon the plants one by the side of the other, and they remain attached there by means of a gluten which cements them. When the little larvæ issue from the eggs, they spread themselves over the leaves to seek their food. Some find it in the plants themselves, from which they draw the juice with their proboscis, others by making war upon insects which they suck until nothing remains but the skin. It is not only in the larvæ state that the corei are carnivorous. The nymph and the perfect insect equally live on such insects as they can catch, and it is not rare to find them sucking a caterpillar larger than themselves, often even in very considerable numbers.

The species are tolerably numerous. M. Latreille found the *coreus histrix* in 1780, in a garden in Paris, on an elm. This is a very curious species. It is five lines in length, grey, with some obscure shades. Its body has so little thickness, that it appears to be composed only of a membrane. It is entirely bristling with rough and grey hairs, from which its scientific name. The antennæ are spinous, and terminating in a knob. The eyes are reddish. The sides of the corslet are raised in rounded and ciliated lobes. The abdomen is boat-formed, and its edges are festooned.

M. Latreille remarked that before catching the specimen

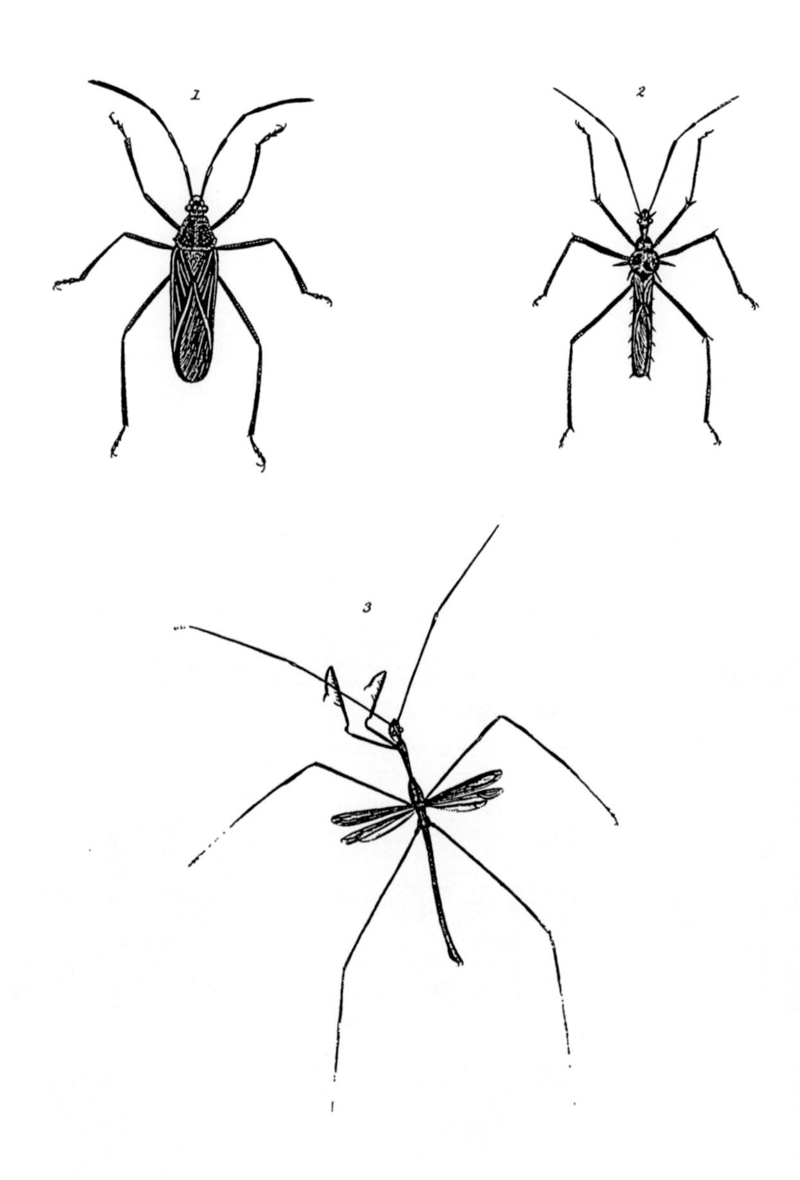

J. O. Westwood del.

1. *Mematopus rufoscutellatus.* – G. R. Gray.
2. *Zelus rufescens.* – G. R. Gray.
3. *Emesa filum?* – Fab.

London, Published by Whittaker & Co. Ave Maria Lane 1832.

alluded to, it agitated its body with much celerity, and sent forth a little noise. It is so scarce in the neighbourhood of Paris, that M. Latreille doubts if another individual has been found there since. It is not rare, however, in the south of France. The same eminent entomologist caught a good number of these insects in the province of Angoumois. He has received others from the neighbourhood of Lyons and Bordeaux.

We insert a figure of an inedited species of Coreus, which might indeed be appropriated to a new sub-genus, from the peculiarity of the third joint of the antennæ, which is palette-shaped, and all the rest simple. The species we call *bicolor*. It is fulvous, the antennæ, the legs, and the tip of the upper wings blueish-black; this insect is from the Cape of Good Hope.

To the genus Alydas we have also added a new species, which from the transparency, and from the nervures being confined to the base of the upper wings, may also be formed into a separate sub-genus. The species we name *pellucidus*. It is black, with the base of the terminal joint of the antennæ, and some spots on the head, thorax, and wings, yellow; the legs and abdomen fulvous; we believe it is from South America.

Our *Nematopus rufoscutellatus* is of a black colour, the thorax margined, with a line down the middle, and femora green; the head and scutellum rufous; the elytra margined with yellow. This species is from Mexico. The sub-genus was formed by Latreille, from the antennæ being slender, and as long as the body.

The Neides, a genus established by M. Latreille under this name, in his General History of Crustacea and Insects, and which Fabricius changed into that of *berytus*, in his "System of Rhyngotes," is composed of small hemiptera, very much approximating to his Corei and Lygæi.

They are found on the leaves of plants, the trunks of trees, &c. But a small number of species are as yet known. The most common is *neides tipularia*, the *cimex tipularius*, of Linnæus. It is of a reddish grey, with antennæ of the length of the body, blackish at their extremity. The hood is advanced into a point. Three raised lines, two of which are marginal, are remarked upon the corslet; the elytra are charged with nervures, and punctuated with blackish. It is found throughout all Europe.

The Lygæi are found on plants, but subsist less on the juice of their leaves than on other small insects. The *lygæus apterus* and some other species assemble in great numbers under the bark of trees, and in the crevices of walls. The first-mentioned species is very common, and usually prefers the mallow for its residence. It has no disagreeable odour, like the majority of the bug family.

We have figured a fine species of Lygæus, but which we consider to form a new sub-genus, from the length of the antennæ, and the prolongation of the abdomen The species, from its size, we name *grandis*. It is red, with two spots on the elytra; antennæ, tibia, and tarsi black, the apex of the anterior femora dentated interiorly. This species is from India, and is in the cabinet of Mr. Children.

We have added a figure of a new species of Syrtis, under the name *fusciatus*. Is light brown, with a broad band across the middle of the abdomen; the last joint of the antennæ and the wings are dark brown. This insect is from North America.

The majority of the insects of the genus Tingis are very remarkable for the semi-transparence of their corslet and elytra, their reticulation, their nervures, and the projection of their lateral edges. They suck vegetables, and some species occasion thereby such a derangement in the organization of the plants on which they live, that they form there

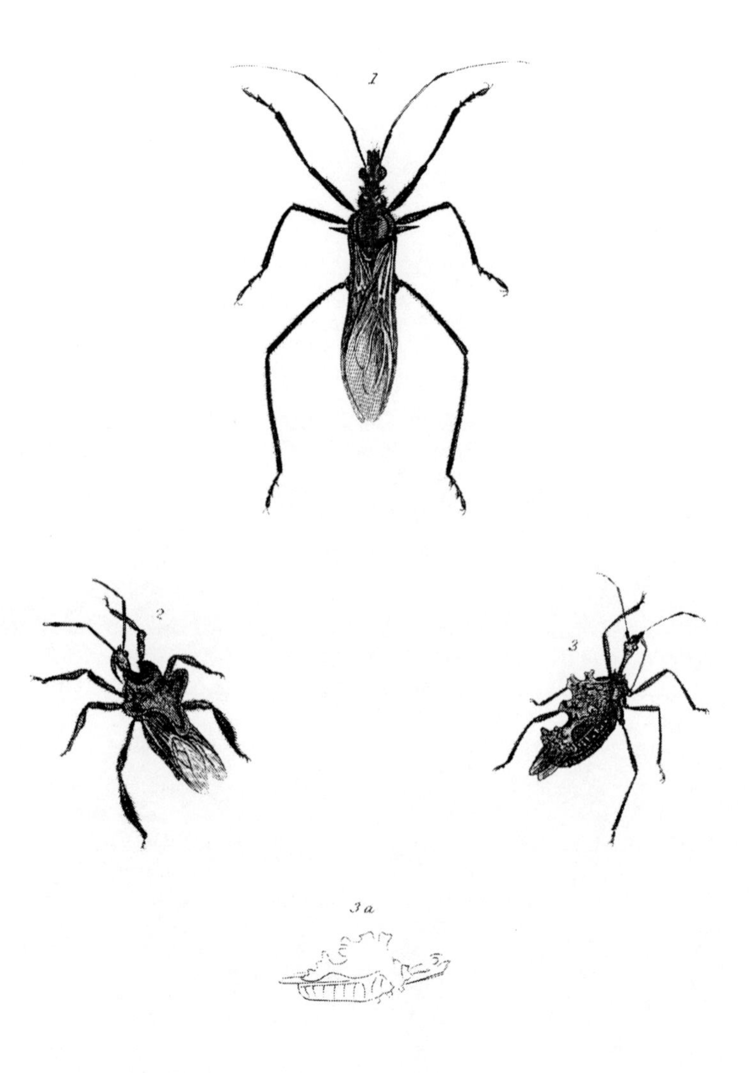

I. O. Westwood del.t

1. Reduvius spinidorsis. _ G. R. Gray.
2. Reduvius dorsalis. _ G. R. Gray.
3. Reduvius tuberculatus. _ G. R. Gray.

London. Published by Whittaker & C.o Ave Maria Lane, 1832.

monstrosities or appearances of galls, as has often been observed on the flowers of the *tencrium chamædrys.* This is especially done by the larva of *tingis clavicornis,* which inhabits the interior of these flowers, and produces, before they open, this species of gall, causing them to acquire an extraordinary volume.

The ARADI remain under the barks of different trees, such as birch, oak, cherry-trees, &c. They pass the winter there, and are sometimes found assembled in great numbers in such places. It is there that they undergo their metamorphoses, which differ little, or not at all, from those of the other hemiptera. It is principally in the spring time that they must be sought.

The genus REDUVIUS is of the creation of Fabricius. These insects live on rapine, as well under the form of larvæ and nymphs, as after having become perfect insects, and undergo the same metamorphoses. They form a genus tolerably numerous, of which but few species are found in Europe. Among these is the *Reduvius Personatus,* Fab., *Cimex Personatus,* Lin., vulgo *fly-bug.* It is found in Europe, often in houses. Its flight is rapid, it pricks strongly with its beak, and spreads a very disagreeable odour. When it is held between the fingers, it sends forth a sound, which is produced by the rubbing of the edges of its corslet with the elytra. Its larva is equally found in houses. It does not differ from the perfect insect, but that it has no elytra or wings. It is usually covered with ordure and dust, which renders it hideous, and causes it to be mistaken. It feeds on insects, and even on the bed-bug.

We are enabled to insert figures of three new species of Reduvius, viz :—

Dorsalis. Of a dirty fulvous; the anterior part of the thorax obscure brown; which is terminated by a paler trans-

R 2

verse fascia; and is considerably elevated and cruciform, the sides being dilated, the hinder part produced, covering the scutellum and base of the wings. This species is from Carthagenia.

Tuberculatus. Brown; the back of the thorax obscure buff, of a singular form, being considerably elevated and covered with tubercules; it is nearly the length of the abdomen, which it almost covers; it is, however, very much contracted beyond the middle, and the margins are very irregular.

Spinidorsis. Black; with a spine on each side of the thorax, also an erect spine on the middle of the scutellum; the second of the antennæ red, and the wings obscure yellow. It is from Demerara.

The ZELI are all exotic and their habits are unknown. The species inhabit the West India islands.

The *Zelus*, which we name *rufescens*, is rufous; with the antennæ and legs brown, with four erect spines on the thorax; spines on the margin of the abdomen. This is from Demerara.

We have figured a species which we consider with doubt to be the *Emesa filum* of the above mentioned author. The specific characters are,—It is brown; with a pale line on each side of the hinder part of the thorax: the abdomen is red at the base; the legs pale, testaceous, varied with darker. This species is from North America.

The PLOEIRÆ have but few species in this genus. The best known is the *Cimex vagabundus* of Linnæus, which is found on trees, where it vacillates, and balances itself, after the manner of the tipulæ.

The insects of the genus HYDROMETRA (divided by our author into three sub-genera) had been confounded by Linnæus and Geoffroy with the bugs. Fabricius at first placed them with his *Gerris*, but afterwards united them with other roving geocorisæ, which composed in M. Latreille's original method, the genera *Gerris* and *Velia*, now sub-genera, according to the system of the "Regne Animal."

The hydrometra frequent the edges of the water and run with swiftness on their surface. But they do not swim, and do not make use of their feet for rowing like the gerris.

The name of *gerres* has been given to certain small fishes, and in all probability from that comes that of GERRIS, employed by Fabricius. The genus which he thus designated at first was composed of insects very different both in their characters and habits. It became therefore necessary to reduce it, which was done by M. Latreille in his "*Precis des Charactères generiques des insectes.*"

We observe in the gerris an elliptical and elongated body, a triangular head, with prominent eyes, and no apparent simple eyes. An elongated corslet, narrowed in front, with the posterior extremity prolonged to form a shield. There are two narrow elytra crossed one upon the other, almost opaque, and with tolerably thick nervures. There are two membranaceous wings. The two front feet are short, with the leg and tarsus folded under the thigh. The four posterior feet spring from the sides of the body from which they are considerably separated. Their thighs are very long, and the legs and tarsi are confounded together. There is an emargination at the anus with a nipple in the middle.

Few persons but have had an opportunity of observing the gerris. The surface of dormant waters, and even of lakes and rivers, frequently present in summer a tolerable quantity of black insects, with attenuated and elongated body, that swim with an extreme agility, making use of their hinder feet, without diving. They have, nevertheless, a remarkable motion, by which they advance, as it were, by shocks. This is caused by their oars, which they are continually pushing backwards.

These insects belong to the most common species of gerris, *gerris lacustris.*

A great number of birds, and particularly aquatic birds, have the feathers of the lower part of the body satiny and polished, so that the water may more easily flow off. The gerris have something analagous to this. Their sides, and their lower surface are covered with a very fine matter, which may be removed by rubbing, the colour of which is changeable, and which, seen under the most favourable aspect for the light, is of a whitish, or silvery ash-colour, and shining like satin. Its use is probably to hinder the body of the insect from being wet. The water, at least, has no hold on the parts which are provided with this matter, while it has upon the others, that is to say, on the upper surface of the body, as an observer may easily convince himself of, by plunging the insect in water repeatedly.

Degeer is, of all entomologists, the one who has best studied the gerris. He has observed, in Sweden, three species, or at least three varieties of these insects. The first is apterous, and appears in spring, after having, in all probability, according to this naturalist, passed the winter under the ice, perhaps in the mud, to shelter itself from the rigour of the cold. We are led to the belief that these gerris are not larvæ or nymphs, but perfect insects, inasmuch as they couple in this state.

Placed on the water, these insects swim there, usually holding themselves elevated on their feet, so that the body does not touch the water. They are carnivorous, and feed on such insects as they can catch. Degeer has often thrown gnats to them. They leaped up, seized them with the two short feet, and introduced the point of the proboscis into the body of the prey, to suck it. Sometimes, even two or three individuals fought for the possession of the body.

When crushed, these gerris shed a disagreeable odour, like

that of our house-bug. The belly of the female is filled with a great number of white eggs, and of a very elongated form. The males are a little smaller than the individuals of the other sex.

This apterous gerris much resembles those which are found later, and which are winged. The body is a little smaller, and the feet, according to Degeer, proportionally shorter. Its colour above, is a brown-black, bordering upon green. The corslet is browner. The under part of the body, and of the sides are of a whitish ashen, changeable, and satiny. Each of these sides has a longitudinal line of spots, of a clear brown. But this is remarked in almost all the species. The antennæ and feet are of an obscure brown.

Degeer is not of the opinion of Geoffroy with regard to the state in which the gerris are when they couple. The latter thinks that there is a union of the two sexes, before they have acquired wings and elytra.

The second species of *gerris*, met with by the Swedish Reaumur, is the *lacustris*.

The third is larger and more elongated than the preceding. The little ones of this third appear on the waters in the month of July, and run there as fast as those which have undergone all their metamorphoses. At first some are seen which are no larger than a grain of sand. Their figure is oval, their head, eyes, and antennæ are thick. The first segment of the corslet has two black and shiny spots. On the second, which is large, there are two black and double plates, which are the germs of elytra and wings. The abdomen is very short, and, as it were, compressed. Also the posterior feet seem to be situated at the anus. All the body is a little hairy, and of an obscure brown, except the bottom of the corslet, and of the breast, which are of a greenish grey. The body elongates with age; the antennæ, and more especially the feet, are then proportionally thicker than in the perfect insect. The corslet is pretty nearly what it will always be. The rudiments of the elytra are apparent and elevated, but the belly is always very short. The animal is then in the nymph state.

Our *Belostoma marginata* is obscure brown, with margin of the thorax and elytra nearest the base obscure white. It is from China.

With regard to the apterous species of Degeer, M. Latreille does not venture to pronounce decisively, but the other two he clearly regards as distinct.

General Hardwicke, in the Linnean Transactions, describes and figures a remarkable insect, which he names *Gerris laticauda*, of which we insert another figure. It is of a testaceous brown, with the margin of the thorax silvery, with darker lateral line; the abdomen darker, the anterior femora with black longitudinal stripes; the antennæ, tibiæ, tarsi, and the pubescence on the middle femora black: this insect is from Java.

Of the VELIÆ all we can say is, that they simply run upon the surface of the water, and with amazing swiftness. In this differing from the gerris which, as we have seen, row and swim by jerks. They seem properly to constitute an intermediate genus between *hydrometra* and *gerris*.

We now come to the second family of hemiptera, the HYDROCORISÆ, which is formed of the genus NEPA, or *water-scorpions*, of Linnæus. The first sub-division of which, that we shall notice, is NAUCORES.

Geoffroy separated these insects from the nepa of Linnæus with great justice. They are aquatic, and have many relations with the *corisæ* and *notonectes;* but they are distinguished from them by their anterior feet. These in some measure resemble the sort of claws, which the *aranëides* have in the front of their head. They make use of them as pincers, to seize and retain the insects on which they feed, while they are sucking them. This character is common, it is true, to the other nepæ of Linnæus; but the naucores differ, because their anterior tarsi have but a single articulation, and because their other feet are proper for swimming. The labrum of the nepæ, properly so called, and of the *ranatræ*, is, moreover, not uncovered. The naucores are very agile, and swim with much swiftness, by means of their posterior feet, which perform the office of oars. They frequently issue forth from the water during the night, to fly in the country. They are extremely voracious. Of all aquatic insects those are they which commit the greatest havock in the waters. The larva and nymph differ from the perfect insect only in the want of wings. The larvæ have only on their breast two very flat pieces, which are sheaths enclosing the elytra and wings, which are developed after the first moulting. These larvæ and nymphs are equally carnivorous with the perfect insect. This genus is not numerous in species.

Linnæus named the hydrocorisæ, whose two anterior feet perform the office of a pincers, NEPA. By the establishment which Geoffroy made of two new generic sections, those of *naucoris* and *corisa*, the genus *nepa* of Linnæus, became more simple and natural. Geoffroy, who designated them under the name of *aquatic scorpions*, not having discovered the antennæ of these insects, took for these organs the two anterior feet, and on this error allowed them but four feet. Fabricius has since separated from the nepæ those which have a linear form, and which constitute his genus *ranatra*.

But the genus nepa, though more restrained, was yet susceptible of being still further reduced. Thus the *nepa cinerea*, and some others of an oblong oval form, and all remarkable for the length of their tail, have but three very distinct articulations to their antennæ, the second of which, only, presents a lateral projection in the form of a tooth.

Their tarsi have but a single articulation, as Geoffroy had already observed. The other species almost all exotic, with broader forms, and tail very short, or not existing, have their antennæ composed of four distinct articulations, and of which the last three are prolonged internally in the form of the tooth of a comb. All their tarsi are formed of two articulations, and almost similar.

The nepæ are aquatic insects, whose anterior feet are in the form of pincers. They are heavy, swim slowly, remain generally at the bottom of the waters in the mud, but fly very well, particularly in the evening. They are carnivorous, as well as their larvæ, and feed on small insects, which they pierce and tear with their proboscis.

The females lay eggs which, seen through the microscope, resemble a seed crowned with seven small threads, the extremities of which are gnawed. They deposit them in the stalk of some aquatic plant. The larvæ issue forth from them towards the middle of summer. They differ from the perfect insect, in being deprived of wings and elytra, and having no threads at the abdomen. They swim very slowly, and walk at the bottom of the water over the aquatic plants. The nymph carries its wings enveloped in sheaths, placed on each side of the body.

These insects are tormented by the *hydrachne* of Muller. There are often found on them some red eggs, which hold there by a pedicle or bill, serving as a sucker, and which grow there.

The Ranatræ are heavy, and swim slowly. They remain usually at the bottom of the waters in the mud. They fly very well, especially in the evening. They are carnivorous as well as the larvæ. Their habits, in fact, are precisely the same as those of nepa.

The eggs of these insects are white, elongated, and have at one of their extremities two threads, or two hairs. They remain fifteen days at the bottom of the waters. The larvæ and nymphs are just the same as those of nepa.

The CORIXÆ remain usually suspended by the hinder part of the body at the surface of the water. But on perceiving the slightest motion, they precipitate themselves to the bottom with great swiftness. They can remain there a certain time, by attaching themselves to some stones. They fly sometimes, but walk badly and slowly upon the ground. In the water, however, they are very agile. They live on aquatic insects, which they suck with their proboscis, after having seized them with the pincers of their anterior feet. When they swim, the under part of their body appears silvery; an effect produced by the air which attaches there. The larvæ and nymph differ little from the perfect insect. They live in the same manner in the water.

THE NOTONECTÆ, thus named because they swim upon their back, live in the water, both in the larva state, and in that of the perfect insect. They always swim on the back, having the belly upwards. The larva does not differ from the perfect insect, but in the want of wings and elytra. In their different forms, the notonectæ are carnivorous. They seize their prey with their anterior feet, and suck it with their proboscis. They attack insects larger than themselves, and do not even spare their own species. The larvæ of the ephemera also, often become their victims. They constitute a genus not very numerous in species. They are almost all found in Europe. The most common are the *glanea* and *minima*. They are frequently met with in stagnant waters.

The *Notonecta glauca* is found in waters, swimming at their surface. It stings very powerfully with its proboscis. During sexual intercourse, the male and female swim together with great swiftness. Afterwards, the latter lays a very considerable number of white, elongated eggs, which it deposits on the stalks of aquatic plants. In the commencement of the spring, there proceeds from these eggs some little larvæ, which in passing to the nymph state, acquire the rudiments of elytra and wings. Both the larvæ and nymphs swim upon their backs like the perfect insect.

The insects of the section HOMOPTERA, have received this denomination, because their elytra are of the same consistence in their entire extent, and they are sometimes half coriaceous, and sometimes almost entirely similar to wings. These hemiptera feed only on the juice of vegetables. Most of the females have an auger, or ovipositor, often composed of three denticulated laminæ, and lodged in a groove with two valves. They make use of this like a saw, to make incisions in vegetables, and insert their eggs there.

The first genus of this family is that of CICADA. It is composed of the division of the *cicada mannifera* of Linnæus, because one of the species, the *cicada orni*, by pricking the wild ash, causes to flow from that tree the honeyed and purgative juice which we term *manna*. Stoll designates the same insects under the name of *cigales chanteuses*, to distinguish them from many other analagous hemiptera, which are also called *cigalæ*, such as the *fulgoræ*, and *cicadellæ*.

The cicadæ are insects which have been known for a very long time. Their size, and the monotonous song by which the male is distinguished and which is heard during a part of the summer, have caused them to be easily discovered. They inhabit warm climates, and usually sojourn on trees. Their mode of flying is very light. While the heat continues they are very lively; but the cold throws them into a state of lethargy, or destroys them altogether.

What most particularly merits attention in these insects is the organs of song. It was for a long time believed, that the females alone possessed this faculty of singing; but this is an error, as they are deprived of the parts which are proper for the production of this effect.

These singular organs, which answer to the males the purpose of calling to their females in the season of reproduction, are lodged in the cavity of the belly, and covered by two scaly plates, placed underneath the corslet, at the origin of the abdomen. These two plates, which attach to the corslet without any articulation, cover each other to a certain extent, and reach almost as far as the third ring of the abdomen. On raising them, we perceive a cavity formed in the belly, divided into two lodgments or cells, the bottom of which is occupied by two small, extended, thin laminæ, as transparent as glass, which Reaumur has compared to two small mirrors, and which some authors have regarded as drums calculated for the returning of sounds. Besides these laminæ, the cavity also contains other parts. Reaumur, on opening a cicada at the back, found there two large muscles, each of which is composed of a prodigious bundle of fibres, applied one upon the other. These muscles lead to two membranes, turned in the form of a kettle-drum, which occupy two nooks, placed in the great cavity. On the side of the belly, we only see the apertures of one and the other, which are covered. These apertures are for the voice of the cicadæ, what our larynx is for our voice. The sounds which proceed through them are modified by the opercles, by the mirrors, and by the great cavity. Each timbal or kettle-drum has a convex and a concave part. The first is folded, and full of rugosities. When the insect causes the two large muscles, which are attached there, to move, these muscles, contracting and relaxing with rapidity, act upon the timbals, whose surfaces, by becoming successively convex and concave, produce the sound, which is termed the *song* of the cicadæ.

The females, though not possessed of the faculty of singing, have yet the rudiments of the opercles. They are, moreover, furnished with an auger, which, in the large species, is about six lines in length. This auger, composed of two pieces, denticulated at the sides, and pointed at their extremity, serves them to notch the wood in which they deposit their eggs. In making a hole in the branch, the cicada causes these two pieces to act alternately, which perform the office of a file. The branches in which the females have placed their eggs are easily recognised by the little inequalities at their surface placed one after the other. Each hole contains from five to eight eggs.

In the body of the female, the eggs are contained in two ovaries, and are sometimes six or seven hundred in number.

On the insects of the genus of which we are now writing, Mr. Kirby has the following remarks:

" The species of the other genus *cicada*, called by the ancient Greeks, by whom they were often kept in cages for the sake of their song—*Tettix*, seem to have been the favourites of every Grecian bard, from Homer and Hesiod, to Anacreon and Theocritus. Supposed to be perfectly harmless, and to live only upon the dew, they were addressed by the most endearing epithets, and were regarded as all but divine. One bard entreats the shepherds to spare the innoxious tettix, that nightingale of the nymphs, and to make those mischievous birds, the thrush and blackbird, their prey. 'Sweet prophet of the summer,' says Anacreon, addressing this insect, 'the muses love thee, Phœbus loves thee, and has given thee a shrill song; old age does not wear thee out; thou art wise, earth-born, musical, impassive, without blood; thou art almost like a God.' So attached were the Athenians to these insects, that they were accustomed to fasten golden images of them in their hair, implying at the same time a boast, that they themselves, as well as the cicadæ, were *Tenæ filii*. They

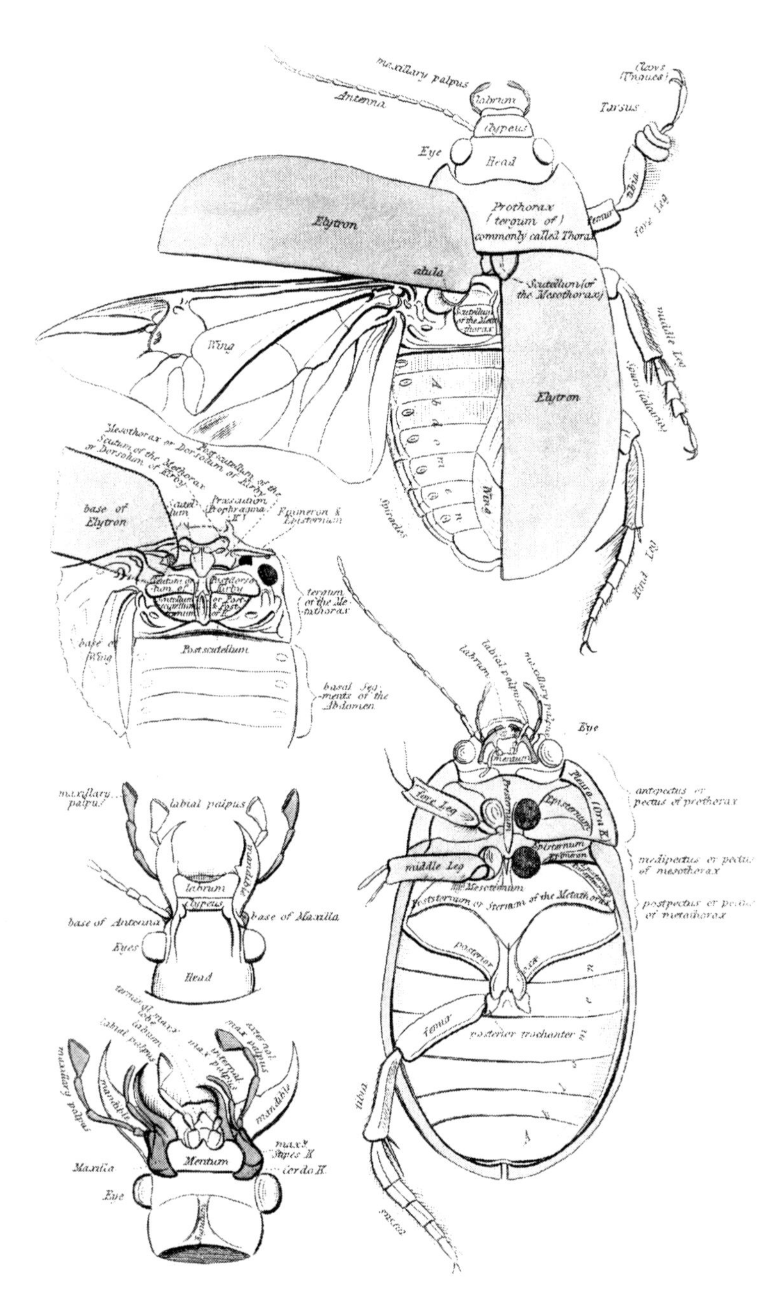

J.O. Westwood F.L.S. &c. del.

London, Published by Whittaker & Co. Ave Maria Lane, 1832.

were regarded by all, as the happiest, as well as the most innocent of animals, not, we will suppose for the reason given by the saucy Rhodian Xenarchus, when he says,

> "'Happy the Cicada lives,
> Since they all have voiceless wives.'

"If the Grecian *Cicada* or *Tettix*, had been distinguished by a harsh or deafening note, like those of some other countries, it would hardly have been an object of such affection. That it was not, is clearly proved by the connexion which was supposed to exist between it and music. Thus, the sound of this insect and of the harp were called by one and the same name. A cicada sitting upon a harp, was the usual emblem of the science of music, which was thus accounted for;—when two rival musicians, Eunomus and Ariston, were contending upon that instrument, a cicada, flying to the former, and sitting upon his harp, supplied the place of a broken string, and so secured to him the victory. To excel this animal in singing seems to have been the highest commendation of a singer; and even the eloquence of Plato was thought not to suffer by a comparison with it. At Surinam, the noise of the *cicada tibicen* is supposed so much to resemble the sound of a harp or lyre, that they are called there harpers (*Lierman*). Whether the Grecian cicadæ maintain at present their ancient character for music, travellers do not tell us."

"Those of other countries, however," (adds Mr. Kirby) "have been held in less estimation for their powers of song; or rather have been execrated for the deafening din which they produce. Virgil accuses those of Italy of bursting the very shrubs with their noise; and Sir J. E. Smith observes, that this species, which is very common, makes a most disagreeable, dull chirping. Another, *cicada septendecim*, which, fortunately, as its name imports, appears only once in seventeen years, makes such a continual din from morning to evening, that people cannot hear each other speak. They appear in Pennsylvania, in incredible numbers, in the middle of May. " In the hotter months of summer," says Dr. Shaw, " especially from mid-day to the middle of the afternoon, the cicada, τεττιξ, or grass-hopper, as we falsely translate it, is perpetually stunning our ears, with its most excessively shrill and ungrateful noise. It is in this respect the most troublesome and impertinent of insects, perching upon a twig, and squalling sometimes two or three hours without ceasing; thereby, too often disturbing the studies, or short repose, that is frequently indulged in these hot climates at those hours. The τεττιξ of the Greeks must have had a quite different voice, more soft, surely, and melodious, otherwise, the fine orators of Homer who are compared to it, can be looked upon no better than loud, loquacious scolds." " An insect of this tribe, and I am told a very noisy one, has been found by Mr. Daniel Bydder, in the New Forest, Hampshire. Previously to this, it was not thought that any of these insect-musicians were natives of the British Isles. Captain Hancock informs me that the Brazilian cicadæ sing so loud, as to be heard at the distance of a mile. This is as if a man of ordinary stature, supposing his powers of voice increased in the ratio of his size, could be heard all over the world: so that Stentor himself becomes a mute, when compared with these insects."

Mr. Kirby in concluding this account of the vocal structure of these animals, has the following most just remarks:

" And now, my friend, what adorable wisdom, what consummate art and skill are displayed in the admirable contrivance and complex structure of this wonderful, this unparalled apparatus! The great Creator has placed in these insects an organ for producing and emitting sounds, which in the intricacy of its construction seems to resemble that which he has given to man, and the larger animals, for receiving them. Here is a *cochlea;* a *meatus*; and as it should seem more than one *tympanum.*"

The larvæ of the cicadæ are white, have six feet, and their form is compared to that of the flea. They issue from their nest to bury themselves in the ground, where it appears that they live in the roots of plants. They change then into nymphs which take nourishment, are active and grow. Their wings are enclosed in sheaths, attaching to the corslet, which resembles that which they are to have in their final form. But we do not discover in those which are to become males, the organs of song, nor the auger or ovipositor in those which are to become females. The anterior feet of these nymphs are very remarkable, and adapted to open for them a way under ground, where they are sometimes found sunk to the depth of two or three feet. When they have acquired their entire growth, which, according to some writers, does not take place until the year after they have been changed into nymphs, and as soon as the heat begins to be felt, they issue from the earth, climb on the trees, strip themselves of their envelope of nymph, and pass into the perfect state. At first these young cicadæ are almost entirely green, but by little and little they become of a blackish brown. It appears that under this final form, they live on the juice contained in the leaves and young branches of the trees, into which they plunge their proboscis.

According to the report of Aristotle, the Greeks used to eat the cicadæ, and have the larvæ of these insects served up at their tables. Before the coupling, they preferred the males, and afterwards the females, because then the belly of the latter was full of eggs, which the Greeks esteemed no small delicacy.

The FULGORÆ are remarkable for the beauty and variety of the colours which adorn the elytra and wings of the greater number, and by the form of the head in some species. This part is as singular as varied. In some it presents a saw, and in others a proboscis similar to that of the elephant, and in some others a kind of muzzle; so that one is astonished to find in insects of the same genus such remarkable differences, and this has given rise to the sub-divisions of this genus made in the text.

A species which inhabits Cayenne (*F. Lanternaria*) has, according to the report of Madlle. de Merian, the singular property of shedding during the night, a light so considerable, that the finest characters can be read by it. But this fact is contradicted by many naturalists, who have lived in the country where this fulgora is found, and who assert that it emits no light whatsoever. M. Richard is mentioned, as having reared this same species, and was unable to observe any luminous point about it. Repeated observations would be necessary to remove the doubts which must of necessity be produced by these contrary assertions. It is possible that this insect may be luminous only at certain periods of its life, and at will, like the lampyrides, which cause their phosphoric points to appear and disappear when they think proper.

This is Madlle. Merian's account of this phenomenon:

"The Indians once brought me, before I knew that they shone by night, a number of these lantern-flies, which I shut up in a large wooden box. In the night they made such a noise that I awoke in a fright, and ordered a light to be brought, not knowing from whence the noise proceeded. As soon as we found that it came from the box we opened it; but were still much more alarmed, and let it fall to the ground, in a fright at seeing a flame of fire come out of it; and as many animals as came out, so many flames of fire appeared. When we found this to be the case, we recovered

J. O. Westwood del.

1. *Fulgora punctata*._G. R. Gray.
2. *Fulgora pallida*._G. R. Gray.
3. *Urophora Hardwickii*._D°.
4. *Pœciloptera antica*._Westw.
5. *Lystra pulchra*._G. R. Gray.
6. *Fulgora nigromaculata* D°.

London, Published by Whittaker & C° Ave Maria, Lane. 1832

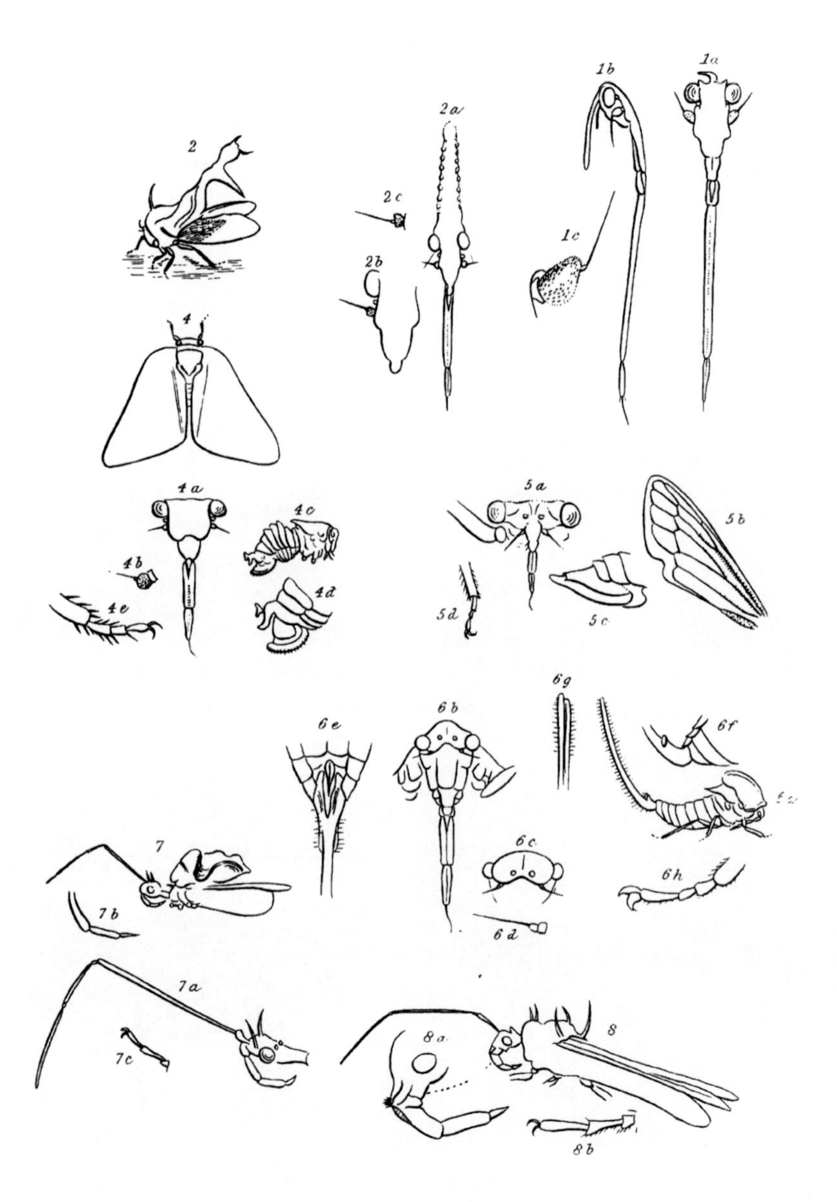

1 Fulgora nigromaculata.
2 Fulgora punctata.
3 Centrotus furcatus.
4 Pœciloptera antica.
5 Darnis Camelus.
6 Urophora Hardwickii.
7 Reduvius dorsalis.
8 Reduvius spinidorsis.

London; Published by Whittaker & Cº Ave Maria Lane 1832.

from our fright, and again collected the insects, highly admiring their splendid appearance."

The veracity of this lady as to facts of which she was an eye-witness has never been brought in question, and it is equally certain from her figures and description, that the species of which she speaks was no other than the one of which we are treating. We cannot therefore, on the authority of others, who assert that they never have observed the phenomenon, dismiss our belief in its existence, or even doubt that the lantern-fly is a luminous insect. The only mode of reconciling these discordant testimonies is the supposition which we have ventured to give as a solution of the difficulty

The luminous points in the lampyrides are placed towards the extremity of their body, whereas it is from the head of the *fulgora* that the light is emitted. Reaumur, who has attempted to discover what could produce this phenomenon, has found in the vesicle which constitutes a part of the head of this insect, only a considerable cavity, and absolutely empty. But this observation made upon an individual which had been for a long time dead, proves nothing, because it is possible that in the living insect this cavity may be filled by a substance, which dries up and evaporates when the insect dies.

The largest fulgoræ are brought into Europe from South America generally, and especially from Cayenne and Surinam. They live on trees. Those which inhabit Europe are very small. They are found upon shrubs and bushes. Their larvæ are unknown. Originally they constituted a numerous species, but they are greatly sub-divided in the "Regne Animal."

We insert figures of three new species of fulgora. First *F, punctata*, which is very pale brown; with black punctures, the lower wings white, the abdomen black, and margin with light brown. This is from the Cape of Good Hope.

The second species we call *nigromaculata*. It has the head and thorax brown; the horn black, short and flexible; the upper wings brown with some small scattered black spots, the exterior margin with a broad black border, the hinder wings, with the basal half, silvery grey, spotted with black; the other half black, with the interior margin brownish black; the body silvery grey, with transverse narrow bands of black. This fine insect is from China, and forms the sub-genus *Aphœna* of M. Geurin.

The third species we name *pallida*. It is yellowish white, the horn long, erect, strong, with the tip black; the suture of the elytra near the base fuscous, apex with some longitudinal brown punctated lines; the legs pale, minutely spotted with black. This species is from India.

We are also enabled to add the following figures of new species which belong to several of the divisions of the genus Fulgora of the text, viz.

To the genus *Lystra* we have added a new species, which we name *pulchra*. It is greenish brown, with the abdomen bright red, the tip with whitish cotton; the upper wings with the base green, spotted with white opaque colour, with a transverse arched band of fulvous yellow; the tip pale yellowish brown; the hinder wings covered with a white wavy matter, with brown spots near the base, and pale yellowish brown at the apex; the nervures are delicate. This species is from India.

To the genus *Pœciloptera* we have added a species, which Mr. Westwood names *antica*. It is dark brown, with a small white spot near the middle of the anterior margin; the lower wings are covered with a bluish powder at the base. This insect appears to belong to a new subgenus of M. Geurin, under the name of *Euryptera*.

In *Darnis* we have figured a species we name *camelus*. It is brown, with the scutellum very large, arch covering the

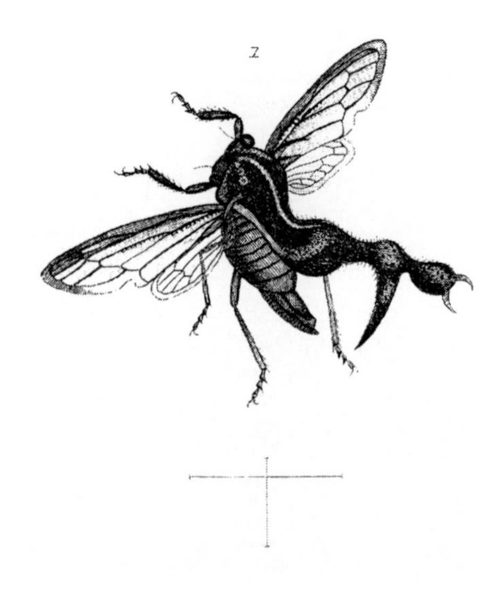

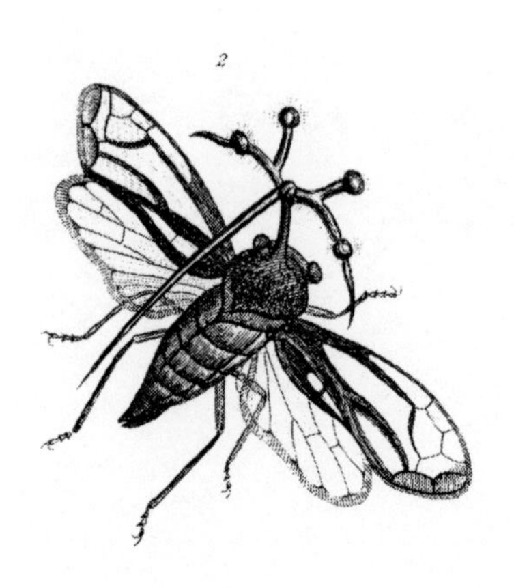

J. O. Westwood del.t

1 Centrotus furcatus. _ G. R. Gray.

2. Centrotus globularis _ Pallas.

London. Published by Whittaker & C.o Ave Maria Lane June 1832.

head, round anteriorly, with the sides projecting, and acute, then narrowing, the posterior end forming a circle, the sides of the latter and legs are yellowish; it is from Mexico.

And to *Centrotus* we have added two species, the first is named *globularis* by Pallas. Its head is black, with the thorax not produced behind, but with an erect cylindrical horn anteriorly, which has four arms, each of which has a small hairy globe at the apex; those on the sides are armed with a small spine, also a longer spine on the body placed posteriorly; the abdomen fulvous, and feet yellow. The second species we have named *centrolus furcatus*. It is brown, with a spine projecting from the body, hanging down, then enlarging again, and bifurcated at the apex. This insect is from Brazil.

To these we have added a figure of a new insect, which must be considered the type of a new subgenus allied to that of Ætalion of Latreille, which we have called *Urophora*, under the specific name of *Hardwickii*. Its characters are: the thorax produced over the head; the posterior legs simple, and with the ovipositor forming a long tube. It is fulvous, with the upper wings spotted between the nervures with brown; the legs and ovipositor obscure; it is from India. Discovered by Major-General Hardwicke.

In the subdivision of CICADELLA of the text, *Cercopis*, we have to notice the species *spumaria*, which former naturalists made a cicada. The following is Dr. Shaw s description, which we shall borrow here:—

"Among the smaller European cicadæ, one of the most remarkable is the *cicada spumaria*, or cuckow-spit cicada, so named from the circumstance of its larva being constantly found enveloped in a mass of white froth, adhering to the leaves and stems of vegetables. This froth, which is popularly known by the name of cuckow-spittle, is found during the advanced state of summer, and is the production of the included larva, which, from the time of its hatching from the egg deposited by the parent insect, continues at intervals to suck the juices of the stem on which it resides, and to discharge them from its vent in the form of very minute bubbles, and by continuing this operation completely covers itself with a very large mass of froth, which is sometimes so overcharged with moisture, that a drop may be seen hanging from its under surface. The included larva, or pupa, (for no material difference can be observed between the two states when arrived at its full growth,) is about the fifth of an inch in length, of an oval shape, with broad head and thorax, and slightly pointed abdomen; its colour is a beautiful pale green, and the trunk or sucker with which it extracts the sap of the plant, may be observed by examining the under part of the thorax, where it will be seen pressed down in a straight direction from the head. When the time arrives in which the animal is to undergo its change into the complete insect, it ceases to absorb any longer the juices of the plant, and to discharge any longer the protecting froth, which at this period forms a vaulted canopy over the insect, instead of entirely investing it as before; the skin of the larva is gradually thrown off, and the animal in its complete form emerges from its concealment. Its size is scarcely superior to that of the larva, but its colour is brown, with a pair of broad, irregular, pale, or whitish bands across the upper wings. If disturbed it nimbly springs to a great distance, and is commonly known by the name of the Frog-hopper, from some fancied resemblance to the colour and shape of that animal in miniature. These insects breed during the month of September, and towards the beginning of October deposit their eggs, which are not hatched till the succeeding spring."

We insert a new *Cercopis*, named *fulvo scutellata*. It is black, with the anterior margin of the band across the

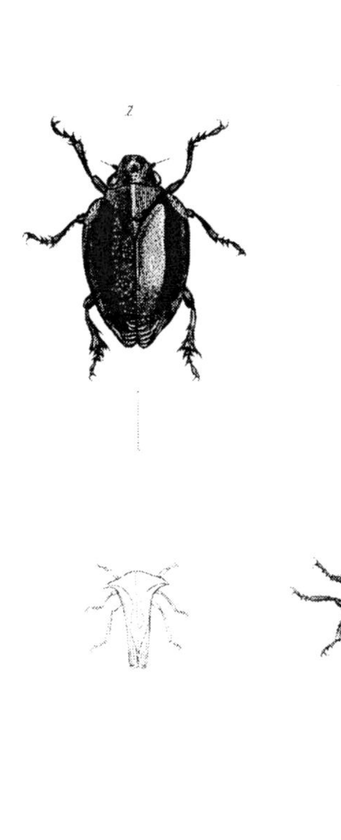

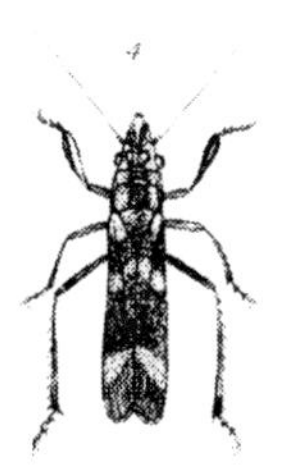

J.O. Westwood del^t

1. *Orthoraphia cassidioides.* — Westwood.
2. *Cercopis fulvoscutellata.* — G.R. Gray.
3. *Darnis Camelus.* — G.R. Gray.
4. *Cicada fulvofasciata.* — G.R. Gray.

London Published by Whittaker & Co Ave Maria Lane 1832

thorax scutellum, and three quarters of the elytra fulvous, but the latter with some black spots.

We have also figured a new Brazilian species, which Mr. Westwood has named *Orthorapha cassidioides*. It is of a dark green colour, with a yellow spot at the base of each elytron; the tip with some short wavy lines, also with a red spot on each side, near the middle; the legs reddish. This insect differs from the genus Cercopis, by the body being very convex and rounded, also by the want of ocelli and the straightness of the suture of the elytra.

We shall now notice the CICADELLÆ proper, or TETTIGONIA.

The tettigonia are leaping insects of small size, and which feed, in all their different states, on the juice of the leaves of different vegetables. Many species, especially among the exotics, have very agreeable and very various colours. The larvæ and nymphs do not differ, the wings excepted, from the perfect insect. The females have at the posterior extremity of their abdomen, an auger which enables them to sink their eggs in such parts of vegetables as are proper for the nourishment of their posterity. The species of this genus, which are peculiar to Sweden, have been published by M. Fallen, but much yet remains to be done towards its complete illustration.

One of these species, the *tettigonia rosæ*, is very small, being scarcely a line and a half in length. Its body is altogether yellow, or of a yellowish green, and sometimes almost white. How slightly soever the yoke-elm, or the hedges formed of that tree may be touched in summer, an immense number of these insects are seen to jump, or flit about them. The female deposits under the leaves of the rose-bush about three hundred eggs, from which larvæ spring, which feed upon their juice. Being thus exhausted, these leaves assume upon their upper surface, the colour of a silvery white. It has been observed, that the mining caterpillars, which form meandering galleries in the tissue of the leaves, will not attack those in which the larvæ of the tettigoniæ are lodged. The little moth, then, which is proceeding to lay her eggs on these leaves, is well aware that another insect has already taken possession of the aliments of which she is in search for her young.

We now come to the family of the Aphidii, all of which are more or less distinguished by the havock which they commit on vegetable substances.

The Psyllæ of Geoffroy, which Linnæus names *chermes*, and Degeer and Reaumur *faux-pucerons*, are little insects which are found on different plants, such as the box-tree, the fig-tree, the alder, the broom, and the nettle. At the first glance, they have some resemblance to the aphides, and they jump with tolerable liveliness, by means of their hind legs, which act as a sort of spring. When it is attempted to seize them, they escape quickly, rather by jumping than flying; it is from this circumstance that they have been named *psyllæ*, which is the Greek word to signify *flea*. In all their different states they feed upon the juice of the leaves, which they suck up with their proboscis. The body of their larvæ is greatly flatted, the head broad, the belly very flat, and rounded at the end. Their six feet are terminated by a sort of bladder, and two crooks. These larvæ change into nymphs, which have, towards the sides of the breast, four broad pieces, serving as sheaths to the elytra and wings. These nymphs are capable of walking. Many of them, as well as the larvæ, have the body covered with a white and cottony substance, which hangs in flakes. Their excrements are in the form either of threads, or masses of a gummy matter. To undergo their final metamorphosis, the nymphs attach themselves under a leaf. There they remain tranquil, until their skin, which divides in a portion of its length, gives a passage to the perfect insect.

Many females are provided with an auger, which serves them to prick the leaves in which they deposit their eggs. These incisions, like those which the *Cinips* make in plants, produce excrescences or tuberosities. They are often to be seen on the summits of the branches of the fir-tree. They are formed by the extravasation of the juices which accumulate in this part. The larvæ and the nymphs live in sorts of galls, which contain a great number of little cells. The leaves of the pine-tree nourish larvæ of the same genus. These last are not enclosed like the preceding. They have merely on the body a white down, which forms, as it were, a sheath, under which they are sheltered.

The psyllæ which live on the box-tree, do not produce excrescences such as we have been describing; but their prickings cause the leaves of the extremities of the branches to turn up like a hood. and many of them thus to unite together, to form a sort of ball, in which the larvæ remain enclosed. These larvæ render, by the anus, a white and saccharine matter, which softens under the fingers. This substance, according to Geoffroy, has some resemblance to manna. They have often long threads of it behind, and little grains of it are to be found in the balls which they have inhabited.

An insect much approximating to the *psyllæ*, which may as well be noticed here, and which M. Latreille at first placed with this genus, produces a remarkable monstrosity on the *junicus articulatus* of Linnæus. Its prickings destroy the parts of the flower of this plant, cause them to acquire a triple or quadruple development, more than they would have naturally, and make them assume the form of little balls, or flowerets of gramineous plants. The resemblance is so much the more striking, as the extremities of the divisions of the corolla terminate in an elongation imitating beards. These sorts of galls enclose a tolerably great number of these insects, of different ages, which feed upon the juices of the plant. The larvæ send forth from the anus, a farinaceous and very white matter. They resemble those of the *Psylla ficus*; of these insects, M. Latreille has formed his subdivision LIVIA.

The psyllæ do not, as far as it appears, produce more than one or two generations at the most in the year. The females survive the winter.

The larvæ of the *psylla alni* live in societies, composed of about a dozen individuals each, upon the alder-tree. If we observe at the commencement of May the shoots of this tree, the pedicles of its leaves, and even the under part of them, we shall perceive a very soft, white, and cottony matter, which seems to be attached to the tree. But if touched ever so slightly, it begins to stir, and divide into several parts, and we discover that these little flakes are only the habits, or the covering of several insects. This cottony down occupies more room than their body, and renders them hideous. It is composed of very fine threads, curved or frizzled behind, towards the head, and many of which are united in the form of pencils, floating over the body. The extremity of these hairs is fine, while that of the hairs of the larvæ of some other psylla is thick, and rounded at the end. This substance increases with the age of the insect, and easily attaches itself to the bodies with which it meets. Although it covers the entire body, it only takes its origin from the posterior rings, or the region of the anus. There no doubt are some excretory glands, and species of spinnerets. The reproduction of this down is very rapid. If it be removed from the upper part of the animal, a new, and tolerably long coat of it will make its appearance at the end of a quarter of an hour. It often happens in the moulting, that the old skin, loaded with its down, remains engaged in the new substance, which begins to form upon the insect as soon as it is stripped. The excrements issue by little and little from the anus, always remain attached to the hinder part of the body, and from these one or two small masses of a yellowish white, a little transparent. This mass is sometimes elongated, irregular, and a little curved. Sometimes it resembles a ball, in the form of a transparent drop. These excrements are at first like a thick syrup, and afterwards become hard. They dissolve in water, and have a saccharine and somewhat acrid taste. The excrements of the *Psylla buxi* are in the form of tortuous threads, and resemble vermicelli.

The insects of the sub-division THRIPS are very small. They live on flowers, and on the barks of trees, where their larvæ are also found. They do not differ from the perfect insect, but by the deficiency of wings and elytra.

We have now to speak of a very destructive race of insects, the APHIDES, *pucerons* or *plant-lice*. They are small insects, commonly found assembled in immense quantities upon almost every species of plant. They are dull, and walk but little. They are to be seen, motionless, and forming great masses upon the stalks and leaves. The most celebrated naturalists have written the history of these insects, which exhibit some singularities well deserving of attention. The first which may be remarked without any great train of observations, is that in the same species, females are to be found both winged and apterous. These last, which might be taken for nymphs, are perfect insects, equally in a state for reproduction, as those which are winged. Another singularity of these insects is, that during one period of the year, both these kinds of females are viviparous, and during another, they lay eggs, which appear destined to perpetuate the species, as the eggs last through the winter, while the living individuals perish. These females couple in autumn, and after this they are oviparous; during the whole summer they are viviparous. The winged, and the apterous females equally produce young ones, which become winged, and others which are to remain always apterous. The fecundity of these females is prodigious. They produce fifteen or twenty young ones in the course of the day.

The third peculiarity of these insects, the most astonishing of all, and the one which has caused them to be remarked with the greatest attention, by Bonnet, Reaumur and Lyonnet, is that they can reproduce without sexual intercourse, and it appears that the female who has received one impregnation transmits the influence of it to her female descendants for a series of generations. The observers just cited, have taken young ones when they had just proceeded from the belly of the mother, have brought them up in the most perfect solitude, and have seen them produce others, which afterwards being reared separately and successively, have been fruitful during many generations, without having had any communication with any individual of their species. Bonnet, who has studied these insects with the greatest degree of attention, has witnessed the birth of nine generations in three months, from a single act of sexual intercourse. However extraordinary this fact may be, that animals should be able to perpetuate their species without the union of the sexes, we can nevertheless, not entertain the slightest doubt of its truth, attested as it is by many witnesses worthy of the highest credit. Reaumur has proved, that in five generations one aphis may be the parent of 5,904,900,000 descendants.

As soon as the aphides are born, they walk, and proceed to seek upon the plant a place in which to fix themselves and suck. As they are fond of living in society, they place themselves always one beside the other. They remain about twelve days under the form of nymph, during which they change their skin four times. After having quitted the last skin, they are in a state to reproduce. Assembled on the leaves or stems of trees, the aphides appear to be in a state of inaction ; but they are busily occupied in extracting the juice with their proboscis. Frequently, their prickings cause very sensible alterations and damage to the leaves, and even to the stems themselves of the trees. Those which live on the linden-trees, attach themselves to the young shoots, on which the little ones arrange themselves in proportion as they are born. They place themselves in file one after another on one of the sides of the shoot ; they cause the new stem to assume different curves, and lodge themselves in the cavities which it forms. We often see on gooseberry-bushes, and apple-trees, some leaves covered with tuberosities. It is the aphides which produce all this. On the leaves of the elm, they produce vesicles, or sorts of hollow galls, commonly of the size of a walnut, and sometimes as thick as one's fist. These galls are not inhabited by the little ones only, as are the galls of the cinips. They also enclose the mother, which lodges there to lay her eggs.

Almost all the aphides are more or less covered with a cottony down. Those which live on the cabbage and the plum-tree, have but very little of this substance, which resembles flour. Those of the vesicles of the elm, are entirely covered with it. This same substance is found on the aphides of the poplar, in the form of cottony threads ; but no species is provided with so large a quantity of it, as that of the beech-tree. These threads are sometimes an inch in length, and are floating on the body of the insect, to which they are but slightly attached. They may be easily removed by friction.

Every where, where the aphides are found, they are sure to be found accompanied by ants. These are attracted by their taste for the saccharine fluid which flows continually from the two horns which the aphides have at the abdomen. So great a quantity of it issues forth, that the vesicles of the elm, and the tuberosities of the leaves of the gooseberry and currant-bushes, contain drops of it of the size of a pea. This fluid, which is limpid and transparent, grows thick in the air. Reaumur tells us, that it is as sweet as honey, and of a more agreeable flavour.

The aphides are extremely numerous, and would be still more so, were it not for the terrible enemies which devour them daily by hundreds. The larvæ of the *hemerobiu*, and those of some diptera of the tribe of *syrphiæ*, in gratifying their own appetite, deliver the agriculturist from a dreadful scourge; for, otherwise, these fruitful insects would multiply to such a degree, that they would finish, by totally drying up the plants, which as it is, they injure and disfigure to so great an extent. The aphides being very soft, may be removed with a moistened brush, and thus trees of no great elevation, may be easily purged of them. But a more expeditious and facile method is, to burn under the trees some sulphur, or tobacco, and to conduct the vapours or the smoke to the affected parts, with a bellows or a funnel.

The *Aphis ulmi* live assembled in great numbers, in a vesicle attached to the leaves of the elm, by a very short pedicle. This vesicle is produced by the extravasation of the juices of the leaf, pricked by these aphides.

The *Aphis populi* is found in large quantities, on the leaves of the black poplar, enclosed in a leaf folded in two, which forms a sort of vesicle. Each leaf, besides, is covered with reddish tuberosities.

The *Aphis quercus* is very small, and of a reddish brown. The great peculiarity of this insect is its proboscis, which is three times longer than its body. It carries it under the belly, and its extremity is raised upon the back; it shortens and elongates it at will, and sinks it so deeply into the bark of trees, that it cannot be drawn out without taking away with it a small fragment of the wood. This aphis has no horns.

"In the quality of the excrements," says Mr. Curtis, in

the sixth vol. of the Lin. Trans., "voided by these insects, there is something very extraordinary. Were a person accidentally to take up a book in which it was gravely asserted that in some countries there were animals who voided liquid sugar, he would soon lay it down, regarding it as a fabulous tale, calculated to impose on the credulity of the ignorant; and yet such is literally the truth. The superior size of the *Aphis salicis* will enable the most common observer to satisfy himself on this head. On looking steadfastly for a few minutes on a group of these insects, while feeding on the bark of the willow, one perceives a few of them elevate their bodies, and a transparent substance evidently drop from them, which is immediately followed by a similar motion and discharge, like a small shower, from a great number of others. At first, I was not aware that the substance thus dropping from these animals, at such stated intervals, was their excrement, but was convinced of its being so afterwards; for on a more accurate examination, I found it proceed from the extremity of the abdomen, as is usual in other insects. On placing a piece of writing-paper under a mass of these insects, it soon became thickly spotted: holding it a longer time, the spots united from the addition of others, and the whole surface assumed a glossy appearance. I tasted this substance and found it as sweet as sugar. I had the less hesitation in doing this, as I had observed that wasps, flies, ants, and insects without number, devoured it as quickly as it was produced; but were it not for these, it might no doubt be collected in considerable quantities, and if subjected to the processes used with other saccharine juices, might be converted into the choicest sugar, or sugar-candy. It is a fact also, which appears worthy of noticing here, that though wasps are so partial to this food, yet the bees appear totally to disregard it.*

* This, however, is contradicted by Mr. White, in his History of Selborne.

"In the height of summer, when the weather is hot and dry, and aphides are most abundant, the foliage of trees and plants (more especially in some years than others) is found covered with, and rendered glossy by a sweet clammy substance, known to persons resident in the country by the name of *honey-dew:* they regard it as a sweet substance falling from the atmosphere, as its name implies. The sweetness of this excrementitious substance, the glossy appearance it gave to the leaves which it fell upon, and the swarms of insects which this matter attracted, first led me to imagine that the honey-dew of plants was no other than this secretion, which further observation has since fully confirmed. Others have considered it as an exudation from the plant itself. Of the former opinion we find the Rev. Mr. White, one of the latest writers on natural history that has noticed this subject. But that it neither falls from the atmosphere, nor issues from the plant itself, is easily demonstrated. If it fell from the atmosphere it would cover every thing indiscriminately; whereas we never find it but on certain living plants and trees. We also find it on plants in stoves, and green-houses covered with glass. If it exuded from the plant, it would appear on all the leaves generally and uniformly; whereas its appearance is extremely irregular, not alike on any two leaves of the same tree and plant, some having none of it, and others being covered with it but partially. But the phenomena of the honey-dew, with all their variations, are easily accounted for, by considering the aphides as the authors of it. That they are capable of producing an appearance exactly similar to that of the honey-dew, has already been shewn. As far as my observation has extended, there never exists any honey-dew but where there are aphides; such, however, often pass unnoticed, being hid on the under side of the leaf. Wherever honey-dew is observable about a leaf, aphides will be found on the under side of the leaf or leaves immediately above it, and under no other circumstances whatever. If by accident any thing should intervene between the aphides and the leaf next between them, there will be no honey-dew on that leaf. Thus then we flatter ourselves to have incontrovertibly proved that the aphides are the true and only source of the honey-dew."

"We have found that where the saccharine substance has dropped from the aphides for a length of time, as from the *Aphis Saliois* in particular, it gives to the surface of the bark, foliage, or whatever it has dropped on, that sooty kind of appearance which arises from the explosion of gun-powder, which greatly disfigures the foliage, &c., of plants. It looks like, and is sometimes mistaken for, a kind of black mildew. We have some grounds for believing that a saccharine substance similar to that of the aphis drops from the coccus likewise, and is finally converted into the same kind of powder."

We now come to the last, and not the least, remarkable genus of this order, namely, Coccus, or the *cochineal* insects. These are very small and delicate insects, and singularly remarkable for the difference of form in the sexes. The body of the males is elongated, the head round, the eyes small, the antennæ tolerably long, and with eleven distinct articulations. They have no apparent organs of manducation. The wings are long and horizontal, with remarkably fine nervures. The females, in their earlier age, have the body ovaliform, apterous, flat, and the antennæ short, with articulations very little distinct. Their mouth consists in a small conical bill, which is formed of a sheath, and a sucker of three silken threads or hairs. It is by means of this bill that they feed upon the juice of vegetables, to which they are very hurtful, in consequence of their great multiplicity. These individuals fix themselves at the period of their amours on the plant or tree, which serves them as a habitation. Their body swells prodigiously, assumes the form of a gall, which covers the little ones, and ceases to be animated.

This extraordinary change, which takes place in the individuals of this sex, has caused these little animals to be named *gall-insects.* Properly speaking, this denomination was applied by Reaumur and his followers only to the insects of this family, the females of which, in the gall state, present no appearance of a ring, and whose form in this respect is still more remote from that of an insect. The females, which do not preserve, when they are arrived at this metamorphosis, any vestiges of their primitive figure, such as the females of the cochineals, are called by Reaumur *pro galles-insectes*, or false gall-insects. This distinction has served as a basis to Geoffroy for his two genera, *Kermes* and *Cochineal*, or *Coccus*. We must observe, however, respecting this denomination of kermes, that the insects so called by Linnæus, are not the same as those so called by Geoffroy and Olivier. The *chermes* of Linnæus, are the *psyllæ* of the latter, and of the " *Regne animal.*"

M. Latreille remarks, that in all probability fresh observations will produce fresh proofs of the natural distinction between the kermes and cochineal—but that in the actual state of the science, it is far from easy to trace with precision the line of demarcation. The gall-insects, which are very much tumefied, will not naturally present annular segments and other vestiges of their primitive form. The female gall-insects, whose brood shall be less numerous, will also be less voluminous. They will exhibit less expansibility in their skin. Their form will be less gibbous, and approximate more to that of their early age. Their wings will not be wholly obliterated. If we examine the figures of the different species of kermes and cochineal, we shall easily see that the variety of forms is very great. But this rather increases the difficulty of finding clear and distinct characters. The males of the kermes and cochineal are so very similar, that a generic identity may be presumed with respect to them, at least, until new observations shall throw some additional light upon this subject.

The cochineals, as well as the kermes, pass a great portion of their lives attached to the bark of trees, from which they extract the sap with their proboscis, without making any sensible motion. Once fixed in a place, they no longer quit it. There they couple, increase in size, lay their eggs, and die. The young ones remain some time under their body, as the young kermes do under that of their mother. What distinguishes, however, as we have already mentioned, these insects from the kermes is, that the female cochineals in acquiring growth, always preserve the figure of an animal, whereas the kermes entirely lose the insect form, and assume that of a berry, or a gall.

Cochineals are usually found in the bifurcations, and under the small branches of trees. They have acquired their full bulk towards the end of spring, or the commencement of summer. Then they resemble a little convex mass, more or less oval, in which it is impossible to distinguish, even with the microscope, either head or feet, but merely the segments which divide the body. Some species are covered with a cottony down, which forms a sort of nest, in which a part of the body of the insect is lodged. This down also serves to receive the brood. The eggs, of which each female lays many thousands, proceed from the body of the mother, through an aperture placed at the extremity of the abdomen, and they pass under her belly to be hatched there. After the laying, the body of the mother dries up, its two membranes become flat, and form a sort of shell or cocoon, in which the eggs are enclosed. If these eggs are bruised on white paper, the greater number will colour it of a red, more or less deep. After the death of the mother, the little cochineals very soon proceed from under her body. As soon as they have sufficient strength, they spread themselves over the leaves to extract the juice with their proboscis. In their youth, or so long as they are under the larva form, they are lively enough, and change place but to pass into their final form ; the females fix themselves, remain motionless, pass the winter thus, and couple in spring.

The males are much less numerous and much less known than the females, which they resemble before they undergo their metamorphosis. Fixed like them on the plant, without taking nourishment or acquiring growth, their skin hardens and becomes a shell, in which the change takes place which distinguishes them when they become perfect insects. Under their new form they are very different from the females. Their body is one half smaller, and they have two tolerably large wings. They are found but seldom, as they do not survive the sexual intercourse for any length of time. As soon as the male has acquired wings this intercourse takes place, and when he has fulfilled the grand end of his being he perishes. The females then grow large, and very speedily deposit their eggs.

Of all the cocci, there are but two species which are employed in the arts. The others are known only by the injuries which they cause to divers plants, more especially to orange-trees, fig-trees, and olives.

It is to the New World that we are indebted for the most precious cochineal, that with which the finest colouring is produced, of all the shades of scarlet and purple. This insect furnishes a branch of commerce so considerable, that in 1736 seven hundred thousand pounds weight of it was imported into Europe. It was employed for a long time before its true nature was discovered ; and it is demonstrated from many passages of Pliny, that that naturalist believed along with the vulgar, that the coccus, or rather the kermes, which was derived from Portugal, from Sardinia, from Asia Minor, and from Africa, was the fruit of a tree. But those who subsequently observed it with attention, soon began to suspect that it was an animal.

The cochineal of commerce, or the *coccus cacti*, is brought up in Mexico, the only known country in which it is gathered. It is brought into Europe in the form of small grains, of an irregular figure, commonly convex on the one side, on which little channels or flutings are perceptible, and concave on the other, with depressions more or less deep. The colour of that which is most esteemed, is a slate-grey, mingled with reddish, and covered with a white dust. Two species of cochineal are distinguished, the finer kind, known under the name of *Mesteque*, because it is gathered at Meteque, in the province of Honduras, and the *woodland*, or *wild cochineal*. The first is obtained, if we may so express ourselves, only by cultivation, or by rearing the insect on plants cultivated for that especial purpose. The other is gathered on plants which grow naturally, as the kermes is gathered upon shrubs, which multiply without the intervention of man. We are still ignorant whether these are in reality two distinct species or not; we only know that the latter is less dear, because it furnishes less colour. This is attributed by some writers, not to any natural inferiority in its colour, but to the quantity of cottony matter with which it is covered, and which at once augments its weight, and absorbs a portion of its colour.

The plant on which the fine cochineal is reared is called nopalli by the Indians, and is the *cactus opuntiæ*. Its articulations are but little spiny, they are oblong-oval, compressed, and fleshy. Its flower is small, and of a blood-red. It is to the juice of this plant that the colour of the cochineal is attributed. The Indians of Mexico eat its fruit, and that of the majority of the cactus tribe, as well as the germs of their flowers. This cactus is reproduced by slips, and issues from its leaves, which are put into the ground. Its culture consists in weeding out the injurious herbs which surround it. It may be planted in argillaceous or in gravelly soils, or in such as are filled with flints; but it thrives better in a good soil, especially when it is sheltered from the northern winds. This shrub grows fast: in six years it becomes several feet in height, and it is in a fit state to nourish the cochineal eighteen months after it has been planted. But at the end of six years it must be renewed, as the younger it is, it is the better suited for the insect. This cactus is seen no where in the open country from Teguahacan to Guaxaca; it is only found in the gardens of these countries, and at St. Juan del Rey. In 1787 it also existed at St. Domingo, as well as that which the Indians name the *Nopal of Castille.*

The Indians of Guaxaca and of Oxaca, who devote themselves to the cultivation of the cochineal, plant these shrubs near their habitations. The most considerable of these plantations does not contain more than an acre and a half, or two acres at most. A single man is sufficient to keep one in a proper state. The cochineal is sown (if we may be allowed such a phrase) on the nopals towards the fifteenth of October, which is the epoch of the return of the fine season in Mexico. This operation consists in placing on the plants the females which have already some young ones. These females are cochineals of the last brood, which the Indians preserve on branches of the cactus, and keep them in their houses during the rains, which would destroy them if they were left abroad. Nevertheless, in some districts, they remain in the plantations, where they are carefully preserved with mats, from the inclemency of the atmosphere.

The manner of setting the cochineal, is to put eight or

ten females into a little nest, made with a sort of flax, taken from the petals of the leaves of the palm-tree, or some other cottony matter. The nests are placed between the leaves of the cactus. They are attached to the thorns with which these leaves are armed, and the bottom of the nest is turned towards the rising sun, that the little family may be thė sooner disclosed. From these nests proceed a great number of these cochineals, for each female produces some thousands of them, which are not larger than the point of a pin, of a red colour, and covered with a white dust. The young cochineals spread themselves quickly over the leaves, and soon attach themselves there altogether. When they are once fixed, should they be deranged by any accident, their proboscis, which is sunk into the plants, breaks, and the insects perish.

The females live about two months, and the males only half as long. Both remain ten days in the larva form, and then become perfect insects, proper for reproduction. The females, in changing their state, do not change their form. They only lose their skin to get another, whereas the males quit their envelop of nymph winged. Up to this period there is nothing to distinguish them from the females, except their being one half smaller. Once become winged insects, they couple and die. The females, which still survive another month after fecundation, acquire their large growth during this time, and perish after they have given birth to their young ones.

There are, according to M. Thierry, six generations of these insects in the year. They may all be gathered, if the rains do not disturb and destroy their posterity; but all authors agree concerning the number of crops which are gathered, and which are three every year. The first takes place towards the middle of December, and the last in the month of May. In the first the nests are taken away from the upper part of the trees, to take out the mothers which have been put there, and which are dead. They wait, to make the second gathering, until the cochineals begin to produce their young. In this operation they employ a knife, the edge and point of which are blunted. That the plant may not be damaged, the blade of the knife is passed between the bark of the nopal and the cochineals, to cause them to fall into a vessel. After this they are dried.

The Indians employ many processes to destroy these insects, which they are in a hurry to kill for fear of losing a part of their harvest. The mothers, although detached from the plants, may still survive some days, and produce young ones. These young ones would soon disperse, and be so much deducted from the weight of the cochineal which has been gathered. Some Indians put these insects into a basket, plunge them afterwards into boiling water, and after having taken them out, expose them to the sun to dry. Others put them into a hot oven, or on chafing dishes; but the boiling water appears to be the most efficacious method. On those different methods of destroying the cochineals, depend principally the different colours of those which are brought into Europe. The living insects being covered with a white powder, those which are destroyed in the boiling water lose a portion of this powder. They afterwards appear of a red brown, and are called *renagrida*. Those which are destroyed in the ovens do not lose this powder. They remain of an ashen-grey, and are named *jarpeada*. Those which die upon the plants become blackish, as if they were depilated. They are called *negra*.

The dead mothers which have been taken from the nests placed upon the nopals, lose more of their weight in drying than do the cochineals which have been taken alive and full of young ones. In drying four pounds weight of the first, they are reduced to one pound; and three pounds of the first, lose only two-thirds in desiccation. When the cochineals are dried, they may be kept shut up in coffers of wood for ages, without being deteriorated, or losing a particle of their tinctorial property.

The wild cochineal is a less bulky insect than the other All its body, except the under part of the corslet, is covered with a cottony, fine, and viscous matter, and bordered with hairs all around. Eight days after it is fixed, the hairs and the cottony matter elongate, and are cemented to the plant, so that one would think that they saw there as many little white flakes as there are insects. In one place some are separated from the others; in another, a hundred of them may be seen grouped together. The group augments in volume in proportion to age, and sticks so close to the plant that when it is desired to detach the cochineal, a part of the cotton which covers it is left upon the plant.

Although this cochineal grows naturally on a thorny cactus, the Indians cultivate it as they do the other, and bring it up upon the nopal of the gardens, because it is more easily gathered. The most skilful workman cannot gather from the thorny opuntium, a sufficient quantity of it each day to make ten ounces when it is dried; whereas he may obtain three pounds dried, when he gathers it on the nopal of the gardens. The cultivators also find another advantage in this; it is, that when reared on this plant the wild cochineal becomes almost as large as the fine one, and in proportion as it is reproduced it loses a part of its cottony substance. As this nopal, and that of Castille, thrive in the western colonies, and as we are assured that the wild cochineal is also found there, in many districts, it would certainly be desirable that its cultivation should be regularly established there.

There is a species of cochineal which lives only in cold countries, which it seems to prefer to temperate climates. It inhabits Poland. Formerly, before the Mexican species was known, it was employed in the business of dyeing. The crops were neither so abundant, so regular, nor so easily gathered as those of the true cochineal. This cochineal is known under the name of *coccus tinctorius Polonicus*. It is found on the root of the plant, which Ray has named *polygonum cocciferum*. Some authors assert, that the same or a similar cochineal is found on the root of the *scleranthus perennis*, of the mouse-ear, of the pimpernel, and of the pellitory of the wall. This grain of scarlet, as it is sometimes called, is gathered at the commencement of summer. Each grain is then pretty nearly spherical, and of a purple colour. The largest are almost of the bulk of a grain of pepper. Each of them is said to be partly lodged in a sort of calix, as an acorn is in its calix. The external part of this envelope is rough; the internal part is polished. Sometimes not above one or two of these grains are found upon the plant, sometimes more than forty. Observations have shewn that from those little grains insects issue forth, which have two antennæ and six feet; that at the end of some days these insects grow shorter, and cease to walk; and that when they have become immoveable, their body is covered with a cottony down, similar to that which surrounds the body of the *coccus ulmi*. The males of this species are similar to the males of certain species of kermes, and couple like the other cochineals. It has been observed, that the females do not cover themselves with down until after they have been fecundated, and those which have not been so remain naked. Both, however, lay eggs, but it is only from those of the first kind that young ones issue. What distinguishes this cochineal from the other species is, that after having been round and immoveable, it again moves its feet and changes form, from round becoming oblong. This insect is gathered only every two years, immediately after the summer solstice, because then it is full of a juice of a purple colour. For this operation a sort of spade is employed, with which the plant is raised, with some earth to detach the cochineal from it. Afterwards the plant is replaced in the same situation, for fear of destroying it. When the cochineal has been separated from the earth by means of a sieve, it is wetted with vinegar or hot water, and afterwards exposed to the sun to cause it to die and dry up. It is said that the Turks and Armenians purchase this drug, and make use of it to tincture silk, wool, leather, morocco, and the tails of their horses, and that the women extract a dye from it with lemon-juice or wine, and use it to redden the extremities of the feet and hands. It is also reported that formerly these people purchased this coccus very dearly, and that they employed it, mixed with half the quantity of the Mexican coccus, to dye cloth scarlet; and that of the dye of this insect, extracted with lemon-juice, or a lixivium of alum, it is possible with chalk to make a kind of lac, capable of being employed in painting, and that if gum-arabic be added, it will be as fine as Florence lac. Finally, it is said that the juice expressed from the cocoons of the polygonum serve in medicine the same uses as the kermes. Notwithstanding, however, all the properties of this cochineal, no other is, at present, used for fine dyes but that of Mexico.

In Russia the inhabitants also extract a crimson dye, from an indigenous species of cochineal. In France no attempt has been made to procure any from the insects of this genus which are so hurtful to orange and other trees. It is by no means improbable that they would furnish a colour which, without being so beautiful as that of the American cochineal, might yet prove of some utility. According to some writers, it is also to a cochineal that lac is owing, a species of gum which comes from the East Indies; but of this there is no very positive proof.

The elm nourishes a species of cochineal which has much resemblance to that of the nopal. It is principally found in the bifurcations of branches which are a year or two old. Towards the middle of summer, the cochineals which have acquired their full growth, resemble a little oval, convex mass, of a brown red, and about a line in length. They are surrounded with a sort of white and cottony cord, which only allows the upper part of the body to be discovered. This substance contains the belly of the insect, and serves as a nest for the young. Reaumur thinks that the females are viviparous; but according to Geoffroy they are oviparous. Towards the middle of July a great number of living young ones are found in the nest, of a yellowish white. They have two antennæ, and six short feet, with which they walk tolerably fast. It would appear, that a day or two after its birth, each little one quits the nest to run upon the branches of the elm, where a great quantity of them are discovered. But it is not a very long time before they all become fixed there. Their growth, as in the other species, does not take place until after the winter. At the commencement of spring their body is a little reddish. Each ring is edged with grey and short hairs, which disappear to give place to the cottony matter which forms the nest. It appears likely that this matter escapes from the body of the insect, as it does from that of the aphis and the kermés. The eggs, in proceeding from the body of the mother, pass under her belly as they come out, and the little ones quit her as soon as they have strength enough to repair to the branches. As soon as the female has finished her laying, she dies, dries up, and subsequently falls from the nest.

The *coccus ficus caricæ* is found in the South of Europe, and all the Levant. These insects produce the worst effect on the fig-trees. They dry them up by pumping out the juice of those trees, and occasioning an extravasation of a considerable portion of the sap. Also those which have been for some time infested by them, lose their leaves sooner than the others. In the new shoots the interval of the knots becomes smaller every year. The number of figs diminish, the fruits fall for the most part without ripening; the leaves and branches are covered with black spots; the bark is detached and scales off; in fine, when the trees are thus brought to a certain degree of weakness, the winter finishes their destruction. Many means have been devised for getting rid of these destructive insects; but their inefficiency has been proved by their very imperfect success. Some cultivators rub the branches and leaves with vinegar, and the lye of oil; but the numerous descendants of these insects survive all the means which may be employed for their destruction. It is only during winter that they can be attacked with any advantage, by rubbing with a cloth the shoots on which they are found, and crushing them, or detaching them with a knife, or a piece of wood a little edged. This operation, which would be neither costly, nor long, would be so much the more easy in this season, as the cochineal is then but slightly attached to the tree.

Those which are attached to the figs grow more rapidly than the others. The figs which have been attached by them, can scarcely be ventured upon to eat, as they cannot be gathered without crushing some one of these insects, from which issues a thick reddish matter, offensive and revolting. As for dried figs, as care is taken to stir them well upon the hurdles, the cochineals are more easily detached from them, and more especially, as the connecting substance which attaches them, grows less adhesive, in proportion to the weakness of the insect.

The *Coccus farinarius* is found in Europe on the branches of the alder. This cochineal covers itself almost entirely, when fixed, with a white and cottony stratum, which extends even from the region of the anus, much beyond that extremity of the body. The eggs are deposited upon this soft nest, and accumulated one upon the other. They thus at once find a bed and a covering, which protect their frail existence. The laying being finished, the mother dies and dries up by little and little. Degeer stripped one of these cochineals of the cottony matter which covered its back. A similar bed, but less thick, made its appearance on the following day, a proof that nature has provided these animals with a considerable portion of this substance. It is somewhat glutinous, and when we try to remove it, many of its threads will remain adherent to the leaf on which the insect is fixed, or to the body itself.

Notwithstanding the length to which this account of the cochineals has extended, it is impossible to conclude without some notice of the *Kermes*, which so long as an article of commerce enjoyed that reputation which has since more deservedly devolved upon the South American *Coccus*. The Kermes more resembles a gall, than any of the cochineals, having the body so much distended, that it presents no vestige whatever of an incision. This point excepted, the characters of the two are identified, and we must confess that we see but little reason for the generic separation made between them by Geoffroy and Reaumur.

In their youth, the females resemble little white *wood-lice*, which would have but six feet. They run upon the leaves, and afterwards fix upon the stems and branches of trees and shrubs, where they pass many months in succession. It is then that they assume the figure of a gall, or excrescence.

It is upon such shrubs and plants as survive the winter, hat th ese insects grow. They need a plant which shall nourish them for nearly a year, that being the time fixed for the duration of their existence. Having acquired their growth, some of them resemble little balls attached against a branch, the size of which varies from that of a pepper-corn, to a pea. Others have a spherical form, but truncated or elongated. Some are oblong, and others, by far the greater number, resemble an inverted boat. The colours are diversified.

Fruit-trees, and peaches more especially, are sometimes so much covered with Kermes, whether of that species like the inverted boat, or the other, like small grains, that their branches appear altogether scabby. These insects do not arrive at the term of their growth until the middle, or at latest towards the end of spring. If the peach-trees be observed at this period, we may remark tuberosities on their branches, which are Kermes, some of which are living and immoveable, and others dead from the preceding year. These may be distinguished from each other, in that the first are extremely adherent to the plant, and that the place where their body is attached, is covered with a cottony matter, on which their belly, which is as much inflated as possible, is applied. If these insects are observed a little later, their skin appears nothing but a simple dried shell, containing or covering an infinity of little, reddish, oblong grains, which are eggs. The little ones which come from them, still remain for a few days under the skin of the mother.

It is impossible to observe without admiration, the manner in which the females cover the eggs and the little ones. A great number of insects know how to weave cocoons, in which they enclose their brood, with considerable art. It is with her own body that the female of the Kermes covers her offspring. It answers all the purposes of a very close shell, or cocoon. She does not leave them for a moment exposed to the impressions of the air, places them in perfect shelter, and covers the eggs from the very instant in which they are laid. She is also useful to her young, even after her death, since they remain for many days under her dried up body.

The females die very shortly after having laid their eggs. Those of some species, according to many authors, lay but two thousand eggs, while those of others produce above four thousand. The little ones proceed from under this skin, through an aperture which exists at the lower part of their body. Scarcely have the young Kermes quitted their cradle, than they begin to run upon the leaves. Their growth is very slow, continuing from the end of spring, or the commencement of summer, the time of their birth, until the spring of the following year, but then they begin to acquire bulk rapidly. If those of the peach-tree are observed at the renewal of the fine season, there will be seen upon their back a number of little tubercles and some hairs or threads, tolerably long, which proceed from different parts of their bodies. These hairs, which are placed in different directions, proceed to attach themselves on the wood, tolerably distant from the insect.

For a long time naturalists were ignorant how these females were fecundated. Some authors believed that they were of both sexes, and could lay eggs without any intercourse with the male. But the observations of Reaumur, who has witnessed the union of the sexes, in the species of the peach-tree, prove that the Kermes, in this respect, do not differ from other animals of the same class.

All the young Kermes resemble one another, and do not assume the form which is peculiar to them, until they have grown. The most celebrated species is that whose figure approaches that of a ball, from which a small segment had been excided. This Kermes lives upon a species of small green oak, which is a mere shrub, that rises to the height of two or three feet, and is the *quercus coccifera* of Linnæus. This oak grows in great quantities, in the uncultivated lands of the southern parts of France, in Spain, and the islands of the Archipelago. It is from these shrubs that the peasants proceed to gather the harvest of the Kermes, in the proper season.

The Kermes for a very long time had excited the curiosity

of naturalists, before its true nature was discovered. It gave rise to an experiment, which succeeded, and led Marcilly into an error on this subject. Every one is acquainted with the composition of ink; we know that it is by the mixture of nut-galls that the solution of vitriol assumes a black colour. Marcilly tried if he could make ink with the Kermes and vitriol, and succeeded in so doing. From this he concluded that the Kermes, producing an effect similar to that of the galls found upon the large oaks, was a gall of the little oak; but he was deceived respecting the nature of these insects. This experiment discovers to us a curious fact; namely, that vegetable substances proper for the making of ink, preserve this property after having passed into the body of an animal.

The Kermes which has come to its full growth, appears like a little spherical shell, fixed against the shrub. Its colour is a brown-red. It is lightly crowned with an ashen crest. That which is obtained through the medium of commerce, is of a very deep red, and only owes its colour to the vinegar with which it has been treated.

The inhabitants of the countries where the Kermes is gathered, consider this insect under three different states. The first takes place in the commencement of spring. At this period it is of a very fine red, almost entirely enveloped with a sort of cotton, which serves it as a nest. It has then the form of an inverted boat. The second state occurs from the moment in which the insect arrives at its full growth, and that the cotton with which it was covered is spread over its body in the form of a greyish dust. It then appears to be a simple cocoon, filled with a reddish liquor. Finally the kermes arrives at its third state towards the middle or end of the spring of the following year. It is at this period that there are found under its belly eighteen hundred or two thousand little round grains, which are the eggs. They are as small again as a poppy-seed, and filled with a reddish liquor. In the microscope they appear set with brilliant points, of the colour of gold. Among these eggs, some are whitish, and some red. The first produce little ones of a dirtier white, more flatted than the others, and whose brilliant points have an argentine colour. These individuals, according to Reaumur, are less common than the red. They are erroneously considered, in the countries where they are found, as the mothers of the kermes.

Towards her second state, the female kermes prepares herself for her laying, by approximating the lower part of her belly to the back. She then resembles a wood-louse half rolled up. The vacancy formed by this contraction is filled by the eggs. The mother having acquitted herself of the duties imposed upon her by nature, very speedily perishes. Her carcase dries up. The traits which characterized it as an insect are obliterated, and totally disappear; nothing more is perceptible, than a sort of gall.

The eggs exclude the young; the latter abandon the cradle of their birth, spread themselves over the leaves of the shrub on which they have just been born; and feed upon their juice, which they extract with their proboscis.

The male at first exhibits the greatest possible conformity with the female. He fixes himself in the same manner that she does, becomes metamorphosed into a nymph in his cocoon, becomes then a perfect insect, raises the cocoon, and issues forth from it, the hinder part of his body being foremost.

Scarcely does he see the light, when excited by the desire of love, he hastens to fulfil the grand, and indeed the only end of his existence. As soon as this is accomplished, he ceases to exist.

The harvest of kermes is more or less abundant, according as the winter has been more or less mild. There is every expectation of its being good, when the winter passes without fogs or frosts. It has been remarked, that the oldest trees, and which appear the least vigorous, and are the least elevated, are the most loaded with kermes. The soil also contributes to their bulk, and to the vivacity of their colour. The insect, which comes from shrubs neighbouring to the sea, is larger, and of a more brilliant colour, than that which comes from shrubs more remote from it.

It is reported that pigeons are very fond of the kermes, which renders it necessary to watch them during the period of its gathering. If some species of kermes do injury to trees, we are amply compensated for this by the use which is made of that one of which we have been just speaking. It holds a distinguished place among the animals which are serviceable to us. The peasants of certain districts in France, and those of some other countries, collect in this way every year, a precious harvest without the trouble of tilling the land or sowing the grain. They proceed to collect this insect, which Pliny names *cocci granum*, and which is also called in the south of France, *graine d'ecarlate*, and *vermillon*. It is with this that the syrup of kermes is made. If the advantage derived by medicine from this drug, may be doubted, it cannot be doubted that the art of dyeing, derives utility from the kermes, which serves to dye silk or wool, of a fine crimson. It must be owned, however, that since the discovery of the cochineal, the kermes has ceased to be a matter of so much importance as formerly, and, perhaps at the present day, it is not applied to the production of so much advantage as it might be. It is women who gather this harvest. They remove with their nails the kermes from the shrubs, and one woman may gather two pounds per day. It is not rare to have two harvests in the year. The second is found attached to the leaves. It is neither so large nor so fit for the purposes of tincture as the first. The kermes destined for this purpose is wetted with vinegar; the pulp or red powder enclosed in the grain is removed. Then these grains are washed in wine, and after having been dried in the sun, they are polished by rubbing them in a sack, and they are shut up, mixed with a quantity of their own powder. The dearness of these grains depends on the greater or less quantity of powder which they afford. The first powder is that which comes out of the hole, which is on the side where the kermes is fixed to the tree; that which remains attached to the grain, issues, it is said, from a very small hole.

Vinegar changes the colour of the kermes. It is also used to destroy the posterity of the insect.

There are found on the large oaks many species of kermes, of different forms, and different colours, one of which being red, very much resembles that of the little oak. It is not good, however, for the purposes of dyeing; but it is considered as good for the confection of *alkermes*, as that which is attached to the *Ilex cocci glandifera*.

The females of the kermes, finish their laying without its being perceived, because their body covers all the eggs. Nevertheless, there are some species in which it covers but a part. The eggs of these latter, are lodged in a mass of threads of silk or cotton, very white, which causes them to be taken for the eggs of the spider. Those eggs which belong to different species are found on the yoke-elm, the oak, and the vine.

The mass which covers the nest of eggs, is usually of a rounded form above. When touched, or deranged ever so little, the envelope attaches itself to the fingers, which remove an infinity of threads parallel one to the other. The kermes do not spin this cottony matter; it escapes from under their shell, in the same manner as it escapes from the body of certain aphides, and of larvæ which eat them. It is not from spinnerets, like those of the caterpillars and spiders, that this substance issues forth. The kermes have underneath the belly a great number of imperceptible apertures analogous to the spinnerets of other insects, which give it passage. The principal are around the body. The species which make these cottony nests, are those, which previously to their laying, have the form of an inverted boat.

In the *kermes ulmi*, neither proboscis nor any organ as a substitute for mouth is to be discovered. There is only visible at the place which it usually occupies in other insects, some small grains or nipples about ten in number, very close, five on each side, that is, two which are larger, in front, two others of the same size behind, and three small ones in a triangle on each side. These grains are polished, lustrous, and resemble little simple eyes.

THE EIGHTH ORDER OF INSECTS.

THE NEUROPTERA.

ODONATA, *and a great part of the* SYNISTATA, *of Fab.*

THE two upper wings are membranaceous, usually naked, transparent, and like the lower in their consistence and properties. Their mouth is proper for mastication, being furnished with mandibles and jaws. The surface of the wings is composed of a very fine net-work. The lower wings are generally of the same size as the upper, sometimes broader, and sometimes narrower, but longer. The jaws and chin are never tubular. There is no sting, and seldom an ovipositor to the abdomen. The antennæ are usually filiform, and composed of a great number of articulations. Two or three simple eyes. Trunk formed of three segments closely united forming one body distinct from the abdomen. The first is short, forming a collar. The number of the articulations of the tarsi is variable. The body usually elongated, or slightly scaly. Abdomen always sessile.

Some undergo but a semi-metamorphosis, others a complete one. The larvæ have always six hooked feet.

I shall divide this order into three families, presenting the following natural relations. 1st. Carnivorous insects; semi-metamorphosis; larvæ aquatic. 2d. Carnivorous insects; complete metamorphosis; larvæ terrestrial or aquatic. 3d. Carnivorous, or omnivorous insects; terrestrial; semi-metamorphosis. 4th. Herbivorous insects; complete metamorphosis; larvæ aquatic, constructing portable domiciles. We shall finish with those whose wings are less reticulated, resembling phalenæ, or tineæ.

The first family,

SUBULICORNES, Lat.

Antennæ awl-shaped, little longer than the head, seven articulations at most, the last a thread. Mandibles and jaws entirely covered by the labrum and labium, or by the anterior and advanced extremity of the head. Wings separated, sometimes horizontal, sometimes perpendicular; lower of the size of the upper, or very small and even nullified. All have reticulated eyes prominent, with two or three simple eyes situated between them.

Some have horny mandibles and jaws; very strong, and covered by the lips. Three articulations to the tarsi. Wings equal. Abdomen terminated by hooks, or foliaceous appendages. They form the genus

LIBELLULA, Lin. Geoff.

Head thick, rounded, or broadly triangular. Two large lateral eyes, and three simple on the vertex. Antennæ inserted in the forehead; five or six articulations, or three at at least, the last attenuating into the form of a stylet. Scaly mandibles, strong and denticulated, jaws terminated by a piece of the same consistence, denticulated, spiny, and ciliated internally, with a palpus of a single articulation, applied on the back, representing the galea of the Orthoptera. There are three leaves to the labium, of which the two lateral are palpi. An epiglottis, or long vesicular tongue, in the interior of the mouth. The corslet thick and rounded, abdomen much elongated, and terminated, in the males, by two lamellary appendages, varying in form. Feet short, and curved forwards.

Fabricius divides the libellulæ into three genera:

LIBELLULÆ (proper). Wings in repose horizontal; head nearly globular, eyes very large and approximated. A vesicular elevation with a simple eye on each side on the vertex. The third by much the largest. The middle division of the lip much smaller than the lateral. The abdomen ensiform.

ÆSHNA. Two hinder simple eyes situated on a transverse keel-formed elevation. Intermediate lobe of the labium larger, the two others separate, armed with a strong tooth, and a spinous appendage. Abdomen narrow and elongated.

AGRION. Wings perpendicular. Middle lobe of the labium divided in two. Third articulation of the lateral lobes in the form of a membranaceous tongue. Antennæ, four articulations. No vesicle on the forehead. Simple eyes, equal, and disposed triangularly on the vertex. Abdomen almost filiform. Serrated laminæ to that of the female.

The other SUBULICORN NEUROPTERA have the mouth entirely membranous or very soft, and composed of parts not very distinct. Five articulations to the tarsi: lower wings much smaller than the upper, or nullified. Abdomen terminated by two or three threads. The genus

EPHEMERA, Lin.

Body very soft, long and attenuated. Antennæ very small, with three articulations, the last long, and in a conical thread. Front of the head, hood-like, often carinated and emarginated, and covering the mouth, whose organs are not distinguishable, in consequence of their softness. Wings perpendicular or but little inclined backwards. Feet slender, legs short,

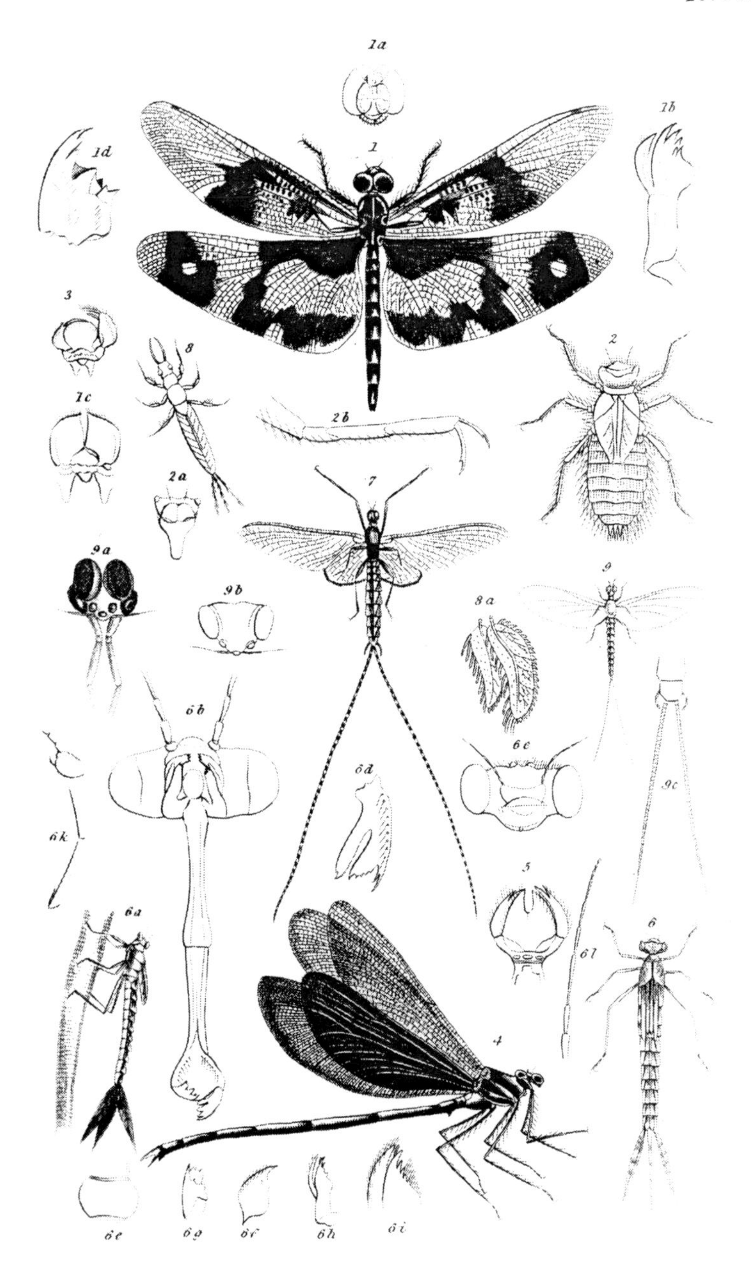

1. Libellula Indica. Fabr.
2. Nymph of Libellula depressa. Lin.
3. Lower lip of Æshne of Egypt.
4. Agrion Chinensis Guer. Fab.
5. Lower lip of Agrion virgo. Fab.
6. Nymph & details of Agrion puella. Lin.
7. Ephemera limbata. Serv.
8. Larva of Ephemera vulgata. Lin.
9. Ephemera bioculata. Fabr.

London. Published by Whittaker & Co. Ave Maria Lane, 1832.

and frequently but four articulations to the tarsi. The two hooks of the last are compressed in the form of a palette. The two anterior feet much longer than the others.

The second family,

PLANIPENNES,

Have the attennæ composed of a great number of articulations, and they are considerably longer than the head. Mandibles very distinct, and lower wings almost equal to the upper. Maxillary palpi filiform, or a little thicker at the extremity, shorter than the head, and with four or five articulations.

This family is divided into five sections:

1st. PANORPATÆ, Latreille. Five articulations to all the tarsi, and the anterior extremity of the head prolonged and narrowed like a proboscis. They constitute the genus

PANORPA, Lin., Fab.

Antennæ setaceous, and inserted between the eyes. Hood prolonged into a corneous plate, conical, and vaulted underneath to cover the mouth. The mandibles, jaws, and labium almost linear. Four to six palpi, short, filiform, and the maxillary with but four articulations. Body elongated, head vertical, first segment of trunk generally small and collar-like; abdomen conical. The two sexes differ much in many species.

In the greater number the uncovered part of the corslet is formed of two segments, the first of which is the smallest. Both sexes are winged, and the wings are longer than the abdomen, oval or linear, but not narrowed at their extremity. Such are

NEMOPTERA, Latr. Oliv. Upper wings separated, almost oval, very finely reticulated. Lower very long and linear; no simple eyes. Abdomen same form in both sexes. Six palpi, apparently.

Bittacus, Lat. Four wings equal, and horizontal. Simple eyes, feet very long, tarsi terminated by a single hook, and without cushion.

Panorpa (proper), Latr Abdomen of the male terminated by an articulate tail, with a pincers at the end like the scorpions. That of the females finishes in a point. Feet of middling length, with two hooks, and a cushion at the end of the tarsi.

In the others the first segment of the thorax is large, and the two following covered by the wings in the males. These are oval-formed, curved at the end, shorter than the abdomen, and wanting in the females. In them the abdomen is terminated by a sabre-like ovipositor. Boreus, Latr.

2nd. The Myrmeleonides. Five articulations also to the tarsi, but head not prolonged like a muzzle. Antennæ gradually growing thicker, or terminated by a button. Head transverse and vertical. Only the usual eyes, round and projecting. Six palpi, the labial longer, and swelled at the end. Palate of the mouth raised like an epiglottis, first segment of thorax small. Wings equal, elongated, and roof-like. Abdomen long, cylindrical, and with two projecting appendages in the males. Feet short. These insects form the genus

Myrmeleon.

Divided by Fabricius into two:—

Myrmeleon (proper), antennæ almost fusiform, hooked at the end, much shorter than the body, and the abdomen very long and linear.

Ascalaphus, Fab. Antennæ long and abruptly terminating in a button; abdomen oblong-oval, and but little longer than the thorax. Wings proportionally broader, and less long than in the preceding.

3d. Hemerobini, Latr. Antennæ like threads, and but four palpi. They form the genus

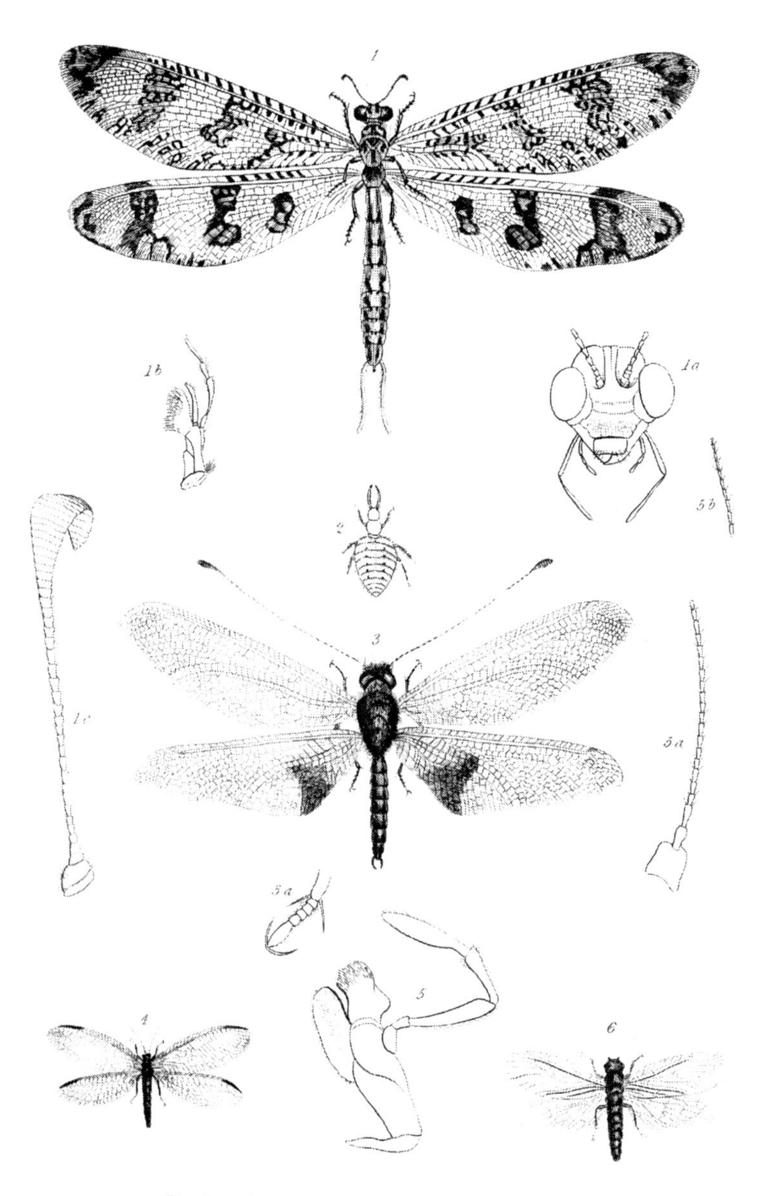

1 Myrmeleo Percheronii _ Guer.

2 Larva of Myrm. formicarius L.

3 Ascalaphus Brasilianus _ Guér.

4 Hemerabius capitatus. Fabr.

5 Détails of Hem. of Egypte.

6 Sialis lutarius. Latr.

London, Published by Whittaker & C.o Ave Maria Lane, 1832.

HEMEROBIUS, Lin., Fab.

Some have the first segment of the trunk very small, wings roof-like, last articulation of the palpi thicker, ovoid, and pointed. They form the genus

HEMEROBIUS (proper), Latr. Body soft, eyes globular, and often adorned with metallic colours. Wings large, and much inclined.

Of *Hemerobius fulvicephalus* (*Maculatus*, of Fab.), Latreille has formed his genus OSMYLUS.

NYMPHES, of Dr. Leach, formed on some insects of New Holland, has the antennæ filiform, and shorter.

The others have the first segment of the thorax large, corslet-formed, wings horizontal, and palpi filiform; last articulation conical, or almost cylindrical. Fabricius gives them the generic name of

SEMBLIS.

This genus is composed of those of CORYDALIS, CHAULIODES, and SIALIS, of Latreille. The first is distinguished by the mandibles, which are very large, and in the form of horns in the males; the second, by pectinated antennæ; and the third, by mandibles of middle size, simple antennæ, and wings, forming a roof.

4th. Another division, that of TERMITINÆ, will comprehend Neuroptera, that undergo a semi-metamorphosis, all terrestrial, active, carnivorous, or rodential, in all states. Except *Mantispa*, they are very distinct from all the insects of this order, resembling even *Mantis*, in the form of their anterior feet. The tarsi have four articulations at most, which removes them from the preceding genera of the same family. The mandibles are always corneous and strong The lower wings are nearly of the size of the upper, and without folds.

Some have from five to three articulations to the tarsi,

labial palpi prominent and very distinct. The antennæ generally composed of more than ten articulations. Prothorax large. Wings equal, and much reticulated.

MANTISPA, Illig. *Raphidia*, Scop., Lin. *Mantis*, Fab., Pall., Oliv. Five articulations to all the tarsi, first two fore-feet formed for seizing. Antennæ short and grained, eyes large, prothorax very long, thickened in front, and wings roof-like.

RAPHIDIA, Lin., Fab. Four articulations to the tarsi. Head elongated, narrowed behind. Corslet long, narrow, and almost cylindrical. Abdomen of the females terminated by a long exterior oviduct, formed of two laminæ.

TERMES, *Hemerobius*, Lin. Four articulations to all the tarsi, wings horizontal, and very long. Head rounded; corslet square or semi-circular. Body depressed, antennæ short, and chaplet-formed. Mouth similar to that of the orthoptera, and labium quadrifid. Three simple eyes, one not very distinct on the forehead, and the two others, situated one at each side, near the internal edge of the usual eyes. Wings transparent, coloured, with very fine and close nervures, not forming a distinct net-work. Two small, conical points, with two articulations at the end of the abdomen; feet short.

The other *termitinæ* have two articulations to the tarsi, the labial palpi not distinct, and very short; antennæ with about ten articulations. First segment of the trunk very small, and lower wings smaller than the upper. They form the genus

PSOCUS, Lat., Fab., *Termes Hemerobius*, Lin. Very small insect, whose body is short, very soft, and often swelled. Head large, antennæ setaceous, maxillary palpi projecting, wings roof-like, but little reticulated, or simply veined.

5. PERLIDES. Three articulations to the tarsi, mandibles almost always partly membranaceous, and small; lower

wings broader than the upper, and doubled on the internal side. They embrace the genus

PERLA, of Geoffroy.

Body elongated, narrow, and flatted; head pretty large, antennæ setaceous; maxillary palpi very prominent; first segment of the trunk almost square; wings horizontal, and crossed on the body, and abdomen usually terminated by two articulated threads.

NEMOURA, Latreille, differ from Perla by an apparent labrum, corneous mandibles, articulations of tarsi of an equal length, and abdomen almost without threads.

The third family of NEUROPTERA,

PLICIPENNES.

No mandibles; lower wings usually broader than the upper, and folded in their length. It composes the genus

PHRYGANEA, Lin., Fab.

Head small; two setaceous antennæ, generally very long and advanced. Rounded and prominent eyes, two simple ones on the forehead, labrum conical, or curved; four palpi, the maxillary very often long, filiform, or almost setaceous, with five articulations; the labial with three, the last a little thicker. Jaws and a membranaceous labium united. Body usually bristling with hairs, and forming with the wings an elongated triangle. First segment of thorax small, wings veined, usually coloured, or almost opaque; silky or hairy in many, and always very much inclined, in the form of a roof. Feet elongated, furnished with small spines, and five articulations to the tarsi.

Some have the lower wings evidently broader than the upper, and folded.

SERICOSTOMA. In one of the sexes the maxillary palpi

are in the form of valvules, covering the mouth like a rounded muzzle, three articulations, and under which we discover a thick and cottony down. Those of the other sex are filiform, and of five articulations.

PHRYGANEA (proper). Mouth similar in both sexes, and maxilary palpi shorter than the head and corslet, and but little hairy.

Some species, as *filosa*, *quadrifasciata*, *longicornis*, *hirta*, *nigra*, with excessively long antennæ, and maxillary palpi equally long, and very hairy, form our sub-genus MYSTACIDA.

Others have four narrow, lanceolate wings, almost equal, and without folds. To this division belongs the genus HYDROPTILA, of M. Dalman. The antennæ are short, almost grained, and of the same bulk.

Another sub-genus might be composed, (PSYCHOMYIÆ), with other *phryganeæ*, with similar wings, but whose antennæ are long and setaceous, as well as in almost all the others. We often meet in gardens a very small species, of great vivacity, the whole body a fulvous brown, and the antennæ annulated with white. This appears to me inedited, or imperfectly described.

SUPPLEMENT

ON

THE NEUROPTERA.

The order Neuroptera was established by Linnæus, and is named from the peculiar character of the wings, which are reticulated, naked, and transparent in the majority, and in all calculated for the purposes of flight. They are in general beautifully clear, and often present very lively reflections. Many, however, exhibit different coloured spots of no great transparency. Their wings are four in number.

The larvæ of these insects are provided with six feet. Many among them live in the water, and do not issue from it but in the state of the perfect insect. Others are terrestrial. Among the latter some live under the barks of trees; others, concealed in the sand, lay snares for ants, and other insects; others make war upon the aphides. These larvæ are in general carnivorous, and live solely on other insects. Some are omnivorous, such are those of the *termes*. Their metamorphoses are not the same in all the species. Some nymphs are immoveable; others stir about, and feed, like their larvæ, on insects, which they catch by divers means.

The larvæ which live in the water, have organs which at first appear analogous to the gills of fishes, but which are only external and tracheal appendages. These are designated by M. Latreille under the name of *fausses branchies* (false gills).

Some of them construct sheathes, or cases, after the manner of the *tineæ*, of different sorts of materials, and carry them with them everywhere. They contrive two apertures in them, which they close up before they change into nymphs, and do not issue forth but under their final form.

Some Neuroptera, in the perfect state, take little or no nourishment, and live but a very short time; such are the *Ephemeræ, &c.* But others are not less carnivorous than their larvæ. The *libellulæ*, for instance, are often seen hovering in the places where they hope to find their prey, and as soon as they perceive it, they dart upon it with rapidity from above, seize, and devour it.

The first family of this order is that of the SUBULICORNES, forming the genus LIBELLULA. Those are the insects, to which, in England, we give the name of dragon-flies.

Notwithstanding the elegant forms of the libellulæ, their inclinations are of the most murderous kind. Far from feeding on the honied juice of fruits and flowers, they hover aloft in the air, ready to dart down on all the winged insects which they can discover, as the raptorial birds come pouncing on their prey. "The name given to them in England," says Mr. Kirby, "seems much more applicable than Damoiselles, by which the French distinguish them. Their motions, it is true, are light and airy; their dress is silky, brilliant and variegated, and trimmed with the finest lace;—so far the resemblance holds; but their purpose, except at the time of love, is always destruction, in which, surely, they have no resemblance to the ladies. I have been much amused by observing the proceedings of a species, not uncommon here, *Anax Imperator*, of Dr. Leach. It keeps wheeling round and round, and backwards and forwards, over a considerable portion of the pool which it frequents. If one of the same species comes in its way, a battle ensues; if other species of *Libellulinæ* presume to approach, it drives them away, and it is continually engaged in catching case-worm flies, and other insects (for the species of this tribe all catch their prey when on the wing, and their large eyes seem given them to enable them the more readily to do this) that fly over the water, pulling off their wings with great adroitness, and devouring in an instant the contents of the body. From the number of insects of this tribe, which are everywhere to be observed, we may conjecture how useful they must be in preventing too great a multiplication of the other species of the class to which they belong."

In fact they devour all the insects which they can lay hold of. Not being very difficult as to choice of species, all come alike to them. They are often seen in the air, carrying off small flies, the blue meat-fly, and even butterflies. It is this taste for insects which brings them into flower-gardens, fields, and especially along hedges, on which many flies and butterflies repose. This same appetite leads them to the banks of waters, where numerous insects flit about, and brings them, in short, into every district which is peopled with their peculiar game.

The libellulæ are born in the water, and there acquire their complete growth. As long as they live there, their form remains tolerably similar to that which they had on issuing from the egg. They change into nymphs while they are yet young and small. This change of state produces no sensible change in their figure. We only perceive on the back of the nymph, four flat, and oblong bodies, which are the sheaths of the wings which the perfect insect is to possess. The colour of these nymphs presents nothing very remarkable. They are usually of a brownish green, often covered with mud; their six feet are attached to the corslet, and differ little from what they are subsequently to be. They have on the forehead a sort of convex, rounded mask, which Reaumur has named *casque;* their mouth is armed with four solid, broad teeth, placed in the middle of its anterior part, and which cannot be made visible without injuring the nymph. They are usually concealed by this mask, which occupies all the front and top of the head. The mask terminates in a sort of solid chin, of a cartilaginous substance. A suture is there distinguishable, which divides it into two parts, the anterior of which, shorter than the other, may be regarded as the forehead, and the other, longer, is what Reaumur calls the *mentonniere*, or chin-cloth. This mask is merely applied against the head; it is not adherent to it. It may be easily removed by means of a fine point: then the mouth and teeth are to be distinctly seen.

The use of the mask is not merely to cover the mouth, it is also to supply it with aliment. Beside its transverse suture, there is a longitudinal one on the front, which divides it into two equal parts, as far as the transverse suture. By means of these different sutures, the nymph opens as it pleases, one or the other of these two parts, or both at once. These nymphs, which are very carnivorous, and continually in ambush for the aquatic insects on which they feed, make use of these different pieces, which Reaumur has called *volets* (shutters), to catch their prey. The edges of these pieces have denticulations, which hold them together when the mask is closed, and they seem to retain the insect when it is seized.

These pieces furnish one of the principal characters which distinguish the larvæ and nymphs of the libellulæ, from those of œshna and agris.

The nymphs of the libellulæ have the body short, broad, depressed, and terminated by a very short tail. Their four teeth, or the parts analagous to the mandibles and jaws of the perfect insect, are covered transversely by the two shutters, which have an almost triangular figure, and are a little vaulted. Their internal sides are denticulated, touch in their entire length, and thus form a suture, perpendicular to the breadth of the mask. The anterior part of the head, closed by the shutters, is properly called the forehead. The mask is in the form of a casque.

The interior of the mouth of these nymphs, presents us, as in the perfect insects which proceed from them, a rounded advancement, almost membranous, situated under the teeth, which M. Latreille calls *palate*, but which Reaumur considers a *tongue*.

These insects, in the larva and nymph state, present a very singular phenomenon in the mode in which they absorb the air which is contained in the water. It is at the end of their body that the aperture is which gives entrance to the water, and by which it is subsequently expelled. This aperture is surrounded by five small, pointed pieces, three of which are larger and triangular. These pieces, when the insect closes the posterior aperture of the body, form a sort of pyramidical tail. Every time that it wishes to respire the water, or eject its excrements, it opens this pyramid, expanding its extremity. In the libellulæ, the three most projecting points are equal; but in the nymphs of the second genus, that of *æshna*, the dorsal piece is truncated, while the two lateral and anterior are pointed.

These triangular points, also, sometimes constitute for the insect an offensive and defensive weapon.

It is easy to see, when these pieces are separated from each other, a round aperture, about half a line in diameter, in the nymphs of the middle size. Jets of water issue forth from it by intervals, and are carried to the distance of two or three inches from the insect. These jets are more or less abundant, according to circumstances. Some may be generally seen to proceed from the anus every time the insect is put out of the water. When deprived, for a quarter of an hour, or longer, of this element, and subsequently put into a flat vessel, where there is scarcely water enough to cover it, its inspirations and expirations become more frequent and more sensible. At other times, there is only perceived, occasionally, a slow circulation of the water round the hinder part of the nymph.

The hole, which is at the end of the last ring, is most frequently stopped by some greenish flesh. But without waiting any length of time, we discover there, by intervals, an aperture, which allows us to see, in the capacity of the body, three pieces, nearly of equal size, formed like shells, cartilaginous, and situated so as to close the aperture at will, and act in some measure as a valve. When these pieces rise, and proceed toward the hinder part, the parts which are above, remove in an opposite direction. We then see through the hole, the interior of the capacity of the body, which appears empty. The last five rings are really so, and form a tunnel, which is filled with air or water. To inspire the water, the nymph separates the parts of its tail, raises the shell-pieces, and forms a vacancy in the last rings of its body, by approaching interiorly to the corslet, a sort of thick bung, or stopple; the water immediately occupies this capacity. If the insect be desirous of rejecting this fluid, the parietes of its body contract, the bung is pushed towards the rump, and the water is squirted out.

This mass, which Reaumur calls *tampon* (bung, stopple), and which performs the office of a piston when the animal inspires and expires the water, is but a net-work of vessels which serve for respiration, tracheæ without number interlaced one with another; four principal trunks, two on each side, extend through the whole length of the body, and throw out, beginning from the middle of their extent, and still more towards the final rings, and of the interior side, a quantity of branches. The posterior extremeties of the grosser vessels are divided, or, as it were, cleft into many small portions. These organs are evidently tracheæ; their tubular form, their contexture, which presents a cartilaginous thread, turned spirally, and from which Reaumur has wound out a length of three inches, their whiteness, their satiny lustre, all convince us of this fact.

The insect has several stigmata disposed longitudinally on the sides of the body. On the corslet there are four very visible, two especially, those which are placed nearer the base of the abdomen. Each ring of this last part of the body, with perhaps the exception of the two at the end, has two of them, but whether water hinders oil from being applied there, or whether by closing with promptitude, they do not permit this last fluid to penetrate to them, the animal does not perish when it is oiled over the external apertures of the tracheæ.

The alimentary canal proceeds in a strait line, from the mouth to the anus. But it has, as it were, three inflations, which Reaumur says may be regarded as three stomachs. The end of this canal appeared to him to remove from, or approach to the anus in the different movements which the insect makes to inspire, or expire the water.

M. Cuvier has seen in the interior of the rectum, twelve longitudinal ranges of small black spots, approximating by hairs, resembling so many leaves, which botanists name *winged.* There are a great number of little conical tubes, of the structure of tracheæ. We observe, that there spring from each of these tracheæ some small branches, which lose themselves in six large trunks of tracheæ predominating in the whole length of the body, and from which proceed all the branches which carry the air into all parts of the body. M. Cuvier suspects that this apparatus of the respiratory organs decomposes the water, and absorbs the air which is there contained. We have seen that Reaumur reckoned but four principal tracheæ: M. Cuvier has discovered two more. This difference arises from Reaumur not having seen the two lateral tracheæ, into which the stigmata almost immediately open. M. Cuvier has also favoured us with some very curious observations on the structure of the eye of the libellulæ, but our limits compel us to deny ourselves the pleasure of inserting them.

The majority of the larvæ of the libellulæ, and perhaps all, live from ten to eleven months in the water before they can undergo the transformation into the perfect insect. During this interval they change their skin several times. It is from the middle of spring to the commencement of autumn, that their final metamorphosis takes place. The nymphs which are about to change their form, may be recognised not only by their size, but also by the figure of the sheaths of their wings. The two which are on the same side, detach themselves from one another, and in some species they change position.

It is out of the water that this grand change from the state of nymph to that of an inhabitant of the air is accomplished. Some nymphs are metamorphosed in one or two hours after they leave the water; others take an entire day to change their form. On issuing from the water, the nymph remains a certain time in the air to dry itself; it subsequently places itself upon a stalk, or branch of a tree, where it hooks itself with its feet, and always fixes so that the head shall be uppermost. The movements by which the transformation is prepared, go on internally. The first sensible effect which they produce, is to cut the sheath over the corslet. This cleft elongates, and the libellulæ disengages its head. Afterwards it puts forth its feet; to complete their disengagement from the envelope, it turns its head downwards. In this attitude it is only supported by its last rings, which have remained in the spoil, and which form a kind of hook, which prevent it from falling. After having remained a certain time in this posture, it turns about, seizes with the hooks of its feet the anterior part of the sheath, fastens there, and completes the emancipation of the posterior part of its body. Then its wings are narrow, thick, and folded like the leaf of a tree, which is about to be developed. It is only at the end of one or two hours that they are entirely unfolded, and sufficiently solid to enable the insect to make use of them.

As soon as their wings are firm, the libellulæ take flight, and, similar to the birds of prey, they proceed to the chase. The males soon have another object in their flight, which is to find the females, with which they may couple. Their amours, and the mode in which their union takes place, constitute one of the most singular phenomena in the history of these insects. From spring to the middle of autum, we frequently observe the libellulæ, either on plants, or flying in the air, in pairs. The one which flies the first, is the male, which has the extremity of its body placed upon the neck of the female; both fly in concert, having the body in a right line.

The females do not retain their eggs very long after they have been fecundated. They come forth from their body, through the aperture which they have near the anus. As these eggs are united, and form a sort of cluster, they lay them all at once, on the same day in which they have coupled. They deposit them in the water, the element in which the larvæ are to grow, and undergo their first metamorphosis.

The colours, in the majority of other insects, usually serve to distinguish the species, but here they most frequently denote only the difference of sex. It is therefore essential to observe, as much as possible, the libellulæ, and also the other insects of this family, at the moment of their amours. Reaumur has seen, in the *libellula depressa*, the most common species, some males which were yellowish, like the females, and others which were of a fine slate-colour. In the *agriones*, especially, he has remarked a great number of differences of colour, among the sexes. He also observes, that the males, or those, at least, of many species, exceed the females, a little, in size, or are not sensibly smaller, a fact of rare occurrence in the insect world.

The insects of the sub-genus *Æshna*, and also those of *agrio*, are called *damoiselles*, by the French, as well as the libellulæ. The difference is in their larvæ, as well as in the perfect state. Those of the libellulæ are short, the abdomen is oval, and terminated by five short points, forming a pyramid. The larvæ of æshna have the abdomen similarly terminated, but it is much longer. Their eyes are larger ; their mask is flat, and provided with two strong talons.

The insects of this sub-genus are remarkable for the rapidity of their flight. They are found, during the fine season, in gardens, fields, and meadows, and near the banks of waters. They go there to seek the insects on which they feed. The species are not numerous.

The Agriones, also, like the preceding insects, live in the water, in their larva and nymph states. But their body is much more attenuated than that of the larvæ and nymphs of the other libellulinæ, and terminated by a tolerably long tail. Their head is broad, with two sorts of hands, crossed over the forehead, above the mask. This mask is long, open, and with two divisions at its extremity. Under their last form they are equally carnivorous, and feed on insects which they catch upon the wing. Like their larvæ and nymphs, they are by no means particular in the choice of food, taking all the insects they can catch, seizing them with their mandibles, and carrying them off to eat them at their leisure. Accordingly, they are usually found in the country, in such places as are most populated by insects, along hedges, the banks of streams, small rivers, &c., where they fly with rapidity, in search of their prey. The females also come to the water, at the time of laying, to deposit their eggs. Their organs of generation, mode of laying, &c., are similar to those of the other libellulinæ.

The name of EPHEMERA has been given to the insects so called, in consequence of the short duration of their lives, when they have acquired their final form. There are some of them which never see the sun; they are born after he is set, and die before he re-appears on the horizon. Many naturalists have made very interesting observations on these insects. Swammerdam speaks of ephemeræ, which issue from the rivers of Holland, for two or three days in succession, in a most surprising abundance. These insects appear at different epochs, according to the species and the country.

The ephemeræ of Holland appear in summer; those of some other countries at the end of spring; towards the middle of summer, there may be seen, in the environs of Paris, clouds of them, which obscure the air. At certain hours of the day, they begin to issue from the water, and the hour is not the same for all the species. Those of the Rhine, the Meuse, the Lech, the Yssel, and the Wahal, commence to fly over there towards six o'clock in the evening, about two hours before the setting of the sun. The most active of those of the Maine and the Seine, which have been observed by Reaumur, do not rise in the air till the sun is ready to set, and it is not until after he has quitted the horizon, that the greatest quantity of them appear. The seasons of the different crops are not better known by agriculturists, than is the time in which the ephemeræ are about to shew themselves on the banks of the rivers which they inhabit, by fishermen. Whatever, during the day, may have been the temperature of the atmosphere, the hour at which the ephemeræ commence to quit their spoil of nymph, is the same with the greater number of them, and another hour appears marked, beyond which it is no longer allowed them to do so.

Those which have been studied in Sweden, by Degeer, are disclosed towards the end of spring, in the evening, in immense numbers, always at the setting of the sun. They assemble by hundreds, hover about continually, rise above some large tree, and seldom remove from it. The duration of the life of these latter, is longer than that of the species observed by Swammerdam, and Reaumur. These ephemeræ commence to fly, an hour before the setting of the sun, swarm together in places but little remote from some river or stream, and remain there until the dew begins to fall in too great abundance. Then they disappear, withdrawing upon walls, and plants, and remaining there in a perfect state of repose, until the following day, when they become re-animated, and rise in the air anew.

In the larva and nymph state, these insects are aquatic. They occupy holes in the banks of streams or rivers, which being below the water, it enters there, and they seldom leave these retreats. Sometimes, however, they swim about, or walk at the bottom, or remain under sticks and stones.

A difference in the temperature of the atmosphere, may accelerate or retard the appearance of the ephemeræ notwithstanding the general regularity of its occurrence. Those of the French rivers, observed by Reaumur, are expected by the fishermen, who call them *manna*, between the 10th and 15th of August. These ephemeræ supply the fish with a great quantity of food, and the latter are always caught in great abundance at this particular season.

"Reaumur first observed these insects in the year 1738, when they did not begin to shew themselves in numbers till the 18th of August. On the 19th, having received notice from his fisherman that the flies had appeared, he got into his boat, about three hours before sunset, and detached from the banks of the river several masses of earth, filled with pupæ, which he put into a large tub, full of water. This tub, after staying in the boat till about eight o'clock, without seeing any remarkable number of the flies, and being threatened with a storm, he caused to be landed, and placed in his garden, at the foot of which ran the Maine. Before the people had landed it, an astonishing number of ephemeræ emerged from it. Every piece of earth, which was above the surface of the water, was covered by them; some beginning to quit their slough, others prepared to fly, and others already on the wing; and every where under the water they were to be seen in a greater or less degree of forwardness. The storm coming on, he was obliged to quit the amusing scene, but when the rain had ceased to fall, he returned to it. As soon as the cloth with which he had ordered the tub to be covered was removed, the number of flies appeared to be greatly augmented, and kept continually increasing. Many flew away, but more were drowned. Those already transformed, and continually transforming, would have been sufficient of themselves to have made the tub seem full; but their number was soon very much enlarged by others, attracted by the light. To prevent their being drowned, he caused the tub to be again covered with the cloth, and over it he held the light, which was soon concealed by a layer of these flies, that might have been taken by handfuls from the candle-stick.

"But the scene round the tub was nothing to be compared with the wonderful spectacle exhibited on the banks of the river. The exclamations of his gardener drew the illustrious naturalist thither; and such a sight he had never witnessed, and could scarcely find words to describe. 'The myriads of ephemeræ,' says he, 'which filled the air, over the current of the river, and over the bank on which I stood, are neither to be expressed nor conceived. When the snow falls with the largest flakes, and with the least interval between them, the air is not so full of them, as that which surrounded us was of ephemeræ. Scarcely had I remained in one place a few minutes, when the step on which I stood was quite concealed with a layer of them, from two to four inches in depth.

Near the lowest step, a surface of water, of five or six feet dimensions, every way, was entirely and thickly covered by them, and what the current carried off was continually replaced. Many times I was obliged to abandon my station, not being able to bear the shower of ephemeræ, which, falling with an obliquity less constant than that of an ordinary shower, struck continually, and in a manner extremely uncomfortable, every part of my face—eyes, mouth, and nostrils, were filled with them.' To hold the flambeaux, on this occasion, was no pleasant office. The person who filled it, had his clothes covered in a few moments, with these flies, which came from all parts, to overwhelm him. Before ten o'clock, this interesting spectacle had vanished. It was renewed some nights afterwards, but the flies were never in such prodigious numbers. The fishermen allow only three successive days for the great fall of the manna; but a few flies appear, both before and after, their number increasing in one case, in the other diminishing. Whatever be the temperature of the atmosphere, whether it be cold or hot, these flies invariably appear at the same hour in the evening, that is, between a quarter and half past eight; towards nine, they begin to fill the air; in the following half hour they are in the greatest numbers, and at ten there are scarcely any to be seen. So that in less than two hours, this infinite host of flies emerge from their parent stream, fill the air, perform their appointed work, and vanish. A very large proportion of them falls into the river, when the fish have their grand festival, and the fishermen a good harvest." (Kirby and Spence, Introd., Vol. I. p. 284, &c.

Reaumur and Swammerdam both consider that earth is the only food of the ephemeræ. It is the only substance found in their stomachs, but yet it may reasonably be presumed that it enters there with decomposed vegetable matter, and remains undigested. *Sand* is found in the stomach of many *testacea*, yet assuredly it does not constitute their food, as Boulli erroneously supposed.

It is singular enough that the ephemeræ, which are never destined to behold the solar orb, should, like moths, have such a propensity to fly round any luminous object. Reaumur gives a very lively description of their gyrations about the flambeau above mentioned. The dullest spectators of the scene witnessed it with astonishment. Innumerable circles of these insects were whirling around the flambeau as a common centre. They intersected each other in all directions, and performed a variety of eccentric evolutions. "Each zone," says Mr. Kirby, "was composed of an unbroken string of ephemeræ, resembling a piece of silver lace formed into a circle deeply notched, and consisting of equal triangles, placed end to end (so that one of the angles of that which followed touched the middle of the base of that which preceded), and moving with astonishing rapidity. The wings of the flies, which was all that could then be distinguished, formed this appearance. Each of these creatures, after having described one or two orbits, fell upon the earth or into the water, but not in consequence of having been burned. Reaumur was one of the most accurate of observers, and yet I suspect that the appearance he describes, was a visual deception, and for the following reason. I was once walking in the day-time with a friend, when our attention was caught by myriads of small flies, which were dancing under every tree; viewed in a certain light they appeared a concentrated series of insects (as Reaumur has here described his ephemeræ), moving in a spiral direction upwards; but each series, upon close examination, we found was produced by the astonishingly rapid movement of a single fly. Indeed when we consider the space that a fly will pass through in a second, it is not wonderful that the eye should be unable to trace its gradual progress, or that it should appear present in the whole space at the same instant." (Ibid. Vol. II., 366, &c.)

We shall oblige our readers with one more description of the picturesque dances of the ephemeræ, from the same admirable writer, without the presumptuous affectation of substituting our language for his own graphic and animated style:

"At the same time of the day, some of the short-lived ephemeræ assemble in numerous troops, and keep rising and falling alternately in the air, so as to exhibit a very amusing scene. Many of these are also males. They continue this dance for about an hour before sunset, till the dew becomes too heavy or tóo cold for them. In the beginning of September, for two successive years, I was so fortunate as to witness a spectacle of this kind, which afforded me a more sublime gratification than any work or exhibition of art has power to communicate. The first was in 1811: taking an evening walk near my house, when the sun declining fast towards the horizon, shone forth without a cloud, the whole atmosphere over and near the stream swarmed with infinite myriads of ephemeræ, and gnats of the genus *chironomus*, which in the sun-beam appeared as numerous and more lucid than the drops of rain, as if the heavens were showering down brilliant gems. Afterwards in the following year one Sunday, a little before sunset, I was enjoying a stroll with a friend at a greater distance from the river, when in a field by the road-side, the same pleasing scene was renewed, but in a style of still greater magnificence; for from some cause in the atmosphere the insects looked much larger than they really were. The choral dances consisted principally of ephemeræ, but there were also some of *chironomé*, the former, however, being most conspicuous, attracted our chief attention—alternately rising and falling, in the full beam they appeared so transparent and glorious, that they scarcely resembled any thing material—they reminded us of angels and glorified spirits drinking life and joy in the effulgence of the Divine favour. The bard of Twickenham, from the terms in which his beautiful description of his sylphs is conceived in *The Rape of the Lock*, seems to have witnessed the pleasing scene here described:

> " 'Some to the sun their insect wings unfold,
> Waft on the breeze, or sink in clouds of gold;
> Transparent forms, too fine for mortal sight,
> Their fluid bodies half dissolved in light;
> Loose to the wind their airy garments flew,
> Thin glittering textures of the filmy dew,
> Dipt in the richest tincture of the skies,
> Where light disports in ever-mingling dyes,
> While every beam new transient colour flings,
> Colours that change whene'er they wave their wings.' "

These insects come out of the water only to couple and to lay. According to Degeer, the first observer who witnessed the coupling of these insects, there are many more males than females. The latter are distinguished by the threads which terminate their abdomen. They have three of equal length, whereas the males have but two, one on each side, and the commencement of a third in the middle. Besides these threads they have four others very short, underneath the belly, and two other parts in the form of hooks recurved into an arch, with which they fasten on the female during the time of coupling. The sexual organ of the females, which consists externally in two apertures, is situated underneath the abdomen, between the seventh and eighth rings—it is through these apertures that the eggs issue forth. The principal assemblages of these insects, according to Degeer, are entirely composed of males; but as soon as a female presents herself among them, these males immediately proceed to the pursuit, and appear to dispute the conquest of her. He who has obtained the preference, flies away with her alone, and the others re-enter the group, to await the coming of more females. The couple proceed to place themselves on a wall or a tree, that they may not be disturbed in their amours. Their union lasts but an instant of time.

As soon as the females are fecundated they deposit their eggs. It is in the water that all should place them; but the majority leave them on the bodies, on which they have themselves been previously situated. There are few insects indeed which produce so great a number of eggs, or lay them so promptly. These eggs are disposed in two clusters as it were, some of which are three lines in length. Each of these clusters contains from three hundred and fifty, to four hundred eggs. Each ephemera has seven or eight hundred eggs to lay, which is an affair of a moment, for she puts forth the two clusters at once. To arrange herself for this operation, she raises the extremity of her abdomen, which she causes almost to form a right angle with the rest of her body. At the same time she pushes forth the two clusters, which issue from the two apertures of which we have already spoken. Those females that lay in the water rest with the threads of their tail on the water itself, while they get rid of their eggs. These eggs, more heavy than the water, fall immediately to the bottom, and are soon separated from each other. We are ignorant of the time which they take in disclosing.

The ephemeræ appear to us to have a very short existence, because they remain but a moment, as it were, in their final form; but they live much longer than a great number of insects, in the larva and nymph states. They pass one, two, and according to some authors, even three years, before they become perfect insects. The larvæ and the nymph live in the water, or in holes below its surface. These larvæ have six feet, the head is triangular, and provided underneath with two scaly and curved parts which terminate in a point.

The body is divided into ten rings; from the extremity of the last three issue three threads, almost as long as the body, and more or less furnished with a fringe of hairs disposed like the barbs of a feather. They are of a brown or yellowish colour, according to the species. The nymphs do not differ from the larvæ, but in having sheaths of the wings upon the corslet.

All the larvæ of the ephemeræ differ among themselves, only in the peculiar inclinations accorded to them by nature. Some pass their lives in fixed habitations, each in its own, which consists of a hole hollowed below the surface of the water, in the earth which forms the bed or basin of the river. Seldom does the larvæ quit this hole for the purpose of swimming. This indeed never happens but under such circumstances as require that it should excavate a new one. Other ephemeral larvæ are as we may say wandering, and swim or walk at the bottom of the water. When these larvæ remain tranquil, there are remarked around their body, tufts of a sensible size, which are continually in a state of extreme agitation. In some the hoops are situated like the oars of a galley, in others they are placed above the body. Some species have them inclined upon the back, and they are directed hindwards. The number of these tufts, which have the appearance of gills, is not always the same in all the larvæ. Some have six on each side, others more. The most common species has these gills couched or inclined upon the body, and does not swim habitually. It is one of those which remain shut up in their holes. These holes are situated horizontally. Their apertures are a little oval. They are very near each other, and communicate with a canal which has two branches, which cannot be better compared than to a tube of glass which has been bent into two divisions. Thus the larvæ inhabit a lodging of two pieces. These holes are never found in banks or beds of gravel, the larvæ living only in unctuous earth. Its lodging is always in proportion to its size. All the vacancies which its body leaves there, are filled by the water with which it is surrounded, as it would be in the middle of the river. It is there in perfect security against the voracity of fishes and aquatic insects of different kinds. These larvæ, according to M. Latreille, feed on earth, the grains of which they reject after having assimilated all that is succulent in it. We have already cited the opinion of Mr. Kirby on this point.

When the ephemeræ are ready to quit their exuvia of nymph, they issue from the water, and proceed to some dry place. They are not long in getting rid of their skin, which is cleft above the head and corslet, and as soon as the ephemera is out of it, it flies off, and proceeds to place itself on some wall or tree. Although it then has wings, and appears to want nothing to complete its perfect state, it has still a moulting to undergo. For this last operation it fastens itself with its feet upon a wall, puts itself for the most part in a vertical position, the head upwards, and remains there sometimes for an hour, until the skin which covers it is cleft over the head and corslet. In proportion as the cleft increases, the insect draws out all its parts one after the other. The wings which are stripped, like all the rest, issue by little and little from a pellicle which covered them, and the exuvia remains attached to the wall, or on the tree on which the insect fixed itself. Swammerdam asserts respecting the species, on which he has given his observations, that this second casting of the skin is incidental to the males alone.

Previously to this last moulting, the body and wings of the insect, were of a dull brown. But subsequently, the skin of the body, and, in many species, the wings, become as it were varnished, dry and friable. The ephemeræ have no very perceptible mouth, and there is no appearance of their taking any food, in their final state. They are so weak and delicate as to be wounded by the slightest touch. Those of the neighbourhood of Paris, mostly every year, towards the end of summer, present to the inhabitants of the banks of the Seine a perfect phenomenon. Such a number of them is born in a few hours, that they form a thick cloud. They hasten to fulfil the functions for which they were born, not arriving at the perfect state, and appearing in the air, for any other purpose than the perpetuation of their species. But after coupling and laying, what becomes of this prodigious quantity of ephemeræ, which hid the face of Heaven as it were, with a veil, and hung, like a dark cloak over the waters? They are for the most part already dead or dying. Some have fallen into the same river in which they received their birth. Those which in falling into the water have not become the prey of the fish, soon perish notwithstanding, being speedily drowned. The others fall on the banks of the river, and sometimes form there so thick a bed, that the earth is not better covered in winter by the snow, than it is by their bodies. The duration of the life of these last is not altogether so short as that of the others. But this is all the worse for them. Heaped upon one another, without having the power to change place, and without shewing much sign of emotion, they die one after the other in succession. Those whose life extends the longest, and which may be considered as centenarians in comparison of the others, do but witness, at the most, the rising of the sun. Thus these insects terminate their life, which is long while they are in the larvæ and nymph form, and so short after they have become perfect insects.

Of the Panorpæ, which commence the second family, all we can say here is that they are found on bushes, in woods, and humid places, and have been named by Geoffroy, *Mouches-scorpions (Scorpion-flies)*. They live by rapine, and their transformations are unknown.

We insert a figure of a new species of this sub-genus, under the specific name of *rufa*. It is red; with the wings yellowish, with some longitudinal irreglur bands, and the tip black; the latter with four minute spots of white. This species is from Georgia.

The Nemopteræ have some approximation to the last mentioned genus. Oliver has described many new species which he discovered in the Levant and Persia. It is to those countries to the South of Europe and the North of Africa that this genus (or sub-genus) appears, hitherto at least, to be confined.

"These insects," says Olivier, "whose metamorphoses, and mode of living we are not acquainted with, fly very badly, and transport themselves slowly, and with a painful agitation of their wings, to small distances, so that they may be always caught with the greatest possible facility. I have seen them in infinite numbers, and they appear to me to enjoy a very short existence. Eight days after their appearance, I could find no more of them, except when I was going from Bagdad into Persia. As I was proceeding from a burning climate to a more temperate region, I have seen during more than twenty days in succession, and almost always in great abundance, the fourth species which I have described" (*Nemoptera extensa*).

To this remarkable genus *Nemoptera*, we have figured a species which is called *Africana*. It is fulvous, the fore wings white, with anterior nervures yellowish, the rest black, the hinder wings long, narrow, with the base fulvous, the apical half whitish, with a band of black. This species is from Egypt.

In the genus Myrmeleon or *Antlion*, the larvæ is an object of far greater interest than the perfect insect. That of the most common species in Europe, is denominated *formica-Leo*, *antlion*, *&c.*, for the same reason that the larvæ of hemerobius has been called by some, *Lion of the aphides.*

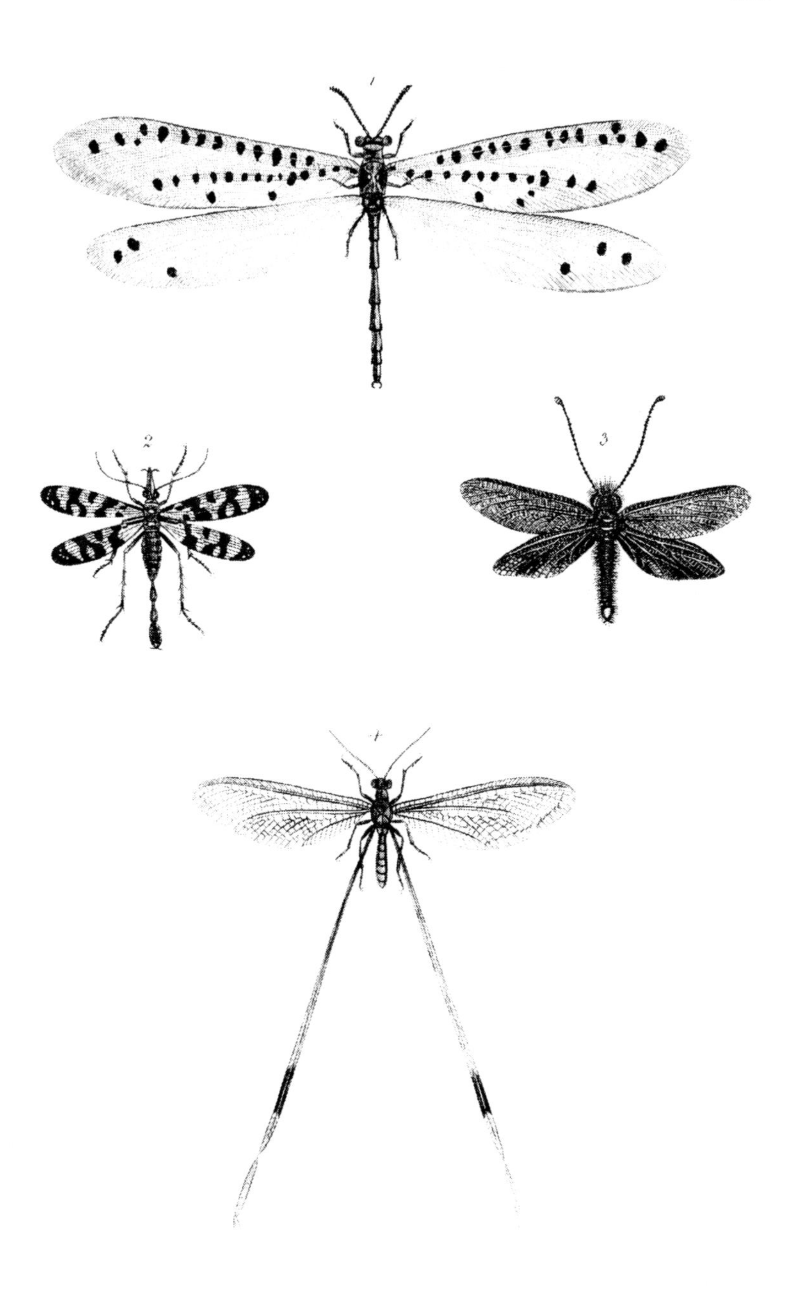

1 *Myrmeleon erythrocephalum* Leach. 3 *Asculaphus barbarus* Fab.
2 *Panorpa rufa* G.R. Gray. 4 *Nemoptera Africana* Leach.

London, Published, by Whittaker & C.º Ave Maria Lane, 1832.

This larva is of a greyish colour, has six feet, and a very remarkable form, the belly being of an extraordinary thickness, in proportion to the corslet and head. The head is very small, flatted and narrow, armed with two horns, tolerably long, mobile, denticulated internally for almost their entire length, curved towards their extremity, and terminating in a point. These two horns answer the purpose of pincers and suckers.

This larva is carnivorous, walks very slowly, and backwards. As it could not catch by running, insects which are much more agile than itself, with which, however, it is absolutely necessary that it should be provided for the purposes of nourishment, nature has instructed it in the means of laying snares for them. It knows how to arrange the place in which it fixes itself, in such a manner as to cause them to fall into its horns which are ready to receive them. It lodges in sand, where it remains quietly at the bottom of a hole which is shaped like a tunnel. There it is entirely concealed, with the exception of the horns which it holds raised up, and separated one from the other. Unlucky is it for any imprudent insect, the ant especially, that in proceeding on its way, should dare to approach this cavern. If one of these insects is so far remote that the larvæ cannot seize it, it showers upon it such a quantity of sand, with its head, which it uses like a shovel, that it becomes completely stunned. It finishes by losing the equilibrium, which it had preserved with difficulty, in walking on a declivity, and tumbles into the bottom of the hole, between the murderous pincers of the larva, which immediately grasp and pierce it, by closing upon it.

When the larva has thus obtained its prey, it drags it under the sand, to suck it at leisure, and after having drawn from it all that is nutritive, it throws out beyond the edges of its hole, the dried carcase, which is no longer of any use to it.

These larvæ are never found but in sandy soils and such as are composed of very fine grained earth. It is at the foot of old walls, in places much exposed to the south, that they most usually establish themselves. One larva does not pass its entire life in the same hole. Whenever any derangement takes place, it changes its domicile, or when it does not find a sufficiency of prey. When it has determined to abandon its habitation it walks forth, and traverses the environs. The path which it takes is marked by a sort of little foss about a line or two in depth. When arrived at the place which suits it, it begins to excavate a new habitation, with indefatigable ardour. To give just proportions to its funnel, it traces its circular extent, by making a foss similar to that which it forms when walking. This foss encloses a space of greater or less extent. Those larvæ which have nearly arrived to their complete growth, sometimes inhabit holes, the diameter of whose entrance is more than three inches. The depth of the newly-formed funnel equals about three-fourths of the diameter of the large aperture. As soon as the larvæ has finished its hole, which it commences and completes sometimes within half an hour, it conceals itself at the bottom to await its prey, where it often remains for a very long time. But as it is capable of supporting a long fast, it will remain for several months deprived of aliment without dying. Notwithstanding this, it is by no means difficult as to the choice of food: all insects come alike to it, even those of its own species.

All the nutriment which this insect takes is assimilated, and contributes to its growth; or, if there be any residue, it escapes from the body by insensible perspiration, for the animal never rejects any perceptible grain of excrement. Accordingly, it would appear, that it has no perceptible aperture at the anus for this purpose.

The larvæ of these insects issue from the eggs in summer,

or in autumn, and do not change into nymphs until the following year. They undergo their metamorphosis in their hole, where they seek in the sand a convenient place to form the cocoon in which they enclose themselves. This cocoon is round. The exterior is composed of grains of sand, which are held together by threads of silk, which the larva draws from some spinnerets, which it has at the extremity of its body. The interior is carpeted with silk, of a satiny white. Some of these cocoons are to be found of four or five lines in diameter. These are the habitations of females. In fifteen or twenty days after the larva has undergone its metamorphosis, the perfect insect issues from the cocoon through an aperture which it has made there, and leaves the envelope of nymph at the entrance.

These larvæ may be easily reared in sand, providing care be taken to supply them with plenty of ants, flies, and other insects.

Bonnet discovered, in the neighbourhood of Geneva, a larva of Myrmeleon, which differed from those which are known, by not walking backwards, forming no funnel, and merely concealing itself so as to catch the insects which come in its way. It may probably belong to the sub-genus termed ASCALAPHUS.

Scopoli, indeed, has placed one species of them with the latter. They inhabit warm and sandy places, remain fastened to plants, from which they fly off when any one approaches them. There is even some difficulty in catching them, their flight being prompt and rapid. Their larvæ are not known, but may be presumed to be perfectly analogous, both in form and habits to the myrmeleons.

We have figured a New Holland species of *Myrmeleon*, which Dr. Leach has called *Erythrocephalum*. Its head, thorax, coxæ, and the four anterior thighs, are reddish brown; the sides of the thorax, antennæ, abdomen, and legs, black; the wings are spotted with fuscous brown, especially the anterior ones; the posterior margin of all four tinted with brownish.

We also figure the *Ascalaphus Barbarus*, a sub-genus of Myrmeleon. It is black, with head and thorax spotted with fulvous; the wings blackish, with nervures yellowish; it is from Sicily.

The name of Hemerobius has been given to that genus, because the little animals which it comprehends, live but a very few days in the form of the perfect insect. They are very pretty insects, usually of a green-colour, and the wings have the fineness and transparency of gauze. The body, observed cross-wise, is of a soft green, and sometimes appears to have a tint of gold. Their corslet is of the same colour. Their eyes, of a fine red bronze-colour, have the brilliancy of the best polished metal. They are frequently found in gardens, where the females go to deposit their eggs, which are very remarkable.

We often see, on the leaves of different shrubs, little stems, about the thickness of a hair, about an inch long, of a white colour, and about ten or twelve in number, placed side by side, and attached above or underneath the leaf. These little stems are seldom straight. They have a small curve, and are terminated in a kind of small, elongated ball, which is the egg. These eggs, which have been taken by some botanists for sorts of mushrooms *(ascaphorus perennis)*, are clothed on one of their sides with a viscous matter, proper to be woven. It is the end which the female applies upon the leaf, where a part of the matter attaches itself, when she elongates the hinder part of her body. This matter, which is elongated, forms a thread, drying up, and becoming hard in the air. This thread serves to draw the egg out of the body of the female, and to sustain and carry it when it is drawn out.

As soon as the larvæ issue from the eggs, they spread themselves along the leaves, in search of aphides, which constitute their ordinary aliment. They seize them with two sorts of little horns, which they have in front of the head, and suck them until nothing remains but the skin. They make so great a carnage of those insects, that Reaumur has named them "*lions des pucerons*" (lions of the aphides). Placed upon a leaf, covered with aphides, the larva is not obliged to use many movements to procure the food for which it has occasion. Accordingly, in a very short time, it destroys an immense quantity of these little animals, which seem to come and present themselves to their enemy. The larva being much more agile than the aphides, it soon gets possession of as many of them as it wants. To seize the largest, and suck them thoroughly, is, for it, but the affair of a moment. These larvæ, so cruel to this species of insects, are not less so to one another. Whenever they meet, they attack each other bitterly, and shew no more mercy than they do to the aphides.

Some other larvæ of the same genus have an instinct of clothing themselves. They form a very inartificial covering of a considerable thickness, in relation to their body, which then appears to be loaded with a little mountain. This is composed of the skins, down, and dried parts of the aphides, which they accumulate one upon the other. All these parts are held together only by a sort of rude interlacement, and this garment is kept upon the back of the larva only by being ensheathed in the furrows and rugosities of the skin which separates the rings. Its construction, however, requires some address on the part of the larva, and especially great suppleness and agility in the head, and the sort of corslet to which it is attached. It is with its two horns that the larva takes up the little mass which it desires to pass over its back. It rests it upon its head, which it then raises briskly. By this movement it throws back the mass. If it has not thrown it to the spot where it wishes to place it, it finally fixes it by making various contortions with its body, and especially with its head. The part to which the head is attached, has so much agility, that when one of these larvæ is placed upon its back, it promptly regains its legs, by turning its head, until it is between its back and the plan of position. In this attitude it is enabled to make a somerset, which replaces it in its natural position.

Reaumur distinguishes three sorts of genera of hemerobiæ. The larvæ of the first two have the body oblong and flatted. Some have tubercles, with aigrettes of hairs on the sides; they form the first genus. Those of the second are destitute of these appendages. The body of the larvæ of the third is less depressed, and covered from the neck to the rump, as we have already described, with the exuviæ of the aphides which they have devoured. An individual, from which Reaumur had stripped its covering, soon formed a new one of some scrapings of paper with which he furnished it. As to other matters, the metamorphoses of these insects are similar.

As they live in the midst of abundance, they soon arrive at the term when their metamorphosis is to take place. It is usually in fifteen days after they have issued from the egg. that they are changed into nymphs. At this period they quit the leaves where they have lived, seek a dry leaf, to withdraw to conceal themselves in one of its folds, and there they spin a cocoon as round as a ball, of a very white silk, in which they enclose themselves.

These cocoons, the largest of which is scarcely of the bigness of a pea, are of a very close texture. The larvæ employ in their construction, the silk of which they have store in their spinnerets, which are placed like those of the spiders, at the

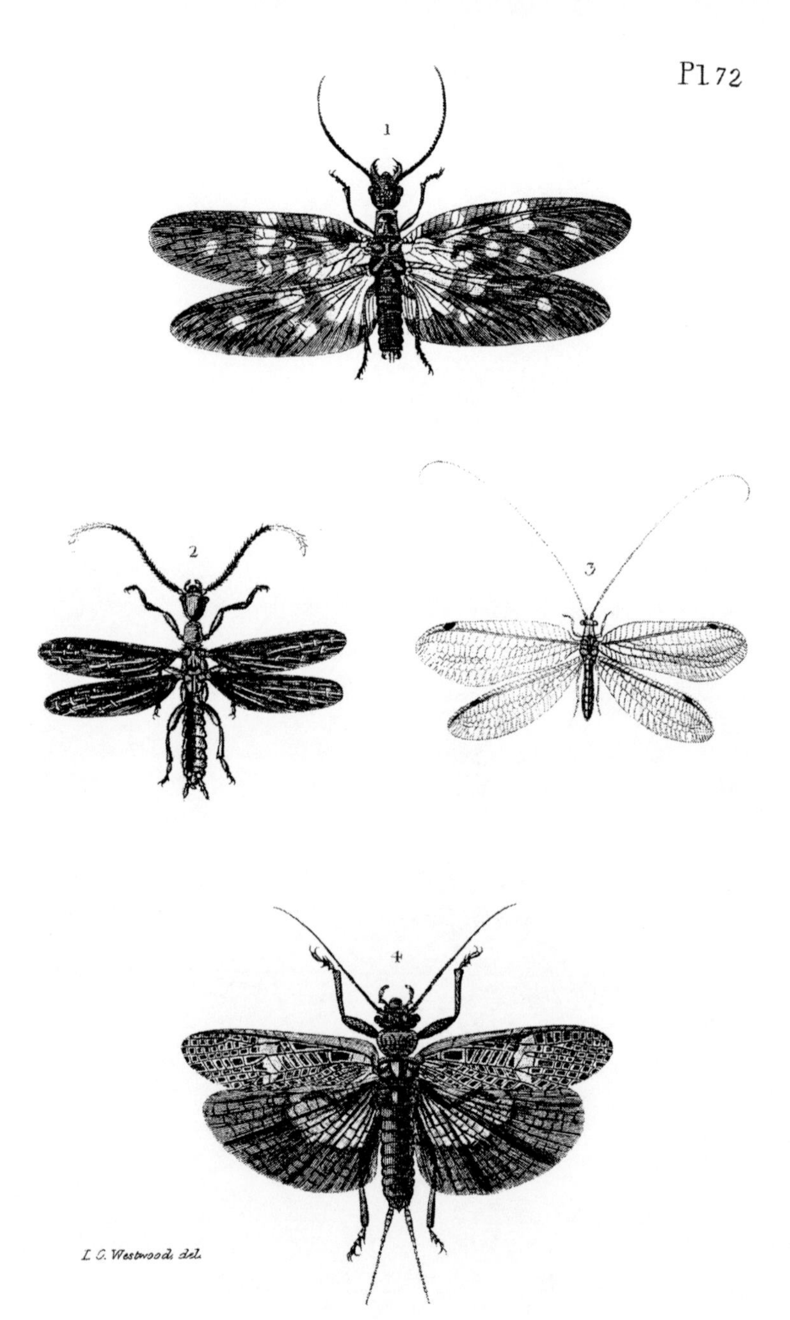

1. Chauliodes *maculipennis*. G. R. Gray. 2. Embius? *Brasiliensis*. G. R. Gray. 3. Hemerobius, *longicollis*. G. R. Gray. 4. Eusthenia *spectabilis*. Westw.

London, Published by Whittaker & Co. Ave Maria Lane. 1837.

extremity of their body. In seeing these cocoons, we can hardly conceive how the body of the larva, curved as it is, and reduced to occupy so small a space, can furnish a sufficient quantity of threads to fill each cocoon, and arrange them with so much dexterity. But if we observe one of these larvæ, when tracing the contour of its cocoon, we shall see the extremity of its body agitated with the most surprising quickness; and the address with which the entire body changes place, sliding over the spherical envelope, which is but sketched out, without deranging the threads which seem scarcely capable of supporting it, being so slender, is not less astonishing.

Soon after it has finished its cocoon, the larva becomes changed into a nymph. If it be in summer that it undergoes that metamorphosis, it becomes a perfect insect in about fifteen days after; but if in autumn, it passes the winter in its cocoon, in the nymph-form, and does not issue forth until the following spring. Although the larva is not large, it is yet difficult to imagine how it can lodge in so small a cocoon; but we are much more surprised when we see the insect which comes out of it.

The flight of the hemerobiæ is heavy Some species walk tolerably fast, but, notwithstanding, they are easily caught. If these pretty insects please the eye by the delicacy of their form, and the beauty of their colour, some of them are extremely disgusting, from the odour of the excrements which they discharge. This odour communicates itself to the fingers, when we touch these insects, and remains there for a long time.

We have figured a new species of this genus, under the name *Hemerobius longicornis*, which is fulvous, the antennæ very long, wings whiteish, with a black spot at summit of each.

In the genus Semblis, we have figured a new species, belonging to the Chauliodes of Latreille, under the name of

Chauliodes maculipennis. It is black: sides of the thorax, and beneath, yellow; the wings blackish, and spotted with white. This insect is from India, and may be formed into a sub-genus, under the name of Hermes.

The larva of RAPHIDIA is very long, narrow, of grey and black mixed, with a scaly head and six feet. It resembles a little worm. Its vivacity is very great. It walks very fast, and turns itself in all directions to insinuate itself between the crevices of trees. It should be carnivorous. The nymph, according to Linnæus, walks, and is active until the moment of its metamorphoses. It carries its wings in an envelope, placed on each side of its body. This insect is found in Europe.

We now come to those most celebrated insects, the TERMITES. They are almost all of them strangers to Europe. The illustrious Linnæus considers them, with justice, the greatest scourge of the two Indies, because they occasion ravages, equally prompt and destructive to the property of man. Under the torrid zone, they pierce and devour all wooden buildings, utensils, furniture, stuffs, and merchandizes of all kinds, and would soon reduce them completely to powder, if not prevented in time. Nothing but metals and stones can resist their destructive jaws.

Although the termites of Africa have attracted the attention of several travellers, by the size and structure of their nests, we are well acquainted with their manners, only through the interesting details given by Sparmann, respecting their industry and mode of living.

These insects, which have been called *white ants*, and many other analogous names, have, in truth, many relations to the ants. Like them they live in societies, composed of three sorts of individuals; like them, they build nests, but still more extraordinary, and, for the most part, on the superficies of the earth. They issue from these through subterraneous passages, or covered galleries, when necessity obliges them, and proceed from them on their devastating excursions. Like the ants, they are omnivorous; like them, at a certain time of their life, they have four wings, and then emigrate and form colonies. The termites also resemble the ants in their laborious activity, but they surpass the bees, the wasps, and the beavers, in the art of building.

Each community, according to Sparmann, is composed of one male, one female, and of workers. He distinguishes these last by the names of *labourers*, and *soldiers*, having observed the first working, and the others combating for the defence of their property. The males and females do not acquire wings, until a short time previous to their being in a state to reproduce their species. The soldiers, which some authors have regarded as neuters, or mules, have a different form from the workers, which have been thought to be males. But, according to Sparmann, this is an error, the soldiers not differing from the labourers, but because they approach within a degree of the perfect state.

Notwithstanding the opinion of Sparmann, it appears very probable that there are, among the *termites*, a kind of individuals, which never acquire wings. The first change which a larva undergoes, in approaching to the perfect state, is nothing else but its metamorphosis into a nymph. Now, as all the nymphs, the perfect insects of which are to have wings, have always the rudiments of these organs, and that the termites of Sparmann have not the least vestiges of them, we may deduce, that not being nymphs, in consequence of the absence of the rudiments of wings, they should form a particular order, and that thus there are three sorts of individuals.

In the nests of the *termes bellicosum*, are found, says our author, a hundred labourers for one soldier. The first are scarcely three lines in length, and twenty-five of them weigh about a grain. Their mandibles appear conformed to gnaw, and retain bodies; whereas, the second, which are much larger, and half an inch long, have very pointed mandibles, in the form of awls, proper only for piercing and wounding, an object which they are found to achieve uncommonly well.

The insect, which after its entire development is provided with wings, differs from the two other individuals, not only in these parts, but also in the form of its body. It is then about eight lines in length. Its wings are as long again, and it has two projecting eyes, which are wanting in the soldiers, and the labourers, or which, if they do exist, are not perceptible. These winged insects are found in the nests, only immediately before the rainy season, a period at which they undergo their final metamorphosis, and after which they commence emigrations, and proceed to form new societies. Thus, one may open twenty nests, without seeing a single one there, because they rarely wait for the second or third shower, before they depart. If the first rain falls in the night, and leaves much humidity behind it, the following morning, all the surface of the earth which neighbours their habitation, is covered with these insects, and especially the waters, because their wings are only formed to carry them for some hours; so that after the rising of the sun, very few of them are found to have preserved these organs, except the morning has continued to be rainy. In that case, they are seen, scattered and isolated, fluttering from one place to another, endeavouring to avoid their enemies, especially one sort of ants, which pursue them even to the trees where they have taken refuge. Those which escape the murderous teeth of these insects, become the prey of birds, of carnivorous reptiles, which make war upon them at that moment; so, that out of many millions that were hovering in the air, there scarcely remain more than a few couples to fulfil the first law of nature, and lay the foundations of a new republic.

Besides these enemies, the termites have others of a different species. These are the inhabitants of many countries of Africa, and particularly those of Guinea, who make them an article of food. M. Kœnig, in his essay on the history of these insects, tells us, that to catch the termites before their emigration, the Indians make two holes in the nest, one to windward. and the other to leeward; at the latter aperture, they place a pot, rubbed with aromatic herbs. On the windward side they make a fire, the smoke of which drives these insects into the pots. By this method they take a great quantity, of which they make, with flour, different pastries, which they sell to the people. This author adds, that in the season in which this aliment is abundant, the abuse of it produces an epidemic cholic, which carries off the patient in twenty-four hours. The Africans are less ingenious in catching and preparing them. They content themselves in collecting those which fall into the water at the time of emigration. They fill large cauldrons with them, and grill them in iron pots, over a gentle fire, stirring them as coffee is stirred. They thus eat them, without sauce, or any other preparation, and find them delicious. This gentleman has several times tasted them cooked in this manner. He has found them delicate, and they have appeared to him nourishing and wholesome.

In the midst of their distress, when thus attacked by such bitter enemies, the termites sometimes forget their danger. The majority have no wings, but they run extremely fast. The males are very ardent after their females; but from the time of their metamorphosis, they are completely degenerated. One of the most active, the most industrious, the most eager for prey, and the most ferocious little animals in the world, becomes at once one of the most indolent and cowardly of created beings. It suffers itself to be dragged along by the ants to their nests, without making the smallest resistance, and it only escapes, when some labouring termites, which are running continually near the surface of the earth, under their covered galleries, perceive it, and come to its assistance. Those which have not been thus protected, infallibly perish.

The workers, that thus save a male and female from the teeth of their enemies, immediately place them in shelter from all dangers, and then shut them up in a small chamber of clay, proportioned to their size. They leave them at first but a small aperture, capable of giving entrance to themselves and the soldiers. They provide for the wants of this couple, and afterwards for those of the young to which they give birth, and defend them until the young be in a state to partake of this task with them. Sparmann, who never witnessed the sexual intercourse of these insects, believes that this is the period when it takes place. A little time after the shutting up of the male and female, the belly of the latter extends by degrees, and enlarges to such a point that in an old female, it is fifteen hundred or two thousand times as voluminous as the rest of the body. Sparmann presumes, that when it is three inches in length, the female must be aged more than two years. It is incessantly putting forth its eggs, to the number of sixty in a minute; and our author has observed some old females lay eighty thousand and upwards in twenty-four hours. If Sparmann was not deceived in this calculation, what an astonishing degree of fecundity is this!

After the male has lost his wings, he changes form no longer, and increases no more in bulk. He usually remains concealed under one of the sides of the vast abdomen of the female, and appears to share nothing of the attention of the other insects.

In proportion as the female lays, the labourers carry off the eggs, and place them in lodges, separated from that of the mother. There the young ones which issue from these eggs, are provided with everything, until they are in a situation to procure for themselves what is necessary, and to take a share in the labours of the society.

Having followed Sparmann in his interesting description of the *termes bellicosus*, the largest and best known species in Africa, and which builds the largest, the most curious, and most numerous nests in the island of Bananes, and in all the adjacent parts of the continent, and whose societies are the most numerous, we must now consider the industry of these singular insects, in the construction of their nests.

Sparmann describes five species of termes, which are—*bellicosus*, of which we have just spoken; *mordax*, *atrox*, *destructor*, and *arborum*. Some build their nests on the surface of the earth, or partly above, and partly below. Others, on the branches of trees, and sometimes at a very considerable height.

The external figure of the nests of *termes bellicosus*, is that of a small hill, more or less conical, the form of which approaches to that of a sugar-loaf. Their perpendicular height is ten or twelve feet above the surface of the earth. If these edifices be compared with those of man, we will find that they are in relation to these insects, whose labourers are scarcely a quarter of an inch in length, what monuments, five times larger than the pyramids of Egypt, would be to us. Each of these edifices are composed of two distinct parts, the exterior and the interior. The exterior is a broad cap, in the form of a dome, sufficiently large, and sufficiently strong to protect the interior from the vicissitudes of the atmosphere, and the inhabitants against the attacks of their enemies. Neither man, nor wild bulls can destroy its solidity by mounting upon it. Each of these edifices is divided into a great number of apartments, which are that of the male and female, denominated by Sparmann the *royal* chamber; those in which their numerous posterity are brought up, termed *nurseries*, by the same traveller, and the magazines. These last are always well stored with provisions, which consist of gums, or the thickened juices of plants, assembled together in small masses. The lodgings occupied by the eggs and the little ones, are entirely composed of parcels of wood, united together by gum. These edifices are extremely compact, and divided into many irregular chambers, the largest of which is not more than half an inch in extent. They are placed around that of the mother. This last is placed at an equal distance from all the sides of the grand body of the building, and directly under the summit of the cone. All the pieces which surround it, compose a complicated labyrinth, which extends on all sides, for more than a foot in distance. The galleries, which are excavated in the loosest pieces, are wider than the calibre of a thick gun. They lead to all the departments, and descend under ground to the depth of three or four feet. It is from these that the labourers get the fine gravel, which they convert in their mouths to a solid and strong clay, or mortar, with which they construct the hillock, and all the buildings, with the exception of the nurseries.

Other nests, again, are seen of a cylindrical form, about two feet in height, each of them covered with a roof, in the form of a cone, the materials of which are the same. Sparmann names them *turretted nests.* These are constructed by the *termes atrox*, and the *termes mordax.* The exterior figure of these nests is more curious than that of the nests of the *termes fatalis;* but the interior is not so well arranged: all are so solidly built, that it would be more easy to overturn them from their foundation, than to break them in the middle.

The nests of the *termes arborum*, differ from those of the other species of this genus, in form and size. They are spherical, and built in the tree. Sometimes they only attach to a single branch, which they surround, to the height of sixty or eighty feet. Some are seen, but rarely, as spacious as a sugar-hogshead. They are composed of parcels of wood, gums, and the juices of trees, with which these insects form a paste to construct their cells. These nests contain an immense number of individuals, of different ages, which the inhabitants seek out to nourish their poultry with. Sometimes the termes place their nests on the roofs, or any other part of houses, and do vast mischief there. But the other species, which are much larger than this, are far more destructive. The *termes bellicosus* and others, advance under ground, descend under the foundations of houses and magazines, penetrate into the stakes which support the buildings, pierce them from one end to the other, and render them completely hollow. The evil is not discovered until it is irremediable, because they never pierce the surface in any part, so that a piece of wood which appears the most entire, falls in powder if the hand be applied to it. When a stake in a hedge has failed to take root, it is their business to destroy it. If it is encompassed with a sound bark, they enter at the lower end, and eat all except the bark, which remains, and preserves the appearance of a solid stake. But if the bark be not sufficiently sound for them to depend upon it, they cover the entire stake, or stump, with a sort of mortar, and it then has the appearance of having been dipped in a thick slime, which has dried over it. They work under this envelope, leaving only as much of wood or bark, as is necessary to support it.

The *termites arborum*, says Smeathman, often enter into a chest or coffer, make their nest there, and destroy everything which it contains. Nothing penetrable is in safety against them; they discover and destroy everything, and, as if in concert with the other species, they ruin a house from top to bottom in a very short time.

The first object with which one is struck at the opening of a nest, is the conduct of the soldiers. They defend, says Sparmann, the common property with fury, and bite away at every thing they meet. If they can reach any part of the body of a man, they fasten their jaws deeply there at the first blow, and never let go. They sooner suffer their bodies to be plucked to pieces than take to flight. As long as the attack continues they are in the most violent agitation. But as soon as the enemy removes, calm is re-established, and in less than half an hour they all withdraw within the nest.

The *termites viatores* are not less curious in the order which they observe in their march, than those already described. This species appears to be much more rare, and much larger than the *termes bellicosus*. Smeathmann was unable to derive any information from the negroes on this subject, from which he concludes that these insects are but seldom seen. He himself only saw them by chance. "One day," says he, "having made an excursion with my gun along the river Caramankes, passing, on my return, through the thick forest, while I was proceeding in silence, in the hope of finding some game, I heard all at once a hissing sound, an alarming thing in this country, where there are an abundance of serpents. The second step which I made caused a repetition of the same sound. I then recognized what it was; but I was surprised to find neither covered ways nor hillocks. The noise, however, conducted me some steps along the path, where, to my equal pleasure and surprise, I saw an army of termites issuing out of a hole in the earth, which was but four or five inches in diameter. They issued forth in very great numbers, moving forward with all the rapidity of which they were capable. At about less than three feet from this place, they divided into two bodies or columns, composed of the first order, which I call *labourers*. They were twelve or fifteen a-breast, and marched as closely as a flock of sheep, describing a right line, without ever deviating to either side. Here and there was to be seen among them a soldier, trotting on in the same manner, without stopping or turning ; and, as he seemed to carry his enormous head with difficulty, I figured to myself a very large ox among a flock of sheep. While these were pursuing their route, a great number of soldiers were spread on one side and the other of the line, some within the distance of a foot or two from each other, posted as sentinels, or roaming like patroles, to watch that no enemies approached the labourers. But the most extraordinary circumstance of this march, was the conduct of some other soldiers, who, mounting upon the plants, which grew here and there in the depth of the wood, placed themselves upon the point of the leaves, about ten or fifteen inches from the ground, and remained suspended above the army in march from time to time. One or another of them was beating with his feet upon the leaf, and making the same sort of noise, or clashing sound, which I had so often observed on the part of the soldier, that performs the office of inspector when the labourers are at work to repair a breach in the edifice of the *termites bellicosi.* This signal among the termites of which I am now speaking, produces an analogous effect. For every time that it was given, the entire army replied by a hissing noise, and obeyed the command by redoubling their pace with the greatest ardour. The soldiers who were perched, and gave this signal, remained tranquil during the intervals. They only turned their heads a little from time to time, and seemed as much attached to their posts as sentinels of regular troops. The two columns of the army united at about twelve or fifteen paces from their separation, having never been more than nine feet distant from each other, and then descended into the earth, through two or three holes. They continued to work under my view for more than an hour, which I passed in admiring them, and appeared neither to increase nor diminish in number, with the exception of the soldiers that quitted the line of march, and placed themselves at different distances on each side of the two columns, for they appeared much more numerous before I retired."

The workers of this species are at least one-third larger than those of the others, and their buildings ought therefore to be more astonishing than those of the other termites. The male and female of the termes viator are unknown.

Such are the principal observations made by Smeathmann on these very extraordinary insects. In his account in the Philosophical Transactions, in Sparmann's journey to the Cape, may be found some other details, which brevity obliges us to omit here. It may also be remarked, that though these observations rest on good authority, they have need of being followed up afresh, and for a considerable time, before the history of these insects shall be complete. The following is the substance of some remarks of M. Latreille on a species which he found in the neighbourhood of Bordeaux, which is the *termes lucifugus* of Rossi.

This insect inhabits, in considerable numbers, in the trunks of some pines and oaks, near the neck of these trees. It always works under cover, gnawing the ligneous portion, which is situated immediately under the bark, without the bark itself being affected, at least externally, and there it forms a great number of holes and irregular galleries. The injured part of the wood appears moist, and presents a great number of little, transparent, gelatinous bodies, similar in appearance to little parcels of gum arabic. This insect appears to be furnished with an acid of a very penetrating odour, which answers the purpose of softening the wood. This odour is a long time preserved in those boxes into which these termites have been put. The societies of these insects are, at a certain period, composed of four sorts of individuals. They present at all times two sorts of individuals, without wings, elongated, soft, of a slightly yellowish colour; with head, corslet, and abdomen distinct; agile, provided with six feet, each pair of which is attached to a proper segment. They have a large head, the essential organs of which are the same as in the winged individuals, the eyes only appearing to be wanting, or, at least, very small. These two sorts of individuals are distinguished by the form of their heads. In one kind, those that compose the bulk of the society, the head is rounded, and the mandibles are not advanced; in the others, which do not form the twenty-fifth part of the community, the head is much larger, elongated, of a cylindrical form, and terminated by projecting mandibles, which cross each other. These last, M. Latreille remarks always remain at the entrance of the cavity, where there is a larger assemblage of the individuals of the first kind. There are found, at least, towards the end of winter, and in spring, some individuals, similar in all respects to the first, but which have moreover certain appendages, in the form of wings, white, and four in number; namely, two on the second ring, and two on the third. The first ring is here, as in all the preceding, and in the winged individuals, in the form of a semicircular plate. It is the first segment of the corslet, that to which the fore-feet are annexed. In the month of June the completely winged individuals make their appearance. They resemble those just mentioned in figure; but their colour is blackish, and they have two very distinct eyes, and four wings, two or three times longer than the body. Some are male, others females. If their colony be visited a month later, we shall meet some of these individuals, but few in number, that have lost their wings. We shall also perceive, in some meanders of the wood, the eggs of these insects, which are almost an impalpable powder.

From these observations, illustrated by analogy, we may be permited to draw the following conclusions: 1st, that the apterous individuals, and short and retiring mandibles, are larvæ; 2d, that individuals, similar in form, but having aliform appendages, are nymphs; 3d, that the individuals, still figured in the same manner, but having large wings, are the insects arrived at their final form, and endowed with the faculty of reproduction; and that the individuals of this sort, though deprived of wings, which are met with later, are females, whose wings are fallen, and which have laid their eggs; 4th, that the apterous individuals, with cylindrical head, and projecting mandibles, and which answer to the soldiers of Smeathmann, form a particular order in the society. These last insects always have the same form, never acquire wings, and do not contribute to the propagation of the species. They are charged, as it would appear, only with the defence of the republic. It is certain that the winged termites have essentially the same form which they had in the larva and nymph states. This is quite in accordance with the march of nature, who, in all cases in this class of animals, of semi-metamorphosis, as in the orthoptera, hemiptera, &c., merely developes the primitive type of the species, which she has established in the larva. Its figure does not change much when this larva passes to the nymph state. The habits being the same in all cases, it would follow that there must be few changes of form. We remark, on the contrary, that the more different the insect is from what it was in the larva state, the more its earlier manners become altered. Nature has been unable to prepare, if we may so express it, these diversities in the modes of existence, but by condemning the insect, when a nymph, to a state of inertia, a species of apparent death. Since then the individuals named soldiers are very different from the winged insects, since the nature of the metamorphosis of the termites excludes great changes of form, and since among these military termites, none are ever found with the appendages of wings, though individuals with such appendages, and similarly figured in all respects, are found in the colonies at all periods of the year, it is fair to conclude that the soldiers compose a peculiar *caste*, and represent, in some measure in this genus, the neuters of the ants and bees.

We may presume that the entire development of the metamorphosis of these insects is not effected in less than the course of two years—since, when the winged individuals appear, a great number of larvæ are found in the nests, which larvæ must belong to a preceding generation, and do not assume wings, at the soonest, until the following year.

To the *termes morio* may be referred the *termes destructor*, of Degeer, and this, perhaps, is the *termes fatalis*, of Linnæus. It is very common in the West Indies, and throughout South America. Rochfort informs us, that to cut off the way against these insects, the inhabitants rub the places where they would pass with oil of that species of *palma-christi*, with which the negroes rub their heads to protect themselves against vermin. The oil of the *manatus* has also the same effect; and if it be poured upon their nest they will abandon it immediately. Rochfort calls them *wood-lice*, and also tells us that these insects will not gnaw the printed parts of books, probably having no particular taste for ink, or peradventure for literature.

Chauvalon, in his "*Voyage à la Martinique*," after speaking of a species of mite, which introduces itself into the flesh, says, "another insect, equally common, and still more hurtful, is that which is called *wood-louse (pou de bois)*. It has, in fact, the size and appearance of a louse. Its colour is a reddish white. It is without wings. They live in troops in sorts of hives, from which they communicate whenever they please, by covered ways, which are made of the same substance as the hives.

"This substance is a sort of paste, composed with a fluid which is natural to them, and which answers for them the purpose of an universal solvent. In whatever place, or on whatever thing they place their hives, and their covered ways, whether on the wood of houses, on the bark of living trees, on paper, on clothes, on stones, or on metals, all is corroded and dissolved by this liquid. Mixed with such materials, it forms, as we have just mentioned, a paste, which they make pretty nearly of the thickness of a card, to form their covered ways and their hives. These hives themselves are but a sort of accumulation of covered ways, assembled one upon the other in all directions. Almost all the houses in our islands being constructed of wood, these insects very speedily destroy the pieces which are most necessary to the solidity of the building, if their labours and multiplication are not very speedily put a stop to.

"A means, however, has been discovered equally prompt and efficacious in arresting their ravages, and destroying themselves, and that is arsenic. Let but a single pinch of this be put into their hives, through a small hole made there, or introduced into one of the covered ways which lead to them, at the end of some hours the millions of *wood-lice* which were assembled in the hive, all perish without exception.

"This insect is a species of ant. It appears to me to be the same which M. Adanson has spoken of in his voyage to Senegal, p. 99, under the name of *vaguevague*. In that part of Africa it is doubtless more maleficent. This academician tell us that they bite the skin, and occasion swelling, and severe pains. They do not bite at Martinique. The only mischief that they do is by their devastations.

"It is astonishing that they have not yet been informed in Senegal of the effect of arsenic on these animals, or have not employed, as in our colonies, where it has been used for so many years. The accidents which may result from its use, are not much to be dreaded, since a very small quantity is sufficient to destroy these insects."

In the southern parts of Europe and in Africa there are

some insects analogous to the termites, but having the head wider than the corslet, the tarsi with three articulations, the wings scarcely reaching beyond the abdomen or altogether wanting, the legs compressed, with the anterior thighs broader; they have no simple eyes and the corslet is elongated. They form the genus *Embia* of M. Latreille. We insert a figure of a singular insect, which bears some similarity to the genus *Embia*, but differs in having the antennæ as long as the body. The thorax much longer and more separate from the head, which is rounded posteriorly, the terminal joints of the palpi rather larger; it therefore may be formed into a distinct sub-genus, which Mr. Gray has named Olyntha.

The species which is figured is from South America, therefore is named *Brasiliensis*. It is black; thorax and anterior femora fulvous; the tip of the antennæ white, the wings blackish, rather lighter between the nervures. In the collection of J. G. Children, Esq.

Many naturalists have confounded the insects of the genus Perla with those of *Phryganæ*, which they resemble in their antennæ and the mode of existence under the larvæ form. But they may be distinguished at once from these insects by their flatted corslet and the threads of their abdomen.

These larvæ, as do those of phryganeæ, live in the water, where they feed on small aquatic insects. Their body is elongated and composed of several rings. The head is scaly, and they have six feet. They enclose themselves in a case of silk, open at the two ends, which they cover with various substances and transport it every where along with them. It is in this sheath that they undergo their metamorphoses. Before changing into the nymph state, the larva opens the apertures of the two extremities of his habitation, with divers threads of silk, which form a sort of grill at each end. This grill, not of a very close tissue, gives passage to the water, which is necessary to the nymph for the purpose of respiration, and it places it in security from those voracious insects, which it could not escape without this precaution. It remains but a short time under this form. Before its final metamorphosis, it breaks one of the grills of its sheath, that it may issue forth with the greater facility when it becomes the perfect insect. On quitting their exuvia of nymph, the perlæ become inhabitants of the air; they do not, however, remove to any great distance from the water, because the females deposit their eggs there, after coupling.

These insects do not form a very numerous genus, and all the species are European.

After the genus *Perla*, we placed *Eusthenia spectabilis*, of Westwood. The body is dark brown, the upper wings pale with a brown spot in each cell; the anterior margin purplish brown, and with a whitish fasciæ rather beyond the centre; the hinder wings red, with a very broad bluish-black margin. Mr. Westwood has established this genus on account of the jaw being horny and very much dentated. The habitat of the species is unknown.

The *Phryganeæ* have been called by Reaumur, "*Mouches papilionacées*," from their resemblance, on the first glance, to butterflies, or rather to the *phalenæ*. The ancients named their larvæ *ligniperdæ*, from a mistaken notion that they were injurious to wood.

The larvæ and nymphs of all the known species live in the water. They are to be found in marshes, ponds, and running streams. They lodge in portable cases which they make with silk, and cover with various substances. They drag them every where along with them.

The interior part of this cave, or tube, in which the body of the larva is lodged, is smooth and polished. Its upper part is covered with fragments of various substances, proper for the purpose of strengthening and defending it. The external parts are often bristly, and full of inequalities. Some larvæ form theirs of different pieces, which they arrange symmetrically side by side. When this sheath or case becomes too short, or too narrow, they make another of a size proportioned to that of their body. Sometimes the new one differs as much from that which they have thrown off, as our modern habiliments do from those of our ancestors, for these insects employ materials that have no manner of affinity among themselves. They make use of leaves, or parts of the leaves of several species of plants, of small cylindrical or irregular sticks, of stems of plants, of reeds, rushes, bits of clay, of aquatic shells, and in fact of all the substances which they can find in the water. Some of these cases are formed but of one substance, and these are the best fashioned. Others are composed of all the discordant materials which we have enumerated, and form a dress equally fantastic and unbecoming.

Each case is internally formed like a cylinder, at each extremity of which is an aperture. The one through which the larva puts forth its head and feet is larger than the other, which is placed in the middle of a circular plate, attached to the end of the case, which is partly stopped by it.

Almost all the cases which are covered with leaves, are flat. But those of this shape are not very common: the cylindrical form being more general. There are some, the entire external part of which is composed of particles of rushes, cemented together. But with whatever materials they may be covered, it is rare to find any without some piece which supports the rest, and this piece is necessary to the perfection of the case. Sometimes it is a bit of stone, a pebble, or a shell. It frequently happens that these cases are entirely covered with the small shells of aquatic snails, or with the shells of mussels which enclosed the living animals.

Cases constructed of such heavy materials would become a burthen to the insect, if it were obliged always to walk upon the ground. But as it sometimes walks at the bottom of the water, and sometimes at the surface, and over the plants which grow there, the case is carried with but little trouble, if the different pieces of which it is constructed, be of a weight pretty nearly equal to that of the fluid. This the instinct of the animal leads it to accomplish, by making use of bodies, the specific weight of which is lighter than that of the water.

These larvæ have six feet; the head is brown and scaly, and the mouth is armed with jaws adapted for cutting the materials which these insects employ in the construction of their cases. Their body is composed of twelve rings. The six feet are attached to the first three rings. On the fourth, there are three fleshy eminences, through which they draw in and reject the water. To the other rings are certain threads, bearing some analogy to the gills of a fish. It is said that they feed upon the larvæ of libellulæ, and tipulæ, and also on the leaves of aquatic plants. But it seems more probable that they are purely herbivorous. When one of these larvæ is stripped of its case, if the latter be left near it the insect will immediately re-enter, putting in its head first.

When the larva, which cannot swim, is inclined to walk, it puts its head, and the anterior part of its body, out of the case.

It is not only in the construction of the case that these larvæ exhibit their industry, but they shew it still more in the manner in which they close the case, before they become changed into the nymph-state. They all undergo this metamorphosis in the water, and in this sort of tube which they have constructed. If nature had not gifted them with the faculty of rendering it inaccessible to aquatic insects which are their enemies, they would become their prey. But they secure themselves from the murderous fangs of their adversaries by stopping the two apertures of this tube. Each larva employs in this operation the silk which it has at its disposal, to form a sort of grating, the meshes of which are close enough to hinder the carnivorous insects from penetrating into the interior of the case, and sufficiently separated to leave a clear passage for the water, which is necessary to the nymph for the purposes of respiration. But before it closes its sheath in this manner, the larva is careful to attach it to some solid substance, that it may quit it with the greater facility when it wants to issue forth.

The nymph is of a citron yellow, and all the parts which the perfect insect is to have are clearly distinguishable in it. Its head is small in proportion to its body, and it exhibits one remarkable peculiarity. This is a sort of bill formed by two hooks, situated one at each side of the head. It makes use of this to detach the grate on that side from which it is to issue forth. This takes place fifteen or twenty days after its metamorphosis. The phryganeæ do not quit their exuvia of nymph in the water. The nymph comes out of its sheath, and withdraws into a dry place. There it remains quietly waiting until the skin which covers it dries up, and cracks, which it does in a minute or two. At the end of this short interval, the perfect insect is in a situation to be enabled to make use of its wings.

These insects usually fly on the borders of streams, marshes, and ponds. The females deposit their eggs on the plants which grow in the water. The eggs are enveloped by a glairy transparent matter, of the consistence of a soft jelly, which adheres to the plant, almost as soon as it is placed there.

During the day-time, the phryganeæ remain tranquil; but towards the setting of the sun, they commence to take their flight. It is not uncommon to see them in apartments. They are habituated to fly round the flame of candles, where, like the moths, they often burn their wings. Their flight is light and lively, and in walking, such is the quickness of their pace, that they seem to glide along. When they are handled they leave a disagreeable odour on the fingers which lasts for a considerable time. Almost all the species are found in Europe, and many are remarkable for the colour of their wings, which resemble those of the phalenæ.

THE NINTH ORDER OF INSECTS.

THE HYMENOPTERA—*Piezata*, Fab.,

AGAIN presents us with four membranaceous and naked wings. A mouth composed of mandibles and jaws, with two lips; but the wings, of which the upper are always the largest, have fewer nervures than those of the neuroptera, and are only veined. The females have the abdomen terminated by an auger or a sting. Beside the composite eyes, they all have three simple ones. Antennæ variable, not only as to the genera, but in the sexes of the same species, filiform nevertheless, or setaceous in the majority; jaws and labium, generally narrow, elongate, attached in a deep cavity of the head by long muscles; they form a semi-tube at their lower part, are often folded at their extremity, more proper to conduct the nutritive juices than for mastication, and are united in the form of a proboscis in many. The tongue is membranaceous, either widened at its extremity, or long and filiform, having the pharynx at its anterior basis, and often covered by a sort of sub-labrum, or epi-pharynx. Four palpi, two maxillary, and two labial. The thorax with three segments united into one mass, the anterior being very short, and the other two confounded into one. The wings crossed horizontally on the body. The abdomen most frequently suspended to the posterior extremity of the corslet by a small thread or pedicle—finally, tarsi, with five articulations, none of which is divided. The auger or oviduct, and the sting, are composed, in the majority, of three long and slender pieces, two of which serve as a sheath to the third, in those which have an auger, and a single one of which, the upper, has a groove underneath to enclose the two others. In those in which the auger is transformed into a sting, this offensive weapon, and the oviduct are denticulated like a saw at their extremity.

This order is divided into two sections. In the first, that of TEREBRANTIA, there is an auger in the females. It is separated into two great families. The first, that of

SECURIFERA,

Is distinguished from the following, by a sessile abdomen, or whose base is united to the corslet in its entire thickness, and appears to be a continuation of it, and to have no proper movement of its own.

The females have an auger, most generally securiform, and which serves not only for the deposition of eggs, but also to prepare a place to receive them. The larvæ have always scaly feet, and often others which are membranaceous. This family is composed of two tribes.

The first, TENTHREDINETA, has long and compressed mandibles; tongue trifid, and as it were, digitated auger composed of two laminæ, serrated, pointed, joined, and lodged in a groove under the anus. The maxillary palpi have always six articulations, and the labial four. The latter are always shorter. The four wings are divided into numerous little cells. This tribe composes the genus

TENTHREDO, of Linnæus.

Abdomen cylindrical, rounded posteriorly, composed of nine rings, so united to the corslet as to seem but a continuation thereof. Their wings appear ruffled, there are two small rounded bodies like grains, usually coloured, behind the scutellum, and the general port of these insects is heavy. The form and composition of the antennæ vary. The mandibles are strong and denticulated. The extremities of the jaws are almost membranaceous, or less coriaceous than the stem. The palpi are filiform or almost setaceous, and of six articulations. The tongue is strait and rounded; its sheath is usually short. Its palpi are shorter than the maxillary, and their last articulation almost ovaliform. There is a double mobile auger to the abdomen of the female, &c.

In some the antennæ have but nine articulations, there are two strait and divergent spines at the internal extremity of the two anterior legs, and the auger does not project behind.

Here, the labrum is always apparent; the internal side of the hinder legs has no spines in the middle, or but a single one.

Sometimes the antennæ, always short, terminate either in a thick inflation, like a cone inverted and rounded at the end, or a button, or in a large articulation, like an elongated knob, prismatic or cylindrical, and forked in some males; there are five articulations besides.

The species in which these organs, similar in the two sexes, terminate in a button or coniform inflation, and in which the two nervures of the upper wings are contiguous or parallely approximate without a large intermediate furrow, compose the genus Cimbex, Oliv., Fab., *Crabro*, Geoff.

On some considerations relative to the articulations of the antennæ, &c., Dr. Leach has divided Cimbex into several other genera, one of which, Perga, peculiar to New Holland, has a moveable spine to the middle of the under side of the four hinder legs. The scutellum is large and square, with the posterior angles advanced like teeth. The valves which receive the auger are furnished, externally, with numerous short, frizzled, and silky hairs. The antennæ are very short, with six articulations, and the knob has no vestige of rings. The genera of the English naturalist, I consider but as simple divisions of Cimbex.

Those in which the antennæ have but three articulations, the last like an elongated knob, &c., and the two costal nervures of the upper wings are very much separated from each other, form the sub-genus Hylotoma, Lat., Fab., *Cryptus*, Jur.

Some, Schyzocera, Lat., *Cryptus*, Leach, Lepel., have four cubital cells, and the antennæ forked in the males. No spines to the middle of the legs.

Others (*Hylotoma* proper), whose wings are like those of the preceding, have their antennæ terminated by a single articulation. Most have a spine to the middle of the four posterior legs.

Sometimes the antennæ have nine articulations at least, very distinct, and do not terminate decidedly and abrubtly in a knob.

There are some, and the greater number, whose antennæ, always simple in both sexes, or at least in the females, have fourteen articulations at the most, and more commonly nine.

Tenthredo (proper), Lat., Fab., nine simple articulations to the antennæ in both sexes. The number of denticulations in the mandibles, varies from two to four. Some difference in the number of *radial* and *cubital* cells.*

Other species, having also antennæ with nine articulations, differ from the preceding, by having them pectinate on one side in the males. Cladius, Klüg., Latr.

* Terms invented by M. Jurine, to express two kinds of little cells, situated near the external edge of the upper wings. Their presence, absence, number, form and connexion, are generic characters employed by this great naturalist.

Some others, having the body short and compact, like the hylotomæ, have from ten to fourteen articulations to the antennæ, and simple in both sexes. ATHALIA, Leach.

The following species are remarkable for antennæ with sixteen articulations at the least, pectinate or fan-like, in the males, and serrated in the females.

PTERYGOPHORUS, Klüg., antennæ with but a single range of teeth, and simply longer, or pectinate in the males, short and serrated in the females; here they are sensibly thicker towards the end.

LOPHYRUS, Latr., whose antennæ, in the males, have a double range of elongated teeth, forming a large triangular plume, and are serrated in the females.

Sometimes the labrum is concealed, or not projecting. The internal side of the four hinder legs, has, before its extremity, two spines, and often a third above the preceding The antennæ are always composed of a great number of articulations. The head is strong; squared, supported by a small neck, with the mandibles strongly crossed. They form the genus *Cephaleia* of Jurine, which has been divided into two others: MEGALODONTES, Latr., *Tarpa*, Fab., in which the antennæ are serrated or pectinated, and PAMPHILIUS, Latr., *Lyda*, Fab., in which they are simple.

The last Tenthredinæ, have the auger protruded beyond the sheath. The internal extremity of the two anterior legs has but a single distinct spine. It is curved, and terminated by two teeth. The antennæ are always composed of a great number of articulations and simple.

XYELA, Dalm , *Pinicola*, Breb., *Mastigocerus*, Klüg. Very distinct, by their elbowed antennæ forming a kind of whip, abruptly narrowed towards their extremity, and consisting of eleven articulations, of which the third is the longest. The maxillary palpi are similarly formed and very long. The thick or callous point in the upper wings is replaced by a cell. The laminæ of the auger are smooth and without denticulations.

Cephus, Lat. Fab., *Trachelus*, Jur. Antennæ inserted near the forehead, and thicker towards the end.

Xiphydria, Lat., Fab., *Urocerus*, Jur., whose antennæ are inserted near the mouth, and more narrow towards the end.

The second tribe, that of Urocerata, is distinguished from the preceding, by the mandibles being short and thick; the tongue entire; the auger of the females sometimes very projecting and composed of three threads, sometimes rolled spirally in the interior of the abdomen, and in a capillary form. This tribe is composed of the genus

Sirex, of Linnæus.

Antennæ filiform or setaceous, vibratile, with from ten to twenty-five articulations. Head rounded, and almost globular, with the labrum very small, the maxillary palpi, filiform, of from two to five articulations. The labial, with three, the last of which is thicker. The body is almost cylindrical. The anterior, or posterior tarsi, and in many the colour of the abdomen, differ according to the sexes. The auger is lodged at its base between two valves, forming a sheath.

Oryssus, Lat., Fab. Antennæ inserted near the mouth, ten or eleven articulations. Mandibles without teeth; maxillary palpi long, with five articulations. Abdomen almost rounded and auger capillary, and rolled spirally.

Sirex (proper), *Urocerus*, Geoff. Antennæ inserted near the forehead; thirteen to twenty-five articulations; mandibles denticulated internally; maxillary palpi very small, almost conical, with two articulations; extremity of abdomen prolonged like a tail, and auger projecting, and composed of three threads.

The second family of HYMENOPTERA,

The PUPIVORA,

Have the abdomen attached to the corslet by a simple portion of their transverse diameter, and even most frequently by a small thread or pedicle, so that its insertion is very distinct, and it moves on this part of the body. The females have an auger which serves as an oviduct.

I divide them into six tribes:

The first, EVANIALES, have the wings veined, and the upper at least areolated; the antennæ filiform, or setaceous, with thirteen or fourteen articulations. Mandibles denticulated internally; maxillary palpi with six articulations; labial with four. Abdomen implanted on the thorax, and in many under the scutellum; auger usually projecting, and with three threads. This tribe forms but a single genus, that of

FŒNUS.

Sometime the auger is concealed, or projects very little, and is like a small sting. The tongue is trifid.

EVANIA, Fab., *Sphex*, Lin. Antennæ elbowed; abdomen small, triangular or ovoïd, abruptly pedicled at its origin, inserted below the shield, at the posterior and upper extremity of the thorax.

PELECINUS, Latr., Fab. Abdomen inserted much lower, a little above the origin of the hinder feet; elongated, sometimes filiform, very long and arched, sometimes gradually narrowed towards the base, and terminating like a knob; posterior legs inflated; antennæ strait and attenuated.

Sometimes the auger is very prominent, and composed of three distinct and equal threads.

Some have the abdomen and hind legs, knob-like; antennæ filiform, or simply marginate. Fœnus (proper), Fab., *Ichneumon*, Lin.

The abdomen of the others is compressed, ellipsoïd, or like a sickle, and all the legs slender. The antennæ are setaceous.

Aulacus, Jur., Spin., in which the abdomen is ellipsoïd.

Paxylloma, Brebisson. When it is like a sickle.

The second tribe, the Ichneumonides, have also veined wings, and in the disk of the upper complete or closed cells: The abdomen originates between the two hinder feet. The antennæ are generally filiform or setaceous (very rarely in a knob), vibratile, and composed of a great number of articulations. In the majority, the mandibles are not toothed internally, and terminate in a bifid point. The maxillary palpi, always apparent, have most usually but five articulations. The auger is composed of three threads. This tribe embraces the totality of the genus,

Ichneumon, of Linnæus.

Maxillary palpi elongated, almost setaceous, with five or six articulations; labial shorter, filiform, and with three or four articulations. Tongue entire, or simply emarginate; body, generally narrow, elongate, or linear; auger sometimes external, like a tail, sometimes very short, and concealed in the abdomen, which then terminates in a point, while it is thicker, and like a knob terminated obliquely, when the auger projects. The middle piece is the only one which penetrates into the places where they deposit their eggs. The extremity is flatted, and sometimes cut like a pen.

The variety and number of the articulations of the palpi, may serve as a basis for three principal divisions. In the first, the maxillary palpi have five articulations, and the labial four. The second cubital cell is very small, and almost circular or nullified.

We shall form a first sub-division with those, whose head is never prolonged in front, like a muzzle, whose tongue is not deeply emarginate, maxillary palpi much elongated, with the last articulations sensibly differing from the preceding. The auger is not covered at its base by a large lamina.

Some species are distinguished from others by their almost globular head; mandibles terminating in an entire point, or slightly emarginate, and metathorax elongated. The second cubital cell is often wanting. Such are

STEPHANUS, Jur., *Pimpla. Bracon*, Fab., whose thorax is very slender in front, and on a level with the origin of the abdomen at its posterior extremity. The hinder thighs are inflated. Many little tubercles on the top of the head.

XORIDES, Latr., *Pimpla*, *Cryptus*, Fab. Metathorax convex, and rounded at its fall; abdomen inserted at its lower extremity, and presenting a very distinct pedicle.

Among the species whose head is tranverse, and mandibles distinctly bifid, or clearly emarginate at their point, some, as

PIMPLA, Fab., have the abdomen cylindrical, with a very short pedicle.

Others have the abdomen almost ovaliform, with an elongated, narrow, and arched pedicle. These are

CRYPTUS, Fab. In some of them the females are apterous, and from this and the division of the thorax into two parts, they might constitute a particular sub-genus.

Sometimes the abdomen is compressed, like a sickle or truncated knob.

OPHION, Fab., whose antennæ are filiform or setaceous. Auger a little projecting.

BANCHUS, Fab. Abdomen in the females narrowed at the end, and terminating in a point.

HELWIGIA. Antennæ thicker towards the end.

Sometimes the abdomen is rather flatted than compressed, either ovaliform, almost cylindrical, or fusiform.

In some the abdomen is remarkably narrowed at its base, like a pedicle.

JOPPA, Fab. Antennæ remarkably widened or thickened before the end, and then terminating in a point.

ICHNEUMON (proper). Head transverse, abdomen ovaliform, almost equally narrowed at both ends.

ALOMYA, Panz. Head more narrow and more rounded, and abdomen more widened towards its posterior extremity.

In others the abdomen attaches to the metathorax in the major portion of its transverse diameter, is almost sessile, almost cylindrical, and simply widened towards its posterior extremity. Such are

PELTASTES, Illig., METOPIUS, Panz. They have a circular elevation beneath the antennæ, and the lateral edges of the scutellum are raised, and sharp.

The second and last sub-division of these has the tongue deeply emarginate, or almost bifid; maxillary palpi with articulations little varying, or gradually changing. Auger projecting, and covered at the base by a large lamina. Hinder legs thick. Head of many advanced, like a muzzle.

ACŒNITUS, Lat. Head not advanced into a muzzle.

AGATHIS, Lat., in which it terminates in the same manner; but in their wings they approximate to the subsequent sub-genera.

Our second division of the ichneumons differs from the first in the number of articulations to the palpi, as there is one less to the labial. The second cubital cell is most frequently as large as the first. The auger projects. The point of the mandibles is bifid or emarginate.

Some have a remarkable vacancy between the mandibles

and hood. The jaws are elongated inferiorly underneath the mandibles. These are BRACON, Jur., Fab.

From them may be detached, under the name of VIPION, the species whose antennæ are short and filiform ; jaws proportionally longer, and forming with the labium a sort of bill, and maxillary palpi but little longer than the labial.

The species with setaceous antennæ, as long as the body, maxillary palpi much longer than the labial, and jaws and labium forming a bill, should be excluded from Bracon.

In the others there is no vacancy between the mandibles and hood. The jaws and labium are not prolonged. The second cubital cell is very small ; auger and abdomen short. MICROGASTER, Lat.

In our third and last division there are four articulations to the labial palpi, but the maxillary have six.

In some the mandibles proceed contracting, and terminate in two teeth, or in a single bifid or emarginate point.

HELCON, Nees d'Es. Abdomen above, presents several rings, terminates with a long auger, and is not vaulted underneath.

SIGALPHUS, Latr. Abdomen vaulted below, but three segments above, the auger is withdrawn, and sting-like.

CHELONUS, Jur., in which this part of the body is inarticulate above.

Sometimes the mandibles are almost squared, with three teeth at the end, one in the middle, and the others formed by the angles of the terminal edge. ALYSIA, Latr.

The third tribe, GALLICOLÆ, *Diploleparicæ*, Lat., have but a single nervure in the lower wings. The upper have some cells, or areolæ, two brachial, the internal one incomplete ; one radial and triangular, and two or three cubital ; the second (in those that have three) very small ; the third very large, triangular, and closed by the posterior edge of the wing. The antennæ are equal, or go on thickening, with thirteen to fifteen articulations. Palpi very long. Auger spirally rolled in the abdomen; hinder extremity lodged in a groove in the belly.

The Gallicolæ form the genus

CYNIPS, of Lin.

They appear as it were humped, having the head small, and thorax thick and raised. The abdomen is compressed, carinated at its under part, and truncated obliquely at its extremity.

IBALIA, Latr., Illig., *Sagaris*, Panz., *Banchus*, Fab. Abdomen greatly compressed; antennæ filiform, radial cell, long, and narrow, two brachial, distinct and complete; first two cubital, very small.

FIGITES, Lat., Jur. Abdomen ovoid, thickened, and rounded above, either simply compressed, or trenchant underneath. Antennæ grained, and proceed thickening; but one brachial cell complete; radial greatly removed from the end of the wing. Second cubital wanting.

CYNIPS (proper). DIPLOLEPIS, Geoff. Antennæ filiform, and not grained. Cubital cells, three in number. Radial, elongate.

Fourth tribe, CHALCIDIÆ. Antennæ (*Eucharis* excepted) elbowed, and proceeding from the elbow, form an elongate or fusiform knob, the first articulation of which is often lodged in a furrow. The palpi are very short. Radial cell usually wanting. Never but one cubital cell, not closed. Not beyond twelve articulations to the antennæ. The genera of this tribe are referred to

CHALCIS, Fab.

Auger often composed of three threads, and projecting. Some, whose antennæ have always eleven or twelve articulations, have the hinder thighs very large and lenticular, and the legs arched.

Sometimes the abdomen is ovoid or conic, pointed at its extremity, decidedly pedicled, and the auger straight, and seldom projecting outwards. Wings extended. Some males have the antennæ fan-like. CHIROCERA, Latr.

Those of the others are simple in both sexes. CHALCIS (proper). *Vespa*, *Sphex*, Lin.

Some have the pedicle of the abdomen elongated. *Sispes* and *Clavipes*, Fab.

The others have the pedicle of the abdomen very short. *Vespa minuta*, Lin., &c.

Sometimes the abdomen appears applied against the posterior extremity of the metathorax, rounded at the end, and compressed laterally. The auger is curved on the back. LEUCOSPIS, Fab.

Others, whose antennæ have often but from five to nine articulations, have the posterior thighs oblong, and the legs straight.

Among those, whose antennæ, always simple, have from nine to twelve articulations, we distinguish at first,

EUCHARIS, Lat., Fab., *Chalcis*, Jur. The only one of this tribe in which these organs are straight. Abdomen pedicled.

THORACANTA, Lat. Insects collected in Brazil by M. St. Hilaire, represent, by their scutellary elongation covering the wings, the hemiptera, called *scutellera* by M. de la Marck.

The other sub-genera, with antennæ always composed of nine articulations, at least, and simple, but elbowed, and wings not covered by the scutellum, may be divided into those whose antennæ are inserted near the middle of the anterior face of the head, or notably removed from the mouth, and into those where they are inserted very near it.

Of the first, some have the abdomen almost ovoid, compressed on the sides, with the auger usually projecting and ascending. Such are

AGAON, Dalm. Very remarkable for the size and length of their head; the first articulation of the antennæ very large, and in the form of a triangular palette, the last three forming, abruptly, an elongated knob. They are furnished with hairs.

EURYTOMA, Illig. Antennæ knotty, and furnished with small hairs in the males. Auger short.

MISOCAMPE, Lat., *Diplolepis*, Fab., in which they are composed, in both sexes, of very close articulations, without small hairs. Auger long.

The others have the abdomen flatted above, rather triangular, and terminating in an elongated point, in the females; or almost heart-formed, or orbicular. The auger is usually concealed.

Sometimes the nervure of the upper wings is always curved, the two hinder feet are the largest, the interior spine of the intermediate legs is small.

PERILAMPUS. Have mandibles strongly denticulated; knob of the antennæ short and thick; abdomen short, and heart-formed; scutellum thick and projecting.

PTEROMALUS, Latr., *Cleptes*, Fab. Thorax short, without anterior contraction.

CLEONYMUS, Latr. Thorax elongated, and contracted anteriorly; abdomen proportionally longer, and antennæ inserted lower.

The nervure of the upper wings is sometimes straight, intermediate feet the longest, and their legs have a long internal spine.

EUPELMUS, Dalm. Nervure curved; first articulation of the intermediate tarsi, large, and ciliate underneath.

ENCYRTUS, Latr. Nervure straight, knob of the antennæ compressed, and truncated at the end.

SPALANGIA, Latr. Antennæ longer, and inserted very near the anterior edge of the head.

EULOPHUS, Geoff., Latr., *Entodon*, Dalm. Only from five to eight articulations to the antennæ, and those of the males branching.

The fifth tribe, OXIURI, Latr., have, in the females, the abdomen terminated by a tubular auger, conical, sometimes internal, and exsertile, sometimes external and caudiform. Antennæ of from ten to fifteen articulations, either filiform, or in the females claviform. Maxillary palpi long, and pendant; it is comprehended in the genus

BETHYLUS, of Latreille and Fabricius.

Some have cells, or brachial nervures, in the upper wings. The maxillary palpi are always projecting. The antennæ filiform, or very gradually thickening in both sexes.

Sometimes inserted near the mouth.

DRYINUS, Lat., *Gonatopus*, Klüg. Antennæ straight, of ten articulations in both sexes, the last a little thicker. Thorax divided into two knots; anterior tarsi terminate in two large denticulated hooks, one of which is folded back; females apterous.

ANTEON, Jur. Ten articulations to antennæ, but thorax continuous. All the tarsi are terminated by common hooks. The upper wings have a large cubital point.

BETHYLUS, Latr., Fab., *Omalus*, Jur. Whose antennæ are elbowed, with thirteen articulations in both sexes. Head flatted, and pro-thorax elongated, and almost triangular.

In some, the antennæ always have thirteen to fifteen articulations, and are inserted near the middle of the anterior face of the head.

Sometimes they are straight, or almost straight.

PROCTOTRUPES, Latr., *Codrus*, Jur. In which they consist of thirteen articulations, in both sexes. The mandibles are arched, and without teeth at the internal side. The pedicle of the abdomen very short and gradual, terminating in the females in a point, or corneous tail, often long, and forming the auger.

Sometimes the antennæ are very distinctly elbowed.

HELORUS, Lat., Jur. The antennæ have fifteen articulations. Mandibles denticulated at the internal side. The first ring of the abdomen forms an abrupt, long, and cylindrical pedicle.

BELYTA. CINETUS, Jur. Their antennæ are of fourteen or fifteen articulations, filiform in the males; more grained, and thicker towards the end, in the females.

The other oxyuri have neither cells, nor brachial or basilary nervures.

These have their antennæ inserted on the front.

DIAPRIA, Latr., *Psilus*, Jur. The maxillary palpi project. The antennæ have fourteen articulations in the males, or twelve in the females.

In *those* they are inserted near the mouth.

CERAPHRON, Jur., Latr. A radial cell. Maxillary palpi projecting, antennæ filiform in both sexes, with eleven articulations; abdomen conical-ovoïd.

SPARASION. Antennæ of twelve articulations in both sexes, but thicker at the end or claviform in the females, and the abdomen flatted.

There are again two sub-genera, that have a radial cell; the antennæ claviform in the females, the abdomen flatted, but all the palpi very short, and making no projection, or not pendant underneath.

TELEAS, Latr. The antennæ of which have twelve articulations.

SCELION. In which they have but ten.

PLATYGASTER. The radial cell does not exist. The antennæ have but ten articulations, of which the first and third are much elongated. The palpi are very short. The abdomen is flatted.

The sixth tribe, CHRYSIDES, Lat., have not the lower wings veined, but their auger is formed by the last rings of the abdomen, like the tubes of a telescope, and it terminates with a little sting. The abdomen, which in the females does not appear composed but of three or four rings, is vaulted, or flat underneath, and can be folded back against the breast. This tribe comprehends the genus

CHRYSIS, of Linnæus.

Body elongated, and covered with a solid dermis. Their antennæ are filiform, elbowed, vibratile, and composed of thirteen articulations in both sexes. The mandibles are arched, narrow, and pointed. The maxillary palpi are usually longer than the labial, filiform, and of five irregular articulations. The labial have three. The tongue is most frequently emarginate. The thorax is semi-cylindrical, and the abdomen semi-oval, and truncated at the base.

Some have the jaws and labium very long, composing a false proboscis, bent underneath, and the palpi very small, with two articulations; such are PARNOPES, Latreille.

The others have no false proboscis; their maxillary palpi are of the middle size, or elongated, and composed of five articulations; there are three to the labial.

Sometimes the thorax is not narrowed anteriorly. The abdomen is semi-oval, vaulted, and has but three segments externally; such are CHRYSIS, properly so called. Fab.

Those whose four palpi are equal, and whose tongue is deeply emarginate, form the genus STILBUM, of Spinola, to which may be united EUCHRÆUS, of Latreille.

Those whose maxillary palpi are much longer than the labial, and which have the tongue emarginate, with the abdomen rounded, and smooth at the end, have been generically distinguished under the name of HEDYCHRUM.

Those resembling the last as to the palpi, but with the tongue rounded and entire, form the genera ELAMPUS, and CHRYSIS, of Spinola.

Sometimes the corslet is narrowed in front, the abdomen has an almost ovoïd figure, with four segments in the females, and five in the males. Such are CLEPTES, of Latreille. The mandibles are short and denticulated. The tongue is entire.

The second section of Hymenoptera, ACULEATA, has no auger. It is re-placed in the females and neuters, of the social species, by a retractile sting, composed of three pieces. Sometimes this does not exist, and the insect defends itself by ejaculating an acid liquor, enclosed in special reservoirs, in the form of glands. The antennæ are always simple, and with thirteen articulations in the males, and twelve in the females. The palpi are usually filiform, the maxillary, often the longest, have six articulations, and the labial four. The mandibles are smaller, and often less denticulated in the males. The four wings are always veined. The abdomen, united to the thorax by a pedicle, has seven segments in the males, and six in the females. The larvæ have no feet. This section is divided into four families. The first, that of

HETEROGYNA,

In which the neuters and females are wingless. In all the antennæ are bent, and the tongue small, rounded, and vaulted, or spoon-like.

In some which live socially, the antennæ gradually thicken in the females (which are winged) and in the neuters. The length of their first articulation equals the third of their total length. The second is almost as long as the first, and like an inverted cone. The labrum of the neuters is large and horny, and falls perpendicularly under the mandibles. They compose the genus

FORMICA, of Linnæus.

The pedicle of the abdomen is like a knot, either simple or double, the head triangular, the eyes rounded, or oval, and entire, the mandibles generally very strong, but varying much in form in the neuters. The jaws and labium small, the palpi filiform, and the maxillary longest. Thorax compressed at the sides, and abdomen ovoid, provided, the females and neuters, sometimes with a sting, sometimes with glands secreting the *formic* acid.

This genus is divided as follows:

1st. FORMICA (proper). No sting: antennæ inserted near the front, mandibles triangular, denticulate, and incisive. The pedicle of the abdomen is always simple.

2nd. POLYERGUS, Lat. Sting wanting; but antennæ situated near the mouth. Mandibles narrowed, arched, or very much crooked.

3rd. PONERA, Latr. Males and females armed with a sting. Antennæ thicker towards the end, mandibles triangular, head nearly so, without any remarkable emargination, at its posterior extremity.

ODONTOMACHUS, Latr., have also the pedicle of the abdomen simple, but terminated above like a spine. The antennæ very slender and filiform in the males. The head of these individuals is a long square, greatly emarginated behind, with long narrow parallel mandibles, and terminated by three teeth.

4th. MYRMICA, Latr. A sting; but the pedicle of the abdomen formed of two knots, the antennæ uncovered, and the maxillary palpi long, with six distinct articulations; the mandibles triangular.

Species certainly similar, but with linear mandibles, compose the sub-genus Eciton, Lat.

5th. Atta, of Fabricius. Like myrmica, but with very short palpi, and the maxillary with less than six articulations.

6th. Cryptocerus, Latr. Sting; pedicle of the abdomen double, but the head very large and flatted, has a grove on each side, where the antennæ lodge.

The other Heterogyna are solitary, but two kinds of individuals, winged males, and apterous females, are always armed with a strong sting. The antennæ are either filiform or setaceous, vibratile, with the first and third articulations elongated. The length of the first nerve equals the third of the total length of these organs. They form the genus

Mutilla, of Linnæus.

Some, of which the males alone have been observed, have the antennæ inserted near the mouth, the head small, and the abdomen long, and almost cylindrical—as in Dorylas, Fabr., Labidus, Jur. Mandibles shorter and less narrow, and maxillary palpi, at least of the length of the labial, and composed at least of four articulations.

The others have the antennæ inserted near the middle of the face of the head, which is stronger than in the preceding. These are Mutilla (proper).

The species whose corslet is almost cubical, without knots, or any appearance of division above, in the females, compose the genera Apterogyna, Psammotherma, and Mutilla, of Latreille. The antennæ of the males in the first, are long and setaceous, in the second pectinate, in the third simple in both sexes.

Those which in both sexes have the thorax equal above, but divided into two distinct segments, the abdomen conical in the females, elliptical and depressed in the males, compose the genus Myrmosa, of Latr. and Jur.

Those in which the thorax of the females is again equal above, but divided with three segments by sutures, and the maxillary palpi very short, form the genus Myrmecoda, Latr.

Scleroderma, of Klüg., have the maxillary palpi elongate, and the second articulation of the antennæ uncovered.

Methoca, of Lat, have the upper part of the thorax knotted or articulate.

The second family of this section, that of

Fossores,

Have stings, individuals but of two kinds, all winged, and live solitarily, but proper for walking, and in many for digging; tongue always more or less wide at its extremity, and never filiform or setaceous; the wings are always extended. They compose the genus

Sphex, of Linnæus.

The jaws and labium are elongated, and of the proboscis form in many.

The numerous sub-genera are divided into seven principal sections.

In the first two the eyes are often emarginate. The body of the males is generally narrow, elongate, and terminates posteriorly in a great number, in three points, like spines or denticulations.

1st. Those in which the first segment of the thorax is sometimes arched and prolonged laterally to the wings, sometimes transversely square, or formed like a knot or articulation; feet short, thick, very spiny or ciliated, thighs arched near the knee; antennæ clearly shorter than head and thorax in the females. These are the SCOLIETÆ of Latreille, thus named from the genus SCOLIA.

Some have the maxillary palpi long, the articulations unequal, and first articulation of the antennæ almost conical. Such are TIPHIA, Fab., and TENGYRA, Latr.

The others have the maxillary palpi short, composed of articulations almost alike, with the first of the antennæ elongated, and almost cylindrical. Sometimes this articulation receives and conceals the following one, as in MYZINE, Latr., mandibles denticulate.

MERIA, Illig., mandibles not denticulate.

Sometimes the second articulation of the antennæ is uncovered, as in SCOLIA (proper), Fab.

2d. The diggers, whose first segment of the thorax is conformed as in the preceding, which have short but slender feet, not spinous, nor strongly ciliate, and whose antennæ are in both sexes as long, at least, as the head and corslet. This sub-division embraces the family of SAPYGATÆ of Latreille, the name of which is taken from the principal genus, SAPYGA.

Some have the antennæ filiform, as in THYNNUS, Fab., eyes entire. POLOCHRUM, Spin., eyes emarginate, and mandibles much denticulated.

In the others, the antennæ are thicker towards their extremity, or even claviform in some males. SAPYGA (proper), Latr.

3d. The diggers, which approach the preceding in the extent and form of the first segment of the thorax, but the posterior feet are as long again as the head and trunk. Antennæ generally narrow, with elongate articulations, and much arched, in the females. They are the SPHEGYDES of Latr., from the dominant genus SPHEX.

Some have the first segment of the thorax square, either transversely, or longitudinally, and the abdomen attached to the corslet by a very short pedicle. Their posterior limbs have usually a brush of hairs on the internal side. The upper wings have two or three cubital cells complete, or closed, and another imperfect and terminal. They form many sub-genera.

Pepsis, Fab. Labrum apparent; antennæ almost strait; composed of close articulations: maxillary palpi little longer than the labial; three cubital cells complete. The legs, and first articulation of the hinder tarsi of the males, compressed.

Ceropales, Latr., Fab. Labrum and antennæ of pepsis; but maxillary palpi much longer than the labial, pendant, and with very unequal articulations.

Pompilus, Fabr. Antennæ of the two sexes arched, and composed of articulations not very close.

Planiceps, head flat, with the posterior edge concave, simple eyes very small, and much separated; ordinary eyes elongated, and occupying the sides. The antennæ are inserted near the anterior edge. The two anterior feet are remote from the others, short, curved underneath, with the haunches and thighs large. The upper wings have but two cubital cells complete.

Aporus, Spinol. Only two complete cubital cells; but the second receives the recurrent nervures.

In the others, the first segment of the thorax is narrowed in front, and also the first ring of the abdomen, and sometimes a part of the second. Three complete cubital cells on the upper wings, and the beginning of a fourth.

Those whose mandibles are denticulated, the palpi filiform, jaws and tongue long, and proboscis-like, *&c.*, are the Ammophilus of Mr. Kirby.

The species whose jaws and labium are much shorter, are the genera Sphex, Pronæus, and Chlorion, of Latreille.

Others with denticulated mandibles, but maxillary palpi

much longer than the labial, compose the genus DOLICHURUS, Latr.

The last fossores of this third division have no denticulations to the mandibles, which are striated. They are AMPULEX, Jur. PODIDUM, Lat., and PELOPŒUS, Lat., Fab.

4th. In the other fossores the first segment of the thorax forms only a simple linear and transverse edge, of which the two lateral segments are removed from the origin of the upper wings. The feet are always short, or moderate. The head transverse, the eyes extending to the posterior edge, the abdomen in a semi-cone, elongate, rounded on the sides near the base The labrum naked, or very projecting. These are Latreille's family of BEMBECIDES, named from the genus BEMBEX, Fab.

Sometimes the palpi are very short, the maxillary have but four articulations, and the labial two. Such is the species called *Apis rostrata*, Lin.

Sometimes the maxillary palpi have six articulations, and the labial four. MONEDULA, Lat.

The others have no false proboscis, and the labrum is short and rounded. Such are STIZUS, Lat. et Jur.

5th. Other fossores differ, by having the labrum concealed altogether, or in great part, and a deep emargination near the base of the mandibles. Three cubital cells in some. These are our LARRATES.

PALARUS, Latr., *Gonius*, Jur. Antennæ very short, gradually thicken. Eyes closely approaching posteriorly. Second cubital cell petiolate.

LYROPS, Illig., *Liris*, Fab., *Larra*, Jur. In which the antennæ are filiform. Third cubital cell, narrow, and crescent-like, and a tooth on the internal side of the mandibles.

LARRA, Fab. Mandibles without teeth, &c.

Sometimes there are but two cubital cells, closed.

DINETUS, Jur., have the two cubital cells sessile. Antennæ of the males maniliform below, and .then filiform. Three denticulations internally to the mandibles.

MISCOPHUS, Jur. Antennæ filiform in both sexes. Mandibles with but a slight projection on the internal side.

6th. Now come fossores with the labrum likewise concealed, jaws and labium forming no proboscis, no emargination at the lower side of the mandibles, abdomen triangular, or ovoïdo-conical, and never on a pedicle. Antennæ filiform. NYSSONIANI.

Some have the eyes entire.

ASTATA, Latr. Three cubital cells closed, extremity of the mandibles bifid, and eyes nearly approaching above.

NYSSON, Latr., Jur. Mandibles terminating in a simple point, and eyes separated widely.

OXYBELUS, Latr , Jur., Oliv. Antennæ short, arched, second articulation shorter than the third; two or three points in the scutellum, like teeth; legs spiny, and a large cushion at the end of the tarsi.

NITELA, Latr. Antennæ longer, almost strait; second and third articulations of the same length. Mandibles terminate in two teeth, no points to the scutel, nor spines to the legs, and cushion of the tarsi very small.

Others have the eyes emarginate. Such are PISON, Spin, Latr.

7th. The last division of fossores, CRABRONITÆ, has the head square; antennæ usually claviform, abdomen ovaliform, broader towards the middle, narrowed at the base.

Some have the antennæ inserted below the middle of the anterior face of the head, with the band short and broad.

Sometimes the eyes are emarginate.

TRYPOXYLON, Latr., Fab., *Apius*, Jur., *Sphex.*, Linn. Mandibles arched and without teeth. Abdomen narrowed at base, in a long pedicle.

Sometimes the eyes are entire.

Here the mandibles are strait, and simply denticulate at the end, or terminate in a simple point, with a single tooth at the internal side. The antennæ approximate at their base.

Gorytes, Latr., *Arpactus, Jur., Mellinus Oxybelus, Fab.* Three cubital cells complete; mandibles moderate, unidenticulate at the internal side. Metathorax, with a sort of false scutel, furrowed.

Crabro, Fab. A single cubital cell closed, mandibles terminate in a bifid point; antennæ bent, filiform, fusiform, or slightly serrated in some; palpi short, almost equal, and tongue entire.

Stigmus, Jur. Thick or callous point on the sides of the upper wings, black; two cubital cells closed. Antennæ not bent, the first articulation a little elongated, and like an inverted cone. Mandibles arched, and terminated by two or three teeth.

Sometimes, the mandibles, in the females at least, are strong, and bidenticulated on the internal side. The antennæ are separated at their base.

Pamphredon, Latr., Fab., *Cemonus*, Jur. Two cubital cells complete, and a third imperfect.

Mellinus. Three cubital cells complete, and often the commencement of a fourth.

Alyson, Jur., *Pompylus*, Fab. Three cubital cells complete, but the second petiolated.

The last crabronitæ have their antennæ inserted higher, or towards the middle of the anterior face of the head. They are usually thicker towards the end. Three cubital cells complete, and two recurrent nervures.

Sometimes the hood is almost square. Abdomen, carried on an abrupt, long pedicle, formed by the first ring. Mandibles terminate with two teeth. Psen, Latr., Jur., *Trypoxylon, Pelapæus*, Fab.

Sometimes the hood is trilobed; first ring of the abdomen

narrowed like a knot; mandibles in a simple point. These insects compose the genus PHILANTHUS, of Fabricius.

Others restrain this generic division to those whose antennæ are separated, and abruptly swelled. Mandibles with no internal projection, and all the cubital cells sessile. They are, PHILANTHUS, proper, Latr., *Simblephilus*, Jur.

Those in which the antennæ are approximate, much longer than the head, and gradually thickening, whose mandibles present at the internal side a tooth, and second cubital cell petiolate, form the genus CERCERIS, Lat., *Philanthus*, Jur.

The third family of the HYMENOPTERA ACULEATA, that of

DIPLOPTERA,

Is the only one of this section, with very few exceptions, which has the upper wings doubled longitudinally. The antennæ are usually bent, and thicker towards the end. Eyes emarginate. Prothorax carried back on each side, as far as the origin of the wings. The upper have three, or two cubital cells, of which the recurrent nervures are received by the second. Many are social, and composed of males, females, and neuters. We divide them into two tribes. The first, MASARIDES, Latr., has for the genus,

MASARIS, of Fabricius.

Antennæ, apparently with but eight articulations, the eighth forming, with the following, an almost solid knob; tongue terminated by two threads, retractile, into a tube formed by its base. The upper wings with but two cubital cells complete. Middle of the anterior edge of the hood emarginate, and receiving there the labrum.

MASARIS proper. Antennæ a little longer than head and thorax. Abdomen long.

CELONITES, Latr. Antennæ scarcely longer than the head. Abdomen hardly longer than the thorax.

The second tribe of DIPLOPTERA (VESPARIÆ) is composed of the genus

The WASPS. VESPA, Linn.

Antennæ, always with thirteen articulations in the males; twelve in the females, and terminating in an elongated knob, pointed, and sometimes crooked (males) at the end—always elbowed. The tongue is sometime divided into four plumous threads, sometimes into three lobes, having four glandular points at the end. The mandibles are strong and denticulated. The hood is large. Beneath the labrum is a small piece, like a tongue. With the exception of a small number of species, the upper wings have three cubital cells complete. The females and neuters have a very strong and venomous sting. Many live in societies composed of three kinds of individuals.

The first sub-genus, CERAMIUS, Lat., Klüg., forms an exception to the general characters of their tribe. The upper wings are extended, and the cubital cells are but two in number. The labial palpi are longer than the maxillary.

Sometimes the mandibles are much longer, than broad, approached so as to form a bill. The tongue is long and elongate; the hood is boat-formed, or oval.

SYNAGRIS, Lat., Fab. Tongue divided into four threads, without glandular points. Mandibles of some males very large, and like horns.

EUMENES, Lat., Fab. Tongue, with three pieces, glandulous at their extremity.

Abdomen in some ovoïd, or conic, and thicker at the base; such are PTEROCHILE, Klüg., with very long jaws, and lip forming a curved proboscis; labial palpi with long hairs, and but three distinct articulations.

ODYNERUS, Latreille. These parts of the mouth much

shorter, and labial palpi nearly smooth, with four articulations.

In the others, the first ring of the abdomen is narrow, elongate, pyriform ; the second bell-like. EUMENES proper ; also *Zethus*, of Fab., and *Discœlius*, of Latreille.

Sometimes the mandibles are but little longer than broad, with a broad and oblique truncation at their extremity ; the tongue shortish ; the hood nearly square. They form the sub-genus, VESPA (proper), Lat.

The fourth, and last family of aculeated Hymenoptera, that of

The BEES. ANTHOPHILA, Latr.,

Is characterised by collecting with the hinder feet the pollen of flowers. The first articulation of the tarsi of their feet, is very large, very compressed, and somewhat like a reversed triangle. The jaws and lips, generally very long, compose a proboscis. The tongue is like an arrow-head, or very long thread, the extremity of which is silky, or hairy. These hymenoptera embrace the genus

APIS, of Linnæus,

Which I divide into two sections :—

The first, ANDRENETÆ, Lat., has the intermediate division of the tongue, heart, or arrow-head-like, shorter than its sheath, and folded upwards in some, almost straight in others. These are *Andrenes* of Fab., and *Melites* of Kirby. The two lateral divisions are very short. The mandibles are simple, or terminated by two denticulations. The labial palpi resemble the maxillary, which have always six articulations.

Some have the middle division of the tongue widened at its extremity, almost heart-formed, and doubled in repose. HYLŒUS, Fab., *Prosopis*, Jur.

Sometimes the body is smooth, second and third articulations of the antennæ almost of a length. But two complete cubital cells. They gather no pollen, and are HYLŒUS (proper), Lat and Fab.

Others have the body hairy; third articulation of the antennæ longer than the second. Three cubital cells complete. COLLETES, Lat.

The other andrenetæ are distinguished from the preceding, by the arrow-head figure of the tongue. In some, this tongue folds back on the upper side of its sheath, as in ANDRENA, and DASYPODA, of Latreille. In the females of the last, the first articulation of the posterior tarsi is very long, and furnished with long hairs. The upper wings in both have but two cubital cells.

In the others, the tongue is straight, or a little curved underneath. Such are, SPHECODES, HALICTUS, and NOMIA, of Latreille. Here the jaws are more strongly bent than in andrena. The closed cubital cells are always three. The males in Sphecodes have knotty antennæ. The tongue is almost straight, with divisions of equal length. The middle one is much longer in HALICTUS and NOMIA.

In the second section of ANTHOPHILA, that of APIARIÆ, Latr., the middle division of the tongue is as long as the mentum, and like a thread, or silky hair. Jaws and labium much elongated, forming a proboscis, folded underneath, in inaction. The first two articulations of the labial palpi, have most frequently the figure of a scaly hair, compressed, and embracing the sides of the tongue. The other two are very small. The third is inserted near the external extremity of the preceding, which terminates in a point.

The apiariæ are either solitary or social. In the first, there are but two kinds of individuals. The females have no brush to their hinder feet, nor particular sinking at the external side of their legs.

A first division of these solitary apiariæ, will be composed of those where the second articulation of the posterior tarsi of the females, is inserted in the middle of the extremity of the preceding. The external and terminal angle of the latter, does not appear more dilated or advanced than the internal, in the following sub-genera.

We may further detach from this group, the *Andrenoïdes*, approximating to the last of the preceding section, in their labial palpi, composed of slender articulations, linear, placed end to end, almost similar to those of the maxillary, and six in number. Labrum always short. The females have no brush to the belly, but their hinder feet are furnished with hairs.

Some have the mandibles narrow towards the end, terminating in a point, and even like the labrum. Such are Systropha, Illig. Three complete cubital cells. Antennæ curled at their extremity, in the males.

Rophites, Spin. Two cubital cells, and antennæ not inflected.

Panurgus, Panz. Mandibles not denticulated. Stem of the antennæ, from the third articulation, fusiform, in the females; but two cubital cells.

The females of some others, have mandibles almost spoon-formed, very obtuse, furrowed, and bi-denticulated at the end. Labrum very hard, ciliate above. Antennæ strongly bent, and filiform; three complete cubital cells.

Xylocopa, Lat., Fab. The male, in some species, differs from the female. Eyes large, and more approached above. Anterior feet, dilated and ciliate.

The labial palpi of some other apiariæ, are like scaly hairs. The first two articulations very large or long, compared to the last two, compressed, and the edges membranous, or transparent. Maxillary palpi always short, and often with less than six articulations. Labrum in a great number elongated, sometimes in a long square, or triangle.

Cereatina, Lat., Spin., Jur., *Megilla*, *Prosopis*, Fab.

Maxillary palpi with six articulations, and three complete cubtial cells; body narrow, and oblong. Antennæ inserted in small fossets, and terminating in an elongated knob. Mandibles furrowed, and tri-denticulated at the end. Abdomen ovaliform, and without brush; labrum shorter than in the following sub-genera. They belong to a little group, which we have called *Dasygastræ*, in which the abdomen of the female (though this sub-genus is an exception) is almost always furnished with a silky brush.

All the other *dasygastræ* have four articulations, at most, to the maxillary palpi, and two complete cubital cells. First come the species evidently provided with a silky brush.

Chelostoma, Lat. Body elongated, and nearly cylindrical. Mandibles projecting, arched, narrow, forked, or emarginated at the end. Maxillary palpi with three articulations.

Heriades, Spin. Mandibles triangular, and but two articulations to the maxillary palpi.

In the following four, the abdomen is shorter, and almost triangular, or semi-oval.

Megachile, Latr., *Anthrophora*, *Xylocopa*, Fab., *Trachusa*, Jur Maxillary palpi composed of two articulations. Abdomen plane above, and capable of retro-elevation.

Lithurgus. Four articulations to the maxillary palpi, like the following sub-genus, but the abdomen depressed above. All the articulations of the labial palpi are placed end to end, and the palpi resemble long scaly hairs, placed end to end. Mandibles narrow in both sexes, with the extremity emarginate in the middle, or bi-denticulate.

Osmia, Panza, *Anthophora*, Fab., *Trachusa*, Jur. Abdomen convex above; antennæ pretty long, in the males.

Anthidium, Fab. Abdomen convex; but maxillary palpi with but a single articulation.

The last two sub-genera of this group, approach the following, by the want of the silky brush. But the labrum is parallelogramic, and the mandibles triangular and denticulate. The maxillary palpi are very short, and with two articulations.

Stelis. Panz. No teeth nor spines to the scutellum, abdomen semi-cylindrical, convex above, and curved at the extremity.

Cœlioxys, Lat. Two teeth, or two spines to the scutellum; abdomen triangular, plane above, pointed at the extremity in the females, and usually denticulate in the males.

Other apiariæ, *Cuculinæ,* sometimes almost smooth, sometimes occasionally hairy, have the labrum like an elongated and truncated triangle, or short and almost semicircular, the mandibles narrow, going into a point, with a single tooth at the internal edge. The paraglossæ are often long, narrow, and in the form of hairs. The scutellum of many is emarginate or bidenticulate ; tuberculous in others.

Some almost smooth, have the paraglossæ much shorter than the labial palpi.

Sometimes the labrum is like an elongated triangle, truncated at the end, and inclined below the mandibles. Never but two cubital cells complete.

Ammobates, Latr. Maxillary palpi have six articulations.

Phileremus, Latr. Only two.

Sometimes the labrum is short, almost semi-circular, or half oval.

Epeolus, Lat., Fab. Three cubital cells complete, and a single articulation to the maxillary palpi.

Nomada, Fab. Six articulations to the maxillary palpi.

Pasites. But two cubital cells complete: four articulations to the maxillary palpi.

The other cuculinæ, whose body is very hairy in places, whose scutellum is often spiny, which have always three cubital cells complete, approach the next apiariæ in the length of the paraglossæ, which almost equals that of the labial palpi.

MELECTA, Lat. Five or six distinct articulations to the maxillary palpi.

CROCISA, Jur. But three, and the scutellum prolonged and emarginate.

OXEA, Klüg. Labrum in a long square, and not semi-oval; and maxillary palpi nothing, or reduced to a single articulation, very small.

The last solitary apiariæ have the first articulation of the posterior tarsi dilated below at the exterior side, so that the following articulation is inserted nearer the internal angle of the extremity of the preceding, than the opposite angle. The external side of this first articulation, as well as that of the legs, is charged with thick and close hairs, forming a sort of tuft: thence called by me *Scopulipedes*. The under part of the abdomen is without brush. The cubital cells, except in a few species, are three.

Sometimes the maxillary palpi have from four to six articulations. In these last, the mandibles have but one tooth at the internal side. In many males there is a bundle of hairs at the first and last articulations of the intermediate tarsi. Others, of the other six, are distinguished either by their long antennæ, or by a more remarkable thickening of the two thighs of the second pair of feet, or of the last.

The species in which the two lateral divisions of the tongue are as long as the labial palpi, and like a silky hair, and whose males have long antennæ, form the sub-genus of EUCERA (proper).

MELISSODES, are Euceræ of America, with but four articulations to the maxillary palpi; three cubital cells.

The other apiariæ of this subdivision, have the paraglossæ

much shorter than the tongue, and always three cubital cells.

There are some of them whose maxillary palpi have evidently six articulations.

MELLITURGA, Latr., whose antennæ are short, and terminate in a knob in the males. All the articulations of the palpi are continuous, and in the same direction.

ANTHROPHORA, Lat. Antennæ filiform, four in both sexes, and the last articulation of the labial palpi in an oblique stem.

Others have but five articulations to the maxillary palpi, and those of the labial are continuous. SAROPODA.

Others, in fine, have but four articulations to their maxillary palpi. The first articulation of the posterior tarsi of the males is very large, curved, and hollowed into a vault, at its internal extremity. ANCYCLOSCELIS. Latr.

In the following, the mandibles have many denticulations at the internal sides. CENTRIS, Fab.

Sometimes the maxillary palpi have but one small articulation, which in some even becomes invisible. The paraglossæ are very short. Mandibles denticulate.

EPICHARIS, Klüg. Last articulations of the labial palpi in the same direction as the preceding; but little distinct, and form the point of these organs, which resemble very long hairs.

ACANTHOPUS, Klüg. Last articulations of the labial palpi form a small oblique stem, lateral.

The last apiariæ are social, composed of males, females, and workers. The hinder feet of these individuals have at the external face of their legs, a smooth depression for the reception of the pollen which they have gathered with their brush on the internal face of the first articulations of the tarsi. The maxillary palpi are small, and have but a single articulation. Antennæ elbowed.

Sometimes the posterior limbs are terminated by two spines, as in EUGLOSSÆ, Lat., Fab., where the labrum is square, and the false proboscis of the length of the body, and labial palpi terminating in a point, formed by the last two articulations.

BOMBUS, Lat., Fab. Labrum transverse; false proboscis much shorter than the body, second articulation of labial palpi ending in a point, bearing on its external side two others.

Sometimes the social apiariæ have no spines at the extremity of their hinder legs. They form two sub-genera.

APIS (proper), Lat. The workers have the first articulation of the posterior tarsi in a long square, and furnished at its internal face with silky down, divided into transverse bands.

The last sub-genus of social Apiariæ,

MELIPONA, Illig. have the first articulation of the posterior tarsi more narrow at the base than in the preceding, and without striæ on the silky brush. Upper wings have but two cubital cells complete, whereas there is one more in *Apis* proper.

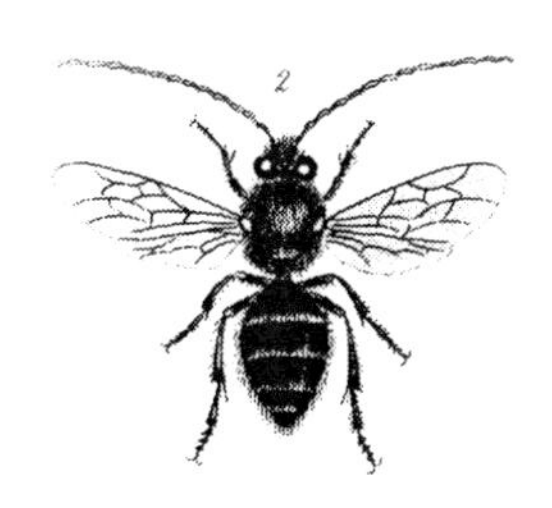

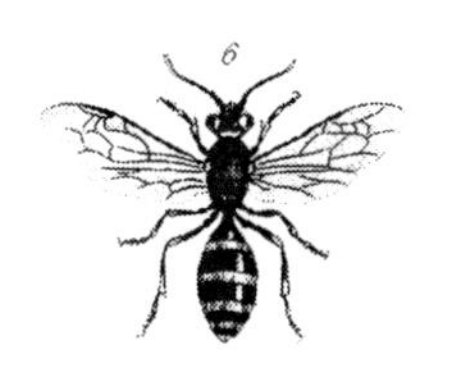

1 *Englossa dentata.* 4 *Andrena pilipes.*

2 *Eucera antennata.* 5 *Bembex rostrata.*

3 *Nomada miniata.* 6 *Hylæus albipes.*

London. Published by Whittaker & C.° Ave Maria Lane. 1837.

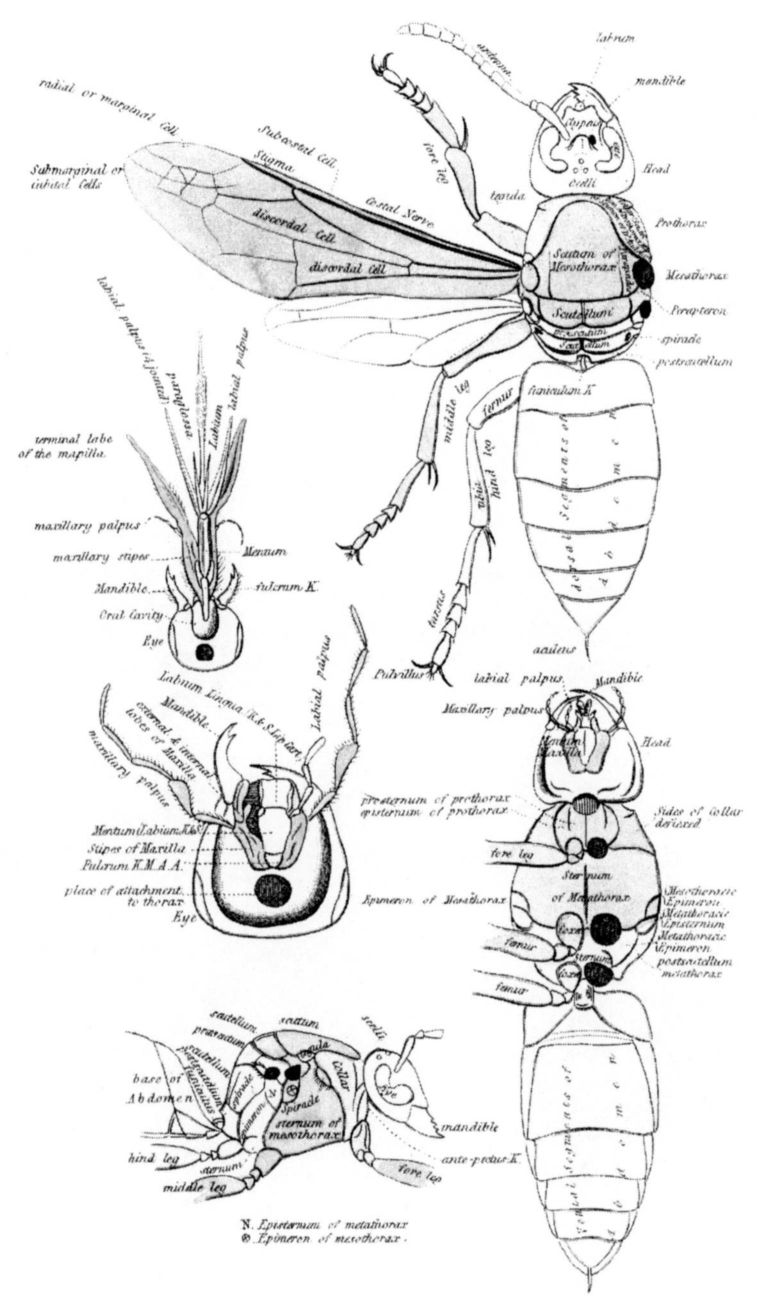

J. O. Westwood F.L.S. del.

London, Published by Whittaker & Co. Ave Maria Lane, 1832.

SUPPLEMENT

ON

THE HYMENOPTERA.

We discover in the writings of the earlier naturalists, the fundamental notion which conducted to the formation of the order on which we are now about to treat. They have remarked, that among the insects with uncovered wings, the *Anelytra*, and in which these organs are four in number, many of these insects, such as bees, wasps, &c. had the abdomen armed with a sting. Linné, in establishing in the first editions of his *Systema Naturæ*, the order sneuroptera and hymenoptera, did not employ this consideration as the principal character. Four *net-work* wings, formed by veins, constituted the distinction of the first order. Membranaceous wings distinguished the second; but the wings of the neuroptera are equally membranaceous with those of the hymenoptera, and we may say that both have equally a network, only with meshes of greater or less size, or more or less numerous. The characters, therefore, of these orders, are not so distinctly marked; and this probably determined Geoffroy, in his history of the insects of the neighbourhood of Paris, to unite both orders into one, under the name of *Tetraptera*. Linnæus subsequently returned to the distinction which had been originally indicated by the earlier naturalists, and made the existence of a sting, a part of the essential character of the hymenoptera. In the first editions of his system of entomology, Fabricius composed with all the insects with four naked wings, as he had in the *branchiopod* and *isopod* crustacea, and the insects of the order *Thysanoura*, an order which he termed *Synistates*. But in the second volume of his systematic entomology, published in 1793, he detached from the *Synistates* the hymenoptera, and formed of them a distinct order, which he characterized in the following vague and insignificant way: *corneous jaw—compressed, and often elongated.*

We have in Degeer, who improved the method of Linné, a more certain and more intelligible guide—"Four wings uncovered, membranaceous, the lower ones of which are shorter, with nervures for the most part longitudinal; mouth armed with teeth; sting or auger in the female:" such, according to him, are the distinctive traits of the hymenoptera.

To the characters which distinguish this order M. Latreille has added another, not noticed before his time, and which may suffice in a system founded solely on the organs of mastication; "a tongue, or labium, enclosed in its base in a coriaceous sheath, which is embraced in the sides, within the jaws."

The order of hymenoptera is so natural that all entomologists have adopted it, and in the same manner as it was circumscribed by Linnæus; the only variation has been respecting its proper situation in the series of insects generally. These changes depend on the greater or less importance attached to characters, derived either from the organs of flight or from the parts of the mouth, considered under the general relation of their forms or action, these animals being either *grinders or suckers*. If, as in the method of M. Lamarck, we place this distinction in the first rank, and that the absence or presence of wings, their number and consistence, form only secondary characters; or if, as has been done by M.M. Clairville and Dumeril, we take the latter organs as a basis, but so as nevertheless to preserve the preceding division, the series of the orders will necessarily differ from that of Linnæus; for, in the other methods, the diptera having but two wings, and being suckers, will be one of the extremes of the division of winged insects, and contiguous either to the hemiptera or the lepidoptera. We shall be unable to prevent the connection of this series of insects, *suckers* with the *grinders*, by the means of the hymenoptera, since they appear to participate, under this relation, one with the other.

It is thus that M. Lamarck, passing from the more simple to the more composite, proceeds from the lepidoptera to the hymenoptera, and that M. Dumeril makes the latter succeed the hemiptera. In the method of M. de Clairville, which terminates with the latter, his division of the winged insects with a sucker conducts us from the hymenoptera, which he names, ridiculously enough, *phleboptera*, to insects with two wings, the first of the series of suckers. But all these arrangements have the defect of conjoining together insects of heterogeneous character, as far as the nature of the organs of flight is concerned. It is thus that the hemiptera, so much approximating, in this respect, to the coleoptera and orthoptera, are placed in the midst of the insects with membranaceous wings. This is the plan, however, that has been pursued by Mr. Kirby, who adopts this grand general division of insects. After all, it does not appear to be of any great consequence how divisions of this kind are made, providing the organs selected as the basis of such divisions be of sufficient importance to constitute essential characters. We may, however, remark, in the insect tribe, that the instruments of flight not only constitute one of the most obvious, but also one of the most essential characters. The very names given by all naturalists to the majority of the orders is quite sufficient to prove this.

We shall here, as the description in the text is so brief, enter a little more largely into the peculiar characters of these insects. The hymenoptera have all composite eyes, often larger in the males. Three small simple eyes, usually in a triangular arrangement, in the vertex; antennæ very various indeed, not only according to the genera, but even in the sexes of the same species; nevertheless, for the most part, filiform, or setaceous, composed of from three to ten articulations, or eleven for the most part in those which have an auger. There are thirteen (in the males) or fourteen (in the females) in those which have a sting: there are two corneous mandibles, also varying according to the sexes. The jaws and labium are generally slender, elongated, attached in a deep cavity of the under portion of the head, by long muscles, in a semi-tube, at their lower part, often folded at their extremity, more adapted to conduct the nutritive juices than for mastication, and in the form of a proboscis in many. The tongue is membranaceous, usually trifid, either widened at its extremity or long and filiform. The pharynx is situated at the upper face of the labium, of which a small triangular lamina, concealed by the labium, the *epipharynx* or *epiglottis*, closes at will the aperture. There are four palpi, two of which, maxillary, are frequently composed of six articulations, and two, labial, have but four. The thorax is formed of three segments, united in a mass, sometimes cylindrical or ovoid, truncated at both ends, sometimes almost globular, the anterior part of which is very short and transverse, and the second, which is properly the thorax, is usually more extended, and intimately united with the metathorax. The wings are crossed horizontally on the body, membranaceous, hyaline, or transparent, and the upper ones larger than the others, having at their origin a little scale, rounded, and convex. In those in which the reticulation is most complicated, there are but three or four principal and longitudinal nervures, united in the direction of the breadth by smaller nervures or veins. The abdomen is most generally narrow at its basis, in the manner of a thread or pedicle, which suspends it to the posterior extremity of the corslet, formed of coriaceous or membranaceous segments, the number of which varies from five to nine, but most frequently six in the females, with either an auger or sting at the end : seven is the most common number in the males. The feet are contiguous, or very much approximated at their base, terminated by an elongated and filiform tarsus, of five entire articulations, with two hooks at the end, and between which there is often a cushion. The two anterior feet are inserted near the neck, with a spine at the internal side of the legs, and an emargination at the same side of the first articulation of the tarsi.

All these parts sometimes exhibit differences according to the sexes.

The auger, or oviduct, and the sting, are composed, for the most part, of three scaly pieces; they are long, narrow, formed like a thread, usually projecting in the manner of a tail, in those which have an auger ; one of them, which is the auger properly so called, is pointed and denticulated, saw-like at the end, and placed between the two others, which form its sheath. These pieces are shorter, circular, and concealed in those species where they form a sting. The upper one has a groove underneath, which emboxes the two others, or the sting properly so called, the extremity of which also presents some denticulations. At the base are two small conical or cylindrical laminæ, in the form of stylets. The auger is sometimes formed by the last rings, and sometimes scaly, projecting in the manner of a pointed tail or sting, and sometimes membranaceous, concealed, and consisting of a series of small tubes, capable of being elongated or re-entering one into another, with a small sting at the end. The sexual organs of the male are composed of several pieces, the most part of which are in the form of pincers, or hooks, surrounding the penis.

The varied reticulation of the upper wings of the hymenoptera has furnished M. Jurine with good auxiliary characters for the distinction of genera.

These insects are all terrestrial, and undergo a complete metamorphosis. The most part of their larvæ resemble a worm, and are without feet. Such are the hymenoptera of the second family, and the subsequent ones. Those of the third have six scaly feet, and with a hook, for the most part, beside twelve to sixteen others, simply membranaceous. In their forms and colours these larvæ are very similar to those of the lepidoptera, and consequently have received the name of *false caterpillars*. Both have a scaly head, and a mouth composed of mandibles, or jaws, and a labium, at the extremity of which is a spinneret for the passage of the silky fluid, which answers the purpose of composing their cocoon, when they are about to pass into the nymph state, and in which they enclose themselves. This substance is elaborated in certain interior reservoirs.

The females of the hymenoptera with auger do not provide provisions for their family, and confine themselves to placing their eggs on the surface, or in the interior parts of animal or vegetable substances, which shall serve as nutriment to their young under the larva form. This section of *terebrant* hymenoptera alone presents us with larvæ provided with feet, and what are called false caterpillars. They are all herbivorous, and most part of the latter, like the true caterpillars, remain uncovered on the leaves which constitute their food. The other herbivorous larvæ of the same section live almost all in those vegetable excrescences, of very various forms, which are called *galls*, and which are formed in consequence of the pricking or wound which the female has made on this part of the vegetable for the purpose of depositing her eggs. The other terebrant hymenoptera feed in the same state either on larvæ, caterpillars particularly, of which they gnaw the interior, without attacking the essential principle of life at first, or on nymphs or eggs of insects, bodies in which they have been deposited under this last form by the mother. The larvæ of the hymenoptera, with stings, are destitute of feet, and resemble worms; some live solitarily in nests or retreats, prepared for them by the mother, and often with a degree of art which must excite our highest astonishment; some on the carcases of insects, or the dust of the stamina of flowers, mixed with a little honey, transported into the habitation prepared for them by their provident parents. The other larvæ are united in society. These last have need of nutritious substances, both vegetable and animal, more elaborated and frequently renewed. Some among them are brought up by females who have survived the winter, and who are subsequently assisted in their labours by other individuals of the same sex, but incapable of coupling and engendering in consequence of the imperfection of their ovaries. Most of the other larvæ, and these constitute the largest number, are confided exclusively to individuals of the same nature, some of which are apterous, others winged, and all of which form very populous societies, and whose works and mode of life are to us a subject of perpetual admiration.

The hymenoptera, in their perfect state, live almost all exclusively on flowers, and generally speaking, are more abundant in southern climates. The duration of their life, from their birth to their final metamorphosis, seldom extends beyond a year.

Of all the orders of which the class of insects is composed that of the hymenoptera is, without contradiction, considered under the relations of manners and habits, the most worthy of attracting our attention. Notwithstanding the numerous and admirable observations of Reaumur, of Degeer, of Huber, and other naturalists, it still presents to the lovers of nature a vast feld of discoveries. A gentleman, named Christ, has assembled, in a special work on this subject, almost all that had been written before his time respecting these insects. But not to mention that the figures which he has inserted are rude, and often not to be recognized, this book, in consequence of the progress of classification, is, at the present day, exceedingly imperfect. The system of Fabricius is a mere specific catalogue, got up in a hurry, without proper preliminary and necessary notices on the sexual differences, without the consideration of natural relations, which is only to be attained by the constant study of the manners of these insects. It is frequently inexact in the exposition of the characters of the genera, and very incomplete with respect to the European species.

M. Jurine has published a most excellent work on the hymenoptera, which was published at Geneva in 1807. He divides these insects into three orders; those of the first have the abdomen sessile; its breadth is equal to that of the corslet. The same division had been also adopted by M. Latreille in his general history of insects. The hymenoptera of the second division have the abdomen petiolate, the petiole being fixed on the corslet; it is inserted behind it in those of the third order. The absence or presence of those cellulæ of the wings, which M. Jurine terms *radial* and *cubital*, their number, their forms, their connections, the mandibles, and finally the antennæ, have supplied him with characters distinctive of the genera. Considering this method merely in a systematic point of view, this philosopher has fully completed the end which he proposed to himself. He has carefully distinguished the sexes: his sections are precise, and there is no mingling of disparate species. The only thing to be desired was, that he had facilitated more the study of the genera of his third order, the most considerable of all, by divisions. The figures with which this work is illustrated are beyond all praise.

Mr. Kirby and M. Klüg have done their part also with respect to this order. The first by his excellent monograph of the English bees, and the other by the establishment of several new genera, and by additional monographs of some of those belonging to M. Latreille's first family of hymenoptera.

The tribe of TENTHREDINÆ have been termed *mouches-a-scie* by Reaumur, Geoffroy, and Degeer, and answer to the genus *tenthredo* of Linnæus. To the general characters given of them in the text we shall add but a few observations.

The instrument which in the females serves the purpose of depositing their eggs, is contained between two scaly plates, or two grooves, from which it issues forth entirely, when these insects wish to make use of it. It is itself composed of two denticulated pieces, similar to a saw: it is with this sort of saw that the tenthredinæ notch the branches of trees, for the purpose of depositing their eggs in them.

The larvæ of these insects have more especially received the name of *false caterpillars*, to distinguish them from the true caterpillars, from which the lepidoptera proceed. The most of these larvæ have not less than eighteen feet, nor more than two and twenty; characters which distinguish them from the caterpillars, which have only sixteen, or even a less number. Such of the false caterpillars as form an exception to the general rule, have but six feet, a character which again distinguishes them from the true caterpillars. In all, the body is composed of twelve rings. Their heads are formed of two caps or heads separated by a channelling: the mouth is furnished with two denticulated jaws, with a *labrum* and a *labium*. Like the caterpillars, they have above this labium a spinneret, through which issues a sort of silk, which they employ in the construction of the cocoon, in which they enclose themselves for the purpose of being changed into the nymph form. Most of them undergo their metamorphoses in the earth. The others spin their cocoon along the branch of some tree. Many of them live in society; but the larger number are solitary.

For the purpose of depositing their eggs, the females notch the branches of trees with the most astonishing dexterity. It is easy to observe the labours of the *tenthredo rosæ.* In the fine days of summer, towards about ten o'clock in the morning, the female is seen traversing with eagerness all the branches of this shrub, one after the other. She usually rests on that one which is nearest to the extremity of the principal stem, and there makes an aperture with her saw, the two pieces of which play alternately. When she has judged that the hole is of suitable dimensions, she deposits an egg in its cavity: she then remains quiet for a few moments, always having her ovipositor engaged in the branch: a moment after, she draws away quickly the largest part of it; and emits at the same time a frothy liquid, which rises as far as the external edges of the notch, and sometimes beyond them. Some authors have thought that the use of this liquid was to bedew and humectate the eggs; but Valisnieri believes that it serves to prevent the aperture from closing. Be this as it may, after the female has emitted this fluid she withdraws her ovipositor, and proceeds to the fabrication of another hole. Sometimes she makes but four in a line, one after the other; most frequently, however, she makes about a score. The part of the branch which is notched in so many places, presents nothing very remarkable the first day of the operation, and it is not until the following day that it commences to become brown; and in the sequel all the wounds become raised, and acquire more and more convexity every day. This growth is owing to the augmented volume acquired by the egg, as it daily grows bigger. It forces the skin of the branch upwards, and the aperture to grow larger. This last finally becomes considerable enough to give passage to the larvæ, which, in coming out of the egg, quits its retreat to seek the leaves of the shrub on which it is nourished.

Some larvæ of these insects present very remarkable peculiarities. Those of the *pine* tenthredo of Linnæus, which live in society on this tree, often to the number of a hundred, after having eaten all the leaves of the branch on which they have been, quit it and proceed off altogether in a march, to seek out another on which to satisfy their appetite. They sometimes make very deep holes in the young shoots of the pine, the bark of which they gnaw. When they are touched, they let forth from their mouth a drop of clear resin, of the odour and consistence of that which issues from the branches of the pine when they are cut. It is this resinous juice which they draw which nourishes them and makes them grow.

Those which live on pear-trees and cherry-trees, &c. have all the upper part of the body covered with a humid, viscous, and shining matter, of a disagreeable odour, which appears intended to preserve them from the rain and the rays of the sun, and especially to assist them in fixing themselves upon the leaves ; for if it be removed, they are observed to attach themselves with difficulty, and appear liable to fall to the ground.

Those of the *tenthredo ovalis* which live upon the alder, instead of this liquid, have upon the upper part a white cottony matter, similar to that which covers the aphides of the vesicles of the elm, those of the aspen, and more especially those of the beech. It is sometimes there in sufficient quantities to form flakes on the sides and back of the larva. This matter, which is soft and light, composed of the re-union of many small flat tufts, which have the figure of a brush, attaches but very slightly to the skin, and is easily removed: but what is most singular is, that if it be removed from the upper part of the larva, at the end of some hours its body will be found covered again with new matter of a similar kind, which issues from many small concave spots, which are perceptible on the skin, and which appear to be so many spinnerets, through which this mass of cottony threads passes. After the last moulting nothing more of the sort is perceptible on the body of the larva, which is then of a bluish green.

We cannot enter here into the detail of all the varieties of form which these larvæ exhibit: every thing of importance in that way will come better a little farther on, in our notice of *tenthredo* proper.

They almost all enclose themselves in a double cocoon, at the end of the summer. Some, and these constitute the greater number, pass the winter in this state. They are changed into nymphs in the spring, and become perfect insects in fifteen or twenty days afterwards. The others undergo their final metamorphoses in a short time after having formed their cocoon.

The genus or subgenus CIMBEX, forms the first division of this tribe. Geoffroy first separated them from the *tenthredo* of Linnæus, under the improper name of *crabro*. Olivier substituted the denomination *cimbex*, which has also been adopted by Fabricius; but it would have been more proper to have preserved to this genus, as M. Jurine has done, the name of *tenthredo*, these insects being the largest of the tribe of which they constitute a part.

They proceed from false caterpillars, with two and twenty feet, whose body is smooth, with lines or longitudinal bands of a colour different from that of the ground.

They feed on the leaves of different trees, especially on those of the sallow, the osier, the alder, and the birch-tree, on which they remain rolled up in a spiral form. Many of them exhibit a phenomenon of a sufficiently singular kind; when they are touched a little strongly, they cause to issue from their sides a greenish liquid, as clear as water, which they shoot horizontally to more than the distance of a foot. These fits only take place when they are caught upon the trees: if they are shut up they produce them no longer, probably because the leaves on which they are fed do not possess sufficient humidity to supply the source which furnishes this fluid. The apertures which give passage to it are situated above the stigmata, at the summit of a fleshy and triangular piece.

Towards the end of summer, the larvæ have acquired their full growth. Almost all of them then quit the trees and retire into the earth. They there spin an oval cocoon, of a rough and thick silk, the tissue of which is similar to that of gum: they pass the winter shut up in their cells, are changed into nymphs at the commencement of summer, or the end of spring, and become perfect insects in a little time after this metamorphosis. Those which do not conceal themselves in the earth fix their cocoons to the leaves, or to some branches.

The flight of the cimbices is heavy, and in flying they make a humming noise, similar to that of wasps and bees. About twenty species are known, almost all of which are peculiar to Europe.

The larvæ of *cimbex amerinæ* is found upon the leaves of the sallow, and usually rolled into a spiral form. It is of the number of those which shoot forth a fluid when they are touched. It undergoes its metamorphosis in a cocoon of silk, shining, and of a yellow-brown, which it attaches to a branch, passes the winter there, and the perfect insect issues forth at the end of the following spring.

We here insert the figure of an insect which belongs to Dr. Leach's section of this genus, which he names perga, mentioned at page 355, with the specific addition *scutellata.* It is of a dark greenish bronze; the scutellum and dilated collar pale yellow; both the wings are fulvous; and the mandibles are reddish.

The larvæ of the HYLOTOMÆ have from eighteen to twenty feet, the first six of which alone are terminated by a conical and scaly crook. The others are membranaceous.

Two celebrated naturalists, Reaumur and Degeer, have followed the metamorphosis of many species of this genus. The *hylotoma rosæ* has more especially fixed the attention of the French naturalist: its larva is particularly remarkable for the strange and singular attitude which it assumes. It usually holds the posterior extremity of its body raised, and often folded in the form of an S. Occasionally it turns it down. It has eighteen legs, the hinder two of which seldom move. The fourth ring, the tenth, and the eleventh, have none. Its scaly legs are terminated by two crooks, which is a character peculiar to this species of false caterpillar. Its body above is of a yellowish, bordering on the colour of a dead leaf, all covered with small black tubercles, from most of which there proceeds a hair. The sides and under part of the abdomen are of a green, which is between sea-green and the colour of water. The under part of the rest of the body is greenish and transparent. The belly thus allows to appear a longitudinal vessel, having a movement like the dorsal vessel, but slower and more feeble. This false caterpillar, to pass into the nymph state, enters in the earth and constructs there a double cocoon, in which it encloses itself. The exterior envelope is a net-work with large meshes, but solid and capable of resisting pressure: its threads, seen with a strong magnifier, seem to be small cords, like cat-gut, having inequalities. They have a sort of elasticity which

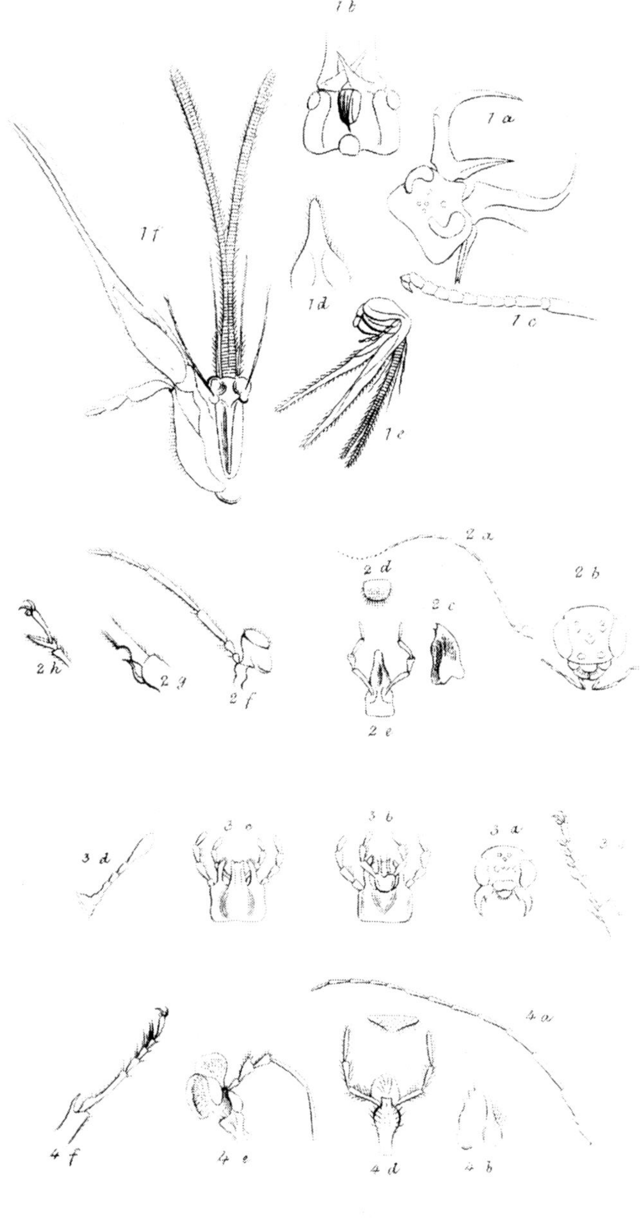

1 *Synargris cornutus* 3 *Perga scutellata*

2 *Stephanus Brasiliensis* 4 *Megalyra faciipennis*

London Published by Whittaker & C° Ave Maria Lane. 1832.

causes them to resume their original position, as soon as pressure is removed. The internal envelope, on the contrary, is of a very close tissue, but without spring, soft and flexible. This internal cocoon is not adherent to the other, of which we may convince ourselves by cutting small portions from one of the ends of the latter, so as to make a passage. Having thus provided for its safety, the false caterpillar spins for itself a second cocoon, the tissue of which is softer and smoother than the finest satin, and which is to constitute a soft bed for the nymph.

The external cocoon is of a reddish cinnamon colour, but the interior is whiter. If earth be refused to the larvæ, it does not the less build for itself the lodging which is necessary for its transformation. The net-work of the external shell is then more perceptible; for when the animal makes it in the ground, it is necessary to clear it that we may be enabled properly to distinguish the meshes.

The larvæ of *hylotoma enodis*, Fab., fastens itself to the edge of the leaves by means of six scaly feet. It keeps the rest of its body stiff and a little raised: it eats much and with avidity: it spins a double cocoon of an elongated oval form, composed of silk of a dirty white, without any mixture of earth: the external part is of the consistence of parchment, the internal is very slender. The perfect insect does not make its appearance until about ten months after, in the summer of the following year.

The other subdivisions of this tribe present nothing worthy of notice here.

Our figure, *Schyzocera Peleterii*, belongs to the subgenus so named by Latr. (see page 356.) It is from Carthagena, and is of a dull pale yellow; with the vertex, the centre of the thorax, the tip of the abdomen, and the posterior tibiæ black; the wings brown, with a pale central fascia.

In the tribe of Urocerata, we shall more particularly consider the genus Sirex, which by some writers, and by our author, formerly has been called Urocerus.

Reaumur, Degeer, and Linnæus, in his earlier works, considered these insects as ichneumons. They were subsequently, however, with great propriety separated from that genus.

These insects inhabit, preferably to others, cold and mountainous regions, abounding in pines and other coniferous trees. The species which is most known *(gigas)* is very much extended in Sweden. Maupertius has met with it in Lapland; and Reaumur, to whom he presented this insect, named it, in consequence, *Ichneumon de Laponie*. It is also very common in the Alps and Pyrenees.

In a work called, "*Ephemerides des curieux de la nature*," is an observation apparently relative to these insects, which would be very extraordinary indeed, if it be true. It is there said, that in the town of Czierck and its environs, there were seen in 1679, some unknown winged insects, which with their stings mortally wounded both men and beasts. They fell abruptly upon men without provocation, and attached themselves to the naked parts of the body: the sting was immediately followed by a hard tumour, and if care was not taken of the wound within the first three hours, by hastily extracting the poison from it, the patient died in a few days after. These insects killed five and thirty men in this diocese, and a great number of oxen and horses. Towards the end of September, the winds brought some of them into a small town on the confines of Silesia and Poland; but they were so feeble on account of the cold, that they did but little mischief there. Eight days after, they all disappeared. These animals have all of them four wings, six feet, and carry under the belly a long sting provided with a sheath, which opens and separates in two. They make a very sharp noise in attacking men. Some of them are ornamented with yellow circles, and others are similar to them in all respects, but they have the back altogether black, and their stings are more venomous. The author of these observations gives an extended description of one species, that with the yellow circles, which he accompanies with figures; and, indeed, the character of *sirex* may be clearly distinguished, and which M. Latreille would refer either to the species *gigas*, or that of *fusicornis*. The other of which he speaks, as being entirely black, may be either *spectrum* or *juvencus*.

Our distinguished author is, however, very far from agreeing with the observations which we have now detailed. Nature has given to the insects of the genus sirex, an auger for the purposes of oviposition in the holes or clefts of trees, and this is the only purpose to which they can turn it. It is impossible to believe that these insects could change their instinct all of a sudden, become the aggressors of man, whom they should naturally avoid, and convert into an offensive weapon a body which is simply an oviduct. Even on the supposition that some persons might have been wounded by them, nothing could result but the common effects produced by the wound of a small sharp instrument, without poison, of a mere thorn or spine. The slight denticulations with which the extremity of the auger of the sirex is armed, could only render the wound a little greater. But ignorance, and her invariable attendant, superstition, are ceaseless perverters of facts, and misinterpreters of nature.

The insects of this genus hum while they are flying. The history of the most common species (gigas) is known through the observations of Rœsel. The female lays in wood some eggs which are very much elongated, and pointed at the two extremities. The larvæ is elongated, radiated, yellowish, cylindrical, with a scaly head, and six very short feet. The posterior extremity of the body is inflated. For the other particulars of its metamorphosis, we must refer the reader to the author just mentioned.

Of the first tribe of the Pupivora, the Evaniales, we can only notice that the insects of the genus Fœnus live on flowers, and their usual attitude is with the abdomen raised. At night, or when the weather is inclement, they remain hooked by their mandibles, and almost perpendicularly, to the stems of different plants. They are frequently observed flying in dry and sandy situations, with solitary bees and spheges; but this is not for the purpose of constructing nests for their young, but on the contrary, to take possession of those which the preceding insects have formed, or at least to destroy their hopes by depositing eggs in the interior of their larvæ or by the side of them. When the young of the fœni are disclosed, they will devour these larvæ, and undergo their metamorphosis in these usurped retreats.

We pass on to the next tribe, the Ichneumonides.

The females of this tribe, when obliged to lay, walk or fly, continue agitating their antennæ, to endeavour to discover the larvæ, nymphs, and eggs of insects, and even those of spiders and of aphides, which are destined to receive and to support their posterity. They exhibit in these researches, an admirable instinct, which leads them to discover the most concealed retreats. It is under the bark of trees, in their clefts or crevices, that the females, whose auger is long, place their eggs: they introduce there the auger, properly so called, or the thread of its middle, in a perpendicular direction: it is entirely disengaged from the other two threads, or half-sheaths, which are parallel with each other, and supported in the air in the line of the body. But the females, whose auger is very short, and but little or not at all apparent, place their eggs in the bodies, or on the skin of larvæ, of caterpillars, and in the nymphs which live uncovered, or which are very accessible.

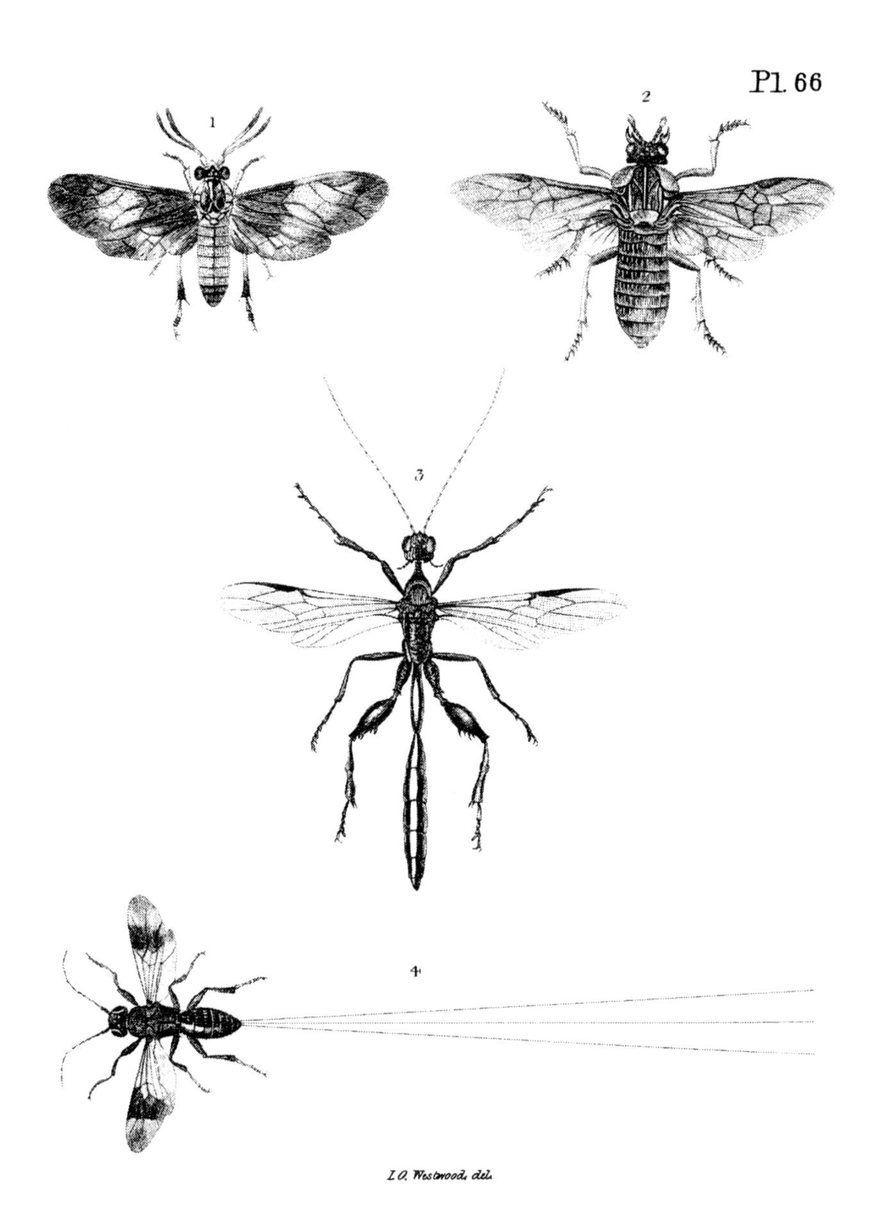

1. Schizocerus *Peletierii*. *G. R. Gray*. 2. Perga *scutellata*. *Westw.* 3. Stephanus *Brasiliensis*. *Westw.* 4. Megalyra *fasciipennis*. *Westw.*

London, Published by Whittaker & C.o Ave Maria Lane. 1837.

The larvæ of the ichneumonidæ have no feet, and resemble small worms. Those which live in the body of larvæ or caterpillars, after the manner of intestinal worms, gnaw at first only their fatty bodies, or those of their internal parts, whose existence is not essentially necessary to their preservation; but when on the point of being changed into nymphs, they pierce their skin for the purpose of coming forth, or otherwise very soon destroy them, and quietly undergo their metamorphosis in the dead body. They all spin cocoons with silk, in which they pass to the nymph state, and then become perfect insects.

The females of the genus Ichneumon have, at the extremity of the abdomen, an auger, which, when it is external, constitutes a long tail. This instrument is an oviduct consisting of three pieces, from which some authors have taken occasion to name these insects *musca tripilis*. It is with the middle piece that the eggs are introduced, which accordingly is more scaly than the other two. Though the apparatus of their ovipositor has the external appearance of a sting, and though the insect, when taken in the hand, attempts to use it for that purpose, there seems no reason to apprehend its effects in general. The ichneumons, indeed, in which this instrument is short, and from that circumstance strong, may be able to, and in fact do sometimes, pierce the skin in the thinner parts, and occasion a sensible degree of pain.

From the shortness of the wings, in proportion to their body, we may conclude that these insects are not capable of any very long or sustained flight. In fact, we observe them continually alighting somewhere, and agitating these organs as well as their antennæ; from this circumstance they have been named by some writers *muscæ vibrantes*.

If we have reason to complain of the number of caterpillars, and of their fatal ravages in the vegetable kingdom, we ought to congratulate ourselves on the existence of the ichneumons, which are their natural enemies. We know that the ancient naturalists designated under this denomination a little quadruped inhabiting the banks of the Nile, and which received divine honours from the Egyptians, in consequence of the opinion that it destroyed the eggs of the crocodile, and even entered the mouth of that reptile, and proceeding into its body, gnawed its entrails. On the latter part of this opinion, or the absurdity of supposing that any *mammiferous* animal could exist within another living body, we have already expressed our opinion in the proper place. But as the light of science has dissipated the night of ignorance and superstition, chasing away the spectral population of credulity, so has it revealed to our astonished view the marvels of truth, not less surprising than the miracles of fiction. But its best result is, that it has taught us to regulate our belief in the permanent exercise of the laws of nature, instead of forming our code of faith on their violation. Poets may "spurn the bounded reign of existence," but philosophers will find ample room to expatiate within its limits to the end of time. Thus the name of ichneumon has been well applied to the insects of which we are treating, since they actually perform, though not exactly in the same manner, what the quadruped, their namesake, was but fabled to do. The *ichneumons* of the entomologists both destroy the eggs, from which the caterpillars—which may be considered as a sort of little crocodile in voracity—proceed, and which we may remark, by the way, are more extensively mischievous, and the caterpillars themselves, and the chrysalids which contain them. Protecting, therefore, the domain of vegetation from such dangerous enemies, they are entitled, if not to Egyptian idolatry, at least to some portion of our gratitude; or rather let us say, that all our gratitude is owing to that beneficent Being, who, for whatever wise purposes he may have permitted evil, has, in all instances, bestowed a compensation of good.

But how, let us inquire, do the females of ichneumons contrive to deliver us from those pernicious insects which despoil nature of her finest ornaments, and restore to us the dreary spectacle of winter in the midst of the finest days of summer? We must not imagine that the ichneumons give mortal battle to the caterpillars: the latter must certainly perish; but it is necessary that they should live for a time, for the purpose of serving both as cradle and nourishment to the posterity of their enemies.

We have mentioned that the female ichneumons are provided with an auger. It may be necessary to add a few details to what we have said on that subject.

We have mentioned that this instrument is an union of three pieces: the two lateral ones serve as a case to the middle piece, being hollowed like a gutter on the internal side, and convex above. The thread of the middle, or the oviduct, properly so called, is smooth, and rounded in the major part of its length; but near the extremity it is flatted and terminates in a point, sometimes formed like the nib of a pen. Observed with the microscope, the part of the stem of this oviduct, which is thick, broad, and flatted, presents on one of its faces a channelling which proceeds from the base as far as the extremity. This gutter is such that the piece seems possible to be divided into two parts, and that the two edges of the cleft are united only by a membrane which permits them to separate at the moment of oviposition. The extremity of the auger allows us to see the opening which gives passage to the eggs; we perceive, at the same time, that some soft and fleshy parts fill the interior of the oviduct. The membrane which unites the two edges of its canal is more apparent at this extremity. The point of the instrument, which to the naked eye appears simple, in reality is not so. Underneath the membrane, and on each side, rises a range of from five to six teeth, similar to those of a saw. This auger, considered in this point of view, has some conformity with that of the cigalæ. This instrument, though delicate and flexible, is nevertheless introduced into very hard bodies. When the ichneumon makes no use of it, it is enclosed in the case, and appears to be composed but of a single piece. Sometimes, again, this case receives but a part of the stem of the auger, and then the instrument appears composed of but two pieces. All this gave rise among the ancient naturalists to the name which they gave to these insects of *musca tripilis*—literally, *fly with one, two, or three hairs.*

Now let us consider how the female, with the long auger, makes use of this instrument.

If any place be favourable for the multiplication of certain insects, it will prove equally so for that of the ichneumons, since these last rear their young at the expence of the others. Consider this ancient wall, with either an eastern or a southern aspect. It answers the purpose of a cradle for the posterity of numerous solitary bees or wasps: its clefts, its plastering, afford an hospitable retreat to their young. This is soon observed by some female ichneumon: she well knows that the larvæ which shall issue from her eggs will there find suitable aliment. She immediately commences roaming round the nests of these insects; she places herself on the plaster which conceals their larvæ, her auger appears to be but of a single piece, but soon she developes it, raises it, lowers it, turns it in the different portions of its length; she contrives to make it pass under her belly, the point being carried forward. The manner in which the insect is placed upon her feet, the difference of length which there is between these parts and the auger, necessitate these movements, and this direction. The point of the oviduct being brought in front, the animal conducts this piece as far as it is possible, applies the extremity of it against the plaster of the wall, making alternate movements from left to right and from right to left.

The operation lasts some minutes, even as far as a quarter of an hour. The point of the auger is then constantly placed in front of the head. Some species have, under these circumstances, the head turned upwards, others downwards.

Some species, more especially those whose abdomen is cylindrical, and terminated by a long tail, know how to find the larvæ, which are under the thick bark of large trees, and in the interior of the wood itself. Their clefts, or external crevices, permit the intromission of the auger. But the situation of this piece, relatively to the body, when the insect sinks it into the wood, is not the same as in the preceding. Here the oviduct is directed almost perpendicularly, and disengaged entirely from its two semi-sheaths, which are parallel, between themselves, and supported in the air in the line of the body.

Some other species of female ichneumons do not experience the same difficulty in placing their eggs. The bodies which their auger must penetrate are less hard and more uncovered—as for instance, the caterpillars and their chrysalids. The cabbage plant nourishes the caterpillars of certain butterflies, which, for this reason, are called *brassicariæ*. The first of these is often devoured by the larvæ of a small species of ichneumon. These larvæ live in a family in the interior of the body of the caterpillar, and spin very handsome cocoons, which they attach one to the other. The assemblage of these cocoons has the appearance of a sort of cottony ball. Goëdart and other naturalists, deceived by these appearances, have said that these larvæ were the true offspring of those caterpillars. They have even attributed to the latter maternal feelings, asserting that they spun the silk for the purpose of enveloping and protecting their beloved progeny. But men who have better pursued the march of nature, and were better acquainted with the constant and invariable harmony of her laws, men such as Swammerdam, Leuwenhoek, Valisnieri, &c. have proved the fallacy of such conclusions. They have demonstrated that the larvæ which live in the body of caterpillars, or in their chrysalids, owe their birth to insects, either ichneumons, cinips, &c.—in a word, perfectly similar to those which these larvæ produce at the final term of their metamorphosis. The only thing which could cause any hesitation on the subject was the explanation of the manner in which these larvæ had been introduced into the body of the caterpillars.

These parasite larvæ live in society, or solitarily. To deserve the attribute of social, they must, according to Reaumur, be in great numbers in the body of the caterpillar, and issue forth together, for the purpose of undergoing their metamorphosis one beside the other. Should there be but one or two, they must be ranged among the solitary larvæ. The greater part of the known larvæ of the ichneumons spin a cocoon more or less silky and ovoid, for the purpose of undergoing their transformation into nymphs.

The larvæ issue sometimes from the body of the caterpillar, sometimes from the chrysalis, according as the caterpillar was more or less advanced in age, when it received the eggs of the ichneumon. The larvæ which live in the interior of cabbage caterpillars, are smooth and without feet. Scarcely have they issued from the body, the sides of which are pierced, than they commence to form their little cocoon. All those which issue from one of the sides descend on the same side, without removing from each other, or from the body of the caterpillar. By means of their spinneret, situated at their under lip, in the same manner as that of caterpillars, they throw out some threads in different directions, and the result is quickly a small cottony mass, on which each larva establishes its cocoon. The tissue of these cocoons is of a fine silk, which differs but little from that of the silk-worm in its texture, and which is either of a very fine yellow, or very white, according to the species.

Reaumur has observed some larvæ which had lived in the body of a caterpillar of the *Aristolochus*. He remarked that those which issued forth all came together, and chose for the foundation of the cocoon which they were about to make, the commencement of another cocoon. The cottony mass which envelopes the whole of these cocoons is only the general interlacing of the stuff which each larva weaves at first. The skin of these insects being very tender, it was necessary that on quitting their cradle they should be promptly sheltered. Accordingly, in less than two hours the cottony mass is completed.

But what an extraordinary phenomena is this! These larvæ have lived for a long time, and in prodigious numbers, in the body of the caterpillar, without the latter appearing to have suffered any consequential detriment. How can we conceive that it could contain within its bosom enemies so multiplied and so terrible, without succumbing almost immediately to their attacks? These larvæ are taught, by their instinct, that their own existence depends on the prolongation of that of the caterpillar. It is therefore incumbent on them to avoid giving it a mortal wound during the whole period in which they are attaining their growth. They do not touch upon the organs which are essential to life. That part, called the *fatty body*, which is of considerable volume, and the use of which appears to be more important to the insect in the chrysalis than in the caterpillar state, furnishes to the larvæ their habitual nourishment; but when they have attained all their growth, it follows of necessity that they must kill the caterpillar in tearing its sides for the purpose of getting out. Accordingly, some other larvæ, whose body more rapidly increases in bulk, still more abridge the existence of the caterpillar in which they have lived. The stems of different plants, and especially granivorous ones, are very frequently loaded with masses of cocoons, pretty nearly similar to those of which we have spoken. The ichneumon which comes from them is very small. The interior of hives also presents, but rarely, a sort of little cake formed by an ichneumon, which probably has lived under the larvæ form in the caterpillar of the tinea. It would almost seem that these larvæ, in spinning their cocoons, were desirous of rivalling the bees, and had taken their industry as a model.

The silk contained in the reservoirs of the spinning caterpillars is sometimes of different shades, which may be referrible both to the quality of the food and to the particular arrangement of the animal; from which it follows that the exterior part of the shell ought then to differ in colour from the internal layers. Cocoons of the ichneumons are accordingly to be found, which are of two distinct colours, disposed in bands. Some are brown, with a white or yellow band in the middle, others have several bands of these colours. This variety does not entirely depend upon the causes which influence the differences of colours of the cocoons of caterpillars; for, were that the case, certain portions of the silky matter would be, some alternately white or yellow, others alternately brown, and these changes would be repeated much more than in the cocoons of the ichneumons. Every thing here appears to be reduced to two causes: first, the first silk which the larvæ of the ichneumon spins, that which forms the external envelope, is white, and the second, or that of the internal layers, is brown; second, the cocoon is more fortified, both by circular spaces in the middle and near the two ends than any where else. This being granted, it is clear that the brown colour of the internal layers will predominate in the places where the layer which is external to the white silk shall be weak; while, on the contrary, all the parts of the external surface which shall have been strengthened with the silk of this last colour will predominate over the brown—from this circumstance arise the brown and white bands. One may convince himself of this by scraping off with the point of a penknife some portions of a white place. The brown will appear then in proportion as the inequality of thickness of the upper layer diminishes. The silk of these cocoons is of an extreme fineness; it possesses a brilliancy and splendour equal to that of varnish, or of a hard body polished in the best manner. These cocoons may be observed on the broom tree in the commencement of autumn. The larvæ is of a greenish white, passes the winter in its cocoon, and is not metamorphosed into a nymph until spring.

Some ichneumons place their eggs in the body of some caterpillars, which are on the point of passing to the state of chrysalis, or are preparing themselves to pass into it. The larvæ issue subsequently from the chrysalis, and spin their cocoons, if they belong to the spinning species, in the interior of the chrysalis, where they are more in safety: other larvæ are transformed into naked nymphs, under the skin of the caterpillar, or chrysalis, which they have devoured.

We meet on the oak the cocoon of an ichneumon singular in many points of view. It is suspended to a leaf or small branch by a thread of silk, which proceeds from one of the extremities of the cocoon. Its form is almost the same as that of the others, but less elongated: in the middle there is a band of a whitish colour. This, however, is not the circumstance which renders it most remarkable. It presents a phenomenon which has fixed the attention of the illustrious Reaumur. The cocoons which he has detached, and enclosed in boxes, were observed to jump there. Placed upon the hand they executed the same movement, and raised themselves to the height of about eight lines, and even occasionally to that of from three to four inches. Reaumur explains this extraordinary fact by supposing that the larva, shut up in the cocoon, acts as a kind of spring, which is let go. With this illustrious naturalist, let us suppose this larva shut up in its cocoon, and lodged there at its ease, and placed upon one of its sides; let us imagine that it curves itself by degrees, so that the middle of its back becomes the middle of the convexity of this curve; that the most convex portion touches the interior and most elevated surface of the cocoon, but that its belly is not contiguous to the interior and inferior surface; that the two extremities alone of its body touch the cocoon; let us grant to this larvæ a sufficient force to make it suddenly assume the same curve in an opposite direction—that is to say, that the middle of its belly, instead of being concave, becomes convex; that the belly is carried towards the bottom of the cocoon, and the hinder part of the head towards the lower part of this same cocoon; in fine, let us suppose that this most elevated point be struck suddenly, before the belly has touched the upper part, the two blows given by the head and the tail will push the cocoon upwards, will force it to raise itself obliquely, and to go forward; and this composite direction results from the obliquity with which the two strokes have been given. But to what end has this larva received from nature this faculty of jumping? There is reason to presume, with Reaumur, that this natural situation of being suspended in the air by means of the thread of its cocoon, is a means of preservation for the insect—that the wind, or other circumstances, happening to displace the cocoon, to bring it in contact with other bodies, it was necessary that the animal should be able to resume its ordinary situation, and for this purpose that it should cause its cocoon to jump. Reaumur, in fact, observed that the larva had recourse to this method when it was disturbed. This naturalist obtained from these cocoons a species of ichneumon, and a fly with four wings, whose body is short, of a blue back, and the abdomen thick, and rather short antennæ (a species of *chalcidites*): This naturalist was thus unable to tell precisely what is the natural inhabitant of these singular cocoons. M. Latreille found in the Bois de Boulogne a small cocoon, suspended, in like manner, to the leaf of an oak by means of a thread; a species of ichneumon proceeded from it, which he has described in the "Bulletins de la Société Philomathique;" but he does not think that this cocoon is of the same species with that of Reaumur: his (M. Latreille's) was of an uniform colour. Muller and Degeer have found similar cocoons, from which also ichneumons were produced. These insects are, therefore, in all probability the true inhabitants of them. The eggs of the lepidoptera are certainly very small; but, nevertheless, they serve as nutriment to one larva of the ichneumon. This may be easily conceived by the smallness of its volume.

The female ichneumons are endowed with so surprising an instinct that they discover the insects in the body of which they are to place their eggs, however concealed they may be. The larvæ of the mason-bees, the leaf-rolling caterpillars, the wing caterpillars, the tineæ, the gall-insects, and even the spiders, cannot guard themselves against them, and become the prey of their larvæ. It was worthy of that Supreme Wisdom which regulates the universe thus to oppose a barrier to the prodigious fecundity of hurtful insects.

The larva of the *ichneumon rufus*, as we are told by Degeer, lives in the body of a caterpillar; it is white, without feet, and with a scaly head: it shuts itself up in an oval, yellowish cocoon, pointed at the two ends; it suspends itself by means of a long slender thread of some lines in length. The cocoon from which Degeer obtained the ichneumon which we have already mentioned, was fixed to a paper covering of the box in which was the caterpillar which had nourished its larva, and near it. The ichneumon, in coming forth, made a small piece jump out, in the form of a cap, at the lower end of the cocoon, and this piece remained attached there.

The females of the *ichneumon aphidum* deposit their eggs one by one in the body of an aphis. The larva there finds a sufficient nutriment, pierces underneath the empty skin of the dead body of the insect, attaches it by means of a thread of silk on the leaf where it is, then lines with a bed of white the interior of the skin of the aphis, and there becomes transformed into a nymph. It passes the winter there, and issues forth in spring, forming a circular aperture, towards the posterior extremity of the skin of the aphis.

When the eggs have been deposited at the commencement of the summer, the metamorphoses take place quicker, and the insect appears with its wings a short time after.

The females of *ichneumon lutedus*, one of the *ophions* of Fabricius, deposit their eggs on the bodies of caterpillars, particularly on those which are named *forked tails.* The eggs are implanted there by means of a pedicle, tolerably long and slender; they are black. The larvæ live and grow on the exterior of these caterpillars, and at their expence. Their skin is tight, smooth, and shiny, as if it had been wetted. Their colour is of a dirty white, with a broad band of obscure green on the back, and some shades of the same colour on the sides. Degeer has remarked that their posterior extremity remained engaged in the cocoon in which they were born.

The *Stephanus* we have figured from Mr. Westwood, and which that gentleman considers to be new, under the name of *Braziliensis,* is of a shining black colour, with the posterior tibiæ dilated in the centre and at the tip: it is from Brazil. The insect which Mr. Westwood distinguishes generically under the name *Shegalyra,* belongs to the family Ichneumonidæ: the species he names *Fasciipennis,* is a dull black colour and hirsute, the wings whitish, transparent, with a brown band near the apex;—this insect is from Melville Island.

We shall now pass on to the next tribe, the GALLICOLÆ, in treating of which we shall confine our observations to the genus CYNIPS, and say a few words respecting the nature of *galls* in general.

Geoffroy injudiciously designates under the name of *diplolepes,* the insects comprised by Linnæus in his genus *cynips,* and applies his last denomination to other hymenoptera, there ranged with the ichneumons. He supposes that those vegetable excrescences, called galls, are equally produced by one and the other. Naturalists, however, of other countries have followed the Linnæan nomenclature, and M. Latreille very properly imitates their example, reserving to the insect whose pricking gives rise to the galls, their primitive designation of *cynips.* The name of DIPLOPUS he has thought proper to suppress, because, in consequence of its injudicious employment by Fabricius, it has encreased the confusion of nomenclature.

The cynips are distinguished from the other hymenoptera by many characters. They have no nervures to the lower wings: their antennæ are straight, filiform, or scarcely thicker towards their extremity, and usually composed of from thirteen to fifteen articulations. Their palpi are always very short. The auger of the females, very analogous to that of the ichneumons in its form, is lodged either entirely, or at least towards its origin, in a cleft or external furrow, formed along the belly. Their upper wings have a complete radial cell, long, almost triangular, and two or three cubital cells, the third of which is very large, and reaches to the end of the wing.

Passing over other details of organization, we shall make a few remarks on the auger or ovipositor of these insects. The abdomen of the female cynips is disposed in a manner very favourable for the play of this instrument. The upper laminæ, each of which forms usually a little more than half of a ring, or segment, make here almost an entire circle, and envelope the laminæ of the under semi-rings. That of the base is the largest. The others are short, and form, by the dilatation of their extremities, or at least of some of them, and by their sharp and oblique convergence, the lower keel of the belly. From the origin of the abdomen, and from the middle of its inferior base, proceeds a small piece, more or less long, almost cylindrical, scaly, flowing longitudinally between the upper laminæ of the belly to their point of re-union, and serving as a crest to the keel of the abdomen. The auger, which Geoffroy, injudiciously enough, calls *aiguillon* (sting), is lodged, previously to its issuing forth, in the furrow or groove of this piece. This instrument, intended to deposit the eggs, is slender, capillary, of the nature of a scale, of a length which, in many of them, equals that of the circumference of the abdomen, measured in its length. As this auger is but little projecting externally, or is even not at all so, it is necessary, seeing its length, that it should be folded back within the abdomen. In fact, it forms there a sort of arch, the curve of which is removed but little from that of a circle a little elongated ; it forms one or two spiral turns. If we figure this instrument, taking its origin from the middle of the upper and lower part of the abdomen, directing itself towards its origin, applying itself on the upper edge, and almost circular, of any powerful muscle, compressed, in the same manner as the cord of a pulley, let us suppose the auger prolonged so as that it shall reach the lower part of the abdomen, and shall glide into the canal of the crest; and then figure to ourselves, at the extremity of the belly, underneath the anus, two small semi-cylindrical pieces, hollowed interiorly into a gutter, to receive the point of the auger, and we shall have an idea of the relative direction and disposition of this oviduct. The two small stems which we have mentioned are much longer in many female ichneumons. These are the two lateral threads. They are, on the contrary, very short in the insects with a true sting, such as wasps and bees; in fact, they are nothing but two small pieces, in the form of a stylet, and are inserted on the muscle of the ovipositor.

This ovipositor, or auger, is not simple, as might be supposed at the first glance. It is hollowed like a gutter, and its extremity is furnished with small lateral teeth, resembling the iron point of an arrow. It is with these teeth that the insect enlarges the notches which it makes on the different parts of vegetables, for the purpose of placing its eggs there. The juices of these vegetables expanding through the vessels which are open in this place, form there an excrescence or tuberosity, which is called a *gall*, in which the egg is enclosed, and which by little and little acquires volume and consistence. From these eggs come small larvæ, without scaly feet, but often provided with a dozen or fourteen membranous feet or nipples. These larvæ find in their retreat the means of nutriment: they suck and gnaw the interior of the gall, which grows and acquires solidity in proportion as they eat the interior.

Many of these galls, considered in general, have a cavity which encloses a certain number of larvæ living in society. Others have many small cavities, between which there are communications. In some others, we see more than a hundred cells, each of which is inhabited by a single larva. In fine, other species of galls have but a single cell, which is inhabited by a larva that lives in a solitary state.

Galls present great varieties in their form. A little further on we shall enter more at large into their description. Here a few general views may suffice. The most common galls are rounded. That, which is the best known of all, which enters into the composition of ink, and which is employed in dyeing, is an excrescence produced by an insect of this genus. The colour and the figure of some galls have caused them to receive the name of certain fruits to which they bear some degree of resemblance. Some are found on the oak, which are called *apple, gooseberry,* or *pippin* galls. Some are seen which resemble fruits in their spongy texture. Among those which are of a round form, some are attached upon the plant, others only hold there by a short pedicle. There are irregular and composite ones, and some of a very curious structure. Some appear to be a portion of the plant tumefied and thickened: such are those observed upon the willow and the osier. Various vegetables, and their different parts, exhibit galls differently figured. There is a species of fibrous gall which is a very singular production; its nut is hard, and solid, and bristled all over with long filaments detached from one another. It is found on the eglantine, which sometimes has three or four of these galls. The same shrub presents another species, more rare than this last; it grows at the end of its branches, where it forms a mass composed of a dozen little galls of different forms. These two species appear to owe their origin to the same insect. The bristly gall of the rose-tree usually encloses ichneumons, chalcidites, and cynips. For a long time it was a matter of doubt to which of these three insects the production of this excrescence was to be attributed. It is now clearly ascertained that the species of two of these genus are parasites, which have lived at the expense of the real inhabitant, namely, the cynips.

Of all trees, the oak is that on which the greatest number of these tuberosities is found. Some have the form of little apples, isolated or joined together; others are prickly: there are some which are branched, and some which resemble little artichokes or mushrooms. Some leaves are loaded with little rough galls, which have the appearance of buds. Other leaves of the same tree possess a species which resembles a small open goblet: these are flatted, even, or crisped. Some are ligneous, and others spongy, according as the eggs have been inserted in a ligneous or pulpy portion of the wood. We shall take no further notice here, however, of the variety of their figures: simply observing that the form and consistence of these excrescences, seem mainly to depend upon the insect which they enclose.

It is not easy to throw much light on the causes of the varieties presented by galls, on their first formation, or on their growth. The majority of them increase in volume with a surprising rapidity. Those of the largest species grow in a few days, and even, as it appears, before the larva has issued from the egg, so that when it is born its lodging is ready made, and has no necessity of further increase. But a singularity worthy of remark, and one which is observed only in these kinds of insects, is that the eggs of cynips and tenthredo, encrease in volume. This proves that they are surrounded by a flexible membrane, rather analogous to that which envelopes the human fœtus, and that of quadrupeds, than to the shells of eggs in general.

The larvæ acquire their growth very speedily, but they afterwards remain five or six months in the gall, before they become changed into nymphs. Some undergo this metamorphosis in the gall itself, from which they do not issue forth but under the form of the perfect insect, after having made a small hole in it. Others quit the gall to enter into the ground, until they have taken their final form. A little time after they have quitted their exuvia of nymphs, these insects couple, and the females deposit their eggs. All do not place their eggs on plants.

There is a very curious operation performed at the present day in the Levant, with one of those insects, which is termed *caprification.* The object of it is to hasten the maturity of figs; and the species employed for that purpose is the *cynips ficus caricæ,* or *cynips psenes* of Linnæus: it consists in placing on a fig-tree, which does not produce flowers, or early figs, some of these last strung together with a thread: the insects which issue from them, full of fecundating dust, introduce themselves through the eye into the interior of the second figs, fecundate by this means all the grains, and provoke the ripening of the fruit. The first figs, as is well known, appear a month before the others: the second ripen successively from the month of August to October, and even later.

This operation, of which some ancient and modern authors have spoken with admiration, appeared to M. Olivier, who had resided for a long time in the islands of the Archipelago, to be nothing but a tribute paid by man to ignorance and prejudice. In many parts of the Levant caprification is unknown. It is not used in France, Italy, or Spain: it has even begun to be neglected in some islands of the Archipelago where it was formerly practised. Notwithstanding this, good figs are to be every where obtained without this operation. If it were absolutely necessary, either that fecundation should be produced by the seminal dust which spreads itself abroad, and is introduced solely through the eye of the fig, or that nature should employ to transmit it from one fig to another, a small cynips, as has been commonly believed, it is clear that these first flowering figs could not fecundate at the same time, those which are arrived to a certain growth, and those which hardly appear, or do not appear at all, as yet, and which do not ripen for two months after the others.

Putting aside all the marvellous in this business of caprification, which will hardly go down now-a-days, and sticking simply to our observations of nature, we must agree that this operation cannot be of any utility; for each fig contains some small flowers towards its eye, capable of fecundating all the female flowers in the interior, and moreover, this fruit will grow, ripen, and become excellent to eat, even when the grains are not fecundated.

Bernard, on this subject, has given some observations equally interesting and instructive. He has remarked that the figs which are cultivated in the south of France, are never attacked by cynips, while they are constantly found in the grains of the wild figs. When the figs are large enough for the female flowers to become sensible, cynips penetrate into the interior through the eye, and proceed to deposit on each seed the germs by which these insects are to be reproduced. A month is sufficient to bring the larva to its final metamorphosis. The cynips issue from each grain by an aperture which constantly follows the direction of the pistil.

We shall conclude this notice of cynips with a few more observations on galls in general; a subject of considerable interest in natural history, and for an account of which this place appears to us to be the most fitting.

Most of these galls are merely objects of curiosity; but there is one of them which is the object of a considerable commerce, namely, that of the oak of Asia Minor, known under the name of *nut-gall,* of which great use is made in dyeing and other arts.

It is to Reaumur, exclusively, that we owe the few notions we possess concerning the nature of galls in general. The more modern naturalists have been well employed in the description of the insects which produced them, but they have given us little or nothing concerning their formation.

There has been much disputation on the means which nature has employed to produce galls so different from one another, from the wound made by an insect, on such or such a part of a plant; but we must aver that the result of all these speculations is far from being satisfactory. Respecting the cause of the regularity of the growth of these singular productions, we can as yet (and it is to be feared that such will be the case for some time) do little more than avow our ignorance.

Galls may be divided into *true* and *false*. The first are those which form an excrescence, exactly closed on all sides, and in which one or many larvæ of insects live, which quit this habitation either before or after their metamorphosis. The second are those which are formed, contrary to nature, on a part of the plant produced by the wound of an insect, but in which the cavity is often open, or but incomplete.

The true galls are subdivided into *simple* galls; that is to say, those in which there is but a single lodging for the animal, no matter whether there be one or many insects there; and into *composite* galls, formed by the union of many lodges which grow together. In both divisions the forms are various; they are globular and smooth, globular and more or less rough, foliated, hairy, osseous, fungous, &c.

In the majority of galls, it is a difficult matter to obtain perfect the insects whose larvæ they contain. Many of these larvæ die the very instant that the gall is removed from the plant to which it was attached; and others require for their transformation, conditions which are either unknown or difficult to be produced.

A great number of insects spring from galls. Coleoptera, hemiptera, lepidoptera, and diptera, are produced from them. The genus, however, particularly consecrated by nature to their formation, is that one of the hymenoptera which we have been just considering. All the species which compose it are born in galls. These two are the only *true* galls. The most remarkable among them are the following:

The *gall of the rose-tree,* which is sometimes called *bede-*

guar : it is as large as an apple, and covered with long reddish and pinnated filaments; it grows on the stem of the eglantine, and is produced by the *cynips rosæ*. It has been placed among the remedies which may be successfully employed against diarrhœa and dysentery, and useful in cases of scurvy, stone, and worms.

The *fungous gall of the oak*. It is as large as the preceding, but smooth externally: it grows at the extremity of the young branches of the oak, is composed of a great number of osseous lodges, enclosed in a fungous matter, and is produced by *cynips terminalis*.

The *grape-like* gall of the oak, which grows on the peduncle of the male flowers of this tree; it is about the size of a grape, semi-transparent, and has but a single lodge: its insect has not been described by Fabricius.

The *artichoke-like gall of the oak*. This comes on the buds or germs of the oak, which acquire a monstrous growth, similar to an artichoke, or a cone of the fir-tree. It is produced by another *cynips*.

The *gall of the oak-leaves*. It grows on the lower surface of the oak-leaves: it is of the size, form, and colour of a cherry; it contains but a single lodge which is inhabited by *cynips quercûs folii*.

The *gall of the toza-oak*, which comes on the young branches of this tree in the Pyrenees. It is round, and as thick as an ordinary apple: and at two-thirds of its height are seen a range of pointed tubercles.

The common gall used in commerce, or the *nut-gall*, which grows on the branches of the gall-oak in Asia Minor, is very hard, and most frequently tuberculous. It is to Olivier that we owe the knowledge both of the insect which forms it, and of the oak on which it is produced. It is much more esteemed when gathered before its maturity, that is to say before the insect issues from it. The galls which are pierced, are of a clearer colour and less heavy. The orientals are particular to gather the galls at the precise moment which experience has taught them to be the most advantageous. This is the time when they have acquired their full growth. In consequence, the agas are careful that the cultivators shall traverse, in the commencement of July, the hills which are covered with oaks. The first galls are set apart, and known in commerce under the name of *black* or *green* galls. Those which have escaped the first search are called *white* galls, and are sold at a less price.

The galls of the neighbourhood of Moussoul and Tocat are inferior to those of Aleppo and of all the interior of Natolia.

The nut-gall is of very great usage in dyeing, to produce black colours and all the shades which depend upon them. It is also employed in the preparation of leather, in the fabrication of ink, and in medicine, as an astringent, either internally or externally. It in general possesses, but to a higher degree, the properties of the oak, that is, it contains a certain quantity of tannin, or of the astringent principle.

The *gall of the roots of the oak*, is ligneous, and composed of a great quantity of lodges united. It grows on the roots of the old oaks, which come out from the ground. It is the hardest of all that grow in our climates.

The *gall of the field-cirsium*, which is only a swelling of the stem of the same plant, formerly enjoyed a very great reputation, because it was considered, when carried simply in the pocket, as a sovereign remedy against hemorrhages, a virtue which it owed only to its resemblance to the principal sign of this disease, the swelling of the vein. It is composed of many lodges, almost ligneous, and produced by a cynips which has not been described.

The *gall of the ground-ivy*, grows upon the stalks and

leaves of this plant. It is hairy, and encloses a small number of ligneous lodges, in the centre of a spongy and spheroidal sort of flesh. It is produced by the *cynips glecome.* These galls have been sometimes eaten; they have an agreeable taste, and to a high degree the odour of the plant which produces them.

The *sage gall,* which much resembles the preceding, is found on a species of sage (*salvia pomifera*). The inhabitants of the island of Crete, where this plant grows, gather these galls every year, as an article of food, according to the report of Poumefort, confirmed by Olivier, who adds, that at Scio it is confected with honey, and that this confection is very agreeable, and of great stomachic virtue.

The *gall of the birch,* is produced by the *cynips fagi,* and sometimes covers the leaves of the birch-tree, under the form of small, very shining, and very hard cones.

All these galls, independently of the insect which produces them, often furnish to those who preserve them in very close boxes, insects of other genera, such as the *ichneumon fly,* &c. These last have been nourished at the expense of the larva of the insect which has produced the gall: they have contributed in no manner to its formation.

Among the galls not produced by cynips, species so well known as those which we have just mentioned cannot be cited; but they are not less abundant in nature. The flowering buds of the common broom are pierced by a gnat, approximating to the tipulæ, and belonging to M. Latreille's genus *Cecidomye.* These buds are not developed, and form a pointed gall, which is sometimes so abundant, that in some years almost half the flowers of the *genista* in the forest of Montmorency, in the neighbourhood of Paris, have proved abortive from this cause. There is never more than a single larva in each gall. The branches of the brier may be seen towards the end of summer charged with tuberosities, in which there are many cells inhabited by larvæ, which change in spring into dipterous flies.

The leaves of the *viburnum* are often loaded with galls, which traverse them from part to part. These give birth to a coleopterous insect, apparently belonging to the genus *Crioceris*.

There are some places, in which the leaves of willows and osiers are furnished with oblong galls, which, like the preceding, project on each side, and which are so abundant, that there are few leaves which have not got one or two of them. These galls, tolerably solid, furnish a retreat to a false caterpillar, which, when it has arrived to certain bulk, pierces the gall, and proceeds to undergo its transformation in the ground. From this proceeds an insect of the genus *Tenthredo*.

It is probable that the hot countries of the old and the new world contain a quantity of galls, proportioned to the number of plants which grow there. Travellers have, however, hitherto but little troubled themselves with them. M. Bosc is perhaps the only naturalist to whom we are indebted for any account of them. He described and designed sixteen species of them during his short residence in Carolina, but he was unable to obtain the insects of any of these species. There, as here, the oak is the tree which furnishes the greatest quantity of them; for among these sixteen species eight appertained to it alone. Two of them were especially worthy of attention.

The first comes on the buds of the red oak. It is spherical, muricated, similar to the fruit of the styrax-tree, but very downy. It is composed of a considerable quantity of galls united. The moment it is touched its hairs sink down, and no more resume their former position. It is necessary to see it to form any idea of it; the effect which it produces cannot be given in description.

The other grows upon the leaves of an oak figured by Michaux, in his superb work on the American oaks, an oak which he regards as a variety of that with willow-leaves, but which certainly forms a distinct species, since it never rises to more than two feet in height, and its thickness rarely surpasses that of a goose-quill, while the true oak with willow leaves, is one of the largest trees in the country, and arrives to the thickness of a man's body. This gall is round, green, about the thickness of a pea, and is formed upon the principal nervure of the leaf. It is hollow in its interior, and its parietes are so thin as to be semi-transparent, and allow us to see in the interior a little ball which rolls there, and which is not more bulky than a grain of millet. It is in this ball that the larva lodges of the insect which has produced the gall. Though M. Bosc has opened hundreds of these galls, he could never conceive how the little ball could remain free in the great one, grow there, or at least retain sufficient freshness to give nutriment to the larva which inhabits it. This fact gives rise to many reflections.

The false galls are not less common in nature than those of which we have been speaking. They are found upon a very great number of plants, and some are very remarkable, both for bulk and abundance. They appear, however, on fewer parts of the plant or tree, being almost solely confined to the leaves or flowers, or the neighbouring and delicate parts. There are few persons who have not remarked hollow, reddish vesicles, which grow in bunches on the branches of the elm, and which sometimes cover entire branches. These, as we have already seen, are the productions of the aphides. When they are young, but a single mother is found within them. But in the middle of the summer some hundreds of insects are found there. Sometimes these galls are entirely closed; sometimes they have a communication with the exterior.

We find in the southern parts of Europe, and in Turkey, on the terebinthus, some galls analogous to these last, which in Spain, Syria, and China, enter into the composition of scarlet dyes. In Syria they are termed *Baizonges.*

The leaves of the black poplar are also often deformed by vesicles of the same nature, as well as those of the willow.

The flowers of the germander are sometimes swelled and entirely closed; they acquire neither the form nor the colour of the others. An insect of the genus *Acanthia (clavicornis)* is the cause of this.

There are observed in the flowers of some other didynamous plants, analogous concavities, which, without doubt, are owing to insects, but which have as yet been but little studied.

Many species of tipulæ, of flies, all the psili, &c. produce monstrosities on the leaves and flowers of a great number of plants, which may also be considered as galls. It may also be observed, that there are many excrescences on trees which are termed galls, but which are not the production of insects. With such, of course, we have nothing to do here.

The genus *Epistenia* of Mr. Westwood, is nearly allied to cleonymus of M. Latreille; but it is distinguished by the penultimate joint of the antennæ, and by the posterior segments of the abdomen being contracted, which have the appearance of a tail. The species we have figured he names *Cœruleate.* It is of a rich blue colour, and punctured, the basal segments of the abdomen shining and iridescent: its habitat is unknown. The specimen is in the collection of the British Museum.

We have also figured a species of which Mr. Westwood has established a new genus, under the name of *Phasgonophora,* which is also allied to cleonymus; but it differs in having the ovipositor exserted and as long as the abdomen, which is subsessile: the antennæ is more slender for the sex;

Pl. 77

1. Stilbum *princeps. G.R. Gray.* 2. Phasgonophora *sulcata. Westw.* 3. Epistenia *cæruleata. Westw.*

London, Published by Whittaker & C.º Ave Maria Lane 1831.

the posterior trochanters are elongate, and the femora compressed, and oval: the species is named *sulcata*. It is black, with the thorax punctated; the abdomen sulcated at the base, and the posterior femora are red. The habitat of this also is unknown; the specimen is in the collection of the British Museum.

On the next tribe, the CHALCIDITES, we can only observe in general, that they are small insects, often adorned with very brilliant colours, and possessing the faculty of jumping. The auger of the females is composed of three threads, like that of the ichneumons, and answers the same purposes, namely, that of depositing their eggs in the bodies of caterpillars, larvæ, or in chrysalids and nymphs. Some females, extremely small, lay in the interior of the eggs of other insects; and even in those of some small ichneumons, already placed in the eggs: others place them in galls, and their young ones destroy the natural inhabitants, the cynips. Geoffroy appears to have been mistaken, in attributing to the chalcidites the formation of these galls, as well as to cynips; for, from all that we have already recounted of the observations of Reaumur and others, it seems incontestible, that the production of the true galls is owing to the cynips alone. Furthermore, the caterpillars or larvæ of the last, are frequently destroyed by the chalcidites, proving that while those feed on vegetables, these, on the contrary, are carnivorous. Degeer, indeed, had already remarked, that the chalcidites, which he, as well as Linnæus, had placed with the ichneumons, were the enemies of the natural inhabitants of the galls, and never produced the latter. He even says, that the females lay but one egg in each of them, because the larva of cynips living there solitarily, its carcase would not suffice for the nutriment of several individuals of these parasite insects.

The larvæ of the chalcidites resemble those of the ichneumons, and are deprived of feet.

On the Oxiuri we can add nothing here.

The Chrysides are hymenoptera, remarkable for the richness and brilliancy of their colours. They may in this respect well compare with the colibris and humming-birds. They are seen moving, but always in a continual state of agitation, and with great swiftness, over walls, and old wood exposed to the heat of the sun. They are also very frequently found on flowers. They deposit their eggs in the nests of the solitary apiariæ, and in those of some other hymenoptera, where the larvæ feed on the offspring of those insects. When they are touched they put themselves into a ball, curve their belly underneath, bringing its extremity into contact with the head; at the same time they apply their feet and antennæ against the corslet, and enclose all the other parts in the cavity of their belly. The females having only a false sting, may be touched without any danger.

In the sub-genus of chrysis *Stilbium*, we have established a species under the name *princeps*. It is of a bright rich green colour, entirely punctured, with a purplish tint in the middle of the thorax; the wings pale brown, with the nervures rather darker. This insect is from Melville Island.

We now arrive at the first family of the second section of hymenoptera, the Heterogyna, and have to consider those most admirable and well-known insects the Ants (genus Formica). In treating of them, we must dismiss here every consideration of structure, as their habits are too interesting to be passed very briefly over.

These insects live in a state of society, like the wasps and bees. Their society, like those of the others, is composed of three sorts of individuals; of males, of females, and of workers or neuters; but these neuters are apterous. It may be also remarked from the great analogy between the female organs and those of the neuters, that it may be presumed that the latter, like the working bees, are barren females,

13

whose organs of generation have not acquired an entire and perfect development.

It is known that the Greeks called these insects *myrmex* or *myrmica*. The ancient naturalists distinguished many species of them, under different names; but their early history, as was to be expected, is intermingled with a variety of fables. It was discovered, however, that they drew their origin from a worm, which at first very small and rounded, becomes elongated, developed by little and little, and receives the suitable form; and that some of these ants had wings.

Leuwenhoek, Swammerdam, and Degeer, were the first who presented the world with sure and positive notions respecting these animals. But of all naturalists, there is none who has observed them with so much attention and sagacity, as M. Pierre Huber, son of the celebrated philosopher of the same name. From him we shall borrow most of our observations on this subject.

The societies of ants are either simple or mixed; that is, solely composed of individuals of the same species, or having moreover, some neuter individuals of one or more species of the same genus. The first six chapters of M. Huber's work are devoted to the history of ants which are united in simple societies; and such are those which most frequently come under our inspection. The author considers these insects successively, in their mode of building, their reproduction, their metamorphosis, and their other peculiar habits.

A species, one of the most multiplied in Europe, and whose larvæ and nymphs are given as food to young partridges and pheasants, is the *formica rufa*. M. Huber distinguishes two varieties of this, according to the difference of the colours of the back, or upper part of the corslet, which is black in one and red in the other. The latter inhabits woods in preference, and its habitation is larger; the former establishes itself along hedges and in meadows: their habits are otherwise but little different.

The habitation of these ants is composed of blades of stubble, of ligneous fragments, of pebbles, and shells of small volume, and of all objects which they meet with of easy transportation; and as they often gather, for the same purpose, grains of wheat, barley, and oats, it has been believed that they laid up provisions for winter, and a period of want. Their laborious life and their foresight, have been celebrated throughout all antiquity; and from the wise Solomon down to the amiable La Fontaine, the sluggard has been referred to the ant to "learn her ways and be wise."

The habitation of the *formicæ rufæ*, is in the form of a hillock or rounded dome, the base of which is often covered with earth and small pebbles, and above which the ligneous materials are raised, like a sugar-loaf. All at first appears disposed without order; but an attentive eye will soon discover that all is arranged, so as to keep off water from the ant-hill, to defend it against the injuries of the atmosphere, and from the attacks of its enemies, to procure for it the heat of the sun, and to preserve that of its interior. The most considerable portion of the nest is concealed, and extends more or less deeply into the earth. Avenues in the form of rather irregular tunnels lead from the summit of the edifice into its interior. Their number is proportioned to the population, and their aperture is more or less wide. Sometimes a principal one is formed at the upper part; there are also often many pretty nearly equal, and around which are placed, circularly, from the base of the hillock to its extremity, many more narrowed passages. Very different from some other species of the same genus, which remain voluntarily in their nest, and sheltered from the sun, the red ants seem to prefer living in the open air, and carry on their labours without any apprehension of our presence. The habitations, like a dome, of many other ants, are closed with earth on all sides, and have but one issue, rather small, near their basis, the way to which is frequently only by a tortuous gallery, which winds through the turf. One would be tempted to believe that the red ants have less foresight, since their dwelling is pierced with a great number of doors, where the rain-water and the enemies of these insects find an easy entrance. But they take care at the decline of day, or on the approach of bad weather, to close the passages and barricado themselves. They bring, at first, little beams near the galleries, whose entrance they are desirous of diminishing, and even sometimes sink them into the mass of the stubble; they then go in search of others, but weaker ones, which they place upon the preceding, but in a contrary direction: finally, they employ pieces of dry leaves, or other materials of a broader form, to cover the hole. The last gates being shut, some individuals are placed behind them, to guard and watch over the safety of the rest. On the return of the morning sun, the barricadoes are removed, and the ordinary passages re-established. These labours are renewed every day, evening and morning, during the fine season; if, however, the weather should be rainy, the gates remain closed.

The ancients believed that the ants, when the moon being too nearly in conjunction with the sun, gives no light, were not in the habit of working.

These ants commence their habitation by hollowing in the earth a cavity more or less spacious. Some then proceed to search in the environs for the materials proper for the construction of the exterior frame-work, and dispose them in an order of no great regularity, but which nevertheless covers the entrance of the dwelling. Other workers bring parcels of earth which they have detached in forming the excavation, mingle them with the substances which have already been employed in the work, so as to fill the vacancies, and strengthen the edifice. To judge from its external part, one might suppose it to be massive; but it is not so: its interior is divided into several stories, and presents galleries and spacious rooms, which, though low, and rudely constructed, are convenient for use. The larvæ and nymphs are transported thither at certain hours of the day. The largest room is almost at the centre of the edifice: it is much more elevated than the others, and traversed only by beams supporting the ceiling. All the galleries lead thither, and it is there also that the greater number of the ants sojourn. The earth being moistened by the rain-water, and afterwards hardened by the sun, forms a sort of mortar which gives solidity to the edifice. The water itself, even after long rains, penetrates there but little beyond a quarter of an inch from the surface. when it is inhabited, and has not been deranged The subterraneous portion cannot be observed, but when the edifice is situated against a declivity. If the hillock of stubble be removed, the interior division of the building may be seen: lodges worked horizontally in the earth, compose these subterraneous dwellings.

M. Huber then describes the architecture of those ants which he terms *masons*, because their nests, always in the form of hillocks, like those of the red ants, are composed merely of earth, without a mixture of other materials, and their interior, divided in the manner of a labyrinth, presents lodges, vaults, and galleries, more artificially constructed.

Many kinds of mason-ants are distinguished. The earth which some species of a certain size employ, such as the ash-coloured and mining ant, is of a coarser mould than that of which the habitations of some other smaller mason-ants are formed, as the yellow, the brown, and that which he names microscopic.

The hillock raised by the ash-coloured ant always exhibits

thick walls, composed of a gross and rough earth, and in the interior very well defined stories, as well as broad vaults, sustained by solid pillars, the strength of which is proportionate to the breadth of these vaults. Every where about may be seen great vacancies and rude masses of earth-work. There are no paths or galleries, properly so called, but passages, in the form of a bull's eye.

The brown ant is much more industrious: its nest is constructed of stories of four or five lines in height, the partitions of which are no more than half a line in thickness, and the material of which is of so fine a grain that the interior linings of the walls appear to be extremely smooth. These stories follow the slope of the soil, and are not always arranged with the same regularity, or on a well determined plan. But the upper one always covers the others, and this concentric disposition continues as far as the subterraneous lodges. At each story we observe cavities carefully worked, lodges more narrow, and elongated galleries, serving for the purposes of communication; small columns, and very slender walls—in a word, real buttresses, support the most spacious places. Here the chambers have but a single entrance, and the origin of which corresponds with the lower story; there we discover broader spaces, which form a kind of cross-roads. The chambers and the broader spaces are inhabited by the adult ants, but the nymphs are always united in lodges more or less approaching to the external surface, according to the hours and temperature; for these insects appear to be very sensible to the impressions of the atmosphere, and to know the degree of heat which is suitable to the family which they are bringing up. If this heat be too strong, they transport the young ones into the lower stories. When the ground floor becomes uninhabitable, in consequence of rain, or moisture from any other cause, they ascend to the most elevated part of the habitation.

In this part there are often more than twenty stories, and there are at least as many more beneath the surface of the soil.

The ant-hill which these insects often place in grass, on the edges of pathways, has a rounded form. Dreading the heat of the sun, they shut themselves up there during the day, or do not issue forth, although the nest has two or three apertures at its surface, except through subterraneous galleries, the issue of which is at the distance of some feet. They do not walk on their habitation except when the weather is fresh, when the dew covers the earth, or after the setting of the sun. To cement the terrene particles which they employ exclusively in the fabrication of their works, having no other resource but water, they do not devote themselves to labour, but at the moments of the day when they are induced to do so by a gentle shower. They particularly avail themselves of the spring rains, and at that period of the year even the night does not suspend their activity: entire stories are constructed from the evening to the morning. M. Huber has often induced them to work by means of an artificial shower.

The ants scrape with their mandibles the earth from the bottom of their domicile, detach the molecules from it, unite them into a little pellet, carry it off with their teeth, and apply it to the place where it is to remain. They divide it, and push it on with these organs, so as to fill the little irregularities of the walls or pillars which they at first commence to construct: this is always the beginning of their work. They feel every moment with their antennæ the bits of clay; and after having given them the suitable form, they strengthen them with their anterior feet. This labour goes on very quickly. The foundations of the pillars and partitions being laid, they throw them more into relief by the superposition of new materials. Often, when two little walls, destined to form a gallery, raised opposite one to the other, and at a little distance, are at the height of from four to five lines, they occupy themselves with the construction of the ceiling, now working in a horizontal direction. They attach against the superior and interior edge of the wall certain bits of moistened clay, and thus form a border, which, extending by little and little, comes at last to meet that of the opposite wall. The breadth of the gallery is often more than a quarter of an inch, and the partitions are almost half a line in thickness. The ceiling is arched. The summits of the pillars, the angles produced by the meeting of the walls, the upper edges, are always the resting points and foundations of the vaults and ceilings, or of the lodges, the halls, and the places, which divide the interior of the stories. We cannot avoid admiring their activity in carrying the mortar, the order which they observe in their operations, and the harmony which reigns between them. The rain augments the cohesion between the parts, and causes the inequalities of the masonry to disappear. When at times it is too violent it may destroy chambers, the vault of which is not yet finished. But the ants very speedily raise them again: often a complete story is finished in the space of from seven to eight hours. M. Huber has, nevertheless, seen these insects destroy chambers that were not yet covered, and distribute the materials on the last story of the habitation, after a violent north wind, which, by drying up the masonry too quickly, diminished the adherence of its parts, and consequently its solidity. These ants are as well acquainted with mining as with building, and their labours proceed in concert equally in their excavations underneath as in their fabrications in the upper part of their habitation, which is raised above the soil. They also construct with earth, and after the manner of the termites, covered ways, which they carry from their nest as far as the roots of trees, and even to the origin of their branches, so that they may be more in safety in the excursions which they make in search of food.

The ash-coloured ants possess but a very rude and simple style of building, in comparison with the brown ants. To give more elevation to their dwelling, they commence by covering with a thick layer of earth, brought from the interior, or from the soil which is near the aperture of the ant-hill, the top, or ridge of the edifice. They hollow passes there, more or less approximating, and of a depth pretty nearly equal. The masses of earth which separate them serve as a foundation to the walls and partitions of the apartments. The useless earth is employed with the most sage economy, and furnishes the materials for this as well as those of the ceilings, which cover the chambers. M. Huber describes with much interest the manœuvres of one of these ants, which he had observed for a long time. Each individual working on his own side, their constructions do not always exactly coincide; but these insects discover the error and repair it. Thus this naturalist has observed one of these ants destroy a tottering vault, and rebuild another, after having given to the wall which composed the preceding more elevation, and that because its architect had not put it on a level with the parallel wall, destined to support the other extremity of the vault.

"I have convinced myself," says M. Huber, "that each ant acts independently of its companions. The first that conceives a plan easy of execution immediately traces out a sketch of it: the others have only to continue what she has commenced. These last judge by the inspection of the first labours concerning those which they ought to undertake: they know how to rough-sketch, continue, polish, or retrench their work, according to occasion. Water furnishes all the cement of which they have need; the sun and air harden the materials of their edifices. They have no chisel but their teeth, no compasses but their antennæ, no trowel but their fore-feet, which they employ with the most admirable dexterity to work and consolidate their moistened earth."

These insects derive the most advantageous part from the local and accidental circumstances which may happen to favour their labours. Blades of grass, chaff of straw, or any other bodies which they may meet with in their nests, are skilfully employed in the construction of the lodges, or the other portions of their building.

The sanguine ant composes with earth, dry leaves, and other materials, a close tissue, difficult to be broken, and impenetrable to water.

The details which we have now presented may serve to give a sufficient idea of the industrious proceedings of the mason-ants; but there are some, such as the red ant of Linnæus, and the yellow, which develope, if they establish themselves in the hollows of trees, another species of talent. The yellow ant, for instance (and our observations will equally apply to the other, or the *myrmica*), chooses the finest parcels of the rotten wood of these trees, mixes them with a little earth, and the webs of spiders, forms a substance of the consistence of *papier maché*, and with it constructs the entire stories of its habitation.

It serves as a sort of compass to the inhabitants of the Alps when they have wandered out of their way during the night, or are environed by thick fogs or mists. The ant-hills of these insects, much more multiplied, and much more elevated in the mountains than any where else, have an elongated, regular form, and are constantly directed from east to west. Their summit, and the most sudden part of their declivity, is turned towards the east in winter; but they proceed in a slope on the opposite side. M. Huber has verified, by observation on thousands of these ant-hills, the facts which were communicated to him by the mountaineers. He found no exceptions, except in the case of such hillocks as had undergone some alteration from the intervention of men or animals.

These nests do not preserve this form in the plains, where they are more exposed to accidents.

Other ants exhibit an architecture of a different kind. These are commonly termed the carpenter ants. Such is the *F. fuliginosa*, called the emmet in this country, where it is less common than those we have already noticed, though occasionally to be met with. The republics of this insect, composed of a great number of individuals, inhabit the interior of many trees, but especially those of the oak and the willow. "Let us figure to ourselves," says M. Huber, "the interior of a tree, entirely carved out, stories without number, more or less horizontal, the floors and ceilings of which, at five or six lines distant from each other, are as thin as a playing-card, supported sometimes by vertical partitions, which form an infinity of chambers, sometimes by a multitude of small slight columns, which allow to be seen between them the depth of a story almost entire, the whole composed of a blackish and smoke-coloured wood, and we shall have a tolerably just notion of the cities of these ants."

"The majority of vertical partitions which divide each story into compartments, are parallel; they follow the direction of the ligneous layers, always concentrical, which gives an air of regularity to the work. The floors considered altogether, are horizontal. The little columns are from one to two lines in thickness, more or less rounded, of a height equal to the elevation of the story which they support, broader above and below, than in the middle, a little flattened at their extremity, and ranged in a line, because they have been cut into parallel partitions."

"What numerous apartments, what a multitude of lodges, of halls, and corridors, do not these insects form for themselves, solely by their own industry? and what labour must not such enterprizes have cost them?"

Without our being able positively to assign the cause, the wood worked by those insects is always blackish, externally, and even internally, where it is most thin. The vegetation of the tree does not appear to suffer, yet M. Huber has found at its foot a black, liquid, and very abundant juice. The same fact has been observed by M. Latreille. The natural colour of the wood, which serves as an habitation to the other species of carpenter-ants, does not change in this manner.

Not having been enabled to accustom the fuliginous ants, or emmets, to work under his inspection, M. Huber attempted, by carefully decomposing the different parts of their edifices, to conceive the order of the labours which they demanded. He thus describes the fragments, the distribution of which he had studied :—

" In one place are horizontal galleries, concealed in a great measure by their parietes, which follow the ligneous layers in their circular form. These parallel galleries, separated by very thin partitions, have no inter-communication, excepting through some oval holes, formed within certain distances of each other. Such is the sketch or outline of these light and delicate fabrics.

" Elsewhere, these avenues open laterally, still preserving between them some fragments of parietes, which have not been thrown down; and it may be remarked, that the ants here and there have contrived transverse partitions in the interior even of the galleries, to form chambers there, by their uniting with other partitions. When the labour is more advanced, we always see round holes, encased by two pillars, cut out in the same wall. In the course of time these holes will become square, and the pillars, at first arched at their extremities, will be changed into straight columns, by the chisel of our sculptors. This is the next step in their architecture, and it is probable that a portion of the building is destined to remain in this state."

But there are fragments worked altogether differently, in which these same parietes or walls, now pierced in all parts, and artificially hewn, are transformed into colonnades which support the stories, and leave a free communication throughout their whole extent. We may easily conceive, that parallel galleries, excavated on the same plan, and of which the walls have been thrown down, leaving only, from space to space, what is necessary to sustain their ceilings, should form together a single story; but as each has been pierced separately, the flooring cannot be very level; it is, on the contrary, very unequally hollowed, which is a great advantage for the ants, since the furrows are better adapted to retain the larvæ which they deposit there.

The stories which are excavated in thick roots are less regular, but of a more light and delicate construction. We sometimes see fragments of from eight to ten inches in depth and height, divided into an infinite number of chambers, the partitions of which are as thin as paper. At the entrance of these apartments, worked with so much care, are presented spacious apertures, formed by arcades pierced in the strata of the wood, and which may be considered as vestibules of the lodges. They leave a free passage in all directions to the insects.

The Ethiopian ant, and that which by reason of its size has received the name of *Hercules,* excavate in the trunks of old trees, and especially of chesnut-trees, long galleries, and large lodges, forming sorts of labyrinths; but their works are much less perfect than those of the *F. fuliginosa,* and scarcely represent the infancy of art. What is most remarkable, according to M. Huber, in the Ethiopian ant is the use which she makes of decayed wood and of its dust. She employs it in stopping chinks at the bottom of the chambers, stopping useless passages, and dividing into compartments the portions of her dwelling which are too large.

To observe the ants in their domestic occupations, M. Huber made use of glass apparatus. But the extreme reluctance of these insects to suffer the light of day to penetrate into their nest, their continual state of alarm and disquietude, at first rendered useless all the means which he employed. He was obliged to call into requisition all the resources of his mind, to invent an apparatus which should finally answer his expectations. He compared the conduct of the yellow ants, which he retained as prisoners, with that of these same ants, when they enjoyed their entire liberty in the fields, and was never able to observe, in this respect, any sensible difference.

In one place the nymphs are heaped in spacious lodges: in another, the neuters surround a heap of larvæ. A little further, the eggs are accumulated together. We also see females who are laying as they walk along, and whose eggs are immediately raised and caught up by the workers. They carry them in a little heap to their mouth, turn them over and over perpetually with their tongue, and moisten them. These eggs differ. The smallest are cylindrical, white, and opake. The most bulky are transparent, with one of their extremities slightly arched. Those of a middle size are only semi-transparent. In their interior, a sort of white cloud is visible, more or less elongated. In some, nothing is observable but a transparent point at the upper end. Some present a clear zone, both above and below the little cloud. In the largest we only observe a single opake and whitish point. There are some, in fine, whose interior is perfectly limpid, and in which we may already perceive very marked rings. M. Huber has observed these last disclose their young. The shell breaks, and the larva appears. Those which are just laid are constantly of a milky white, entirely opake, and one-half smaller. These eggs then acquire a sensible growth, and the longest are the only ones which this naturalist has seen disclosed. Whether the workers communicate to them a necessary degree of humidity, or whether they have any other secret for preserving them, it is certain that the most advanced eggs dry up and perish, if they are removed from the care of these ants.

The larvæ issue from their shells in fifteen days after the laying, and their body is then perfectly transparent: many workers, raised upright on their feet, with the belly forward, watch for their defence; others are occupied in clearing the conduit; others are in a state of inaction. But if the sun should happen to enlighten the external part of the nest, the ants which are at its surface descend immediately with precipitation to the bottom of the ant-hill, strike their companions with their antennæ, even sometimes seize with their teeth the antennæ of such individuals as appear not to comprehend them, drag them to the summit of the habitation, and leave them there, for the purpose of returning to those who are guarding the young. All is immediately in motion; the larvæ and nymphs are transported in all haste to the top of the nest, and receive for some hours the influence of the sun. The larvæ of the females, which, in consequence of the greater volume of their body, and greater weight, give more trouble, are placed alongside of the others. At the end of a quarter of an hour, the ants withdraw them into lodges proper for their reception, under a layer of stubble, but which does not entirely intercept the heat. Many of these ants take advantage of the sunshine themselves, extend themselves before it, and, heaped upon each other, appear to enjoy repose. Among the rest, some are employed in working on the ant-hill, and others, when they perceive that the sun is beginning to decline, bring back the little ones into the interior of their habitation.

The larva, according to the observations of M. Latreille, resembles a small white worm, without feet, thick, short, and of a conical form; its body is composed of a scaly head, and twelve rings; its anterior part is more slender and curved; its mouth presents two small, separated crooks, which are the rudiments of mandibles, and underneath, four small points, two on each side, besides a nipple, almost cylindrical and retractile, through which the animal receives its nutriment: the workers supply it with this every day. If the larva be sufficiently grown, it raises up its body, and sucks with its mouth that of the neuter ant, which then separates its mandibles, and allows the larva to suck in the nutritious fluid. M. Huber does not believe that this has undergone any preparation in the body of the working ants, because he has often seen these insects present to the larvæ the nutriment which they had themselves just taken in, and which consisted of honey, or sugar dissolved in water. But they have not always at their command similar materials, so easy of deglutition. It is known, moreover, that they live, and that very frequently, on animal substances, such as caterpillars, coleoptera, the carcases of small quadrupeds, birds, &c. Now they cannot feed their larvæ but with the most essential and fluid extract of these substances, or with the sort of chyle which they derive from them, which evidently appears to require a particular elaboration. It is probable, as M. Huber conjectures, that the working ants proportion the quality of the regimen of the larvæ, to their differences of age and sex. They keep them in a state of the most extreme cleanliness, constantly cleaning their bodies with their tongue and mandibles. Near the epoch of their transformation, they pull off their skin, which is softened and relaxed.

The larvæ of certain ants pass the winter heaped together at the bottom of their chambers. The body of those which have this destination, is hairy during this season,—a new proof of the foresight of Nature. M. Huber has observed, at this epoch, some very small larvæ in the nests of the yellow ant, and of some other species. But those of the wood-ant *(rufa)*, of the ash-coloured and mining ants, had none; according to him, the larvæ of males and females are only to be found in spring, and they do not metamorphose until the beginning of summer. All these differences are to be explained by those of the epochs in which this last change takes place; for the wood-ants, the ash-coloured, and the mining, are developed in spring, while the others appear much later.

The larvæ of the ants proper, and of the *polyergi*, or formicariæ, without sting, all spin a cocoon of silk, of a close tissue, very smooth, cylindrical, elongated, and of a pale yellow, when they are on the eve of being changed into nymphs. Having enclosed themselves in this, they are stripped of their skin, and the black spot, in the form of a point, which is often remarked at one of the extremities of the cocoon, is formed by the residue of the aliments which these insects have rejected previously to this operation. The ant, in the nymph state, presents the form and size of the perfect insect; but it is weak, of a consistence yet tender, and its limbs, incapable of action, are each enclosed in a particular sheath, which is composed of a pellicle of silk. Their body, a little time after this transformation, becomes immoveable, and its colour passes from white to pale yellow, then to red, and in some species becomes brown, or almost black.

These nymphs have still occasion for the assistance of the working ants. They would perish in their cocoons, if the latter did not open a passage for them at the proper moment, by tearing, or even cutting, this cocoon with their mandibles, in a longitudinal and band-like scissure. They bring them forth from their cells, remove the satiny pellicle which envelopes the parts of their bodies, draw it away delicately, especially contribute to extend the wings of such individuals as are provided with them, and second with all their own efforts, the efforts of Nature. They then hasten to impart to them some of their provisions, or give them aliments of some description. The new-born, whether males, females, or neuters, now enjoy their full liberty, and the active faculties with which they are endowed. The debris of their cocoons are sometimes placed by the nursing ants in the lodges most remote from the centre of the domicile, or are heaped up together in some other lodges; sometimes they are transported to the exterior surface of the habitation, or even to a considerable distance from it.

The nests of the ash-coloured ants present naked nymphs, or nymphs shut up in their cocoons. M. Huber has often seen the neuters open the cocoons, a little time after the larvæ had been metamorphosed into nymphs. The mining ants do the same. But wherefore do the neuters anticipate this operation for a certain number of individuals? This is a question, which our knowledge does not yet enable us to answer. It would be necessary to study the nature of those individuals, thus stripped, before the epoch of their final transformation: for perhaps this operation takes place only with regard to the males and females. M. Huber has often drawn from their cocoon larvæ which had but just spun it. They have changed their skin, but have not been able to disengage their feet. These have remained attached to the abdomen, and the nymphs have speedily perished. It appears then, that for them, the cocoon is as it were a necessary prop, or support.

The neuter ants continue during some days to watch over the individuals which have been just developed; they nourish them, accompany them into all places, appear to teach them the paths and labyrinths of their habitation, assemble in the same chambers the males which are dispersed, keep them in, when they are desirous of issuing forth before the weather be favourable, and when it is so, conduct them out of the ant-hill.

M. Huber has never observed the working ants to lay, and the approaches of the male have always proved the forfeit of life to those whom he has surprised in the fact. He asks, what can be the object of Nature in condemning those neuter individuals, as well as those of the wasps and bees, to eternal sterility? May it not be, he says, for the purpose of augmenting the number of individuals of the same family, without producing a multiplication which should be proportionate to that number, and thus keeping the race within proper limits. But a better cause may be assigned, to which we have already alluded in our general article on insects, —this is, the proper attention to, and preservation of, the young. If the neuter ants are deprived of wings, they are well compensated for their absence, as this judicious observer remarks, by this natural maternal sentiment, and this unlimited power which they enjoy over the other individuals of the society at large.

M. Latreille has observed, as well as M. Huber, that the winged individuals do not issue forth from their domiciles, but when the temperature is elevated to at least fifteen or sixteen degrees of the thermometer of Reaumur; accordingly it would seem that most of our indigenous species of ants do not undergo their final metamorphosis until summer, or even autumn. If the weather be favourable, the neuters open many issues for the other individuals, at the external entrance of which, the latter come to respire. Many of them are sometimes even observed parading over the ant-hill; but if it be touched at that moment, the neuters hurry forth to drive the others in. It is usually in the after part of the day that the males and females abandon their cradle, and take their flight. The neuters remain alone in their habitation, and, according to M. Huber, carefully close up all its avenues.

M. Huber has depicted, with a charm and interest peculiarly his own, all the details of these emigrations. After having described what occurs in relation to the turf-ants (*Myrmica*), the solicitude of the workers for the winged individuals, the cares which they lavish on them, the efforts which they make to retain them, and finally, the sort of farewell which they seem to address to them, he expresses himself after this fashion:—

"But what brilliant objects are those which we observe on that other hillock, which rises above the verdant turf? These are the male ants, which are issuing by hundreds from their subterraneous retreats, and parading their silvery and transparent wings on the surface of the nest. The females, fewer in number, are dragging along, in the midst of them, their broad and bronze-coloured abdomen, and also unfolding their wings, whose varying brilliancy increases the gratification derived from this pleasing assemblage. A numerous train of workers accompany them over all the plants which they traverse. Already disorder and agitation prevail upon the ant-hill; the bustle increases every moment; the winged insects mount with vivacity along the blades of grass, and the workers follow them, run from one male to another, touch them with their antennæ, and present them with nutriment. The males at last quit the paternal roof; they rise into the air, as it were, by a general impulse, and the females quickly follow them. The winged troop has disappeared, and the workers return for a few moments, over the traces of these cherished beings, whom they have fostered with so much perseverance, and whom they are destined never to behold again."

"The variety of the colours and forms of this multitude of insects sometimes presents very striking pictures. In some, the body has but a single tint throughout. The workers are yellow, the males are entirely black, and the females of a light gold-colour. Their glittering wings exhibit all the colours of the rainbow. In other species, all the workers are of an ashen- black, with red spots upon the corslet; the males, whose body is black, have the feet of a fine yellow, and the wings whitish, while their females have the corslet and abdomen spotted with fawn-colour, on a brown ground, and the wings transparent, and blackish at their extremities."

We have now seen the ants make the first essay of their wings, and begin to spread themselves through the air. This is the theatre of their amours. But speedily these insects fall, two by two, like rain, either upon the ground or on plants, where many of them consummate their union. The object of nature being accomplished, some individuals take flight again, and rejoin a cloud of winged ants, assembled on some tree, or sporting round its summit.

The males and females of another species of ants, assemble together in the manner of a swarm, which balances itself in the air, rising and sinking alternately, in the space of about ten feet, but at no great height above the soil, and but little distant from the ant-hill from which it issued forth. These movements are executed with great slowness. The males which form the mass of the swarm, fly obliquely, and in zigzag, with great rapidity. The females, suspended like balloons, and turned against the wind, appear motionless; nevertheless, they follow the movements of the swarm, until the males take hold of them, and draw them away from the crowd.

These swarms are common in the month of September. They disperse upon the least wind, but are not long before they re-unite, and are even frequently confounded with others. The general humming produced by these insects does not equal that of a single wasp. These swarms are sometimes placed above our head, and follow us in our walk. We find in the memoirs of the Berlin Academy, a description of a prodigious swarm of these insects, observed by M. Gleditch, which from afar produced an effect somewhat similar to that of an Aurora Borealis, when from the edge of the cloud shoot forth, by jets, many columns of flame and vapour, many rays like lightning, but without its brilliancy. Columns of ants were coming and going here and there, but always rising upward, with inconceivable rapidity. They appeared to raise themselves above the clouds, to thicken there, and become more and more obscure. Other columns followed the preceding, raised themselves in like manner, shooting forth many times with equal swiftness, or mounting one after the other. This phenomenon continued for the space of half an hour. Each column resembled a very slender net-work, and exhibited a tremulous and undulating movement. It was composed of an innumerable multitude of little winged insects, altogether black, which were continually ascending and descending in an irregular manner.

Some authors have advanced, but without proof, that the male ants, after fecundation, undergo the same fate as the drones among bees, and that the female ants return to their habitation to confide to the same neuters which have taken care of their infancy, the germs of their posterity. But the first individuals, far removed from their native home, deprived of the assistance of their nurses, almost incapable of providing for their own sustenance, exposed to the inclemency of the weather, and to many other dangers, very speedily perish after the epoch of their amours, from these causes alone. The bees, humble bees, and wasps, can find their habitation again, according to M. Huber, by an instinct peculiarly their own. They turn round about it, examine its position, and, in fact, adopt every measure which prudence can suggest, to prevent them losing their way, before they issue forth, for the first time, from their dwelling, The queen bee does the same, when she flies out of the hive for the purposes of reproduction. But the winged ants, when they quit the nest, turn their backs upon it at once, and proceed forward in a right line, until it becomes totally out of sight.

The female ants, after having been fecundated, and only then, cause their four wings to fall off, weakening at first, by divers movements, the muscles which serve as an attachment for them, and afterwards compressing them by means of their hinder feet, which they pass over them at a little distance from their basis. This fact has been confirmed by many observations, both of M. Huber and M. Latreille. These organs are so slightly attached to the corslet, under these circumstances, that they almost always come away, when these insects are laid hold of.

All the females, however, do not quit their primitive dwelling. Such are those that couple in the ant-hill, or round about it, and which are retained or brought back by force, by the workers. They tear off their wings, guard them with the greatest assiduity, never permit them to go out, feed them with the greatest care, and conduct them into those parts of the habitation, the temperature of which is most suitable for them. These female prisoners become accustomed by degrees to their confinement. Their belly enlarges, and a single sentinel, who is regularly relieved, watches over them. But when every thing indicates that they are on the point of becoming mothers, they receive, according to M. Huber, the same attentions that are paid by the bees to their queen. They are accompanied every where by a dozen or fifteen of the working ants, who lavish on them caresses of all kinds. They lead them along with their teeth, and even carry them about into the different quarters of the habitation. The body of the female is then crooked, and suspended to the mandibles of the ant that carries her, and rolled up like the proboscis of a butterfly. She is so well fixed under the corslet of the latter, that she does not, by any means, embarrass the motions of her carrier. Sometimes they are contented with dragging her along; but if the ant which is charged with this burthen is too much fatigued, another takes its place, and in this moment of repose, the female is environed by her suite, which pays her all kinds of attention. The eggs, from the moment of their being laid, are gathered and united around her.

The fuliginous ants, among which the departure of some individuals takes place more slowly, are more easily observed in operations of this kind. M. Huber ingeniously compares the general movement which goes on in their hillock, to a sort of national festival, in which all the individuals of the population take an active part. Many females can live without experiencing any rivalship, or suffering any injury, in the same nest. They contribute equally to the increase of the society, but are possessed of no power, or influence: authority exclusively belongs to the neuter ants. The naturalist whom we have just cited, gives to the females the title of *queen*. He preserved, from the month of November to the end of April, some yellow ants, with one of their females and a number of small larvæ. The box which enclosed this nest, having remained in his chamber, these insects were not benumbed by the cold, and therefore he had ample opportunities of observing the attentions of the workers towards this female. They surrounded her, or rather covered her so completely, that her existence could only be ascertained from a very slow movement of this living mass. They constructed for her, at different times, a particular lodge, inviting her, by all kinds of caresses, to come and take possession of it. M. Huber has also been the first to inform us, that some females, independently of the assistance of the workers, will lay the foundations of a new colony. Animated by maternal love, they construct lodges, which are sometimes common to other similar individuals, lay their eggs there, take care of them, and bring up their family. By repeated experiments and observations, M. Huber convinced himself of the truth of these facts which we have just recited. Some virgin females, and provided with wings, have opened the cocoons of the nymphs of working ants, which he had placed with them in an apparatus where these females were isolated. He has surprised them occupied in delivering other workers from their last envelope, and without appearing at all embarrassed by the part which they were playing for the first time, and one contrary to the presumed intention of nature.

The attachment of the working ants to their females appears to extend even beyond the existence of the latter; for, according to M. Huber, when a fecundated female perishes, five or six of the workers remain about her, brushing and licking her body incessantly, for several days, as if they wished by their cares to restore her to existence.

Such is the political economy of the ordinary societies of ants, considered as to the principal end of their institution. We have contemplated them engaged in the construction of their habitations, in rearing their family, and in perpetuating their race; let us now exhibit a few of their more peculiar habits.

All the relations which exist between the individuals of their different orders, their harmony, the union of all wills for the completion of one grand design, all presuppose, in the opinion of M. Huber, the intervention of some kind of language, or some mode of expressing their designs, their wants, in a word, all the impulses of their instinct.

If any ants on the external part of the nest be attacked, some of them, while the others are defending themselves, hurry down to the bottom of their galleries and spread the alarm. The individuals set a guard over the young, in the upper stories, and hasten to carry off their precious deposit into the deepest recesses of the habitation, for the purpose of placing it in shelter from the dangers by which it is threatened.

The Hercules ants, which are the largest belonging to these countries, which form their dwelling in the excavations of the trunks of trees, and which do not issue forth from it, but in spring, to accompany their males and females, have furnished to M. Huber many curious observatious of this kind. When he disturbed the individuals who were most remote from the others, either by examining them too closely, or breathing slightly upon them, they immediately ran towards the former, gave them little blows with the head, against the corslet, went from one to the other, traversing in a semicircle, and jostled several times against those that, by their inaction, manifested no signs of fear. When they became acquainted with the danger, the latter also set forward, describing in their turn different curves, and stopping occasionally to strike with their heads the other ants which they met with on their route. The alarm became general: the ants of the interior quitted their retreats, and augmented the tumult; but the males appeared to be more insensible, or less terrified. They sought no asylum, nor retired precipitately into the interior of the trunk, excepting when the working ants, having approached them, had given them over and over, the signal for flight. All these facts, related by M. Huber, are of the most exact truth, and have been further verified by M. Latreille, who has often had occasion to observe this species, which is extremely common in the southern departments of France.

M. Huber placed the feet of an artificial ant-hill in troughs full of water, so as to prevent the ants from getting away. This water proved to them a source of great enjoyment; for like the bees, butterflies, and other insects, they are fond of quenching their thirst during hot weather. He disturbed some ants one day, that, assembled at the foot of their habitation (which was constructed of wood, in the manner of a hive, and surmounted with a bell), were eagerly licking the little drops which filtrated through the wood, and which they appeared even to prefer to the water in the troughs: most of the ants re-ascended directly; the rest, without any appearance of fear continued to drink. But one of the first re-descended, approached one of its companions that was still drinking, pushed it repeatedly with its mandibles, lowering and raising its head by jerks, and at length succeeded in making the other depart. It then addressed itself to another drinker, and finding that there was no use in striking its abdomen, it gave it a few smart blows on the corslet, with the end of its jaws, and finally succeeded in forcing it to return to the upper part of the habitation. A third, warned in the same manner, also yielded to the suggestions ; but a fourth remaining alone at the edge of the water, was so obstinate, that though seized by one of its hind legs, it still resisted, and even seemed by opening its mandibles to testify, so that its adviser passing to the front, was obliged to take it up with its mandibles, and drag it back into the nest.

It has been remarked that the ants, when they are proceeding along in file, may be stopped by simply passing the finger several times across the line of their march. From this it has been concluded that they conduct each other reciprocally by the smell, since their march cannot be suspended, but by the interruption of the odorant emanations which they leave on their route. The majority, however, soon get over this obstacle, and resume their direction. A fosse of several inches deep excavated around them, will at first disconcert them, but nevertheless does not hinder them, within a certain time, from returning to the path which leads to their nest. Impetuous winds, rain, and other causes, may efface the traces of their steps, yet, notwithstanding this, they always arrive at their habitations. Certain sensations may otherwise embarrass them; for instance, if the traces of those emanations by which they are guided, should form various lines, crossing each other in divers directions. Without denying the utility of this sense of smell to the ants, M. Huber presumes that other means, such as the inspection of objects, the memory of localities, or some other causes to us unknown, may enable these insects to recognize their way. He has sometimes amused himself, by disposing in the middle of a chamber, the debris of a small ant-hill of earth. The ants spread themselves on all sides, and wandered in every direction at hazard; but when one of them had discovered some cleft in the floor, or some other place of safety, it returned into the midst of its companions, and indicated to them, by certain motions of the antennæ, the route which they should take, or even accompanied them to the entrance of the retreat it had discovered. These, in their turn, conducted others thither. Every time they met, they struck each other with their antennæ, in a very perceptible manner, and all the ants thus repaired into this asylum. Those which penetrate into closets where confections, sugar, or fruits are kept, of all which they are remarkably fond, receive information in the same manner, from one of their companions who has made the discovery. It is observed, either in going from or returning to the domicile to inform them of their good fortune, to hesitate, and pause at intervals, as if it wanted to recognize the localities, and assure itself of the right road. Another ant will frequently put it right by a touch of its antennæ. There are many species which are thus directed by the use of these organs.

These insects, if their habitation be too much shaded, too humid, exposed to the insults of passengers, or near a hostile ant-hill, proceed to lay elsewhere the foundations of a new city, and this is what M. Huber terms migration. It is not a colony, for the entire nation transports itself to a new settlement. Many naturalists had already noticed a usage very common with ants—namely, that of carrying each other from place to place—but they were ignorant of the cause of this. M. Huber having one day deranged the habitation of a horde of yellow ants, perceived that they were changing their domicile, and saw, at ten paces distant, the new ant-hill, which communicated with the old one by a path beaten through the grass, and along which these ants were passing and repassing in great numbers. He remarked that all those that were going towards the new establishment were loaded with their companions, while those who were proceeding in the opposite direction marched one by one. This circumstance proved a ray of light to our observer. By dint of torturing, with repeated demolitions, these republics of ants, he succeeded in obliging them to change their locality. It was natural to expect that the scene which he had witnessed would be repeated, if the migration of those insects had been the cause of it. The number of ants which carried others was, however, but small at first: he observed but two or three in the path; but it soon increased, and many of the colonists, thus transported, returned again to recruit in their turn. Sometimes they invited the others to desert, simply by caresses; sometimes they seized them by surprise, dragged them out of their ancient habitation, and carried them off with celerity. If these latter ants were disposed for the journey, the carriers caught them with their mandibles, and when they turned round to carry them off, the others suspended and rolled themselves round the neck of their bearers. The number of recruiters augmented in such a rapid progression, that the path of communication became full of them, and the external surface of the primitive ant-hill was perpetually exhibiting examples of these abductions. In certain glass apparatus, in which M. Huber had shut up some ants, he again witnessed a similar spectacle: he observed that by taking away the first recruiter, the emigration was stopped until another had discovered the place proper for the new establishment. This sort of recruiting lasted for many days, and only ceased when all the ants had acquired a perfect knowledge of the road to their new habitation. The locality being prepared, and the young family transported, the road of the old ant-hill is abandoned altogether. The tranquillity which prevails in the old ant-hill at the period of migration, seems to prove to M. Huber that the design of expatriation is by no means generally known, but that begun by one individual, it is executed gradually. Sometimes, indeed, several working ants form at the same time the plan of establishing a new city, and this gives rise to the momentary existence of several ant-hills. But they are very soon confounded into one, by means of a final enlistment. It also happens, that if the ants be discontented with the site which they have chosen, they will change three or four times, and even return to the first: but the last emigration has always the advantage over the others. In case that the new habitation is very much removed from that which they have abandoned, intermediate lodgings, or kind of relays, which clearly represent, with smaller dimensions, and a less considerable population, true ant-hills, are prepared upon the road, and become points of repose: sometimes these are so many little colonies dependent upon the metropolis. M. Huber has even seen in fir-woods some large ant-hills in the proximity of each other, and communicating together by beaten routes, sometimes a hundred feet in length, and several inches in breadth, and excavated by the ants themselves.

But this art appertains exclusively to the wood-ants (*F. rufa*).

The species called Hercules, Ethiopian, ash-coloured, sanguine, and mining, also carry on the recruiting system.

The turf-ants (*myrmicæ*) have, like the brown ants, the *F. emarginata*, &c., an instinct of directing each other by means of signs, and of carrying each other, but not precisely in the same fashion. The companions with whom they charge themselves, and also seize with their mandibles, have their body in the air, and the head downwards. Boinut has transformed this act of duty into one of discord and combat.

The brown and fuliginous ants, which have not the habit of carrying one another in their migrations, nevertheless know how to employ this means with the males, the females, and the workers, which have been but just metamorphosed. This would prove, according to M. Huber, that these latter individuals are not yet well acquainted with their language, and the art of directing themselves. M. Latreille is more inclined to think that this kind of transportation only takes place in time of danger, and because these individuals are not yet sufficiently strong to evade it.

The ants do not yield to the bees in devotionate affection to their family. They defend it with the grèatest courage and pertinacity. Their abdomen might be separated from the rest of their body, before they would let go the larva or nymph which they are carrying between their mandibles; and they would drag themselves along, as well as they were able, so as to put it in a state of perfect safety. The courage and audacity which they exhibit in defence of their property, are well known, and have often excited the repentance of the aggressor. M. Latreille tells us a very singular story in his history of these insects, illustrative of their mutual affection. He deprived an ant of its antennæ; one of her companions, that no doubt perceived the nature of the evil which had occurred to her, distilled upon the wound a drop of yellowish and limpid fluid from her mouth.

M. Huber reports two other traits of similar attachment:—

Having taken, in the month of April, from the woods, an ant-hill, for the purpose of peopling one of his glass apparatus, he set a portion of the ants at liberty. The freed ants fixed themselves at the foot of a large chesnut-tree, in the garden of the house which he inhabited. The others remained four months prisoners in his cabinet. At this period he transported them into the garden, within ten or twelve paces of the preceding. Some of the prisoners succeeded in escaping, met, and recognized their ancient companions. "They were observed," says M. Huber, "to gesticulate, and caress each other mutually with their antennæ, and catch each other with their mandibles. The ants of the chesnut tree led the others into their nest, came soon in crowds to seek other fugitives, even ventured to glide under the bell of those which were captive, and produced a complete desertion, by carrying them off successively."

Some yellow ants, which this naturalist had assembled in another artificial ant-hill, the base of which, instead of being perpendicular to the table which supported it, was inclined by some degrees, not liking this disposition, went to establish themselves under a bell which was above the table. M. Huber hoped to make them return into the chest, by warming its glass work by means of a flambeau. Some ants which were in this part, very much pleased with this temperature, and manifesting their delight by brushing their head and antennæ with their feet, adopted the resolution, after hesitating a few moments, of re-mounting to the upper stage, or above the bell. Two of these ants re-descended very soon into the chest, bringing in their mouth two of their companions, which they deposited in the warmest place, and returned immediately into the upper part. Those which had newly arrived, after having been warmed, also mounted under the bell, and brought back others with them. This manœuvre soon became general, and not a single ant remained above.

When M. Huber ceased to warm the chest, these insects remounted under the bell, and he made them repeat this trait of sociability every time that he brought back the flambeau. "These observations," says he, "recall to our minds those ideal republics, where all goods were supposed to be in common, and where public interest is the rule of action to all the citizens. It appertained, however, only to Nature to realize this chimera, and it is only amongst insects, exempt from passions like ours, that she has established such an order of things. She has given to the ants the faculty of communicating together by the touch of their antennæ. By this means, they can mutually assist each other in their labours, succour their companions in danger, find their way again, when they have wandered from it, and make known their wants to their consimilars. The insects which live in societies are then in possession of a language, and should not this relation, which they have with us, though in an inferior degree, elevate them in our eyes, and does it not even contribute to embellish the grand spectacle of the universe?"

The ants give chace to the insects which they meet with, especially to caterpillars, cock-chafers, &c. If they are not sufficiently numerous to arrest their prey, they go in search of succour; and the little animal, overcome by so many aggressors, incapable of delivering itself from them, and often even of walking, succumbs, and is quickly dragged within the ant-hill. They attack by open force; their mandibles, like hooks or pincers, and an acid and irritating fluid, which chemists call *formic acid*, a sort of poison, which they pour into the wounds produced by their bite, curving the posterior extremity of their abdomen, where it is contained, and applying it against the injured part, are very formidable and dangerous weapons with which Nature has provided them. Other formicariæ, as the myrmices, make use of their sting, and though but equal in size, are superior in power to the species which are not provided with this organ. We know that ants are very dexterous dissectors of the carcases of quadrupeds, birds, and reptiles of small size, and that we may avoid the trouble of preparing their skeletons, by leaving the dead bodies in certain ant-hills, especially those of the yellow ants.

If they perceive one of their enemies, but at a distance to which they cannot reach, they stand up on their hind feet, pass their abdomen between their legs, and shoot simultaneously, and with force, some jets of their acid. Its odour is so penetrating, that its action is felt, particularly in hot weather, at a considerable distance from the ant-hill. In the woods, where the yellow ant is very common, an agreeable lemonade may be procured, by placing a piece of sugar in their nest, and leaving it there some minutes. It soon absorbs a certain quantity of formic acid, which is a perfect substitute for that of lemon.

Some other species of the same genus are those of all their enemies that our indigenous ants are most afraid of. The smallest, by hooking themselves to their feet, and to other parts of the body, and often in great numbers, are far from being the least formidable. The strongest species, in the combats which they have with the latter, are even obliged to act by surprise; for when the others provoke the attack, they proceed to advertise their companions, and reinforcements come to determine the victory in their favour. These insects are so bitter against their enemies that they will sooner suffer themselves to be torn to pieces than let go their hold. The Hercules ants sometimes wage a cruel war against the sanguine ants, which they proceed in search of even to the gates of their habitation. The latter, only one half as big, but considerably superior in number, remain on the defensive, and to avoid such dangerous enemies, in general prudently determine to remove their habitation to a greater distance. This they perform by a regular levy, such as we have already described, while small bodies of troops are posted within a little distance from their nest to cover their retreat. "But," says M. Huber, "if we want to see armies in presence of each other, and a war in all its regular forms, we must go into the forests, where the wood ants have established their domination over all the insects which are found within their tracks. We shall there see populous and rival cities; beaten roads, proceeding from the ant-hill like so many rays, and frequented by an innumerable crowd of combatants; wars between rival hordes of the same species; for they are naturally hostile and jealous towards every neighbouring territory. There have I observed two of the largest ant-hills engaged in arms against each other. I cannot say what might have kindled the torch of discord between these republics. They were of the same species, similar as to size and population, and situated at a hundred paces distant one from the other. Two mighty empires do not possess a greater number of combatants. Figure to yourself a prodigious crowd of these insects filling all the space which separated the two ant-hills, and occupying a breadth of two feet. The armies met half way from their respective habitations, and there gave battle; thousands of ants, mounted on the natural projections of the soil, struggled two by two, holding by their mandibles, and opposite to each other. A greater number were brought out, attacked, and dragged along as prisoners. The latter made vain efforts to escape, as if they had foreseen that when arrived at the hostile ant-hill, they would experience a cruel fate.

"The field of battle was two or three feet square: a penetrating odour was exhaled from all quarters. A number of ants were to be seen there dead, and covered with poison;

others, composing groups and chains, were hooked together by their feet or pincers, and dragging each other by turns in opposite directions. These groups were formed successively. The struggle had begun between two ants, who seized each other by the mandibles, raised themselves on their legs to let their belly pass in front, and mutually spouted venom against the adversary. They squeezed so close that they tumbled on the side, and fought for a long time in the dust; they soon rose again, and tugged away at each other, each endeavouring to drag off his antagonist. But when their strength was equal, these athletæ remained motionless and crooked to the ground, until a third ant came up to decide the advantage. Most frequently one and the other received assistance at the same time: then all the four, holding by one foot, or one antennæ, again made vain efforts to gain the victory: others united themselves to the former, and sometimes these were again seized in their turn by new comers. In this manner were formed chains of six, eight, or ten ants, all hooked to each other. The equilibrium was broken only when several warriors belonging to the same republic advanced at once; they forced those that were enchained to let go, and the private combats were recommenced.

"At the approach of night each party re-entered gradually into the city, which served it as an asylum, and the ants which were killed or taken prisoners not being replaced by others, the number of the combatants diminished, until at last no more remained. But the ants returned to the fight with the dawn of day, the groups were formed, the carnage recommenced with increased fury, and the place of the encounter was six feet in depth and two abreast. The success was for a long time doubtful; nevertheless, towards the middle of the day the field of battle was removed ten feet from one of the hostile cities, from which I concluded that that one had gained ground. The bitter obstinacy of the ants was so great that nothing could distract them from their enterprize: they did not perceive my presence, and though I was immediately at the edge of their army, none of them climbed upon my legs: they had but a single object in view, that of finding an enemy that they could attack."

Nevertheless, the habitual labours of the two rival societies were not interrupted. Order and tranquillity predominated in them. The war terminated without any very fatal result: rains of long duration put an end to it; and the belligerent ants no longer frequented the road, which might have proved the occasion of new combats. M. Huber has witnessed many battles, similar to that of which he has given us so admirable a description.

The sanguine ants, which are often attacked by the wood ants, when their nests, though sufficiently remote from each other, are placed along the same hedge, and the paths lengthened into their respective territories become a subject of discord, defend themselves like partisans, and carry on a little war, very amusing to the observer. The two parties put themselves in ambuscade, and suddenly rush out upon each other. If the sanguine ants observe that they are inferior in force, they call in succour, and immediately a considerable army issues from the gates of their city, advances in mass, and envelopes the hostile platoon.

Let us not envy the inhabitants of the equatorial countries of the new world the enjoyments which the riches and beauty of their climates seem calculated to procure for them. The gifts of nature are too well counterbalanced there, by calamities without number, and we may felicitate ourselves on not having such compatriots as those ants which Malouet observed, in visiting the forests of Guiana, and of which he has spoken in his travels into that part of the globe. He perceived in the midst of a level savannah, and as far as the eye could reach, a hillock which he would have attributed to the hand of man, if M. de Prefontaine, who accompanied him, had not informed him, that in spite of its gigantic construction, it was the work of black ants of the largest species. He proposed to conduct him, not to the ant-hill, where both of them would infallibly have been devoured, but to the road of the workers. M. Malouet did not approach within more than forty paces of the habitation of these insects. It had the form of a pyramid truncated at one-third of its height, and he estimated that its elevation might be about fifteen or twenty feet, on a basis of from thirty to forty. M. de Prefontaine told him that the cultivators were obliged to abandon a new establishment, when they had the misfortune to meet with one of these fortresses, unless they had sufficient strength to form a regular siege. This even occurred to M. de Prefontaine himself, on his first encampment at Kourva. He was desirous of forming a second a little further on, and perceived upon the soil a mound of earth, similar to that which we have just described; he caused a circular trench to be hollowed, which he filled with a great quantity of dry wood, and after having set fire to it in every point of its circumference, he attacked the ant-hill with cannon-shots. Thus every issue was closed to the hostile army, which to escape from the invasion of the flames, and the shaking of the ground, was obliged to traverse, in its retreat, a trench filled with fire. The largest formicariæ, which come from Cayenne, are of the genus *ponera,* and it is presumable that it is of those that M. Malouet speaks.

M. Huber has further observed some singular facts relative to the wood ants, one of which presents to us a sort of gymnastic scene. Having one day approached one of their habitations, exposed to the sun, and sheltered on the north side, he saw these insects heaped in great numbers upon its surface, and in a general movement, which he compares to the appearance of liquid in a state of ebullition; but having set himself to follow separately the manœuvres of each ant, he discovered that they were sporting together, in a sort of mock fights, similar to those in which we often observe young dogs to indulge. Our observer suspected that local and favourable circumstances, such as the happy situation of the nest, abundance of food, and a solitude which places these ants in security from ordinary perils, had disposed the animals to sports of this kind.

The other ants, half warlike, and half social, rarely gave M. Huber an occasion for similar observations. Initiated in the mysteries of the life of these insects, he has made us acquainted with two of their maladies: one is a species of vertigo, occasioned, as he thinks, by a too great heat of the sun, and which transforms them for two or three minutes into a sort of bacchantes. The other malady, much more severe, causes them to lose the faculty of directing themselves in a right line. These ants turn in a very narrow circle, and always in the same direction. A virgin female, enclosed in a sand-box, and attacked by this mania, made a thousand turns by the hour, describing a circle of about an inch in diameter: it continued this operation for seven days, and even during the night.

In their habitual relations with their companions, the ants make such continual use of their antennæ, that M. Huber, further developing those ideas which we have already noticed, and bringing further facts to their support, endeavours to persuade us of the existence, in these insects, of a language of tact, which he names *antennal.*

It has been known for a long time, that many ants are extremely fond of that fluid which the aphides eject through two horns at the posterior extremity of their body, or through the natural passages, and which is a secretion from the juice of the vegetables on which they feed. Accordingly we see a great number of ants spread over the trees and plants which abound in aphides, and other insects from which a similar secretion proceeds, hurrying to get this sort of honey, or manna, at the moment when these animals eject it in little drops. Their abdomen, much more voluminous then than it was before, shows us how partial they are to this species of aliment. But a fact unknown before, and discovered by M. Huber, is that these ants in flattering and caressing, as it were, those insect providers of theirs with their antennæ, and striking alternately with one of those organs the posterior extremity of their body, obtain directly from them, and voluntarily, this secretion which is so precious to them. M. Huber has many a time seen the brown ant, and some other species, though not so often, employ these singular processes to procure themselves this provision, and always with success. They are soon, however, satisfied with these libations. The red ant catches dexterously the drop of fluid with the swelled extremity of its antennæ, which it then carries to its mouth, and causes it to enter by pressing alternately with one end the other of these organs, as if with real fingers.

The yellow ant, very different from the others in the relation of their habits, scarcely ever issue from their subterraneous abodes: they are not to be met with on trees and fruits: they do not even proceed to the chase of other insects, and nevertheless they are greatly multiplied in meadows and orchards. How then do they sustain themselves? This is a difficulty which naturally presents itself to the mind. To find the explanation of it has led M. Huber to the discovery of many other facts far more strange. Having one day turned up the earth of which the habitation of these ants was composed, he found some aphides in their nest. The roots also of the gramineous plants which shaded the ant-hill, presented some others of different species, and assembled in families tolerably numerous. The ants seemed to watch near them for the moment of their melliferous secretion, or otherwise they were endeavouring to produce it by the means which we have above described. It was important to know whether this kind of cohabitation was general; M. Huber proceeded to search in a great number of the nests of yellow ants, and he always found aphides there, especially after warm showers. He was soon a witness of the interested affection which the ants have for them, which even proceeds to a degree of jealousy. They often took them in their mouth, and carried them to the bottom of the nest; at other times they assembled the aphides in the midst of them, or followed them with great assiduity and solicitude.

The establishment of one of these hordes of ants, with their aphides, in a glass case, gave him a facility of again corroborating these observations, and of convincing himself that they preserve the aphides with the same diligence, and treat them with the same care as if they formed a part of their own family. As the body of these aphides is remarkably soft, what delicate precautions must these ants employ, when they are desirous of detaching them from the vegetable to which they are affixed by their proboscis, so as to be able afterwards to transport them into their dwelling! It is always by caressing them with their antennæ, that they engage them to withdraw the instrument which is infixed in the plant, and which answers the purpose of pumping up its juices. Some neighbouring ants will often attempt to carry them off; but the proprietors are too well acquainted with the value of these little animals, and stoutly defend their possession.

"An ant-hill," says M. Huber, in his agreeable way, "may be considered more or less rich according to the number of aphides which it possesses. These are its cattle, cows, or goats. We should never have guessed that the ants were a pastoral people."

It might be believed that the aphides had come voluntarily

to establish themselves in the ant-hill; but it seems more probable that the ants have transported them thither, at least in a great part.

There are four or five species which thus possess aphides, but in a fewer number, and less constantly than the yellow ants. Those species now alluded to being more active and wandering, can climb upon the trees which are loaded with aphides, and milk them in the manner we have described, without displacing them. There are even some that construct with earth a tunnel which leads from their domicile to the tree, in which their nurses, if we may so express ourselves, dwell. They are there under cover, out of the view of other ants, and can without any fear bring back the aphides to their lodging. The red ant, the turf, the brown, and another species which is nearly microscopic, have always in autumn, in winter, or in spring, a supply of these insects. Those which inhabit with the last mentioned species, are of a size proportioned to its littleness.

Other ants, still more provident and ingenious, build with earth around the stalks of plants, little houses destined for the reception of the aphides which they assemble there. Sometimes these are of a spherical form, smooth and hard within. Such was one which M. Huber found in the middle of a stem of *tithymalus*. It had, in the lower part, a very narrow aperture, through which the brown ants, the proprietors of this fold, came in and out, and were always in the neighbourhood of their own proper habitation. Sometimes this dwelling of the aphides (as was the case with one which the same naturalist observed at the foot of a thistle, and the construction of which he attributes to the red ants,) has the form of a tunnel, two inches and a half in length, and one and a half in thickness. Having opened it at the bottom, he perceived that the ants were living there, with their larvæ and some aphides.

The brown ants, of which we have just spoken, avail themselves, sometimes, of the arrangement of the leaves of the tithymalus, to construct around each branch so many elongated lodgings. M. Huber having destroyed one of these buildings, the insects immediately transported the aphides into their nest, which was situated around the roots of the plant. But in a few days afterwards, the lodge was repeopled and repaired.

M. Huber has seen a habitation destined to the same use, which, at about five feet above the soil, environed a little branch of poplar, at the end of the trunk. It was composed of rotten wood, of the vegetable earth of the tree itself, and formed a blackish and rather short tunnel. The ants arrived thither from the excavations in the interior of the tree, and an aperture formed at the origin of the branch allowed them to introduce themselves into the habitation of the aphides, without appearing in the light.

The aphides of the common plantain withdraw, when its stem is dried up, under the radical leaves. Ants will follow them thither, and shut themselves up along with them, walling in with humid earth all the vacancies which they find between the soil and the edge of the leaves. Then by excavating the earth which is underneath, they give themselves more space to approach the aphides; and they contrive subterraneous galleries, which lead from thence to their own proper habitation.

The ants do not fall into a lethargic state until the temperature is two degrees above congelation in the thermometer of Reaumur, and when the winter is not rigorous, the depth of their nest preserves them from it, and their activity continues uninterrupted. Without some particular resources they would be then exposed to perish. The aphides supply their wants, and what is most remarkable, the latter insects fall into a lethargic state exactly at the same degree of cold that the ants do, and also recover from their lethargy at the same time. The ants which do not possess the instinct of appropriating these animals, are at least acquainted with the places in which they are concealed, and bring back to their companions what little honey they have been able to gather from them. The evaporation of this fluid at such times is very slow, and almost insensible. A bed of earth, and their close aggregation, which perhaps augments the interior heat of their domicile, preserve them against the cold, if its inclemency should increase.

The preservation of aphides is an object of so much interest to the ants, that even the eggs of these insects are objects of their solicitude. This is what M. Huber has observed in relation to the yellow ants. They assemble and preserve these eggs with the greatest solicitude. They lick them continually, invest them with a gluten which cements them together, and in a word, perform all the conditions which are necessary to their preservation and fecundity. In consequence of all this, these eggs disclose the young in the habitation of the ants, just exactly as if they had been abandoned to the cares of nature. M. Huber observes, after Bonnet, that the aphis, in the egg form, in consequence of the want of nutritive aliment, not being capable of growing like other germs similarly enclosed, we cannot strictly pronounce these insects to be oviparous, at least in the same sense as we do the others.

Most of these very curious facts which we have now detailed, have been verified by M. Biot, a gentleman who stands in the very foremost rank of the philosophers of Europe, and who has devoted a large portion of his leisure to the study of the insect world.

To complete the history of indigenous ants, it remains for us to speak of those which form mixed societies. By this we mean societies where there are to be found workers of one or two other species, beside the original founders. These last individuals have been taken away by force from their own societies, in their early age, by the workers of these mixed hordes, and when arrived to the perfect state, become the auxiliaries of the others, or are even exclusively charged with the labours of the ant-hill, and the education both of the family of their ravishers, and of the young of their own species, which have undergone the same fate. All the neuter warrior ants have neither the same form, nor the same functions. Some, which have been distinguished by M. Huber under the name of Amazons, have long, narrow, and arched mandibles, without denticulations, and formed like fangs, and which, in consequence of this conformation, are very little adapted for the transportation and arrangement of the materials of their habitation. They are more properly weapons than tools for mechanical operations; accordingly, these ants are purely warriors. Their trade is war, and the construction of the nests which they inhabit, and the education of their little ones, are exclusively confided to the strange workers which they have carried off: the latter even feed them. The other workers of the mixed ant-hills do not give themselves up to rapine of this kind, except under certain circumstances, and when pressed by extreme necessity. Having received from nature organs entirely analogous to those of ordinary ants, having all their habits, and occupying themselves, like them, in all the labours which are necessary to the support and well-being of their society, they confine themselves to taking some assistants, who partake their cares, and defend their young family at those times when they are themselves absent from their family. The amazon ants, when the heat of a serene day begins to decline, and regularly at the same hour, at least during many consecutive days, quit their nests, advance in a serried column, more or less considerable, in proportion to the extent of their population, and proceed thus in battle array, towards the ant-hill which they intend to spoliate, and the local situation of which has most probably been indicated to them by some one of their companions, who has made the discovery of it. They penetrate thither in spite of the prompt opposition and obstinate defence of the proprietors, seize with their teeth the larvæ and nymphs of the neuter ants of these societies, and transport them, still observing the same order, to their habitation. It is on the species called ash-coloured and mining that they exercise rapine of this kind, always, however, taking care to make choice of those of the species already established in their domicile. Their societies present in alternation one or the other sort of these expatriated workers, and which are compared by M. Huber either to helots, slaves, or negroes: the males or females of such species are never found there. He has observed that they receive the amazons with pleasure, when the latter bring home the trophies of their victories, and when the reverse is the case they testify their disappointment. Having shut up thirty amazon ants with the larvæ and nymphs of their own species, and twenty of the ash-coloured larvæ in a glass hive; having again put a little honey into a corner of their prison, but without associating any auxiliary ants with them, most part of them died of hunger in less than two days. The others were languishing and weak; but having given them one of their companions of the ash-coloured species, the latter, although alone, re-established order, formed a lodging in the earth, assembled the larvæ there, unswaddled many nymphs, and preserved the life of the amazons which were still existing. It then appears that the fate of these latter insects is altogether dependent on the presence of the ash-coloured workers in their nests. In educating their posterity the latter pay equal attention to the larvæ and nymphs of the members of their own species, and thus strengthen the establishment of their conquerors: but more of this anon.

The sanguine ants present us a second example of the mixed societies, but different from the last, inasmuch as all the workers have forms essentially similar, and concur in the same labours.

They have great relations with the wood ants (*F. rufa*), both in the form and colour of their bodies, and in their mode of building. Their ant-hills, composed of earth, mixed with pieces of leaves, blades of grass, moss, and small stones, the union of all which constitutes a very solid mortar, are usually along hedges with a southern aspect. These appear to be very common in the cantons of Switzerland, which were the theatre of M. Huber's observations. The sanguine ants often proceed in pursuit of another small species of their own family, of which they make their prey, and they are never assembled but in small bands. They lie in ambush, according to the report of this naturalist, near an ant-hill, and dart suddenly upon the individuals which are sallying forth: such other insects as they meet upon their route also become their prey. Nevertheless, they also go in search of aphides: but this is rather the province of the auxiliaries. The latter open the gates in the morning, for all the external passages of the ant-hill are closed every evening. The sanguine ants moreover provide themselves with ash-coloured workers, by military expeditions similar to those of the amazons. M. Huber thus describes their hostile operations.

" On the fifteenth of July, at ten in the morning, the sanguine ant-hill dispatched forward a handful of warriors. This little troop marched in haste as far as the entrance of a nest of ash-coloured ants, situated at about twenty paces from the mixed ant-hill: it disperses itself around the nest. The inhabitants perceive the strangers, rush out in a crowd to attack them, and take several of them prisoners; but the sanguine ants advance no further, they appear to be waiting for assistance. Every moment I behold arrive little bands of these insects, which issue from the sanguine ant-hill, and come to reinforce the first brigade. They then advance a little further, and seem more disposed to risk an engagement. But the more they approach the besieged, the more eager they appear in sending back couriers to their own nest. These arrive in haste, spread an alarm throughout the insect ant-hill, and instantly a fresh swarm sets out to join the army. Even still the sanguine ants do not press forward much to seek the combat: they alarm the ash-coloured only by their presence. The latter occupy a space of two feet square in the front of their habitation, the great majority of the nation having issued forth to meet the enemy. All around the camp we begin to see frequent skirmishes, and it is always the besieged that attack the besiegers. The numbers of the ash-coloured being very considerable bespeak, a vigorous resistance; but they are distrustful of their strength, think beforehand of the safety of their little ones, and present us, in this respect, with one of the most singular traits of prudence, of which the history of insects can furnish an example.

Long before success can possibly become doubtful, they bring out their nymphs from their subterraneous recesses, and heap them together in a part of the nest opposite to that side on which the sanguine ants are making their attack, so that they may be able to carry them off more easily, if adverse fortune should oblige them to retreat. The young females take to flight on the same side. The danger now approaches; the sanguine ants finding themselves in sufficient number, rush into the midst of their opponents, attack them in all points, and get even as far as the dome of their city. The ash-coloured, after a sharp resistance, renounce all idea of defence, take up the nymphs which they had assembled outside of the ant-hill, and carry them off to a distance. They are pursued by the sanguine ants, who endeavour to rob them of their treasure. The whole body of the ash-coloured is in flight, nevertheless, some are seen to throw themselves with the utmost devotion into the midst of their enemies, and returning to the nest, penetrate into its caverns, from which they withdraw some larvæ, and carry them off in haste. The sanguine ants, however, proceed into its interior, possess themselves of all the avenues, and appear to establish themselves in the devastated city. Small troops then arrive from the mixed ant-hill, and they begin to carry off what remains of the larvæ and nymphs. A continuous chain is established from one dwelling to the other, and the day passes in this manner. Night arrives before they have transported the entire booty. A good number of the sanguine ants remain in the city which has been taken by assault, and on the following morning, at dawn of day, they recommence the transference of their prey. When they have taken off all the nymphs, they carry one another into the mixed ant-hill, until but a very small number remains outside. But there may be perceived some couples going in an opposite direction. Their number increases. A new resolution has doubtless been adopted among these truly warlike insects. A numerous enlistment is set on foot in the mixed ant-hill, in favour of the pillaged city, and the latter becomes the abode of the sanguine ants. All is transported thither with promptitude; nymphs, larvæ, males, females, auxiliaries, and soldiers, all that the mixed ant-hill contained, is deposited in the conquered habitation, and the sanguine ants renounce for ever their ancient country. They establish themselves in the settlement of their ash-coloured congeners, and from there undertake new invasions."

M. Huber remarks, that the ash-coloured ants, when attacked by the sanguine, conduct themselves differently from what they do when their business is with the amazons. The impetuosity of the last leaves them no time for defence. The tactics of the besiegers being different, those of the besieged must also be different.

The invasions of the sanguine ants are much more rare than those of the amazons. They attack but five or six ant-hills of the ash-coloured race in the course of a year, and content themselves with a certain number of domestics. Besides, as it is necessary that they should assemble in a single month all the nymphs which they want, and which are all developed in August, the period of their depredations is very limited. Very carnivorous, and always occupied with the chase, the sanguine ants cannot do without these auxiliaries, for their little ones would then be without defence. The mining ants, also carried off from their nests when young, render them the same service. But it is very remarkable, that there are some sanguine ant-hills in which these two species of auxiliaries are to be found. Such is the substance of M. Huber's observations on these subjects: a naturalist from whose authority there can be no appeal, and whose name does equal honour to his country and to science. M. Dupont de Nemours has also treated of these insects, but there is more of imagination and wit, than of authentic fact, in his speculations.

We are far from being in possession, relative to the foreign ants, of knowledge as extensive and as certain, as that for which we are indebted to M. Huber, with respect to indigenous species. The others are known only by their ravages. It is for enlightened men, resident on the spot, armed with courage and with patience, enthusiastic and persevering in their studies, and not for travellers, who stop scarcely a day in one place, and who adopt popular prejudices and errors, that the task is reserved of preparing the materials for the history of these most interesting insects. The ants are, unhappily, a scourge, even in Europe. They cause considerable ravages in gardens, injure fruits, attack them before their maturity, and communicate to them a disagreeable odour. They also damage the roots of many useful plants, by excavating the galleries which lead to their habitation, and transporting thither a considerable quantity of corn. All the mischief, however, which is done by the European ants, is nothing in comparison of the devastations of the ants of America, and the West Indies. They are sometimes so numerous, according to the report of Mr. Castles, that they lay waste entire plantations of the sugar-cane.

These insects, according to this observer, appeared for the first time nearly fifty years ago in Grenada; it was believed that they came from Martinique. They soon destroyed the sugar canes, and all other vegetable productions. Their multiplication was so prodigious, and their ravages became so alarming, that the government offered, but without effect, an immensely large reward, for the discovery of some means of their destruction. These ants are of a middle size, elongated form, deep red colour, and remarkable for the vivacity of their movements. They are particularly distinguished by the impression which they make upon the tongue, by their infinite number, and the choice which they make of particular places in which to construct their nests. All the other species of ants which are found at Grenada have a bitter, musky taste; these, on the contrary, are acid in the highest degree, and when many of them are crushed between the hands, they emit a strong sulphureous odour. Their number is prodigious; Mr. Castles has seen roads of several miles in length covered with these insects; they were so numerous in some places, that the traces of the horses' feet were perceptible only for a few moments, that is, until the ants which were around had taken the place of those which had been crushed. The common black ants make their nests around the foundations of houses, or of old walls, some of them in the trunks of hollow trees; one large species selects the savannahs, and enters the earth through a small aperture. The sugar-cane ants, of which we are now speaking, place their nests between the roots of the canes, of the lemon, and the orange-trees. It is in making their nests between the roots of plants, that these insects become hurtful. Castles informs us that there was much difficulty in preserving cold meat from their attacks. The largest dead animals were carried off by them, as soon as they began to be in a state of putrefaction. The negroes who had sores and ulcers, found the greatest difficulty in defending themselves against the approaches of these ants; they had entirely destroyed all the insects, and even the rats, from the sugar plantations. It was with the greatest difficulty that fowls could be reared; the bodies of these birds, when they were dying, or dead, were in an instant covered with these insects. Two methods were employed of destroying them, poison and fire;—arsenic and corrosive sublimate, mixed with animal substances, such as salt fish, crabs, &c., were immediately carried off by them. In this manner, thousands of them were destroyed; it was even remarked, that such of these insects as had touched the corrosive sublimate, fell into a frenzy, previously to their death, and killed the others. The contact of their bodies was also sufficient to cause the destruction of many. But these poisons could not be spread in sufficient abundance to cause a very sensible diminution of these insects. The employment of fire appeared at first to be more efficacious; it was observed that wood burned into charcoal, but which emitted no flame, when placed upon their path, attracted them immediately, and that, precipitating themselves upon it by thousands, they speedily extinguished it. Mr. Castles tried this experiment himself;—he put some burning coals into a place where there was at first but a small number of ants: in a moment, he beheld thousands arrive, and cast themselves upon the fire; and they continued to come, until the fire was totally extinguished by the dead insects, which entirely covered the coals. In consequence of this, hollows were formed in the earth at certain distances from each other, in which fires were made; the ants immediately threw themselves upon them, and when any of the fires was extinguished, the mass of insects which had perished in this manner was so great, that it formed a hillock elevated above the level of the soil. Though a prodigious number of these insects were thus destroyed, yet, after all, there did not appear any very sensible diminution of them. This scourge, which had resisted all the efforts of the planters, finally disappeared, and was replaced by another, the hurricane of 1780. Without this accident, which destroyed the ants, the people would have been obliged to abandon, at least for some years, the cultivation of the sugar-cane in the best parts of Grenada. These happy effects, as Mr. Castles tells us, were produced by the rain, which disturbed the nests. It appears that these insects cannot multiply, except under ground, or under the roots which shelter them from lesser rains and disturbances.

We read in the History of the Insects of Surinam, by Mademoiselle de Merian, that there is a species of ant in America (*Atta Cephalotes*, Fabr.), which travels in troops. In the country it bears the name of the *visiting ant*. On its appearance, the people open all the cupboards, closets, &c., in their houses: the ants enter, and exterminate rats, mice, *kakerlacs* (a species of blatta peculiar to this country), in fine, all hurtful animals, as if they had received an especial commission for the purpose of delivering mankind from nuisances of this nature. Some historians of these insects pretend, that if any one was so ungrateful as to annoy them, they would set upon him, and reduce to atoms his shoes and slippers. The misfortune is, that their visits are not frequent;

they are sometimes three years without making their appearance in the habitations.

They do not always make so good an use of the large jaws with which they are armed: they often despoil, in a single night, the trees of their leaves, to such a degree, that in the morning they might more readily be taken for brooms than trees. Some of these insects cut the leaves, while others receive them on the ground and carry them into the nest.

These ants occasionally excavate a sort of cellars in the earth, which are sometimes eight feet deep, and they fashion them precisely as men might do. When they are desirous of passing from one branch to another, they form a sort of bridge, in the following manner: the first places itself on the nearer branch, and then attaches itself to a piece of wood, which it holds closely between its teeth; a second attaches itself behind the first, and so on. In this manner they suffer themselves to be carried by the wind, until the last of the string finds itself on the other side. Immediately a thousand ants pass over these which serve them as a bridge. Such is the account of Mademoiselle de Merian; but it is by no means confirmed by Capt. Stedman, who on the contrary, declares, that in traversing the places frequented by these insects, he could never gain any knowledge or information of its truth or falsehood. We shall here cite two curious passages from this last-mentioned writer, concerning the *exotic ants.*

"During the day we were continually assailed by entire armies of little ants, termed in this country *fire ants,* in consequence of the burning pain produced by their bite. These insects are black, and remarkably small; but they are accumulated in such vast numbers, that, from their thickness, their hillocks in some measure obstruct the way; and if unluckily, you should chance to stumble over one of them, your legs and feet are immediately covered by these animals, which seize the skin so sharply with their pincers, that they will suffer the head to be separated from the body, sooner than let go. The kind of smart which they occasion, does not, in my opinion, proceed solely from the very acerated form of their pincers. I think that it may also be produced by some poison which they cause to flow into the wound, or which is drawn in by the latter. I can assure the reader that I have seen them cause such a starting among an entire company of soldiers, that you might have said that they had been just scalded by some boiling water.

"After having passed the Cormœtibo Crique, we went south-west by south as far as the Cottica, on the banks of which we embarked. We saw nothing remarkable the first day of our march, except a great number of ants of an inch at least in length, and perfectly black. The insects of this species despoil a tree of its leaves in a very little time: they cut them into small pieces about the size of a sixpence, to carry them under the earth. It was very pleasant to see this army of ants, each with its piece of green leaf, perpetully following the same route. We are so much prone to believe in the marvellous, that some persons have pretended that this devastation was made for the benefit of a blind serpent. The truth is, that these leaves serve as food to the young of the ants, which have not strength to procure it for themselves, and which sometimes are lodged in the ground at the depth of more than six feet."

In this last conclusion, Captain Stedman is quite mistaken. From the conformation of the mouth of the larvæ, it is impossible that they could eat these leaves. The ants carry them off to employ them in the construction of their nests.

The ants have many formidable enemies: among the quadrupeds are the ant-eaters, the tatou, the pangolin; natives of both Indies. Among us, many birds, and some insects, as the ant-lion. The wood-pecker especially feeds on ants. It introduces its long tongue into their nest, and does not withdraw it till it is completely covered with these insects, which are then swallowed. Other birds also destroy a great quantity of ants: they carry off the nymphs and larvæ, to present to their young. But the most terrible of all their enemies is man; he overturns and destroys their habitations, to obtain their larvæ as food for the birds which he is rearing,—more especially for pheasants and partridges. These insects also furnish him with an acid: this is shed so sensibly by the *formica rufa*, (wood ant) when an ant-hill is stirred, that it can occasion an inflammation. If a living frog be fixed upon an ant-hill which is deranged, the animal will die in less than five minutes, even without having been bitten by the ants. Many experiments have proved that this acid could occasion very serious accidents. This acid, which the chemists have named *formic*, will answer all the same purposes as the acetous acid. It is obtained in two modes; 1st. by distillation: the insects are introduced into a glass retort, distilled by a gentle heat, and the acid is found in the recipient. 2dly. By the process called lixiviation: the ants are washed in cold water, spread out upon a linen cloth, and boiling water poured over them, which becomes charged with the acid part.

Fourcroy has published an article on the chemical nature of the ants, in which he tells us that these insects are formed of a great quantity of carbon, united to a small quantity of hydrogen, and doubtless also to a little oxygen. This composition is mixed with phosphate of lime, which constitutes the solid part of the body of these animals. It also appears that the formic acid is composed of the acetous and the malic acid in a state of very considerable concentration.

Many modes of destroying these insects have been devised. The most common, and which is known to gardeners, is to put water and honey in a bottle, which is suspended to the trees, which are attacked by the ants. The odour of the honey attracts them, and entering the bottle they are drowned. This mixture must be boiled, that the honey may be the better dissolved, and the water prevented from floating above it, and that the odour of the honey may spread with more force, and attract a greater number of ants. The bottles should never be more than half full. There have been many other methods pointed out, which it would be but a waste of space to dilate on here, as the best method, after all, appears to be diligent cultivation. In lands that are properly attended to, ant-hills are seldom found.

The remarks which we have now made upon the ants in general, apply more or less to all the sub-divisions of this family, on which our limits will not permit us minutely to dwell. But that to which we have already slightly alluded, namely, the amazons, POLYERGUS of the text, is worthy, from its peculiar interest, to detain our attention a little longer.

M. Huber has made us acquainted with facts respecting these insects, of so marvellous a nature, that we might at first be tempted to take them for mere reveries of imagination. But independently of this gentleman's being utterly incapable of deceiving others, or of suffering himself to be deceived by fallacious appearances, there were many other men of science, among the rest M. Jurine, who were witnesses of his discoveries. M. Latreille, in the neighbourhood of Paris, has had frequent occasions of verifying the observations of M. Huber; and that no doubt might remain, he invited many naturalists to accompany him in his researches, among these were MM. Bosc, Monges, and Olivier. To all these authorities, we shall add one which, singly, is sufficient to dissipate every shade of doubt,—we mean that of our distinguished countryman Mr. Kirby, who in company with M. Latreille, was a witness of one of the military campaigns of the amazons, or, as he terms them, *rufescent* ants.

M. Huber thus relates the first discoveries he made respecting these animals.

"On the 17th of June, 1804, walking in the environs of Geneva, between four and five o'clock in the afternoon, I saw at my feet a legion of tolerably large ants, red or reddish, which were traversing the road. They were marching in a body with rapidity; their troop occupied a space of from eight to ten feet in length, and about three or four inches in breadth. In a few minutes, they had entirely evacuated the road: they penetrated through a very thick hedge, and repaired into a meadow, whither I followed them. They took a serpentine direction over the turf, without losing themselves, and their column always remained unbroken, in spite of the obstacles which it had to surmount.

"Soon they arrived near a nest of ash-coloured ants, the dome of which was raised in the grass, at about twenty paces from the hedge. Some ants of this species were at the door of their habitation. As soon as they discovered the army which was approaching, they darted forth on those which were at the head of the cohort. The alarm was spread at the same instant in the interior of the nest, and their companions sallied forth in crowds, from all their subterraneous caverns. The rufescent ants, the bulk of whose army was but two paces distant, hastened to arrive at the foot of the ant-hill. The entire troop precipitated itself thither at once, and overturned the ash-coloured ants, which after a very short, but very sharp combat, retired to the bottom of their habitation. The rufescent ants clambered up the sides of the hillock, collected on its summit, and introduced themselves in great numbers into the avenues. Other groups of these insects were working with their teeth, to procure themselves an opening in the lateral part of the ant-hill. This enterprize succeeded, and the rest of the army penetrated through the breach into the besieged citadel. They made no long stop there; three or four minutes afterwards, the rufescent ants re-issued through the same passages, each holding in his mouth a larva or a nymph, belonging to the invaded ant-hill. They resumed precisely the route by which they had come, and proceeded, without order, one after the other. Their troop was easily distinguished on the turf, by the peculiar aspect of this multitude of cocoons and white nymphs, carried by so many red ants. These last a second time traversed the hedge and the road in the same place where they had passed at first, and finally directed their course into grass fields in full maturity, whither I regretted that I had not the power of following them.

"I returned towards the ant-hill which had suffered this assault, and there I found a small number of the ash-coloured workers, perched on blades of grass, holding in their mouth some larvæ which they had saved from pillage. They were not long before they brought them back into their habitation."

Having on the following day returned to the same place, and at the same hour, M. Huber was eye-witness to a new expedition. But the rufescent, or amazon ants, which had not been so fortunate as the rest, or which had brought back no booty, formed a particular column, which, while the other division was triumphantly regaining its domicile, directed itself against a second ant-hill, where it was amply compensated for the inutility of its first attempt. This time our observer had the pleasure of following the entire march of the army to the place of its encampment. But he was not a little astonished to find, at the surface of the habitation, and within the exterior layer which he raised up, a multitude of ash-coloured ants; to see them receiving the conquerors without the least opposition, approaching them, flattering them, as it were, by touching them with their antennæ, giving them food, and relieving some of these amazon ants of their burthen, by seizing the larvæ which they were carrying, and descending with them into the interior of the nest, where the others having arrived, were also depositing those with which they had been charged. The amazon ants came out no more for the entire day. The ash-coloured remained outside for some time, but they re-entered the nest before night.

One of these ant-hills having been opened, the last-mentioned individuals immediately busied themselves in re-establishing it, after having taken the precaution to transport into the subterraneous parts the larvæ and nymphs which had been thus uncovered. But the amazons took no part in those labours, and after having wandered some time on the surface of the nest, retired for the most part into the interior of the habitation. These annoyances, however, did not hinder them from trying on the same day a new enterprize. "But," says M. Huber, "at five o'clock in the afternoon the scene changed; all on a sudden I saw them issue from their retreat. They were agitated, and they advanced to the external part of the ant-hill: some of them wandered from it, but in a curved line, so that they speedily returned to the edge of their nest. Their numbers increased every moment: they traversed larger circles; one gesticulation was repeated continually among them. All these ants went from one to another, touching with their antennæ and forehead the corslet of their companions. The latter in their turn approached those which they observed coming, and communicated to them the same signal: it was that of departure. It produced no equivocal effect. Those that received it were observed immediately to put themselves in motion and join the troop. The column was soon organized; it advanced in a right line, and directed itself into the turf: all the army proceeded on, and traversed the meadow. No amazon ant was seen to remain near the ant-hills. The head of the legion seemed sometimes to wait until it was rejoined by the rear-guard: it then spread itself right and left without advancing. The army assembled afresh into a single body, and set forwards again with rapidity. No particular chief was to be remarked: all the ants were first by turns: they seemed to seek to outstrip each other. Nevertheless some amongst them I observed to go in an opposite direction: they came down from front to rear, then retraced their steps, and followed the general movement. There was always a small number returning back, and it was probably by this means that they directed their course. Having proceeded to about thirty feet from their habitation, they stopped, dispersed, and felt the ground with their antennæ, just as dogs scent out the traces of game. They soon discovered a subterraneous abode of ants. The ash-coloured ants had retired to the bottom of their habitation. The legionaries finding no opposition, penetrated into an open gallery. All the army entered the nest in succession, took possession of the nymphs, and re-ascended through several apertures. I saw them immediately resume the road back to the mixed ant-hill. They no longer constituted an army disposed in column, but an undisciplined horde. They issued forth in file from the scene of their pillage with rapidity. The last which came out from the besieged ant-hill were pursued by some of its inhabitants, who attempted to deprive them of their prey: but they very seldom succeeded.

On their return to their domicile the amazon ants deposited their burthens at its entrance, and resumed their way to the habitation which they had robbed. Their ash-coloured auxiliaries hurried out to raise up, one after another, the nymphs which were heaped together, and to transport them into the subterraneous apartments. They often even unburthened the amazons, after having saluted them in a friendly manner with their antennæ, and the latter delivered up to them, without difficulty, the nymphs which they had carried off.

During this interval the inhabitants of the invaded ant-hill had time to recover themselves, and to place guards at every portal. Intimidated by these means of defence, those of the amazon ants which had taken the front for a new attack, fall back several times, and wait until the majority of the troop be arrived. All the army then casts itself in mass on the habitation, penetrate thither after having put the proprietors to the route, carry off their nurselings in haste, but without taking any adult prisoners. When returned to their domicile the amazon ants again experience from the ash-coloured ants that live along with them the best reception, and on their side they confide to their faithful auxiliaries the fruits of their conquest. M. Huber has seen them return a third time to pillage. But they then experienced more resistance, and were forced to undertake a formal siege. Instructed by so many reverses, the ash-coloured ants had taken new precautions to defend their dwelling. They had taken care to barricade their gates and strengthen the interior guard; but these measures did not more than retard their defeat. Their enemies, after some moments of hesitation, gave the signal, poured in upon the habitation with extraordinary impetuosity, separated with their teeth and feet all that could bar their passage, and penetrating by hundreds into the ant-hill, soon issued forth again with new trophies of their victory. But this time they did not deposit their spoils with their associates; the amazons themselves entered the subterraneous apartments, and appeared no more for the rest of the day.

According to the account which M. Huber has made of another expedition of these ants, which took place the twenty-third of the same month, at a quarter to five o'clock in the evening, and also according to the observations of M. Latreille, their project was indicated by some precursory signs. Within about an hour before their departure, some individuals quit their retreat, to parade about the neighbourhood, and very soon re-enter; others, but few in number, come to respire at the entrance of the galleries. Many of those which first go forth return, after having advanced a little, to their domicile, approach, by turns, all the ants which are found above it, and determine this rear-guard to join the body of the army. The troop sometimes halts, and the warrior ants then extend themselves in all directions, in search of booty. If their search is not fortunate, they continue their march, and M. Huber has seen them remove from their domicile to the distance of fifty paces. M. Latreille, from his own personal observations, informs us, that in cases of urgency, they will go double that distance. The ash-coloured and mining ants being the only ones they press into their service, the amazons will not stop if they meet with ant-hills of another species. It is M. Latreille's opinion, that they do not commit themselves wholly to chance in campaigns of this kind: he has frequently observed them go in a right line, and in spite of many obstacles which they had to surmount, towards the ant-hill which they were about to attack. It is doubtful whether they send forward scouts to take cognizance of the territory, or whether their sense of smell is their only guide; at all events, it is quite certain, from the observations both of M. Huber and M. Latreille, that some of these ants are to be met with, a little time before the departure of the troop, wandering here and there, at a greater or less distance from their residence, seeming to look out for chances, or to proceed to discoveries, and that the expedition takes place immediately on their return. Arrived under the walls of the place, the column stops, and is concentrated, that it may rush forth simultaneously, and with greater force.

Although it does not preserve, on its return, the same discipline, and the same regularity of march, as in going; and though it then hurries on, yet still it preserves, according to M. Latreille, some sort of order in file, or in approximating groups. Their most distant expeditions generally last an hour, or a little more. M. Huber has sometimes seen their column go in one direction, return immediately to the nest, and set out again from an opposite side. It also occurs, though rarely, that it is divided into two bodies, which go in different directions. That which is inferior in number, perceiving the inequality of their forces, will return back, and join the other. If the two troops are pretty nearly equal in number, they will go separately to plunder, and each return loaded with booty: they succeed more by the impetuosity of their attack, and the terror which they inspire in the ash-coloured ants, than by their actual degree of strength. M. Huber has seen an army, composed, at the most, of but a hundred and fifty of these amazons, attack an ant-hill, take it by storm, and carry off the plunder. According to him, the signal of departure is various; sometimes it is with the mandibles, sometimes with the forehead that they strike each other, and sometimes it is merely the play of the antennæ. What is especially remarkable in their order, is, that none of the ants, which compose the troop, run constantly in the same direction. In proportion as they arrive at the head of the column, they make a little circuit, and re-enter the body of the army; then, as we have already mentioned, they return to the rear-guard, to give directions to those that they find behind. The front of the army is always composed of from eight to ten ants, that appear to endeavour to outvie each other in the advance; but as soon as they have passed their companions they return into the crowd. Thus their van-guard is continually renewed.

M. Huber has never seen any female in the ranks of the amazons; and the neuters only, as in other species of this family, are exposed to the hazards of war. These insects do not go on feeling their way, but run in the suite of their companions in arms, without any apparent fear of losing their way. If one of them should wander, which, however, very rarely happens, it is brought back to its habitation by one of the ash-coloured ants, which has perceived its embarrassment. M. Huber says, that he never perceived but once an army of these legionaries to be deceived respecting their route. After having put itself in movement, according to custom, instead of following a right line, it described a curve. It went on more than fifty paces, stopping several times; and after having spread itself on all sides, without finding any ant-hill, it re-assembled and returned by the same road to its domicile, without having reaped any fruit from its expedition. The amazons, upon this occasion, were very ill received by their auxiliaries. The latter assailed them individually, worried them in all ways, dragged them out of the nest, and forced them to defend themselves. But these hostilities soon ceased, and peace was re-established. The ash-coloured ants, says M. Huber, were surprised to see them return without the cocoons which they usually brought, and which, as he somewhat fancifully implies, might be a sort of passport for them with their auxiliaries.

The amazons are not carnivorous. M. Huber has often thrown amongst them caterpillars, worms, and cooked meat, without their ever touching any of them; but the ash-coloured ants immediately took possession of these viands. He also presented them with honey and fruits, but equally to no purpose. When they are hungry, they approach their auxiliary ants, and the latter disgorge into their mouth the juices which they have derived from their intercourse with the aphides. This observer never saw them take nourishment in any other manner. He has often put his hands across the army in its march, and the amazons passed quietly between his fingers, without being alarmed at his presence, or attempting to bite.

M. Latreille, however, informs us that they pinch very strongly when they are seized ; and the fear of death will not induce them to let go their hold, for their head, separated from the trunk, has often, in cases of this kind, remained attached to his fingers.

In almost every fine day in summer, the spectacle of those warlike expeditions which we have described may be enjoyed upon the continent, for the amazons are not natives of our country. The hour at which they take place varies, according to the length of the day and its temperature. Thus, in the climate of Paris, and for the first fortnight in August, the amazons do not enter on their campaign until about four o'clock in the afternoon. M. Latreille has seen them set out half an hour sooner, but then the season was more advanced. As we mentioned above, M. Huber has seen them undertake those expeditions in the day, but it was during the summer solstice, and the ant-hills which they invaded were, in all probability, not very remote from their own habitation. It may be well supposed that when the days are less long, and the heat is abated, that they have no longer the same facilities for their operations. M. Latreille has never observed them make, towards the end of summer, more than one expedition of this kind in the day. They are never seen, according to M. Huber, to issue from their retreat, unless the temperature of the atmosphere be above sixteen degrees of Reaumur in the shade. The general rendezvous is usually about five in the afternoon. They sometimes, however, says M. Huber, set out sooner, but never before two o'clock, or later than five. They return about six, or half-past, and never go out unless the weather be fine.

The amazons sometimes procure, and by the same means of violence, another kind of auxiliaries, namely the neuter mining ants. But this is only in case of the want of the ash-coloured species, for only one or the other of these species is ever to be seen in their cells. The mining ants are a little larger than the ash-coloured, and although approximating to them in the mode in which they construct their habitations, differ, nevertheless, in their manners. They are lively, carnivorous, and extremely courageous, while the others are timid and pacific. Accordingly the mining ants, when attacked by the amazons, defend their property obstinately, and do not fear to measure weapons with their aggressors. They attack them with fury, fight body to body, dispute the ground to the very last extremity, and frequently snatch from them the nymphs and larvæ which they are carrying away. Notwithstanding, however, all this resistance, the amazons come off finally triumphant. Knowing also that they have to do with courageous enemies, that will continue to harass them to a considerable distance, they resume their route in good order, with close ranks, and forming but a single legion, in fact, in an arrangement perfectly similar to that of their first march.

During the contest, some hundreds of the mining ants have the prudence to remove from their nests, carrying the larvæ, the nymphs, and young females with them, which they are desirous of saving from pillage. The most part of those which are thus loaded, climb on the surrounding plants. Others are assembled under thick bushes. When the danger is passed, all are transported into the city, the gates of which are barricaded, and a great number of sentinels placed over them.

The amazons re-enter their dwelling peaceably, and are received as its mistressesby the auxiliaries. But often the hope of a new conquest engages them in another enterprize. "I see them," says M. Huber, in his animated style, "set forward again in close column; they direct their course towards one of the most considerable of the mining ant-hills, in sufficient force to be able to grapple with the proprietors. They precipitate themselves in a crowd into one of the galleries which they find carelessly guarded. But their number not permitting them to enter all at once, the mining ants which were upon the top of the ant-hill precipitate themselves on the strangers. While they combat in a state of desperation, their fellow-citizens, in innumerable crowds, having probably abandoned the hope of being able to defend their habitation and their young, issue forth from the nest, carrying with them the nymphs and larvæ of the youngest ants. They are seen flying on all sides, and their multitudes cover the entire surface of the soil, for a distance of several fathoms from the ant-hill. Every moment the fight grows hotter. Here, the amazons are attempting to seize the nymphs which the others are carrying away; there, the besieged are despoiling the victors of the fruits of their rapine. All is in confusion: the legionaries and the mining ants attack each other with impetuosity, and often in their fury are deceived in their object, and fall upon their companions, which, however, they immediately let go. All this takes place at the rear-guard of the legionaries. In the mean time, a great part of their army issues forth from the subterraneous recesses, loaded with booty, and return in a square battalion to their native city, always assailed by the mining ants, that follow them to a considerable distance from their habitation. It is only by their address, the rapidity of their movements, and the use of their sting, that the amazons are enabled to disengage themselves from the pursuit. I have often observed, during these combats, some females of the mining ants in flight, and carrying the nymphs in their mouths; but they do not meddle with the defence of the nest. These affairs did not last for any length of time. In less than a quarter of an hour, the amazons resumed the road to their domicile, and, in spite of the courage and obstinacy of both parties, but a small number of ants were destroyed."

M. Huber is erroneous in supposing that the amazon ants have a sting. M. Latreille discovered, on examination, that these insects are possessed of no such organ.

We have hitherto considered the amazon ants only in their external relations, or military manœuvres. It remains for us to speak of their form of government, their internal economy, and domestic life. To do this, it will be requisite for us to explain briefly the physical and moral characters of the various members of this singular union, which M. Huber has judiciously named the *mixt* society of ants, or, to translate him literally, the *mixed ant-hill.*

During the greater part of the year, the interior of this ant-hill is peopled only by the neuter and wingless individuals of the amazon and ash-coloured ants, or of the mining ants, though occasionally females are found there. A greater or less considerable quantity of insects of the same family, but not yet entirely developed, or even very young, compose the rest of the population. If the neuter amazons be compared with those of the ash-coloured or mining ants, there can be no hesitation in pronouncing the former to be a particular species, perfectly distinct from either of the others. Some of these specific differences are of importance to be acquainted with, because they serve to explain to us the diversity of habits in these insects.

The head of the amazon is square, almost vertical, with very small and round eyes. In the auxiliary ants it is triangular, advanced, and with large and oval eyes. The mandibles of the amazon are long, narrow, arched, and pointed, without denticulations, and, in a word, exactly formed like large hooks. Those of the ash-coloured and mining species are broad, thick, hollowed like a spoon, and denticulated at their interior edge. The first can only act, like crooks, or harpoons; the second have a conformation which renders them proper for various labours, such as those of the pioneer, the mason, and the carpenter. The other parts of the mouth of the amazon ants are extremely small, so that their action on the solid substances on which they ought to feed, must be very slow and feeble. It appears that these insects have need of aliments altogether prepared or masticated. In the auxiliary ants, these same organs have the ordinary proportions and distinctions. M. Huber has observed, but rather rarely, among the amazons, some individuals of the size of the females, larger than the neuters, but otherwise similar to them in the forms of the body and other characters. He has never seen them mix with these latter individuals in their excursions, and he suspects that they are females, but with peculiar modifications, so that these individuals form the passage from the neuters or workers to the ordinary females, or such as are provided with wings. He does not, however, know if they are capable of laying eggs, or what are their peculiar functions.

The societies of some other species of ants present a similar anomaly, but M. Latreille thinks that those extraordinary individuals ought not to be the less classed among those which are designated under the name of neuters. A more abundant nutriment, and other favourable circumstances, may have extended their development beyond the ordinary proportions, and this appears more probable, as the neuters of the *myrmice hamata*, of Fabricius, a foreign species of the same family, vary much, not only in this respect, but also in the relation of the relative length of the mandibles. Thus the ash-coloured and mining ants, though united in society with the amazons, are perfectly strangers to their race. It would appear that they must have been born in the habitation of the latter, since their number predominates there, and no observation has tended to prove that they ever came there of their own accord to settle; every thing, on the contrary, goes to show, that the amazons carry away from the ant-hills, which they are continually invading, nothing but nymphs and larvæ.

If, on the other hand, we compare these auxiliaries with the neuters, which are the proprietors and natural inhabitants of the plundered ant-hills, we shall soon perceive that their physiognomy is perfectly similar, and that they must have had a common origin. We surely are authorised to deduce from all these facts, that the larvæ and nymphs which have been kidnapped from their native soil by the amazons, and transported into their habitation, have undergone their metamorphoses there, and are the same auxiliary ants which we now find there in such numbers. The very striking dissimilarities which we observe, on a comparative examination of these insects in their perfect state, ought also to be again observed in their nymphs; for the organization of these nymphs, though under contracted forms, is essentially the same as that of the insect in the full enjoyment of all its faculties, or arrived at its complete developement. Now the fact is, that all the nymphs found in the nest of the amazon ants are only of two sorts: the one presents us with the type of the amazon, but under three sexual modifications; the other is the perfect mould of the ash-coloured ant, and all the individuals of this species of nymph are entirely similar.

If we examine the interior of the formicary, from the end of July to the middle of August, we shall find winged individuals in which we cannot possibly mistake the characteristic traits of the amazon ants. Among these individuals, some, much smaller, of a shining black, are the males, and cannot be confounded with those of the ash-coloured ant. The others, of an orange-yellow, larger than the amazon neuters, but having great relations with them, are the females: but at this time, as well as at every other favourable time of the year, we should vainly seek to find any males or females of the ash-coloured or mining species. All the individuals of those two species, developed in the nest of the amazons, are exclusively of the cast of neuters or workers. The same parallel might be extended to the larvæ of these insects; but that would be quite superfluous.

Let us now consider what has been the object of Nature, in the institution of these mixed formicaries; and what we shall say respecting the functions of the ash-coloured auxiliaries (or, as Mr. Kirby calls them, the *negroes*), will be equally applicable to the mining species. The destiny of the auxiliary ants is to construct the habitation, to keep it in order, to augment it, if necessary, by excavating new galleries, to guard and defend its entrance, to bestow all kinds of attention on the young family of the ant-hill, to procure provisions for it, to nourish, educate, and protect it, thus laying the favourable foundation for a new generation; finally to provide for the wants and security of their ravishers, and the preservation of their existence. "The amazons," says M. Huber, "in tranquillity at the bottom of their subterranean abodes, wait for the proper hour of their departure, reserving all their courage, strength, and skill to seek in a neighbouring ant-hill for thousands of larvæ, which they confide to their housekeepers, and which are to become in their turn useful to the community. These insects have no other object in their excursions but that of carrying off ants, so to speak, yet in their cradle, from a laborious community, and to make helots of them, which shall perform their work, educate their young, and furnish them with provisions. It is on this account that they never take away any but the larvæ and nymphs of the workers. The males and females would be of no use to them, besides that Nature would not thus permit the destruction of the ash-coloured ant-hills, which would be totally contrary to the interest of their invaders. These warrior ants are acquainted with all the habitations of the others in their neighbourhood, they visit them by turns, they vary their direction every day, and as we have already said, they sometimes pillage the same nest several times; but they never destroy the ant-hills from which they have removed a portion of the young. Very few ash-coloured ants perish in these engagements, the object of which is not to make prisoners, nor to dispute the possession of the invaded city.

M. Latreille has frequently caught many amazons, when

they were returning to their domicile loadedwith the fruits of their victories; he has carefully examined the nymphs which they were carrying, and been perfectly satisfied that they were all of the working class.

From the nature of the architecture which the mixed formicaries exhibit, where the ash-coloured species is domiciliated, we can easily discover that the ants of this species are the sole architects. The general form and internal arrangement of their ant-hills are the same as those of the nests of this species, when its members are in a simple society, or one solely composed of their consimilars. In consequence of the double population, however, the mixed ant-hills are more extended. The ash-coloured ants especially avail themselves of the rains, to raise new stories, and construct saloons and apartments at the edge of the ant-hill. Sometimes in the space of three or four days they have added a complete suburb to the original enclosure. Their number being more considerable than that of the workers of a simple society, they advance their work with very great rapidity, and without having need of overseers, or of being excited to labour. Very different from all this, the amazons are afraid of rain, and continue in an absolute state of inaction. It is curious to observe the activity of the auxiliaries in transferring the others from the old quarter into the new. The emigration of the horde, the urgency of which is decided, and the execution directed, by the workers, presents a still more interesting spectacle. Having once chosen a place proper for an establishment, as, for instance, a soil which is easily mined; they at first carry one another thither. Some of those that have first arrived begin to hollow cavities, while the others return to seek their companions. Without waiting until the new habitation be finished, they occupy themselves with the carriage of the amazon ants. A file of the ash-coloured, loaded with these amazons, and whose colour contrasts with that of their conductors, extends along the road which communicates from the new city to the old. The males and females are brought in succession, and deposited before the gate, where other ash-coloured ants take them up with their mandibles, and conduct them into the interior of the nest. The transportation of the larvæ and nymphs finally terminates this operation. There are to be seen constantly around the ant-hill, ash-coloured ants, some of which go to a distance to seek provisions, and others are returning loaded with them. Many of the latter have their stomachs filled with the honey which they have derived from the aphides; but the amazons are never seen to go in search of this, and rarely do they even issue forth from their retreats in the morning. It will sometimes happen that some of them wander, or appear not to know their way: the ash-coloured ants soon perceiving this, come to their assistance, and bring them back into the ant-hill. Some of them may occasionally be seen, who, appearing to fear that they may themselves be mistaken, leave the amazon in its embarrassment, return to the habitation to make the issues from it, and then go back to recover their companion.

It appears very clearly from all that we have stated now, and from the experiment made by M. Huber, of putting these ants into a glazed box, which we mentioned in our general account of this family, that the amazons cannot subsist without the aid of their auxiliaries. However it appears from another anecdote of their transporting their companions to a deserted ant-hill, and which we have also mentioned before, that they will occasionally exert themselves. From this circumstance, this naturalist conjectures that the amazons which live few in number near their mother, not being at first accustomed to an idle life, do not confine themselves to military operations, but also mingle in domestic labours, and that for a time their society is composed only of individuals of their own species. He was unable to follow the expatriated females, that, having lost their wings, were running here and there, and seeking an asylum. The primitive formation of these hordes, in consequence of the want of proper observation, is extremely difficult to conceive. How could they possibly, at this early epoch, procure auxiliaries, when they are altogether destitute of the strength and numbers that would enable them to undertake invasions? It appears unquestionably that at such a period they must be obliged to shift for themselves as well as they can; and such is the presumption of M. Huber on the subject. M. Latreille thought that he could observe, that these societies are sometimes composed of a small number of individuals, and without any mixture of auxiliaries. But he adds that this observation was but casual, not confirmed by consequent research, and that it is possible the ash-coloured ants might have escaped his notice.

"On the 30th of July," says M. Huber, "at half-past ten in the morning, I saw issue from a mixed ant-hill several small black males: a great number of ash-coloured workers accompanied them. The number of males was continually increasing. Many amazon ants sallied forth also, and paraded in the midst of them, although it was an unseasonable hour for them: they approached the males and licked them, just as the ash-coloured ants do. The latter were the most numerous. Afterwards came the large females. They climbed upon the grass, and met with, in their turn, from the ash-coloured and amazon neuters, the same reception as the males. At eleven o'clock the latter began to grow animated; they climbed along the plants, ran one against the other, clapped their wings, overthrew each other, and ended by taking to flight. The females followed their example, and quitted the mixed ant-hill. I saw more than fifty females depart, and four times as many males. I waited some time at the gates of the city which they had abandoned, to see if they would return; but I saw none of them do so."

As for the preservation of these mixed societies, M. Huber seems to think that it is effected in the same way as that of the other societies of ants. Some female and fecundated amazons are reserved to keep up the population. He has seen very often, and in all seasons, females without wings in ant-hills of this kind. He opened in the middle of April some mixed ant-hills, and saw, in the most elevated parts, some female amazons, having near them their eggs agglomerated together, and a train of auxiliaries. The larvæ of the male amazons do not begin to spin their cocoons to pass into the nymph state until the month of June. The larvæ of the females are more tardy. The auxiliaries do not draw the nymphs from their cocoons but at a little time before their final transformation. This takes place at the end of July, or the commencement of August. The mixed ant-hills exhibit at this period a great number of nymphs of neuter amazons, but there are as yet no ash-coloured or mining nymphs. Those which have been carried off the preceding year have been developed before autumn, since the last invasions take place in the month of September.

The amazon neuters do not commence their campaigns, before the time when they are ready to be metamorphosed; so that they have not above two months and a half to assemble in their habitation the larvæ and nymphs of the auxiliary ants. They sometimes sally forth individually; but then the ash-coloured ants always bring them back, at least at this period, into the interior of the ant-hill.

If the amazons, as M. Huber informs us, could pillage before the above-mentioned time, they would be able to take a great number of the young ash-coloured ants: but in that case they might get hold of the males and females, which would be contrary to the intention of Nature. Accordingly, in the opinion of this naturalist, to prevent the grievous inconveniences which might result from this error of the amazons, Nature has not permitted them to commit their robberies until after the metamorphosis of the males, females, and auxiliary ants. But M. Latreille, with greater probability, attributes still more wisdom to the regulations of Nature in this respect, by supposing that the exclusive choice made by the amazons of the other auxiliary ants is the result of a superior instinct, incapable of any mistake. Besides, the observations both of M. Latreille and of Degeer, are in contradiction to the assertion of M. Huber, on the subject of the metamorphosis of the ash-coloured ants, which he supposes to precede that of the amazons. The first takes place, according to M. Latreille, from the end of July to that of the following month. Degeer, on the 19th of July, detaching the bark from a carious tree, found underneath a large family of ash-coloured ants, accompanied with a great number of *larvæ*, *nymphs*, and *cocoons*. Moreover, it is certain that, at least in the month of September, the nests of the ash-coloured ants contain young larvæ of males and females of this species; and how should the amazons, in a hurried choice, made in the midst of confusion and obscurity, avoid committing such an error, were they not guided by an infallible instinct, with which Supreme Wisdom has endowed them?

For the purpose of having such interesting insects continually under his inspection, and of better observing the internal administration of their society, M. Huber obliged them to establish themselves in a glazed hive, with shutters, and of a very ingenious construction, the model of which he has given.

He filled the lower part of the apparatus with fine earth, slightly moistened, and poured honey upon several parts of it: some holes made in the upper part served to introduce the same substance, or water, when it was necessary. M. Huber then got possession of a great part of one of the most considerable mixed ant-hills, in which he found many males and young. female amazons, and put it into a sack of thick cloth, which he had transported into his cabinet. To the opening of this sack was fitted one of the ends of a wooden canal, glazed above, and communicating by the other extremity with the hive. By means of this apparatus, placed upon the turf, full liberty being allowed to the insects of going backwards and forwards, he corroborated most of the facts which we have already stated; such as the indefatigable activity of the ash-coloured ants, their tender solicitude for the amazons and their nurselings, the neglect and inertia of the latter, respecting all domestic duties and their own wants, their military excursions, and the departure of the winged individuals. Once, however, he observed one of these amazons occupied in removing the last pellicle from an ash-coloured nymph, which was ready to make use of its limbs, and conducting this operation with the same delicacy with which the other ants perform it. He remarked that the cocoons of the amazons, less numerous than those of the ash-coloured species, were one-third longer, and of a brownish kind of silk. The embarrassment of some of these ants to disengage themselves from a heap of earth, by which they had been overwhelmed, presented to his view a scene which was truly touching. He saw the auxiliary ants, whose succour they seemed to implore, run to them, search out in every corner, visit the smallest clods, sweep with perseverance the debris of their nest, and draw out their companions, both of one and the other species, which they carried in their mouth, and then introduced into the hive. They brought back thither the individuals which had wandered, and sometimes, not being sufficiently strong to deliver their unlucky companions, proceeded quickly in search of a reinforcement. "They reminded me," says M. Huber, ingeniously, "of those celebrated dogs, which seem associated in the charity of their masters, for the purpose of getting out those unfortunate travellers who are caught in the Alpine snows." He was always able, by means of his hive, to study the distribution of their apartments. They were excavated in rather irregular stories: some were larger than others, some loftier, some more elongated and narrow, and some were observed, for which the ants rather mine the earth than work it into mortar. The earth which they had thus drawn from the interior was heaped above the last roof, but the work was altogether so massive, that the weight of this matter, thus heaped, could not cause it to sink in. Here the nymphs and larvæ were accommodated together in large rooms, there resided the amazon horde, and elsewhere the majority of the ash-coloured auxiliaries were assembled. The latter, continually and exclusively occupied with their labours, were carrying, according to the hour, and the direction of the sun, the nymphs into different quarters. They divested them of their envelopes, under the very view of our observer. The larvæ of the two species were assembled, at least at that time, in different passages, their wants being relative to their age. The amazons, always fixed in groups to the ceiling of their subterraneous apartments, never stirred, except to approach the ash-coloured ants. The latter lavished on them all kinds of attentions, nourished them, brushed them, and carried or conducted them from one quarter to another, and that principally when the temperature was warmer, and re-united them to their companions. Around a female which had lost its wings were assembled other ash-coloured ants, whose cares for her were of the same description. A numerous train accompanied, outside the ant-hill, some other females and males, which quickly, however, took to flight, and returned no more.

M. Huber saw no amazon worker approach the larvæ and nymphs, (except in the instance above cited,) nor touch any of the provisions of various kinds, with which he presented them, though apparently suited to their taste. He constantly observed, that the amazons, on their return from their expeditions, confided to the ash-coloured ants the booty which they had taken. Annoyed, perhaps, by the pertinacious observations of this naturalist, these latter ants ventured, at the end of some time, to emigrate. He thought to prevent them by placing the hive elsewhere; but they soon found out a new situation, which determined M. Huber to replace his artificial formicary in his cabinet. Finding that he had collected all the facts which this mode of observation could enable him to do, he resolved to make his hive subservient to an experiment, which he had had a long time in contemplation, namely, of engaging in combat two armies of legionaries.

He carried his hive to the front of a mixed ant-hill, situated in his garden, the moment he saw a legionary army issue from it. At the conclusion of a slight combat, which took place at the door of the hive, the amazons of its interior sallied forth in a crowd. The hostile column appeared to wish to avoid the battle, took at first another direction, then returned, and re-entered the habitation. Several ants from the hive rushed to the pursuit, many went as far as the ant-hill, but most of them were kept there, and but two or three succeeded in escaping, and returned in a great hurry—then all the amazons from the hive sallied forth; but when the column had gone on some paces, it went back again. However, a platoon of about three hundred individuals resumed the road, proceeded to the ant-hill, and penetrated into it, in spite of the obstinate resistance of the proprietors. But after having destroyed nearly half of the legionaries of this habitation, the assailants ended by giving way themselves. M. Huber having replaced his hive on the turf, the ash-coloured ants emigrated as above mentioned.

Such is the political economy and truly extraordinary system of these mixed societies of ants, called by M. Huber amazons and legionaries. The workers of the ash-coloured and mining species are, in some sort, their negroes or slaves. Taken from their country at the tenderest age, their instinct appears to be modified to such a point, that these insects identify themselves with the family of their ravishers, which they consider and treat as they would have done their own. They even unfold new industrious faculties for the supply of wants which they would not have known in their native state, namely, to nourish and protect the amazons, at all their ages. This institution of Nature, however, cannot be adduced as an argument or sanction for the existence of slavery among men. The barbarous cupidity which tears the Lybian from his country and his home, to subject him to a life of cruel and thankless servitude, can find neither advocation nor analogy in any of the laws of Nature. Although transplanted to a foreign soil, our auxiliary ants experience neither slavery nor oppression; they enjoy liberty in its fullest sense; live with their companions like sisters, and find, in a sort of maternal sentiment, the recompense of all their labours. In a word, their original destination has undergone no essential change. We shall find in a family of laborious insects, such as the apiariæ, that Nature, in relation to the *nomades melecles*, &c. has deviated from her general plan, by giving parasite habits to the species of these genera. These apiariæ profit by the labours of some other hymenoptera of the same family. They place their eggs in the nests of the latter, and their young ones are nourished and developed there, at the expense of the posterity of the others. The alimentary subtances deposited in these retreats are preserved until the larvæ of these parasite apiariæ are transformed into nymphs.

The food of the larvæ of ants is more liquid, more various, and must be renewed daily. The young ants must be fed by the mouth. Their situation, in other respects, calls for other

Pl. 76

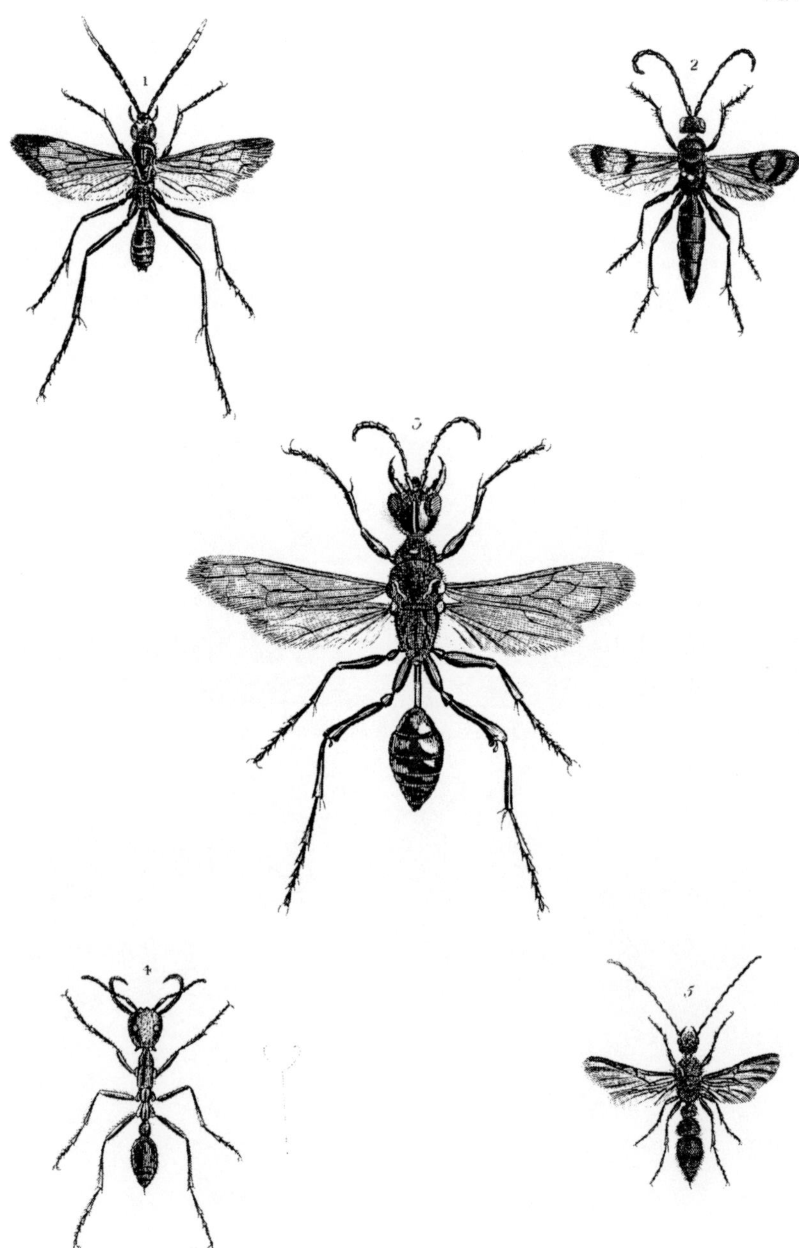

1. Pepsis *apicalis.* *G.R. Gray.* 2. Pompilius *bifasciatus.* *G.R. Gray.* 3. Podium *nigripes.* *Westw.*
4. Camptognatha *testacea.* *Leach. Mss.* 5. Apterogyna *Ehenbergii.* *G.R. Gray.*

London. Published by Whittaker & C° Ave Maria Lane. 1831.

cares, which they cannot receive but from individuals expressly formed for the purpose, charged with this function, and assembled in a numerous society. In establishing in the family of the ants a society of parasite species, Nature has not been able to preserve their posterity from a certain and general extinction, but by instituting a new order of things, such as we have now explained, and which, notwithstanding its apparent anomaly, is perfectly consonant with the wisdom by which all her operations are directed.

M. Latreille, in terminating his history of the amazon ants, very judiciously invites naturalists to make the following experiments, for the purpose of completing it—1st. To transport into their primitive habitation the auxiliary ants, and to introduce the neuters of those species which have remained on their native soil, into the mixed ant-hills.—2d. To take away from these last habitations, the larvæ and nymphs of the auxiliary ants, for the purpose of ascertaining whether the affection which the neuters of these species have for the nurselings of their own kind, be not the sole motive which retains them.—3d. To deprive, by an inverse change, these mixed ant-hills of the young amazons.—4th. To retain and preserve their females, immediately after fecundation, at first solitarily, afterwards in company, with workers either of their own species only, or with the auxiliary ants.

Having dwelt so fully upon the formicariæ, our limits now oblige us to pass over all the subsequent divisions mentioned in the text (respecting the habits of which we could say but little). We have, however, the pleasure first of inserting a short notice of our figures of several new species belonging to these subdivisions.

In the subgenus *Apterogyna*, we have figured a species under the name *Ehrenbergii*. It is black, entirely clothed with fuscous pubescence; the wings reddish brown. This species is from Africa.

We have also figured a species under the name of *Camptognetha testacea,* of Dr. Leach's MSS., but which may probably be the *Eciton hamata* of Latreille. It is of a pale testaceous colour, and slightly hirsute, with the mandibles long, and very much curved.

Our *Mutilla Klugii* is hirsute, black, the abdomen fulvous, with the base black, and is from South America.

In the subgenus *Myrmecodes,* of Latreille, we figure a species which Dr. Leach has named *Australis.* It is fulvous, the head and thorax reddish brown, and the abdomen with transverse arch bands of black. This insect is from New Holland.

Of *Scolia* we have figured a species which we call *Fulva.* It is black, but entirely clothed with fulvous hairs; the basal segment of the abdomen and posterior femora black; the former shining; it is from South America.

Of the genus *Pepsis* we have figured a species under the name of *Apicalis.* It is a bronze green, the antennæ black, with the three last joints pale buff, the wings reddish brown, with the exterior margin dark.

Of the Fabrician genus *Pompilius,* we have figured a species under the name of *Bifasciatus.* It is black, shining; the wings brown; the anterior with two contra arched bands on each, black. This species is from Brazil.

And, finally, Mr. Westwood has added a fine species of *Podium,* of M. Latreille, under the name of *Nigripes.* It is of a shining black colour, with the thorax clothed with greyish brown pubescence; the wings dark, testaceous. This insect is from South America.

We shall proceed at once to those hymenoptera, beginning with the tribe VESPARIÆ.

The WASPS (*Vespa*), like the ants and bees, live in society. They are comparable to the latter for their industry, and

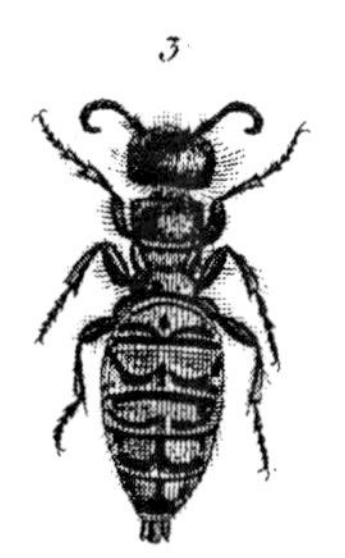

I. O. Westwood del.

1. Scolia *fulva. G.R. Gray.* 2. Mutilla *Klugii G.R. Gray.* 3. Myrmecodes *australis. Leach.*

London, Published by Whittaker & C.o Ave Maria Lane, 1837.

approximate to the former in the extent of their ravages. The bee, continually occupied with its labours, lives only on what it gathers from flowers, and the sting with which it is armed, is merely a defensive weapon, never unsheathed but for the protection of itself or its country. But the wasp, on the contrary, is ferocious, and subsists only on rapine and destruction. Its sting is an offensive weapon, a means of overpowering animals more feeble than itself. Nevertheless it is not less ingenious than the others, nor less attached to its offspring. United in a single republic, the wasps spare neither care nor labour. The works which they perform evince their dexterity, their patience, and the delicacy of their instinct. The peculiar style of their architecture is worthy of admiration.

Among them we particularly distinguish two species, the *hornet* (*vespa crabro*), and the *common wasp* (*vulgaris*). The first makes its nests sheltered from winds and heavy rains, either in barns or the holes of old walls, but most frequently in the large trunks of trees, the interior of which is rotten. There these insects form a large cavity, by detaching fragments of the wood, which is ready to fall into dust. It is in spring that the females, after having passed the winter in a lethargic state, and now re-animated by the heat of the atmosphere, issue from their retreat, to find out a suitable place in which to establish their nest. This place once found, they lay there the first foundations of the edifice, which consist of a thick and solid pillar of the same material as the rest of the nest, but much harder, and more compact. The material of which the wasps make use, is the bark of the slender branches of the ash, which they detach in filaments. Then they grind and bruise it with their mandibles, so as to form a paste, which hardens after it has been employed in building. They collect, at the same time, a clear and saccharine fluid, which drops from the places which have been recently gnawed from the branch which they have been despoiling. This pillar is always placed in the most elevated part of the vault, and the hornets attach to it a sort of cap or covering of the same material, which is to serve as a roof to the edifice, and prevent the dirt, &c., detached from the upper part of the general cavity, from falling on the combs. Within this cavity or vault they place a second pillar, which is in some sort only a continuation of the first. This is to serve for a base to the first comb of the cells. These cells are hexagonal, and their aperture is turned downwards. The mother constructs some of them. As none but females are found in spring, it is probable that they have been fecundated previously to the winter. What is certain is, that the females commence laying as soon as they have constructed a few cells in which to deposit their eggs. These eggs soon disclose the young, and the mother feeds the young larvæ which issue from them, with the products of the chase. When the latter have acquired their full growth, they line their cells with silk, and stop them with a covercle of the same material. Beneath this envelope they undergo their metamorphoses. They do not come forth from it until they are perfect insects. The wasps which are first born are workers. Analogy leads us to the belief that, as among the *bees*, they are only females destitute of the ovariæ. They are designed for the occupation of constructing the nest, and nursing the larvæ. As the female continues to lay, the family increases, and the lodging becomes too small. Then the workers increase the covering and the comb, and when the latter is pushed to the edge of this envelope, they construct another immediately. This last is attached to the first by one or many pillars. Speedily the covering is finished, and filled with new combs. Then there remains but a single aperture to the nest. This aperture corresponds to that of the hole which is the gate through which the wasps arrive at their nest. It is often no more than an inch in diameter.

It is only towards the commencement of autumn that the young females and young males come forth from their nymph state. All the larvæ which could not become perfect insects until the month of October, usually are put to death before this period, especially when the cold begins to be sensibly perceptible. The wasps, instead of continuing to nourish the larvæ, are then solely occupied in plucking them out of their cells, and flinging them out of the nest. The nymphs, or pupæ meet with no more mercy. The males and workers are daily perishing, from the growing inclemency of the season; so that at the end of the winter, none but some females remain, which have passed that season in a state of lethargy, at the bottom of the nest.

In autumn, males and females are to be met with on trees, from which acid and saccharine fluids exude. They return no more to the nest, and perish miserably on the first approach of cold. Thus invariably finishes this society, whose largest population but little exceeds one hundred, or one hundred and fifty individuals.

The *common wasp* makes its nest in the ground, usually at the depth of about half a foot. The entrance to it is a conduit of about an inch in diameter, and very seldom in a straight line. The edges which are at the surface of the earth look as if they were ploughed.

The most usual form of the vespiary is that of a ball. It is thirteen or fourteen inches in diameter: its envelope is a sort of paper or paste-board, which is sometimes more than an inch in thickness. Its colour is a grey of different shades, disposed in bands. This envelope is rough, and appears formed of pieces, in the form of valves of shells placed one upon the other, so that nothing is seen but their convex exterior. When finished, this envelope has two gates, which are two round holes, through which the wasps enter and go out. The interior of the nest is occupied by several combs, parallel, and pretty nearly horizontal: they resemble those of the bees in form, but are composed of a very different material. The vespiary sometimes contains fifteen or sixteen combs of a diameter proportioned to that of the envelope. All these combs are, as it were, so many floors, disposed in stories, which furnish the means of lodging a great number of inhabitants. Free passages are left between them. In these intervals are sorts of columns, which serve to support the combs. The foundations of the edifice, (if we may use such a solecism) are, at its highest part, for the wasps, unlike other builders, begin their work at the top, and descend as they go on with it: these pillars, formed of the same material as the combs and envelope, are massive; their base and capital are of greater diameter than the rest.

These wasps, which work under ground, are concealed from our inspection; they must therefore be drawn, that we may observe the manner in which they construct their nest. When a nest is procured, which is easily done, there is no fear that they will abandon it. It may then be placed under a glass hive, where we can have the pleasure of seeing these insects at their work.

As soon as they are lodged, they begin by repairing the disorders which the vespiary may have suffered, after having attached it solidly to the hive, and they increase the thickness of its envelope.

This envelope deserves a description a little more detailed. Its thickness, which is often more than an inch, is not massive. It is formed of several layers, which have vacancies between them: each layer is as thin as a sheet of paper. In proportion as the wasps thicken this envelope, they build another layer on those which are already formed. The number of these layers sometimes exceeds fifteen or sixteen.

Nothing can be more amusing than to see these wasps

working for the purpose of extending or thickening this envelope. Several of them are engaged at this work, which they perform with the greatest celerity, and without the least confusion. They proceed into the country to find the necessary materials: she that has collected some, returns loaded with a little ball composed of a soft paste; she holds it between her jaws. Arrived at the vespiary, she takes it to the place where she intends to labour, and immediately applies it there. She walks backwards; at each step which she makes, she leaves before her a portion of the ball, without detaching it from the rest, which she holds between her two fore feet. When she has thus applied it all, she unites and smooths it, by repassing over it several times. The materials which she employs are filaments of wood, which she tears off with her mandibles: she moistens and kneads them well previously to use.

These vespiaries contain males, females, and workers: these last, as among the bees and other social insects, are charged with all the labours of the society. Those that go in search of provision are continually employed in the chase; some seize on insects by main force, which they bring back almost entire to the nest; others pillage the shops of butchers, where each one attaches herself to the piece of meat which she prefers, and when she is satiated, cuts off a piece, sometimes larger than herself, to carry it home. Others again plunder the fruits of gardens and orchards: they gnaw or suck them, and bring back the juice. All share their spoils with the males and females, and even with the other workers; and the division is made with the most perfect good will on all sides.

The mothers do not fly into the country, excepting in spring and autumn. During the summer, they are shut up in the interior of the vespiary, occupied in laying, and especially in nursing their larvæ.

A vespiary which has all its combs usually contains fifteen or sixteen thousand cells, each of which is filled by an egg or a nymph. It is the larvæ principally that occupy the attention of the wasps. The latter feed them in the same manner in which birds feed their young, giving them from time to time the *bill-full*, after having softened in their mouth the aliments which the larvæ could not otherwise digest.

Twenty days having elapsed since the eggs were laid, the larvæ are now ready to be metamorphosed into nymphs. Like those of the hornet, they inclose themselves in their cells, and become perfect insects eight or nine days after they have been changed into nymphs. The cell which a young wasp has quitted does not remain a long time vacant. An old wasp cleans it out, and renders it fit for the reception of a new egg.

The cells destined for the eggs which produce the workers are never placed among those which contain the eggs destined to give birth to males and females. The edifice built by the wasps, and which occupies them during some months, is to last no longer than a year. This habitation, so populous during summer, is almost deserted in the winter, and entirely abandoned in the spring: most of its inhabitants have perished the preceding autumn. Some females destined to perpetuate the species pass the winter in a state of numbness, and in the following spring each of them becomes the foundress of a new republic, and the mother of all the individuals which compose it. The workers, as being the most useful, are the first who are born: the males and females do not appear until towards the end of summer, or the commencement of autumn: they couple in the vespiary itself in which they were born.

The occupation of the males in the vespiary is limited to cleaning it out and removing the dead bodies: they are smaller than the females, and larger than the workers, which are the smallest of the three kinds of individuals which compose the society. Like the males of the bees, they are destitute of a sting. The mothers and the workers are alone provided with this organ. The sting of the females is longer than that of the workers, and the wound which the wasps inflict is more severe, and causes a sharper pain than that made by the bees. The violent smarting which it produces is, however, caused in the same manner, by a poisonous fluid, which is introduced into the wound.

Peace does not always reign in the societies of the wasps. Combats often take place among the workers, or between them and the males. The last individuals are more cowardly or weaker than the others; but these combats are rarely fatal.

When the cold weather first approaches, the workers snatch from their cells the larvæ which are not yet metamorphosed, and assisted by the males, turn them neck and heels out of the nest. It appears that they know that the little ones could not support cold and hunger during the winter season, when at this early period they can scarcely find wherewithal to nourish them. To cause them thus to perish is, therefore, an act of mercy, not of cruelty. It is the quick prevention of a long and lingering state of misery.

Notwithstanding all the admirable industry of the wasps, agriculturists are not the less desirous to get rid of these insects, which do most particular damage to fruits, even previously to their maturity. Many means have been pointed out for destroying the species which live in a social state, especially the common wasp. When the places which they inhabit can be discovered, it is easy to dispatch thousands of them in a little time. Some have adopted the plan of putting glue on blades of straw, and placing them in the neighbourhood of the nest, but this method is long and troublesome. Boiling water cast into the hole may be used with success.

But when the nests are remote from houses, a sufficient quantity cannot well be obtained to destroy the wasps: sulphur matches are far more efficacious. The aperture of the hole which conducts to the vespiary must be widened a little, and lighted matches introduced into the hole, after which its entrance must be closed with small stones, so that the wasps cannot get out without mining, which they cannot do in a little time: they will then assuredly be suffocated by the vapour of the sulphur. Care must be taken, however, not to close the hole so exactly as to prevent all access of air, and give no issue to the smoke, for then the matches would be too soon extinguished.

Our history of the wasps would be very imperfect without some notice of one or two of the subdivisions of this tribe, on which, however, we must be as brief as the interest of the subject will allow us to be.

The subgenus Odynerus of the text comprehends those wasps of Linnæus which live in a solitary manner, without constructing hives, and are very remote from the wasps we have been reviewing, and from some which we shall by and by have occasion to notice. All those build nests with combs, are united in societies more or less numerous, and composed of three kinds of individuals, males, females, and neuters, whose labours, vigilance, and care, concur to the preservation and well-being of the state.

Not so the solitary wasps. The species of this genus, whose manners are best known, is the *vespa muraria* of Linnæus, the history of which has been given us by Reaumur.

This insect goes to work in the commencement of June, and continues to operate until July. Its first operation is to excavate a hole in the sand, the diameter of which but little exceeds that of its body, and the depth is a few inches. It raises above this with the grains of earth or sand, which it draws from it, a cylindrical tunnel, the direction of which is at first straight and afterwards turns a little. This tunnel is, as it were, made like a rude kind of filigree work, or rather waved. It is formed of thick, grained, tortuous threads, leaving vacancies between them in some places. These insects soften the sand and separate its molecules with greater facility, by moistening it with some drops of water, which they disgorge. They scrape this paste afterwards with their teeth, when it becomes soft. The feet of the first pair receive what the teeth have thus detached: they knead it, and compose a little pellet, about the bigness of a currant. This pellet is the material which is to serve as a foundation for the tube. The circuit of the subterraneous gallery which the insect desires to excavate being determined, this paste is quickly turned about, flatted, and employed. The gallery is then prepared; and the grains of sand, detached by the raking up, are successively applied on the layers which form the basis of the tube. But as the provision of the fluid which the insect disgorges, for the purpose of moistening the sand, is quickly exhausted, it goes in search of more, whither it gets it, from the water of some neighbouring rivulet, or derives it from plants, or fruits, or flowers. The length of this tunnel, and the depth of the hole, vary a little, according to circumstances, which depend more or less upon the soil, its nature and situation. The tube is about two inches long. The materials which do not enter into its construction are thrown out, and this rubbish falls upon the ground, if this tube is placed in a wall; or falls round the base of this tube, if it is situated in a horizontal ground.

It is easy enough to divine the object of the insect in piercing a hole in a mass of sand; but we do not see so clearly with what design it builds the tunnel, the construction of which requires considerable art, and which, by the way, is destined to almost immediate ruin. Reaumur conjectures, that it is to supply materials at hand proper for stopping up the hole, when the egg of the insect shall be deposited there. This explanation, though plausible enough, still leaves something to be desiderated. Many other hymenoptera excavate, in like manner, holes for the reception of their young in walls, stop them with earth, but form no tubes or tunnels of this description. Many apiariæ of the genus anthophorus do the like. There is but a single species of that genus whose mode of modification is somewhat similar, and that is the *Megilla Muraria* of Fabricius. Wherefore then, may we ask, has Nature thus changed the habits of congenerous animals? Must there not be some sufficient motive which determines this *mason* to perform this additional work, which the others do not? May it not be, says M. Latreille, to render the habitation which it has commenced for its young, more difficult of access to the *Chrysis*, *Cinips*, and to some other apiariæ, which would come thither to deposit their eggs? This is also the conjecture of Reaumur. But why is not this foresight common to all the insects of the same genus? Their interest is surely the same, as they are all liable to similar invasions. It is nothing but future study and observation, that can lead us to an explanation more solid and more certain.

Reaumur says on this subject, " while the wasp is absent, some ichneumon fly might come and deposit in its nest an egg, enclosing an insect fatal to its young. The ichneumon fly will not so readily enter a hole, when, to approach it, it is necessary to make a long journey, and pass a tube which prevents it from seeing if the wasp be absent. I have observed one, which, after much hesitation turned and returned around the mouth of the tube, and at last ventured in; I also saw that this was very *mal à propos*, for the wasp happening to be at home, presented itself before the ichneumon, who had believed it out of the way, so that there was nothing left for the latter, but to take speedily to flight *."

Reaumur has found, in the cavities of these holes, larvæ similar to little caterpillars, but without feet, green, with bands of a clearer colour, and others of a deeper hue, with a brown head. There were eight, and sometimes even ten or twelve in number in each hole, placed in file, one after the other, and rolled together in the form of rings. The larva of the insect, which is the exclusive proprietor of the habitation, will thus have a sufficient provision for its complete growth. It may thus feed, at its ease, on the victuals which are placed within its reach, without fearing any thing from the movements and attack of the larvæ which are shut up with it, and which being so closely packed, are totally incapable of hurting it.

We shall conclude our account of the wasps, by some notice of the operations of the *vespa nidulans*, Fabricius, which will be found under *Polistes*, in the text. It is also called *vespa*

* There is one observation which we will hazard here, before we dismiss this subject. We know, that though the general plan of Nature be the same, at least in essentials, she developes her works in an endless variety of modifications. This diversity of operation is one of the best proofs that all is the result of design and choice, and not of blind necessity. Were all things formed exactly on the same model, were there no diversity or deviation, we are not aware that a very sufficient answer could be given to the pantheist, who maintains the eternal self-existence of the universe. But such phenomena, as the one we have just described, prove that the works of Nature are not works of compulsion; that there is a directing power which can vary when it will, and from the diversity observed here, can doubtless form other worlds, on plans totally different from that of ours. Therefore we may be satisfied with the explanation of Reaumur, and conclude, that if a similar instinct, for the better preservation of their offspring, be not granted to the congeners of the solitary wasp, that such is the *will* of the great directing Mind, who certainly has thought proper not to grant equal advantages to all the beings of his creation.

chartaria, from the material of which its nest is composed. The paste-board nests, which are in the form of a truncated cone or bell, very common at Cayenne, and to be found in several collections, are the production of this insect. Some other species make their nests of a material and form somewhat similar.

The works of these insects are, perhaps, superior even to those of the bees. The latter are obliged to find a retreat, an edifice prepared by Nature, or the hand of man, in which to fix their establishment. They do no more, if we may so express ourselves, than furnish the house—they do not build it. The materials which they employ are easily found, and are fashioned without difficulty, in consequence of their softness. Their nature too, being very frail, the work which results from them is of no long duration. The insects of which we are now treating, and which might be called *paste*-board wasps, construct even the walls of their habitation, raise it in the air, and have need only of a resting-place or support, which is usually a branch of a tree. The material of their edifice is of the same nature as that of a most excellent pasteboard, and will resist the most abundant rains. Let us consider, for a moment, what a degree of labour, what pains, its formation must have required. It was necessary to detach from the different trees an inconceivable quantity of small ligneous particles, to chop, moisten, and knead them, to compose the paste of this papyraceous substance. The bee, besides its mandibles, and the other organs of manducation, possesses instruments proper for gathering, in a more prompt and commodious manner, the substances which enter into the fabrication of its combs. The pasteboard wasp has almost no other means, than those which are furnished by the parts of its mouth. Moreover, the interior of its habitation exhibits as much art, as much symmetry, in a word, as much perfection, as the interior of the hive. It is therefore past a doubt, that if the labours of the paste-board wasps were as useful to us as those of the bees, we should not hesitate to give the preference to the former.

Some of the nests of these wasps are more than a foot and a half in length. This nest is a sort of box, composed of a paste-board, remarkable for its fineness, its polish, and its whiteness, in the from of a bell, more or less elongated, and more or less wide, having no other aperture than a little circular hole, placed in the centre of the lower plan, which forms a convex covercle or lid. It is suspended vertically in its natural situation to a branch of a tree, by means of a ring situated at the upper part, and composed of the same material. The interior is filled with combs, distributed in horizontal stories, like those of other vespiaries. But they are not supported by columns, and are immediately attached to the parietes of the box, on the external surface of which their place may be distinguished, and they may be counted by means of little circular reliefs, or knots, which they form at the points of union. Each comb represents a sort of cap, the convex part of which is underneath, but which, instead of being rounded, is raised in a point towards the summit. The summit is pierced with a small circular hole, which is the external and only door to the habitation. All these holes are thus in one vertical and central line, which permits these insects to go from one extremity to the other, or to get at pleasure at any one of the combs. If the lower lid of the box be removed, it will be easily perceived, that the cells or alveoli, which, like those of other vespiaries, are hexagonal, occupy all the convex part of the combs, which is underneath, and that the insect had previously constructed the floor or foundation of the comb which bears them. These combs are also perfectly alike. It is easy to conceive why their convexity is the same underneath as that of the lid which closes the box: it is that each of these combs has successively served as a lid. We may imagine the box having as yet but two combs; to make a third, the insect lengthens the envelope underneath, by bringing its edges beyond the lid; then it constructs underneath a new lid, and this operation being finished, it raises new cells, on the convex parts of the old. When it recommences the same operation the insect will close the last lid, and make it serve for a similar purpose.

We now come to, if not the most wonderful of the insect race, at least to those which must ever prove most interesting to us, from their utility, and whose works, though perhaps not in all respects comparable to some that we have already reviewed, are yet, from their ingenuity and precision, worthy of the highest admiration. We need scarcely say that we mean the Bees.

Originally, the name of bee was given only to the valuable insect which furnishes us with wax and honey. The application of this name has since become more general, and is extended, in most part of the works on natural history, to insects of the same order, whether solitary or living in society, that collect the fecundating dust of flowers, or as it is usually termed, their pollen.

Although Scopoli, Degeer, and Fabricius, had already restrained the genus *Apis* of Linnæus, it still exhibited a very heterogeneous group, both in the organization and habits of the species which were preserved in it, and of those subsequently referred to it. M. Latreille, and more especially Mr. Kirby, our distinguished countryman, have done much towards its reformation; but in this place we can but afford a mere reference to their labours, contenting ourselves with as brief a notice of the habits of the domestic bee, and some analogous species of the same family, as the interest and importance of the subject will permit.

The societies of these insects present us, like those of other social insects, with three distinct kinds of individuals. The *working* bees, or mules, form the mass of the population, and are charged with all the labours. The *males* (sometimes

Pl. 123.

1. *Xylocopa violacea* — Carpenter Bee.

2. *Bombus ruderata* — Lat.

3. *Apis centuncularis* Leaf cutting Bee.

3a. Nidus of Rose leaves of same.

4. *Apis mellifica* — Worker

4a. Female of same.

4b. Male or drone.

4c. Part of the Honeycomb with the Queen's cell.

Lane. 1832.

called *drones*), whose number varies in each hive, are destined only to fecundate the third sort of individuals, the *females,* often designated under the name of *queens* by the moderns, and under that of *kings,* or chiefs of the society, by the ancients. There is generally but one of them in each hive.

The workers and females are armed with a sting; the males have none. It is by means of a little brush, which invests the internal side of the first articulation of the posterior parts, that these little animals gather the fecundating dust of the stamina of flowers.

We shall now give a short description of the sting. It is composed of three threads, extremely slender, which are enclosed in a sort of case or sheath. The two pieces which form the true sting are scaly, and furnished each, at the extremity, with ten or sixteen denticulations. If the bee be much agitated or enraged, it will leave this sting in the wound, and die in consequence. To little animals this sting is fatal. Man himself is more or less exposed to dangerous consequences from it, according to the delicacy and sensibility of the parts which receive the poison. Many accidents annually take place from this cause. Dioscorides long ago recommended a solution of salt, or merely sea-water, as a remedy for the sting of a bee. This very simple remedy is sufficiently efficacious. The essential precaution, however, for calming the pain, and abridging its duration, is in the first instance to withdraw the sting. Alkali and chalk have also been recommended, and are not without their merit. It is the poison lodged in a bag at the end of the sting, which produces the disagreeable effects experienced from it.

The workers have been, as we long since alluded to, in describing analogous genera, discovered to be only imperfect females. M. Huber has ascertained a difference between them, in consequence of which he has divided them into *wax-*

workers and *nurses,* or large and small. To the former are confided the charge of external affairs, the gathering of the honey, of the wax, its preparation, and the establishment of the cells, which they shape, polish, regularize, and modify at will, according to the end which they have in view, or rathei according to the sort of inhabitant who is to be lodged in these receptacles. The nurses, on the contrary, are formed for retirement; more feeble, and less adapted for the business of carriage, they are scarcely employed except in cleaning the alveoli, and procuring for the young larva the aliment which is suitable to it. The ancients appear to have been acquainted with these facts, for Virgil, in treating of the bees, in the fourth book of the Georgics, expresses himself in a manner to lead us to that conclusion.

> "Namque *aliæ* victu invigilant, et fœdere pacto
> Exercentur agris: pars intra septa domorum,
> Narcissi lachrymam, et lentum de cortice gluten,
> Prima favis ponunt fundamina; deinde tenaces,
> Suspendunt ceras: *aliæ,* spem gentis, adultos
> Educunt fœtus."

The great difference, however, between them, is in point of size.

All the parts which enter into the organization of these insects, have been formed for one determined end. By those of manducation, the harvests are gathered in, the carriage performed, the cells cut out, and the magazines filled. It is to the works of Swammerdam and Reaumur, that we must refer for a full description of these different organs. The mandibles, composed of a solid substance, and moved by powerful muscles, with the assistance of the upper lip, can tear and dig into vegetable substances, and detach from the trees the mastic which is employed for stopping up the chinks of the internal part of their habitation. Their jaws and lips form a very elongated proboscis, which they plunge into the bottom of the calyx of flowers, to extract the honey. Sometimes they will even pierce the tube of the monopetalous corolla, when they are either too much closed, or it is impossible for them to reach to their lower extremity.

Reaumur had already observed, and the observation has since been corroborated by M. Latreille, that the bee, to pump out the nectarious juices, carried to right and left, bent back, and turned the extremity of its tongue, and seemed to lick, or lap like a dog. The liquid matter which the insect extracts, passes near the jaws and the sides of the tongue, flowing under the lateral divisions of the latter.

The corslet offers very considerable resistance; thus it is enabled to support the successive efforts of the wings and feet which are united in it. The feet in bees have some particular uses, independently of the business of locomotion. The disposition of the tarsi and of the hinder limbs, gives them the faculty of gathering and transporting at will, the substances of which the insects make use.

M. Huber furthermore accords to them the faculty of touch, on which hypothesis, the hinder parts being, as we above mentioned, provided with brushes, must possess this advantage in a high degree. This writer also thinks that the wings are employed for the ventilation of the hive, during very hot weather, or when it is necessary that the air should be renewed. The experiments which he has made, prove that there is some degree of probability in his opinion. All the organs of the abdomen serving for digesting, are in their uses similar to what we observe in other animals; but among the secondary functions relating to this primary use, there is one which is so peculiar to the bees, that it is to be met with in no other animated being. It is the secretion of a substance, which is nothing but an elaboration of the saccharine matter of honey. Wax was a long time considered by naturalists as a conversion of the dust of the stamina performed by the stomach of the bee, which, as they thought, disgorged it for the making of the combs. M. Huber, but little disposed to acquiesce in this received opinion, resolved to ascertain the truth by experiment. This was the only way of confirming or destroying the common belief. Reaumur himself was desirous of penetrating into the origin of this business. He concluded, after many experiments, that the dust of the stamina was only one of the principal ingredients, but he never affirmed this in a positive manner. Some time after, a cultivator of Lusace observed some plates of wax under the rings of the abdomen. M. Willemi communicated this to M. Burnet, who, however, did not change his opinion on this subject. In 1793, M. Huber observed similar plates to those which we have mentioned, at the same place, where a long time before they had been discovered. This naturalist made some observations, which confirmed this fact positively. Our countryman, Hunter, was also led to the same results. Nothing could be more interesting than the careful study of this phenomenon.

Accordingly, the Genevese naturalists, after having scrupulously examined the abdomen of some bees, after having made a comparison of the different species of a hive, and those hymenoptera which have the closest relations with the bees, submitted to divers chemical reactions some plates newly formed: all proved to him their identity with wax. He had not, however, discovered any canal of communication between the second stomach and the innumerable little lodges which serve as receptacles for the wax. He was therefore obliged to suppose their existence, founded on a great number of probabilities. Although he had recourse to the assistance of the celebrated Madlle. Jurine, the only fruit of their researches, was the discovery of a soft, transparent, yellowish membrane, of an extreme degree of tenuity. He saw that this lined the wax-boxes, adhering to the bottom of these little cavities, and to the abdominal membrane; it appeared to him destined for the secretion of a humour of no great abundance, which appeared to be intended to prevent the immediate contact of the wax, and of the membrane between which it was situated.

Being overcome by a mass of evidence, M. Huber found himself obliged to conclude that the whitish plates which he had found under the rings of the abdomen in the working bees, were of a nature identical with that of wax, although less elaborated. Thus he overturned the received opinion on the preparation of this substance, while, at the same time, by other experiments he determined its origin. To establish this last truth, he fed some bees for several days; some with honey and water, others with fruits and pollen, having taken the necessary precaution of hindering those insects from going out. The first, after a seclusion of five days, exhibited five cakes of this substance; while the second had formed none, after eight days. All this was decidedly in favour of the new discovery, and well recompensed the observer for the pains which he had taken.

A pound of sugar-candy, reduced to a syrup, and clarified with white of egg, produced ten drachms and fifty-two grains, of a wax less white than that which the bees extracted from honey. Brown sugar of an equal weight, produced twenty-two drachms of a very white wax. The same effects were obtained with maple sugar.

M. Huber has made various experiments relative to the respiration of these insects, the result of which may perhaps prove interesting to our readers.

He found, by comparative experiments, that the various gases acted differently on such bees as he submitted to their influence, either in consequence of their more or less deleterious principles, or in consequence of the diversity of sex. Sometimes the bees perished; sometimes they remained but for a moment in a state of asphyxy, and returned to life as soon as they were placed in contact with the atmosphere. To strengthen these experiments more, the observer made new trials, on benumbed individuals, in whom respiration did not exist. In vain he submitted them for a long time to the action of the same gases which he had employed in experimenting on bees in full vigour. He easily recalled them to life by the heat which he communicated, by placing them upon his hand. If the first experiment left any doubt on what particular part the gases acted, whether it was to their influence over the whole body, that the phenomena must be attributed, or merely to that which they exercised over the organs of respiration, the final trial put an end to the difficulty, and decided in favour of the last opinion. He was the better able to confirm the reality of his experiments, inasmuch as the air contained in two cells, which had been closed, each containing one hundred and twenty workers, produced asphyxy in some, put out candles, &c. He had already convinced himself of the asphyxy, by a complete and long immersion in water. In this experiment he saw four stigmata, which had been discovered by Swammerdam, and two others which had escaped the dissections of this skilful anatomist. One of these stigmata is situated at the place where the corslet is united to the head by a little fleshy pedicle, and the other at the opposite end, quite close to the pedicle of the abdomen.

The immersion of the head did not cause the bees to suffer. It was not the same with the immersion of the corslet, and abdomen. That of the first destroyed them, that of the second only produced in them a degree of agitation. It was therefore easy to conclude in which of these three parts were placed the most important organs of respiration. This first experiment necessarily conducted M. Huber to those which he afterwards attempted with such great success.

This most positive determination of the existence of respiration in these insects, led to the indication of the source of the heat, observable in hives, which had falsely been attributed to the fermentation of the honey. As bees respire, phenomena must result from the exercise of this function, analogous to those which accompany it in other animals. Some trials had already demonstrated the alteration of the air through the influence of this operation. The contact of this fluid with the animal and living parts, arranged to be in relation with it, must also cause them to experience modifications, similar to those which they undergo, in the high degrees of the animal economy, with proportional differences.

The modes of relation of the bees with the other beings of nature, and their method of reproduction, are equally worthy of the researches of those who delight to recognize, even in minute things, the sublime laws of the universe, and of those elevated minds, who possess the faculty of observing, measuring, and comparing all phenomena, and deducing from them felicitous and important results.

The numerous discoveries made concerning the manners of these interesting hymenoptera, suggested the idea of inquiring into the secret springs of their actions. Observations were consequently made; they weakened all doubts which might have been started concerning the existence of a great number of senses in these insects. M. Huber, in fine, came to the belief, that he could not only pronounce with certainty on their existence, but even indicate with precision, the seats which some of them occupied.

It is not possible, however, as yet, to assign their limits: thus, for instance, is sight the exclusive property of the complex eyes; or is the faculty divided between them and the small simple eyes which surmount the head? Taste, bees do not appear to possess in any great degree of delicacy, for indifferent in their choice, they gather their provisions on aromatic and sweet plants, and on those which exhale an infectious and nauseating odour. Now, if the elements which enter into the composition of honey are so different, is it astonishing to find so much variation in their properties? The water with which these insects slake their thirst, is not selected with a greater degree of discrimination. They sometimes appear to prefer that which stagnates, and is in a state of decomposition in the most disgusting marshes.

But if they do not possess taste in any perfection, the delicacy of their smell appears to compensate them for this defect. On presenting odorant bodies to these insects, M. Huber could observe that some appeared to displease, and others to attract them. Besides such trials on their bodies in general, he made others on particular parts, as on the eyes, the antennæ, and the proboscis; and because he thought that he perceived more marked effects on the application of odorant bodies, at the origin of this last organ, he has concluded that the sense of smell exists in the mouth.

The proofs brought forward for the existence of hearing, are far from being convincing, although some facts seem to confirm it. It must, however, be left to future research and observation, to decide the negative or affirmative of this question.

In the bees, one of the most important of all the senses is that of touch: nevertheless, on first consideration, every thing in their composition would seem calculated to diminish its sensibility;—hardness of envelope, abundance of hairs, and deficiency of numerous parts adapted for studying and measuring surfaces. The tarsi, but more especially the antennæ, parts less hairy and more mobile, have been regarded as the depositories of this faculty.

What proof can be stronger in favour of the opinion which would place the sense of touch in the antennæ, than the conduct of the bee, when occupied in the obscurity of its retreat, in seeking out the cell where its honey is to be deposited in the pollen accumulated for the support of the larvæ. There we perceive it directing in all ways those useful and supple instruments which supply the place of sight, which is no longer serviceable, and seem to enable it to recognize that particular cell for which its labours have been called into action. To be perfectly assured of the importance of these organs, we have only to remove them from a certain number of bees. In vain, after this operation, will they attempt to find their own proper hive. The cares of it are abandoned. If some, after many efforts, have succeeded in getting up upon the combs, they very speedily fall down again, and, as if they were conscious that their presence is useless in a place where they can work no longer, they speedily make their escape to return no more.

On data like this, the existence of the sense of touch has been established, and the parts by which it is exercised have been determined.

Provided with all these senses, the bees have different functions to fulfil: their number, which is sometimes as much as fifteen, eighteen, or even thirty thousand, is divided by Nature into four kinds, to each of which she has assigned their particular functions. The queen, which is usually single in each hive, and the males, from two hundred to a thousand and upwards, are charged with the business of reproduction: the others form the multitude of the population, do not participate in the sex of the preceding, and are, as we have already observed, distinct among themselves.

Fecundation takes place in the air. It is difficult to say how the female and the male recognize each other; how one is informed of the presence of the other in the atmosphere. No doubt but that the discovery of this would lead to some further interesting results, concerning these curious animals. Taste, smell, and touch, useless when removed from bodies in which they re-act, do not appear under those circumstances to guide the instinct of the bee. Some noise, perhaps, made by the queen, or some peculiar movement, may strike the hearing, or the sight of the numerous males which are spread through the atmosphere.

However this may be, the young queen, five or six days after her birth, feeling the desire of uniting with an individual of her own species, having issued from the hive, abandons herself to the caresses of the first male she encounters. If, on her first excursion, fecundation should not take place, another soon follows. When the queen returns to her dwelling in a state of fecundation, she becomes a new stimulus to exertion for the labouring population. Every thing is then done with greater care. The workers, encouraged, hasten to finish the alveoli which they have commenced, lay new foundations, and appear to be fully conscious of the importance of their activity. The progress of their labours is the signal by which the female is directed in the laying of her eggs.

According to M. Huber, a single coupling is sufficient to vivify the eggs which the female shall lay for two years, and perhaps for all those which she shall lay during her entire life. The male who co-operates to the existence of so many thousands of lives, never enjoys the pleasure of beholding his posterity; he dies immediately after coupling. By this union he is deprived of the organs of generation, which remain with the female. The laying of the eggs continues about a month. The extraordinary fact to which we have just alluded is sufficient to overturn the opinion of Butler and Swammerdam, who pretended that the impregnation of the female was caused solely by effluvia from the male. The ancients were not less opposed to the truth, when they imagined the generation of bees to be similar to that of fishes.

Occupied with the care of laying her eggs, the queen visits the dwelling destined for the new individuals to which she is to give birth. She enters there, head foremost, traverses the place, and examines the most retired parts of it. It is only after this operation, that she introduces the extremity of her abdomen into the alveolus, to make her precious deposit there. The egg remains cemented to the upper angle of the cell. It is retained there by the viscous humour, which Swammerdam observed in the neighbourhood of the ovaries. Many hundreds of the eggs are often laid duringa single day in spring. Should there happen not to be cells enough, she deposits several eggs in one, leaving to the workers that construct them, the care of making a new division. This is particularly observable, according to Reaumur, in a young swarm.

The laying is over in autumn, because the pollen being then wanting for the existing larvæ, the honey which is gathered becomes necessary for them. It is thus Nature prevents both the exhaustion of the magazines, and the dearth which a great population would induce, at a time when the renewal of provisions is impossible. The suspension of the laying in winter, is a matter of course, for then the queen is in a state of lethargy. On the return of spring, at the appearance of certain early flowers, she awakes. She then regulates the laying of her eggs, which she recommences according to the temperature. Cold diminishes her activity in this respect, and heat increases it. The eggs of the workers appear the first, because they being destined to nourish the larvæ of the other species, require to be developed before them. At the end of two months, those of the males are laid—some eggs of the females follow. All are deposited in the cells which are suitable for them. The queen never lays a male egg in the alveolus of a female, nor a female egg in that of a male. Both are oval, elongated, a little curved, of a bluish white, and about a line in length. The temperature of the hive, which is often considerably elevated, and always above that of the atmosphere, is sufficient to make them disclose in three, four, five, or six days.

The bee is then in its second state. It is a worm, without feet, wrinkled circularly, and always rolled round, at the bottom of its cell, where it moves with difficulty. We are not aware whether it changes its skin like the larvæ of other insects.

The nursing bees take care of the larvæ with the utmost solicitude. They visit the cells incessantly, enter them, remain there a certain time, during which they give to each the matter on which it is fed, or more properly speaking, renew it. This substance is a kind of insipid jelly, thick and whitish, but becoming more transparent and saccharine, in proportion as the worm advances towards its metamorphosis. It is also more or less succulent, and more or less copious, in proportion as it is designed for the nourishment of the females, or of the population. It always forms a bed at the bottom of the cell, on which the worm is placed, which, thus in the bosom of abundance, opens its mouth when it is hungry, and this food, so to speak, enters it.

The nature of this jelly was not known, until M. Huber, the younger, made several experiments, which demonstrated it to be a mixture of honey and of pollen, which undergoes some modification from the workers.

Six days are sufficient for the worm to acquire its complete growth, when the temperature is high. At this period the nurse knows, by a kind of instinct, that other phenomena are about to take place. She no longer furnishes to the larvæ a nutriment that is no longer necessary; but she encloses it in the cell with a convex lid. This is what distinguishes this cell from those which contain honey, and which have a flat lid or cover: both, however, are composed of wax.

The prison being thus closed, the larva thenceforth employs itself in lining it with a fine and delicate silk. When this labour is finished, it changes into the nymph state. All the parts which it is to have in the sequel, are observable through a white and transparent skin. In twelve days after its seclusion, it is sufficiently developed, and sufficiently strong to tear its envelope, and issue from its cell, after having gnawed away the lid. These young bees have scarcely broken the bonds which retained them captive, when the nurses hurry around them, begin to lick and clean them, and present them with food. Many of them hasten to take possession of the abandoned cells, so as to arrange them for the reception of a new egg on that very day. Experiments have been made to determine the influence of air, and of gases more or less deleterious, on the eggs, the larvæ, and the nymphs, as well as on the perfect insect. The same results have been obtained. There was always a consumption of oxygen, and a formation of carbonic acid; only the larvæ consumed a greater quantity of the first than the eggs, and less than the nymphs, whose death almost immediately followed the experiment. The azotic gas and carbonic acid appeared to be less pernicious to the larvæ than to the perfect insects, and considerably less so than to the nymphs. Their effects were altogether nullified during the lethargic state. It was not the same with the influence of the air which had been respired by the bees; for it prevented the eggs from becoming developed.

In seeking to be acquainted with all that took place, at each epoch in the life of the bees, it was requisite to observe the mode in which they escape from prison. This mode is the same with all the individuals that compose a hive. The female, being retained captive a longer time, after the last metamorphosis, is able, in her less confined prison, to lick her wings, and acquire all the vigour necessary for flight, so that she may issue forth on her first excursion, to prevent her destruction; and that individuals may be preserved to the species, the workers instinctively throw obstacles in the way of those efforts made by the queen to approach those who are soon to replace her; they take care to keep guard over the cells while the metamorphoses of the young queens are going on.

Far from being able to follow their companions to the field on issuing from their cells, the workers are obliged to pass an entire night upon a comb, where they dry, wipe, and prepare the delicate organs which are to transport them into the air.

By its grey colour, and the number of hairs by which its body is shaded, a young bee is easily distinguished from an old one, which is redder, and less furnished with hairs. It has been endeavoured, but in vain, to determine the number of years which this insect can live. The opinion of the ancients was, that it could live seven years.

All writers once considered the workers as insects deprived of sex. But M. Huber, the elder, and his observations have been confirmed by his son, was perfectly convinced that they are all originally of the female sex, and consequently provided with ovaries, though Swammerdam and Reaumur were not able to find any. He cites, in support of his proposition, the discovery of M. Schirach, who observed, before him, in certain hives, workers converted into females. But not knowing how this phenomenon, unheard of before, could take place, he endeavoured to find out by experiments. His observations convinced him that such bees are never to be seen except in hives deprived of queens, and that they are indebted for this metamorphosis to the workers; that when these last have lost their mother, a fact which they very promptly discover, if there are found in the combs any eggs, or larvæ of workers not more than three days old, they soon hasten to choose themselves another queen. They immedi-

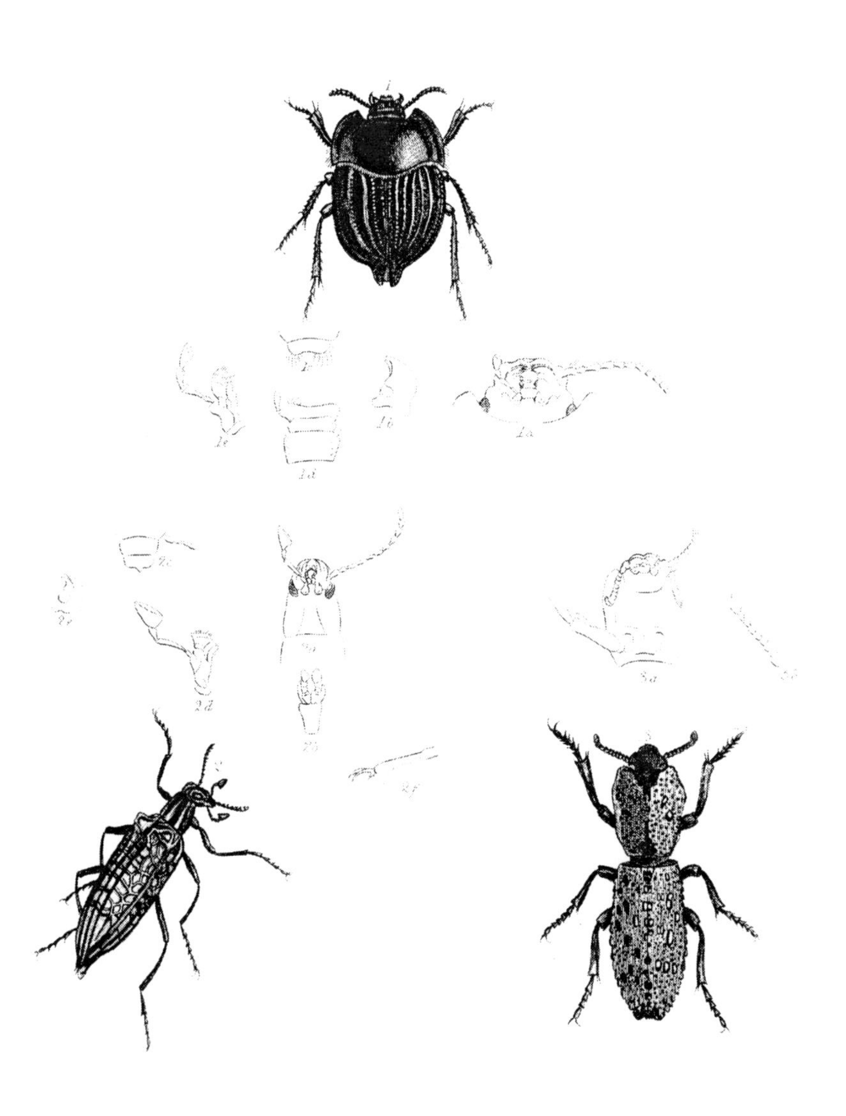

1. *Aulacus chilensis* — G. R. Gray.

2. *Cyphonotus dromedarius* — Guer.

3. *Zophorus chilensis* — G. R. Gray.

London. Published by Whittaker & Co. Ave Maria Lane, 1832

ately enlarge the cells of some of these larvæ, prepare some paste, similar to that which they usually give to the larvæ of the females, and nourish with it those which they destine to the maternal state. Finally, by dint of care and labour, they succeed in producing a female, which replaces the one that they have lost. From these circumstances, this author concludes that if all the workers are not proper for reproduction, it is because in the larva state they have received but a small quantity of a paste, much less active than that given to females, and because they have been lodged in too narrow a cell. These two causes have so much influence over them, that they prevent the development of the ovaria.

M. de Riemps has made a discovery which appears equally astonishing with what we have just related. Having shut up some pieces of comb in several boxes, from which pieces he had removed the eggs, along with some workers, he subsequently found eggs upon these very combs. He judged that those could not have been laid except by the workers; thus, according to him, there are sometimes fruitful workers; a thing the possibility of which was not before suspected. But M. Huber tells us that he has been able to make some working bees fruitful in his hives, whenever he thought proper. His method is to remove the female from a hive. "Immediately the bees hasten to replace her, by enlarging many of the cells which contain the larvæ of workers, and giving to the worms which they contain the *royal* jelly. They also let fall some of this jelly, in a small quantity, on the young worms lodged in the neighbouring cells; and this nourishment developes their ovaria to a certain point. Therefore, fruitful workers are always born in hives, when the bees are occupied in repairing the loss of their queen; but it is very rarely that they are found there, because the young queens brought up in the royal cells, fall upon and massacre them.

To save their lives, it is necessary to remove their enemies; then the fruitful workers finding no rivals in the hive, are well received; and a few days after they will lay male eggs, but never any other."

In the fine season, the number of bees which are daily born in a hive gives rise to the formation of swarms. No equivocal signs indicate their expedition; those which announce it for the same day are when there appears in the interior of the hive a great number of males or drones, when few workers issue forth in the morning to gather in provisions, and when those which return bring back no wax upon their feet; in fine, when an extraordinary noise is heard. If at that moment we examine what is passing, we shall see the female traversing every corner, with a degree of agitation which is communicated to the workers: the last give over their labour, and assemble together. Heaped one upon the other, they experience a heat, which is still further increased by the rays of the sun. Not being able to support the fatigue any longer, they repair in a crowd towards the door, sally forth with precipitation, and often drag out the mother along with them. It is always in calm weather, when the sky is serene, between nine in the morning and four in the evening, that they determine to quit their habitation. If the female is not found among the first which issue forth, she is not long in repairing to them, and in less than a minute she is followed by all the bees which are to compose the swarm.

As soon as they are abroad they disperse in the air, hover about, and seem in search of a place where they can re-unite. By degrees they fix themselves upon a branch, form a group there, by hooking themselves one to another with their feet, although they are exposed. They remain quiet, and often in less than a quarter of an hour, we see scarcely more bees hovering round a swarm than we see around a hive in warm weather.

If at the moment of their sally they direct their flight towards some large trees, there is reason to fear that they may rise too high, and wander from the limits of the hive, which sometimes happens. But a method of making them descend when they are only at a certain height, is by casting handfuls of earth and sand upon them. This kind of rain forces them to get lower, and the nearest shelter appears to them the best. Two or three discharges of a gun or pistol loaded with powder only, will produce the same effect upon them. It does not appear that it is the female that makes choice of the place where they go to settle; for when a certain number of bees fix upon a branch, the female remains with some others on a neighbouring branch, and goes to join the first, when they are assembled in sufficient quantity to form a group.

Although the swarm remain tranquil, it must not be left long in this position, without presenting it with a lodging, especially if the sun be warm; because it would speedily go elsewhere in search of a new situation. Thus, therefore, in the swarming season it is necessary to have hives quite ready, to make use of as occasion requires. If there are none at hand at the moment, a sort of tent must be made for the swarm, with some linen wetted to keep off the heat, and hinder the swarm from quitting the tree before a hive is procured. The interior of a hive should be well cleaned before it be presented to the bees, for they are fond of cleanliness. To render it agreeable to them, the sides are rubbed with flowers of melissa, and bean-flowers, of the scent of which they are fond. Some parts of it are also moistened with a slight layer of honey.

To cause the swarm to enter into the hive, an operation which is not difficult, it is brought near the branch on which the swarm is fixed. It is held up reversed, and with small branches the bees are made to fall in. The hand may even be used with safety, for under these circumstances the bees will not employ their stings. It is sufficient if the greater portion of them be in the hive, to induce the others to enter it. Then it may be turned, taking care to manage the apertures so that those which are outside may be able to get in. If any should persevere in remaining on the branch, they may be forced to quit it and join the others, by rubbing it with leaves of rue and elder, the scent of which displeases them. After the setting of the sun the hive may be taken to the support which is destined for it, providing it be carried cautiously and gently.

The swarms do not always place themselves so that they can be easily obtained. Some seek a very elevated branch, others a thick hedge, and others again take refuge in the trunk of a tree, or the hole of a wall. These last must be watched until the sun has quitted the horizon, and their retreat approached only at the coming of night, because then the bees are less formidable, and may be removed without danger. The hive is brought to the foot of the tree or wall, held there with the aperture upwards, and the bees may be taken with the hands, or a large spoon to put them in. As they are somewhat benumbed by the coolness, they will suffer themselves to be taken in quantities without making the least resistance. If they cannot be all obtained in this manner; the hive may be reversed and placed on the ground, pretty near the others, so that they may enter it on the following day.

A well-peopled hive furnishes two, and sometimes even three swarms during the year. But the last weakens it very much, and puts it in danger of perishing during the winter. The first is always the most numerous and the best, because the bees go to work in a favourable season, which furnishes the harvests of wax and honey in abundance, and they have more time to labour before winter. A good swarm will weigh five or six pounds. Reaumur has found some weighing eight pounds, but this is rare. He has calculated that there were forty thousand bees in that swarm. The last swarms are sometimes not more than three or four thousand in number. M. Schirach, after the discovery which he made of the conversion of workers into females, imagined that he might derive some advantage from this in the formation of artificial swarms. His process consisted in taking away, in the month of May, from the interior of a hive, some pieces of combs containing eggs, larvæ newly issued from eggs, others ready to be metamorphosed, and nymphs. You shut up the combs in as many boxes as you are desirous of having swarms, and in each you put seven or eight hundred workers, with a small provision of honey, so that the bees which cannot get out may have wherewithal to eat. The boxes are to be placed in a warm situation, without approaching them to the fire. The workers, deprived of females, but having the power of producing them, set themselves immediately to the construction of a cell to rear a larva for this state. In two or three days after they have been shut up they are allowed to go out. They take advantage of the liberty which is given them to go into the country, and they return to their new habitation. At the end of fifteen days the box is opened to see the state of the cell which they have made. If it is gnawn on the side, it is a proof that the female is dead: if, on the contrary, it is pierced in the middle, the operation has succeeded. Then the new female should be placed in a more commodious lodging.

This method of forming swarms has had several partizans in Germany; but two very strong objections are made against it. The first is that it is very injurious to the hives thus to subtract any part of their brood; the second, that it hinders them from swarming. To the first M. de Schirach replies that the brood is removed only from very populous and powerful hives, which have subsisted for several years; and to the second, that we are frequently exposed to lose the natural swarms, because they can issue forth without being perceived, and that, by his process, this inconvenience is obviated; that, besides, the artificial swarms are infinitely better than the others, because they are composed of laborious bees, and less disposed to form new colonies, which is a very great inconvenience to the swarms, which are continually weakened thereby.

When the bees are newly in a hive that is agreeable to them, they set themselves promptly to work. They often work more wax during the first fifteen days after their settlement than for all the rest of the year beside. Sometimes, for the first two or three days that they are there, they do not go out in search of provision. During this time they employ the wax which they had the precaution to bring with them, before they go to seek for any more.

Before men began to occupy themselves with their culture, bees lived wild in the woods. But we are ignorant of the places which they naturally inhabit. They are to be found wild in the forests of Russia, in different parts of Asia, and even in Italy, and the Southern departments of France. As they do not know how to make a nest, they retire into some large cavities, to shelter themselves from the inclemency of the weather. They live in very numerous societies, which some authors have named republics.

The hives in which they are usually kept are of different forms, and different materials, in different countries. Some are merely the trunk of a hollow tree; others are made of four equal planks, which form a kind of long box, placed upon one of its ends, with a lid upon its upper part. The form of the greater number is that of a bell or cone. These are generally sorts of baskets. Some are made of osier, or of some other pliant wood; others are composed of twisted straw. We shall mention by and by those invented by M.

Huber. The invention of glass hives is comparatively recent. The ancients, who were unacquainted with glass, had hives into which the light could penetrate. Pliny informs us that a Roman senator had some of very transparent horn.

A well-peopled hive contains one female, from two hundred to eight hundred males, and from fifteen to sixteen thousand workers, and often more. When the bees enter into a new hive, their first care is to close its apertures. They do not make use of wax for this operation: they employ a material which stretches and attaches itself better. This material was known to the ancients, who termed it propolis. They derive it from the young germs of the poplar, the willow, and other trees, before the buds are blown. They also cover with it the sticks which support the combs, and sometimes the interior of the hive.

As soon as the bees have taken possession of the hive, the workers proceed to gathering the dust of the stamina of flowers, and honey, to nourish their larvæ and construct their combs. If it is spring time this labour occupies them the entire day. In the great heat of summer they give it over about ten o'clock in the morning. After a bee has remained on a flower the necessary time for collecting its provision, until all its hairs are covered with this dust, it collects it with its feet, which it uses as brushes, forms two small pellets of it, which it places under each of its hind legs, and returns to the hive.

The bees construct combs parallel to each other, and leave a space between them: these are so many paths by which they arrive thither. Very often their combs are not attached to the top of the hive, but by a sort of foot of small extent. Each comb has its two surfaces of pretty nearly an equal number of cells, of an hexagonal figure, applied one against the other. All are composed of many pieces, fabricated with art and regularity. These pieces are assembled by a considerable number of bees, who work at the same time. Some are placed on one of the surfaces of the comb, or a part are occupied in elongating the sides of the tunnels. Others sketch out the bases of the new cells, while others do the same for the opposite surface. A certain number of cells serve to preserve the honey. The female lays her eggs in the others. Those in which the workers are to be born are well distinguished from the rest: they are smaller than those in which the males are born. That of the female is much larger than the others; it has not the same form; it is usually placed upon one of the surfaces of the comb; it generally hangs to one of its lower edges in the manner of a stalactite, and is suspended only by sorts of pedicles: their form is oblong. A single cell of this kind contains more wax than is required for a hundred common ones. The workers do not begin until they find the female occupied in laying male eggs, which takes place when she has already laid a considerable number of workers' eggs, and, according to the younger M. Huber, only when the hive is sufficiently peopled to furnish a swarm. In some hives there are but two or three of these cells; in others, from thirty to forty.

This skilful naturalist has perfectly distinguished the difference of these cells of the first rank, or, as our naturalists call them, *royal* cells, by the pentagonous form of their tube, and the modification which takes place on the pieces which compose the bottom. All other cells are hexagonal. Their tube is consequently formed of six fronts, which are cut under a right angle at one of their extremities—that which forms the orifice of the cell—while the opposite extremity is exactly adapted to the angular contour of the pyramidal bottom on the edge of which they are raised. Each cell, in whatsoever place it may be found, is formed of two distinct parts. The first, or bottom, is the basis which supports the second. Accordingly, the bees appear to work there with more care than they bestow upon the tube: they seem to foresee, as this observer has demonstrated, that every thing must depend on this first work, as, for instance, the form of the tubes, their close union, the solidity of the combs, and the economy of the wax.

The bottom of the royal cells presents on one of the faces of the comb a slight concavity, and a proportional protuberance on the other. Such are the results of the manner in which the pieces are cut, which compose the cells, and of their respective inclination. Instead of being hexagonal, like all the cells which follow them, the royal cells have but six sides. One of these, the upper, is a horizontal line, formed by the top of the hive, to which the comb is suspended. Four pieces of wax, therefore, alone enter into the structure of the tube; two, oblique to the horizon, united at their lower edge in an obtuse angle, form the entire contour of the alveolus, and two are placed vertically. The cells, of whatever kind, are so inclined, that their raised orifice, and low foundation, prevent the flowing off of the honey which they contain, while those intended for winter magazines are closed by a lid.

It is impossible to witness, without a feeling of the highest admiration, what passes at the moment in which the foundations of these cells are laid. A worker detaches herself from the chain which she had concurred to form, presses herself through the crowd, and proceeds to place, after having cut it properly, the wax, in pentagonal plates, which she carries under the second, third, fourth, and fifth rings of the abdomen. The fragments produced by this operation, far from being abandoned as useless, are carefully collected, and pass into a cavity on each mandible. When the plates, whose diameters are equal, and whose form is similar to the diameters and form of the plates which cover them, are entirely employed, the fragments of the wax re-appear, but in a different manner. Softened and rendered more tenacious by their mixture with the fluid which flows from the tongue of the wax-worker, they issue forth in the form of a narrow riband, which is cut into laminæ, serving as a base for the new cell. Nothing more remains to complete the cells, than to invest them with a certain quantity of propolis. By this the union of the pieces is consolidated. The workers extend this substance with their tongue. For longer details we must refer to the equally instructive and amusing work of M. Huber, or to the not less excellent Introduction to Entomology, by Messrs. Kirby and Spence.

Besides the gathering of propolis and wax, the bees have another store to make, namely, that of honey. After having filled their stomach with this, all that does not serve for the purposes of nutriment is carefully carried to the hive, where they disgorge it into the cells, some of which are destined to receive what is for daily consumption, and others to contain what is laid by for the season in which no honey can be gathered. Sometimes, before re-entering the hive, a worker finds the means of getting rid of some of her store. If she meets another worker, who has had no time to procure any, she lets some drops issue forth, as far as the aperture of her mouth, and the other sucks them in with the end of her proboscis. She also renders the same service to those whose occupations retain them in the hive.

The manner of deriving the best advantage from these insects, is to lodge them commodiously, to place the hives in situations convenient for the bees in gathering their stores, and sheltering them from the effects of a too powerful heat, and still more from the cold, which would destroy them during winter. When there are a certain number of hives an apiary may be constructed, which will obviate these inconveniences. This is a kind of a cabin, raised something more than two feet from the ground, near a wall. Some pieces of wood planks, and some unctuous earth, will answer for this purpose. The roof should be made with straw. The hives placed within this shelter are much better than in the open air. An aspect favourable to the bees must be chosen for this apiary. A northern aspect is fatal to them, nor is an eastern one very suitable. The preference is to be given to a southern position. M. Wildman prefers a western to every other, because the workers that remain late in gathering their stores, will have more light to find their habitation again. But the south is, after all, preferable. The brood is less liable to fail in hives thus placed, because it is not so liable to be chilled by the northern winds, as in those which have an eastern or a western aspect. Moreover, it has been remarked, that the bees in hives thus situated, swarm six or eight days sooner than the others. Hives, thus placed, and not protected from the heat of the sun by an apiary, require a little attention during the summer. It is necessary to cover them with leaves, and wet cloths, on very hot days, to prevent the wax from softening too much, and the honey from running out.

The apiary should be built, as far as it is possible, in the neighbourhood of a meadow, in a garden, and near a stream. The bees will thus find the water, of which they have need. Columella assures us, that when this fails them, it is impossible for them to make wax and honey, and to rear their offspring. There is no place more favourable for the culture of bees upon a large scale, than those plains upon which meadows abound, or where buck-wheat, &c., is cultivated, and which are near woods and mountains covered with aromatic plants. There they find, in sufficient quantity, all that is necessary for them. Though dry, arid, and sandy countries, do not present such advantages, they are still able to gather something in them.

Some agriculturists are in the habit, when the season of

flowers is over in their district, of transporting their hives to some other place, where the season continues longer. This plan was known even to the ancient inhabitants of Egypt. Niebuhr tells us that he met on the Nile, between Cairo and Damietta, a convoy of four thousand hives. The Italians in the neighbourhood of the Po embark theirs upon that river. The inhabitants of Beauce, in the same manner transport their hives. It is equally the practice of the agriculturists who inhabit Mont-Blanc. They reap an abundant harvest in this way, from the *artemisia alpestris,* which flourishes better there than any where else. These methods seem deserving of general imitation.

For a long time back, many have employed themselves in devising means how to render the lodging of the bees adapted to make them work as much as possible, and to favour their multiplication, so as to derive all possible advantage from them. Such motives have induced many celebrated men, amateurs of these useful insects, to construct hives of different forms, and if they have not completely attained the end which they proposed, their labours are yet entitled to the gratitude of mankind.

These newly invented hives present some advantages not to be found in the old ones. If they do not possess all the perfection which could be desired, perhaps by some alterations their defects might be remedied. This, however, can only be learned in the course of practice.

To enter into any description of these different kinds of hives, though many of them are very ingenious, would be totally beside our present purpose, not to mention that mere verbal descriptions of complicated machinery, are in general very useless, from the great difficulty of rendering them intelligible. The things themselves must be seen, or the descriptions accompanied with suitable figures, otherwise no clear ideas concerning such matters can possibly be attained.

We shall only observe that glass hives are very commodious for observing the operations of the bees. Reaumur gave them a variety of forms, and it was by means of such hives, that this celebrated naturalist, and many others, became acquainted with the manners and habits of these insects, on which they have left us such interesting works.

To constitute a valuable hive, it is requisite that it should be well peopled, and its inhabitants should be young and active. Whether they possess these qualities or not, is easily discovered by their colour, by the vivacity with which they issue forth from their habitation, and re-enter it, when the hive is clean and abundantly provisioned. We know that it is well peopled, when, on striking it slightly, in the evening, when all the bees have re-entered, or in the morning, before they come out, we hear a humming noise in the interior, following the blow. When this humming is dull, and repeated at different intervals, the hive is well furnished with bees, and with provisions. The reverse is the case, when the sound is clear, and finishes almost immediately after it has been excited. The whiteness of the wax is also a certain indication of the goodness of the hive.

The most favourable season for transporting hives from one place to another, is the end of the winter, or the commencement of spring, because the bees, not being yet awakened from their lethargy, are better able to support the fatigue of their journey. Two or three days after they have arrived at their destination, they must be allowed to come forth to take the air, and the hives must be examined, for the purpose of removing the combs which may happen to have been broken.

When the bees have constructed a great number of combs, and the hive is so full that they can work no longer, they lose all courage. In spite of the fecundity of the mother, which above all attaches them to their habitation, they become disgusted with it, because they can no longer build cells in which to lodge their eggs. A part of them then determine to quit the hive; but this emigration may be prevented by removing a portion of the combs.

This operation is performed at different seasons. The most favourable, according to M. de Palteau, is the commencement of summer, because the bees have then repaired the losses of the winter, and made considerable stores. No hives should be thus treated in spring excepting those where the provision is superabundant. But he recommends to prune them all in this way in autumn, even though it may have been done before, because in that season the honey is excellent, and it would lose its quality by remaining in the hive during the winter. The wax is also finer, and can be more easily bleached than when it has remained too long in the hive, where it acquires a reddish hue. In spring, which is the time when the queen lays her eggs, care must be taken not to derange the brood, and only to remove the combs in which there are no eggs.

In whatever season this operation is performed, the division must always be made with discretion, so as to leave sufficient for the bees to live on, especially before the approach of winter, when it would be impossible for them to get any provision.

It is not every person who is fit to prune the hives in this manner, because it is necessary to be able to distinguish among the combs those which contain honey, from those which enclose the future posterity of the bees. In the common hives used by the country people, the eggs are usually in the fore part, being the most convenient situation for excluding, and for nourishing the young. In the combs the cells which contain the larvæ and nymphs ready to be metamorphosed, are easily recognized by the lids, which are convex, and somewhat brown; whereas, those which close the cells containing nothing but honey, are flat and white.

The attempt to remove the combs from the middle of the hive, where thousands of well-armed bees are ready fiercely to defend their property, may be considered as a kind of military expedition. He who undertakes it must carefully cover his hands and face, to protect himself against their stings. Nevertheless, M. Bosc thinks such precaution useless, and that firmness alone is sufficient for the task. He advises to seize, for this operation, the moment when a great number of bees are out, to burn a rag at the aperture of their dwelling, and to approach by so much the more as the humming increases. Then, he tells us, that the bees which are sentinels advertise the others of their danger, and speedily all hurry off to envelop their queen, and defend her at the peril of their lives. The evening before the day fixed on for this business it is necessary, at the approach of night, to detach the hive from its support. If there be no fear of cold, it may be turned upon its side, and the following day, before the rising of the sun, it should be smoked for some moments. For this purpose, at a hole made in the top is placed the tube of a funnel: some old linen, or even cow-dung, is burned. The smoke is directed with a bellows into the mouth of the funnel, and spreads through the hive; something may also be burned underneath the hive. The smoke from this causes the bees to mount to the top. In a little time they lose their activity, and are easily driven from the combs that we wish to remove.

Swarms may be managed in the same manner as old hives, but still greater moderation must be used with them.

There are hives, sometimes old as well as new, which are but badly provisioned, because the bees cannot always gather a sufficient harvest to put a portion in reserve. In this case it is necessary to provide for their wants, so that they may suffer no famine. The best nutriment which they can receive, which suits them best, and is most to their taste, is honey and raw wax. If these are not to be had their place may be supplied by the juice of pears or apples cooked, in which a quarter of a pound of honey, or brown sugar, is mixed, which is boiled until it is reduced one-third. When it is in a syrup it is poured into a wooden vessel. On the top some straws are placed to enable the bees to take it more easily, and it is placed in the hive. All fruits baked in the oven in their own juice may also be given to them in time of want. According to M. de Palteau, a pound and a quarter of honey, or syrup, per month, suffices for the consumption of the bees of the best populated hive. But he advises to give them more at the approach of winter, and at the return of spring, as in these two seasons they gather nothing.

When the first cold begins to be felt, it will be necessary to prevent the bees from leaving the hive, lest, excited by the sun which makes its appearance during the day, they may be induced to wander into the country, and proceeding too far, be seized by the cold, and perish. To hinder them from attempting such expeditions, which could be of no use, a grating should be placed at the door of the hive, which, while it leaves a free circulation of air, will force them to remain in the interior.

This precaution alone is not sufficient to protect the bees from the effects of an excessive degree of cold. It is, moreover, necessary, before winter, to shelter them in some close place, such as a cellar, &c. and to cover with mats, or something similar, those hives which are but thinly populated, taking care to leave a passage for the air. If we intend to preserve our hives, we must visit them in the month of February, to see if the bees have still any provisions, and to supply them with some, if they are in want. In spring, when the weather grows mild, and the sun begins to exercise a genial influence, they should be restored to their liberty, but not if the weather be unfavourable.

In this season they are sometimes subject to a malady which has been named dysentery: the excrements which they then pass, instead of being of a yellowish red, are black, and have an insupportable odour. Many writers attribute the cause of this to the flowers of the linden-tree, and the elm, of which they are remarkably fond: others, to the new honey, which they eat to excess, for the first few days after they issue from the hive. But it seems more probable that it is occasioned by the long stay which they have made in the hives, and by their food, which, for want of raw wax, consists entirely of honey. This contagious and mortal malady, may destroy an entire hive, if remedies be not quickly applied. It may be prevented, they say, by putting into the hive, a syrup made with equal quantity of good wine and sugar; and remedied by giving to those bees which are already attacked by it, some cakes containing raw wax. The weakest bees are the soonest to be attacked by this disease.

It appears that the poisonous quality of some plants, occasionally produces in these insects a fatal and incurable malady, which some writers call *vertigo*. Those attacked by it run about, and turn and twist in all directions near the hives. Their hinder limbs and the hinder part of the body, appear uncommonly feeble. This malady generally appears towards the end of spring.

The bees are also subject to another disorder, which is not quite so dangerous. It appears in their antennæ, the extremity of which becomes very yellow, swelled, and resembles the bud of a flower just ready to blow. The front of the head is also of the same colour. The bees attacked by this malady become languid and weak: their vivacity may be restored by giving them Spanish wine, which is placed in a saucer under the hive that they may the more easily reach it. This remedy strengthens and cures them.

A fourth malady by which they are visited, which is contagious, and a kind of plague amongst them, is what M. Schirach, and after him the French naturalists, call *faux-couvain* (false brood). This name is given to the larvæ and nymphs which are dead and rotten in their cells. This accident occurs when the mother has deposited her eggs in the cells so badly, that the larva cannot break its envelope to get out, or when the cold has been sufficiently great to affect the larvæ, or when they have received unwholesome food. The only remedy is to remove the combs which are infected, to clean the hive thoroughly, perfume it by burning under it some aromatic plants, to suffer the young bees to fast for some days to give them time to void their excrements, and finally to give them some Spanish wine to strengthen them.

We are not precisely acquainted with the term of life allotted to bees. Virgil and Pliny assure us that they may live seven years. Some extend this term as far as ten years. But from the experiments of Reaumur, we may believe that it is far from being so long. This indefatigable observer, marked in the month of April five hundred bees with red varnish, and in the month of November, there was not a single one alive. It may be well believed that these insects share the fate of all others, which die soon after they have fulfilled the functions for which they were intended by nature. The female lives longer than the workers. The males are condemned to perish sooner than the latter: not a single one remains during winter in the hives; the workers will not suffer them. These bees, which are so attentive, and lavish their cares indiscriminately on all the larvæ, make in the last two months of summer a horrible carnage of the males. This continues for two or three days, until they have killed all that exist, or may be produced from the existing brood. They carry off from the cells the larvæ of which they had taken so much care, and tear away the nymphs which are about to be metamorphosed. But, according to M. Huber, the males are spared in those hives where there is no female, and in those where there are only fertile workers, and in all where the massacre does take place, it never happens until after the swarming season.

As to the hives, with proper attention to the brood, they may be preserved for a long time. Some have been known to last from twenty-five to twenty-eight years.

All the hives which are so well peopled during the fine season are infinitely less so at the end of autumn. Many bees perish every year, some naturally, others by a violent death. They have many enemies, some of which insinuate themselves into the hives, others catch them on the wing. Mice, rats, field mice, &c. sometimes introduce themselves, and chiefly in winter, into a hive, and destroy a great number of bees, eating only the head and corslet. Spiders spread their webs in the neighbourhood of their habitation, and catch numbers. Toads place themselves near the entrance, and swallow many. Several birds, such as sparrows, swallows, the kingfisher, hens, &c. are partial to these insects. Foxes sometimes turn the hives over in the winter to get at the honey. Ants penetrate into the habitation of the bees, being very fond of honey, and even occasionally attack the eggs. The odour which exhales from some species is very disagreeable to the bees. The *death's head sphinx*, when it enters the hive, occasions great confusion there, and causes many bees to perish.

Among their most dangerous enemies we must not forget to enumerate wasps and hornets, which will make their nests even in the very hives. The most destructive enemy of all is that known under the name of the wax-moth, *galleria cereana*, Fab. This lepidopterous insect is not afraid of depositing its eggs in the best populated hives. From these eggs spring caterpillars, which make a silken gallery in the combs, where they remain secure, and devour the wax in tranquillity. They pierce the cells in different directions, and if their number is considerable, they soon force the bees to abandon the hive.

These caterpillars, for the transformation into the nymph state, envelop themselves in a cocoon, which serves them as a defence. Neither the vigilance of the bees, nor the arms with which nature has provided them, can protect them from these fatal parasites. It is the business of man to destroy them, by visiting the upper part of the hives where they remain by preference, and changing the hives at least every four or five years. These moths especially attack the old wax. A mite (*acarus gynopterorum*) is sometimes found upon the bees, but this insect is not dangerous, or at all events not particularly injurious to the population of the hive.

The best method, in general, of protecting the bees against these dangers is to raise the hives to a tolerable distance from the earth, to isolate them as much as possible, and to watch, by assiduous visits, the enemies which threaten those interesting insects.

"And to what occupation," says M. Latreille, with the enthusiasm of a philosopher, "more innocent and more agreeable can any person dedicate his leisure moments? Here the philosopher may find ample room for meditation, the physician discover secrets important in his art, the geometrician find problems to resolve, the friend of nature and the poet delightful associations; the man of commerce an object of luxury, the agriculturist a powerful resource against the capriciousness of the soil, and the pernicious influence of cold and storms. The latter, confining himself to what is merely necessary, will have but few expences to incur. Some bundles of straw and osiers, a few stakes, useless in the forest, constitute all that is necessary for his operations. During the long evenings of winter, during the rigorous days when the cold or rain shall confine him to his dwelling, he may employ those moments, lost by so many others, in the construction of his hives. The *happiness* of labour, the expectation of the future, of riches to be gained without danger, and enjoyed without remorse, all concur to inspire his heart with gaiety. He is not, however, entirely absorbed by one single care: his hives must be duly visited, the snow which may cover them be removed, and every aperture carefully closed which might admit the rain. Has spring returned with all her genial influence, and is the cold no longer formidable; already the working bees begin to busy themselves with the construction and regulation of their storehouses and habitations. If the meadows are adorned with but few flowers the cultivator must sow his seeds in abundance, in the neighbourhood of his hives, to compensate his bees for long and fruitless excursions, and time bestowed in vain. The recompence of his labours is at hand. A new nation is about to found another colony, and to people the habitation which his winter's industry has prepared for it. What multitudes are yet to spring from this scion of his ancient stock; what advantages must redound to him from their production? Reflections like these will sweeten all his present toils and cares, while in anticipation he enjoys the good which is at least as certain in reversion as any other in this sublunary scene of perpetual mutation."

We could not be expected, even if our limits permitted, to enter into any lengthened details concerning the culture of bees, a subject more strictly appertaining to rural economy than to natural history. We shall merely observe that the cares which they require may be referred to three different periods of the year. The first extends from the moment in which they awake from their lethargy to the destruction of the males, the second from the end of July to the end of October, and the third from this last time to that in which they shake off their winter's slumbers. Each of these periods, which are no further fixed than by the temperature of the atmosphere, require different cares. In spring the winter covering must be removed from the hives. The bees, which are too weak, whose magazines are exhausted, and cannot be replenished for the want of flowers, must be fed. Syrup, composed usually of good wine, honey, and sugar, nourishes them, purges them, and cures them of the dysentery. But an abuse of this resource would be followed by some inconveniences.

When the swarms are once commenced the bees must be left in peace and tranquillity; when they are formed, their flight must be watched and prevented. Cases of this must be prolonged to the end of July, and it must be most particularly remembered that these insects be never suffered to want water. The use of this liquid is facilitated to the bees by some plants growing on the edge of the fountain where they quench their thirst, or some blades of straw placed upon the dish or trough in which this beverage is presented them. The neighbourhood of great lakes is often very pernicious to them. Should they traverse one of these, at the period of a storm, they are liable to be precipitated into the water and drowned. As they often lose their females in the month of August, at which time they cease to lay eggs, a critical era for the queen, they deliver themselves up to pillage. Those to whom this accident occurs unite themselves to the inhabitants of another hive. If the period of their gathering be past, their provisions must be augmented. As the lethargic state puts an end to all their wants, we have nothing to do in the third period but to moderate the influence of cold upon these insects.

The numerous subgenera of the apiariæ will not detain us much longer. To notice them all, or even the greater part, would be to extend this portion of our work far beyond the limits of a volume. We shall therefore conclude with a very brief review of those more immediately allied to the hive-bee, to which we have thought proper to accord the foremost rank, without attending to the order in the text.

The HUMBLE BEES (BOMBUS) have sometimes been confounded with the males of our domestic bee (at least in name), but which they are as unlike as possible. They are insects very common and very well known, especially by children, who often deprive them of life for the sake of the honey which they contain. They live like our bees, in societies, but much less numerous, and which end at the close of autumn to recommence in spring. They make a loud humming noise, from which originate their Latin and French names (*Bombus* and *Bourdon*), and they are still termed in many parts of Britain *bumble-bees*, a name much more appropriate than their other English denomination.

The humble bees live in subterraneous habitations, united in a society of fifty or sixty individuals, and sometimes as many as two or three hundred, and which is dissolved at the approach of winter. It is composed of males distinguished by the smallness of their size, and relative proportions, of females, which are larger than the other individuals, and of mules or workers of an intermediate magnitude. Reaumur distinguishes two varieties of these workers: the first stronger, and of a middle size; the second smaller, and to all appearance more lively and active. The younger M. Huber has verified this fact. According to him, many of the workers born in spring couple in the month of June with the males that proceed from their common mother, lay soon after, but only bring forth individuals of this last sex. These again fecundate the ordinary or later females, which do not appear until the after season, and which in the spring of the following year are to lay the foundations of a new colony. All the other individuals, without excepting even the small females, perish.

Such of the ordinary females as have escaped the rigours of winter take advantage of the first fine days to make their nest. One species (*Bombus lapidarius*) establishes itself at the surface of the earth, under stones, in a manner which we shall presently explain; but all the others place their nests in the ground, and often at the depth of one or two feet. Meadows, dry plains, and small hills are the places which they select for this purpose. These subterraneous cavities, of a very considerable extent, and more wide than high, are in the form of a dome. Their vault is constructed with earth and moss, which they bring in blade by blade, entering backwards. A sort of cap, of a rude and inferior kind of wax, invests the interior walls. Sometimes a simple aperture, formed at the lower part of the nest, serves as a passage; sometimes a tortuous path, covered with moss, and one or two feet in length, conducts to the habitation. The bottom of its interior is carpeted with a bed of leaves, on which the brood is to repose. The female first places there masses of brown wax irregularly, and nipple-formed, which Reaumur calls *pâtée*, and which he compares, in consequence of their figures and colours, to truffles. Their interior vacancies are intended to enclose the eggs and the larvæ which proceed from them. These larvæ live in society until the time when they are about to change into nymphs. They then separate, and spin cocoons of silk, which are ovoid, and fixed vertically one against the other. The nymph is there always in an inverted situation, or with the head downwards, like those of the females of the domestic bee in their cocoons. Accordingly these cocoons are always pierced at their lower part, when the perfect insect has issued from them. Reaumur says that the larvæ live on the wax which forms their lodging; but in the opinion of

M. Huber, it simply guards them from cold and humidity, and the nutriment of these larvæ consists of a tolerably large provision of pollen, moistened with a little honey, of which the workers take care to furnish a fresh supply when it is exhausted. For this purpose they pierce the lid of their cells, and then shut them again. They even enlarge them, by adding a new piece, when these larvæ, from their growing size, are confined within too narrow limits. There are found, moreover, in these nests three or four small bodies, composed of brown wax, or the same material as the *pâtée*, in the form of goblets, or little pots, almost cylindrical, always open, and more or less filled with honey. These form an agreeable refreshment to the mowers, when they light upon any of these nests. The places which are occupied by these reservoirs of honey are not always the same. It has been said that the workers make use of the empty cocoons for the same purpose; but this appears doubtful, as these cocoons are of a silky material, and always pierced underneath.

The larvæ issue from the eggs, four or five days after the deposition, and finish their metamorphosis in the months of May and June. The workers remove the wax which embarrasses their cocoon to facilitate their issue. It was believed that they produced none but workers, but, as we have already seen, males also issue from them. The workers assist the female in her labours. The number of cocoons which serve as a habitation to the larvæ and nymphs increases, and these cocoons form irregular combs, rising by stories, and on the edges of which the brown wax, called *pâtée* by Reaumur, is especially distinguishable. According to M. Huber, the workers are exceedingly fond of the eggs which the female lays, and even, sometimes in her absence, partly open the cells in which they are enclosed, to suck the milky matter which they contain; a very extraordinary fact, since it seems to belie the well known attachment of the workers for the germ of the race of which they are the natural nurses and guardians. The wax which these insects produce is of the same origin as that of the domestic bee, or but an elaborated honey, which also transudes at intervals, between the rings of the abdomen. Several females live in good intelligence under the same roof, and do not testify any mutual aversion. They couple outside their dwelling, either in the air or on plants.

The CARPENTER-BEES (XYLOCOPA, Latr.) so much resemble the humble-bees, that Fabricius has placed some of their species with the latter: these are the *abeille-perce-bois* of Reaumur. The species to which he has particularly given this name is the *apis violacea* of Linnæus, an insect which we only begin to meet with in temperate Europe. It is observed to appear in the first fine days of spring; it flies humming round walls exposed to the sun, round windows which have old outside shutters, window-frames, projecting beams, &c. It thus seeks out a favourable place for the deposition of its eggs, which it always places exclusively in old wood. One individual will prefer a vine-prop, another the pieces of wood which serve for a support to the hedge-rows; some will choose an outside shutter, some an old bench or beam. In all cases it is necessary that the wood be dry, and that it has commenced to rot, the insect having then less difficulty in excavating it. It requires a very persevering strength and courage to enable this little creature to complete its enterprize. The hole which it opens is at first directed obliquely towards its axis. At some lines in depth its direction changes, and it becomes nearly parallel to the axis. The wood is pierced like a flute obliquely, yet sometimes from one end to the other. The cavity should be tolerably spacious, to enable the insect to turn about in it. Reaumur says that he could put his fore-finger into it. These holes are often more than a dozen or fourteen inches long, and are three or four in number, if the thickness of the wood will permit it. The mandibles and teeth are the instruments employed by this insect in the work of excavation.

It is in these tunnels that the carpenter-bee lodges her eggs. Each tunnel is composed of a range of several chambers or compartments. It is generally divided into about twelve of these compartments, which have no communication with each other. Each lodge encloses an egg, and a quantity of paste, necessary for the growth of the larvæ which is to be born there. This paste has the consistence of soft earth, and is tolerably similar to that with which the humble-bees nourish their young. It is doubtless the dust of the stamina of flowers, mixed with a little honey. The larva at first is lodged within a narrow space, as the paste almost entirely occupies the cell. But the consumption which it makes by little and little produces a vacancy which sets it at its ease. In each range of cells the larvæ which are in the lower are older than those in the upper, and are transformed and issue forth first.

We lastly come to the *mason-bees,* which form three subgenera in Latreille—OSMIA, MEGACHILE, and ANTHOPHORA. The history of the first will comprize almost all that we need say about them, and also about what have been called by English naturalists the *upholsterer-bees.*

These insects, like those of the preceding germs, are solitary. Reaumur, Degeer, and Spinola have collected some traits of their history, which have been illustrated or confirmed by M. Latreille. The first, after having made us acquainted with the manners of that species of mason-bee called *muraria,* speaks of some other species of the same family, which are likewise masons. The mortar which the latter make is not so good as that of the former; it is nothing but a fine earth, the grains of which are cemented together by means of a fluid. It would be useless to give to this mortar more solidity, because these insects know how to construct their cells in places where they are not exposed to be injured by the rain. They seek out stones which have cavities, to serve as an habitation to a single one of their larvæ. They cover with earth the walls of this cavity, fill it in part, and leave no vacancy except the space necessary to contain the provisions destined for the nutriment and growth of the larvæ which is to issue from the egg deposited close by. To shorten the duration of their labours, these bees choose such cavities as are not too large, and the apertures of which have scarcely more diameter than is necessary to allow them to pass. If the apertures be too wide they contract them, by attaching some earth to their interior edge, and leaving in the middle a very circular hole, and proportioned to the magnitude of their bodies. The paste which Reaumur drew from some of these holes had the consistence of *bouillie*. The honey which served to moisten the pollen, of which this paste was composed, had a very agreeable flavour. Having provided for the wants of its posterity, the insect closes the entrance of the cell with earth prepared for that purpose. Another species mentioned by Reaumur (*bicornis*), makes in old wood works similar to those which the preceding execute in stone. By no means fearing the presence of man, and being, as it were, quite tame, it nidificates in doors and the frames of windows, when it finds cavities in such places, proper to become the cradle of its young. The individual observed by this great naturalist had taken advantage of a hole in one of the doors of his kitchen. The movements of the persons who were going and coming continually did not alarm it: it never interrupted its labours, even when any one stopped to inspect them. Whether the door was open or shut, it did not the less continue its work, entering into its hole and coming out many times in the course of the day. It invested with earth the walls of this cavity; and after having laid its eggs there, it sealed up the two ends of the habitation of its young with the same materials. Reaumur having waited three weeks or more before he deranged the interior of the nest, when he came to examine it, found it empty.

The females of these two species, and those of some others of this group, present a peculiarity worthy of attention. There are two projections on the head, in front, in the form of horns, which are, in fact, but elongations of the sides of this part; another species (*tricornis*) has even an additional one. The interval which separates them is more or less sunk, more or less extended, and usually smooth and shining; it is a kind of trough, for we often observe there parcels of earth and mortar. It is possible that these horns may answer the purpose of enlarging the cavity in which these insects nidificate, and rounding its walls and centre. The mandibles may perform the office of trowels.

Degeer has given us the history of the *Osmia cœrulescens*, Latr., *Apis venea*, Lin. He had remarked, for many years in succession, in the inequalities of a wall, built with large stones of granite, some oval plates, embossed, and of the colour of dried potters' clay. On examining them closely, he perceived that they were formed of earth and grains of sand intermingled, and forming a tolerably solid mass; but that they were easily detached with the point of a knife, and fell immediately into powder if touched a little roughly. These masses are the nests of this bee. Having opened one in the month of May, Degeer saw two or three cells in its interior, each filled with an oval cocoon of silk, of a dirty white, and which contained one of those insects full of life, and ready to break forth. These nests had been constructed the preceding year. The same observer found another, made of the same materials, in a thick bed of argilla, mingled with lime, with which, in his country, they are accustomed to line the walls of wooden houses. In this nest there was a larva, which passed the winter in that state, and was not transformed into a nymph until the beginning of June in the following year.

A very curious species is that which Reaumur names *Abeille tapissière*, and which we call the *upholsterer-bee.* It at first excavates in the earth a perpendicular hole of some inches in depth, cylindrical at its entrance, more wide at bottom, and resembling a sort of bottle. Some portions of the petals of the flowers of the wild poppy, which the animal cuts in a semi-oval form, and transports with its mandibles, are employed to line the walls of this cavity, and prevent their falling in. To make these pieces enter, it folds them in two, then opens and extends them as smoothly as possible. This tapestry often extends some lines beyond the aperture of the hole, and forms all round it a flame-coloured riband, which attracts the attentive eye of the observer. As this invests all the interior of the burrow, it naturally takes its form, and the nest may be drawn out of the hole without losing its shape. Having placed at the bottom a paste, composed of pollen and honey, the insect lays an egg there, and closes the nest by folding down the superior extremity of the tapestry. If the hole is tolerably deep, it raises a second nest above the first. A little earth closes and conceals the entrance to the habitation. If the flowers of the poppy be rare in the neighbourhood, this insect will use others, as the flowers of rape, which being of a very different colour from the others, it is evident that the animal, in this case, acts more by virtue of its instinct, than from the impression made on its senses by external objects. It is also remarkable that the nymphs of this insect do not disclose, except during the season in which these flowers are in bloom.

Another of these insects makes its habitation in the spherical galls, which have been abandoned by the *cinips.* As

J.O. Westwood del.t

1. *Alphaneura rufescens*. _ Westwood.

2. *Crocisa nitidula*. _ Jurine.

3. *Polistes apicalis*. _ G.R. Gray.

London. Published by Whittaker & C.o Ave Maria Lane, June 1832.

these are not sufficiently spacious, it increases them considerably, and polishes the interior.

The larvæ and nymphs of all these bees are exposed, like those of many other insects, to the attacks of the ichneumon and other parasitical hymenoptera. The OSMIA *fulviventris*, when in its perfect state, is sometimes covered with a prodigious quantity of *acari*.

In this third family of stinging hymenoptera, the diploptera, we are enabled to add figures of several new species belonging to its different genera.

In the singular genus *Synagris*, of M. Fabricius, we have re-figured the *cornuta* of the above-named author. It is of a dark orange colour, with the vertex, the centre of the thorax, and the abdomen, of a dull black, the wings blackish brown, rather shiny. This insect is from Fernando Po, and the specimen is in Mr. Hope's collection.

In *Polistes* we have figured a species we call *apicalis*. It is brownish black; the wings black at the base; the apex much lighter. This insect is from Brazil, and is in the collection of the British Museum.

We insert a beautiful species of *Crocisa*, which is named *nitidula*, by Fabricius. This insect is from India; it has the antennæ and head black, with the front of the latter silvery blue; thorax shining black, with seven spots of silvery blue; scutellum black; the abdomen black, with the base of the first segment and of the sides of each segment marked with silvery blue.

Mr. Westwood has formed a new genus in this family, called *Aphaneura*, which is allied to *Ancyloscelis*, but is distinguished by the flat and dilated form of the posterior tibia in the female: the first joint of the posterior tarsi being large and flat. The species figured he names *rufescens*. It is reddish brown; with the head and thorax dirty yellow, with the centre of the latter blackish; the wings yellow, with the nervures reddish.

Lastly, we have inserted the type of MM. Lepeletier and Serville's genus *Aglæ:* the specific name is *cœrulea.* It is of a very shining violet blue, covered with black hairs; the antennæ black; the sides of the abdomen and wings brown, with a very slight reflection of gold; the labrum and scutellum very smooth; the trunk is testaceous. This genus is from Cayenne.

J. O. Westwood del[t]

1. *Synagris cornuta* _ Fab.

2. *Aglae cærulea* _ Lep. et Serv.

London Published by Whittaker & C° Ave Maria Lane. June 1832

THE TENTH ORDER OF INSECTS *

LEPIDOPTERA. *Glossata, Fab.*

TERMINATES the series of insects with four wings, and displays, exclusively, their characters.

The wings are covered on both sides with small coloured scales, like farinaceous dust, which comes off by touch; a trunk, called a tongue, rolled up spirally between two palpi, beset with scales, or hair, forms the most important part of their mouth, and is the instrument with which these insects abstract the honey of flowers, which is their only food. We have seen, in our general remarks on the class of insects, that this trunk was composed of two tubular threads, representing the jaws, and each carrying, near its exterior base, a very small feeler, called the upper feeler, having the form of a tubercle. The visible, or lower palpi, act as a sort of sheath for the trunk, and represent the labial palpi of the grinding insects; these are cylindrical, or conical, commonly elevated, composed of three articulations, and inserted on a fixed lip, which forms the partition of the portion of the mouth beneath the trunk. Two small pieces, scarcely distinct, corneous, and more or less ciliated, situated one on each side, on the anterior and upper edge of the fore part of

* As this order of insects is much more generally studied than the others, we have translated the text of Cuvier at full length. Want of space alone has compelled us unwillingly to abbreviate the text in the preceding orders. —ED.

the head, near the eyes, appear to be vestiges of mandibles. Indeed there may be found, but quite in miniature, the labrum, or upper lip.

The antennæ are variable, and always composed of a great number of articulations. In most of the species, there may be discovered, but concealed among the scales, two smooth, or simple eyes *. The three segments, of which the trunk of hexapod insects is formed, are united in one body: the first is very short; the other two are confounded together; the scutellum is triangular, with the point toward the head. The wings are simply veined, variable in figure, size, and position. In many insects, the under-wings have some longitudinal folds toward the internal edge. At the base of the upper wings, is a piece, in form of an epaulet, elongated behind, and corresponding to what is called the *tegula* in the hymenoptera; but being more developed here, I shall call it the pterygoid. The abdomen is composed of six or seven rings, is attached to the thorax by a small part only of its diameter, and has neither sting or borer analogous to that of the hymenoptera. In many of the females, nevertheless, as the *cossus*, the last rings can be thrust out, and withdrawn again, to form an ovipositor in the shape of a pointed and retractile tail. The tarsi have always five articulations. There are never more than two sorts of individuals, male and female. The abdomen of the former is terminated by a sort of flat pincers, which inclose the penis. The females lay their eggs often extremely numerous, generally on the vegetable substances on which their larvæ are to feed, and they cease to exist very shortly afterwards.

The larvæ of the lepidoptera are known under the name of caterpillars. They have six feet, scaly, or crooked, corres-

* From the observations of M. Dalman, these do not exist in the diurnal lepidoptera.

Pl. 59

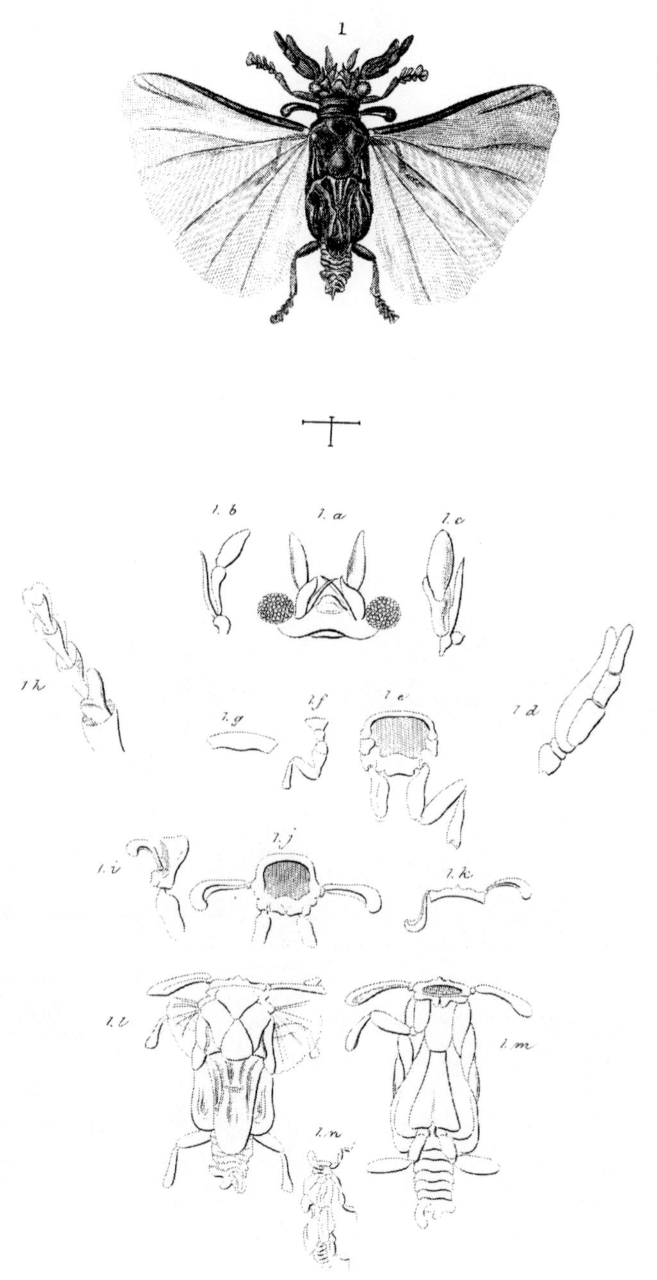

Stylops *Childreni.*

I. O. Westwood del.

Published by Whittaker & C.o Ave Maria Lane London. 1837.

ponding with those of the perfect insect, and, moreover, from four to ten membranous feet, the last pair of which are situated at the posterior extremity of the body, near the anus; those which have altogether but ten or twelve feet have been called, in French, by reason of their mode of locomotion, *géomètres*, or *arpenteures*, geometricians or surveyors. These cling to the plane of position by means of scaly paws, then elevating the intermediate articulations of the body into the form of a ring, or buckle, they approximate the posterior legs to the anterior, disengaging the latter, they hold on by the former, and carry the body forward to recommence the same manœuvre. Many of these caterpillars are fixed, when in a state of repose, to the branches by the hind legs only, and in that state bear resemblance, by the direction, form, and colour of their bodies, to the branch; they keep themselves a long time in this situation without showing the least sign of life. A position so constrained, presupposes a prodigious muscular force; and Lyonet has accordingly reckoned in the *cossus ligniperda* 4041 muscles. Some caterpillars, with fourteen or sixteen feet, but some of whose intermediate membranes are shorter than others, have been named semi, or pseudo-geometricians. The membranous feet are frequently terminated by a coronet, more or less complete, of little hooks.

The body of these larvæ is, in general, elongated, nearly cylindrical, soft, variously coloured, sometimes naked or bare, sometimes beset with hairs, tubercles, or spines, and composed, exclusive of the head, of twelve rings, with nine stigmata on each side. The head is covered with a corneous or scaly dermis, and has on each side six small shining grains, which appear to be little simple eyes; it has, moreover, two very short and conical antennæ, a mouth composed of strong mandibles, of two jaws, a labrum, and four small palpi. The silky matter they make use of is elaborated in two interior, long, and tortuous vessels, the upper extremity of which gradually diminished, terminates at the labrum: a tubular and conical mamma, situated at the end of the labrum, gives issue to the threads of the silk.

The majority of caterpillars feed on the leaves of vegetables; others gnaw the flowers, roots, buds, and seeds; the ligneous, or hardest parts of trees, supply food for some of them, this they soften by means of a liquid which they disgorge. Some species gnaw our cloth, silk, stuffs, and furs, and are very pernicious domestic enemies to us; leather, grease, lard, and wax, are not spared by them. Many subsist exclusively on one sort of matter; but there are some less delicate, which attack many sorts of plants or substances *.

Some unite in societies, and frequently under a silken web, which they spin in common, and which affords them a shelter during the bad season; many make themselves cases, fixed or moveable; some are known which lodge in the *parenchyma* of leaves, in which they form galleries; most of them delight in day-light, but others quit their retreat in the night only. The severity of winter, so opposed to almost all insects, does not reach some phalena, which appear only at that season.

The caterpillars commonly change the skin four times before they pass into the state of nymphæ, or chrysalids: most of them then spin a cocoon in which they enclose themselves. A liquid, generally reddish, being a kind of meconium, which the lepidoptera eject from the anus at the moment of their change, softens one end of the cocoon, and facilitates their exit; one end of the cocoon is, moreover, weaker than the other, or affords, by the disposition of the threads, a more easy issue. Other caterpillars merely tie with silk, leaves, particles of earth, or particles of substances where they have lived, and thus form a large cocoon. The

* One of the most manifest proofs of Providence, is the perfect coincidence in the adaptation of the caterpillar to the vegetable substance on which it feeds.

chrysalids of the diurnal lepidoptera, ornamented with golden spots, which have given rise to the general denomination of chrysalids, are uncovered, and fixed by the posterior extremity of the body. The nymphs of lepidoptera present a particular character which we have discussed in our general observations on the class. They are swathed, or in form of a mummy *. Those of many, especially of the diurnal lepidoptera, open in a few days; indeed these insects often breed two generations in a year. In others, the caterpillar, or its chrysalid, pass through the winter, and the insect does not undergo its last change till the spring or summer of the following year. In general, the eggs which are laid in the autumn do not hatch till the following spring. The lepidoptera issue from their chrysalid form in the common manner, or by a cleft which is made on the back of the corslet.

The intestines of caterpillars consist of a thick canal without inflexions, whose anterior part is sometimes a little separated from the rest like a stomach, and in which the posterior portion forms a wrinkled cloaca. The biliary vessels, four in number, and very long, are inserted very far behind. In the perfect insect there may be seen a first lateral stomach, or gizzard, a second turgid stomach, and a thin, largish intestine, with a cœcum, near the cloaca †.

The larvæ of the ichneumonides and of the chalcides rid us of a great part of these destructive insects.

We shall divide this order into three families, which correspond with the three genera of which it is composed in the system of Linnæus.

The first family, that of

* The sheaths of the feet and antennæ are fixed: a character peculiar to this sort of metamorphosis.

† See, on the anatomy of the caterpillar, Lyonet's admirable work; and on the development of the organs of the chrysalis and caterpillar, that of Herold.

The Diurnal Lepidoptera. Diurna,

Is the only one (a few nocturnal excepted) in which the exterior edge of the lower wings does not afford a rough scaly silk or a kind of fringe, to retain the two upper wings; the former, and, in general, the latter also, are elevated perpendicularly when at rest; the antennæ are sometimes terminated by a swelling in form of a little club, sometimes nearly of the same thickness, or even thinner, and in a bent point at their extremity. This family includes the genus of

Butterflies, Papilio, *of Linnæus*,

Whose caterpillars have always sixteen feet. Their chrysalis is almost always naked, attached by the tail, and in general angular. The perfect insect is always provided with a trunk; they fly only in the day time; the colours of the under side of their wings are not inferior to those of the upper side.

We shall first divide them into two sections.

Those of the first have only a pair of spurs, or spines, to the legs of their posterior extremity; their four wings are all elevated when at rest; their antennæ are sometimes bent at the extremity like a bud or little club, sometimes rounded at the top, and sometimes nearly filiformed.

This section includes the genera *papilio* and *hesperia rurales* of the entymological system of Fabricius.

This division, very numerous in species, may be divided in the following manner.

1st. Those in which the third articulation of the lower palpi is sometimes entirely naked, at others very distinct, but as much covered with scales as the preceding, and in which the hooks of the tarsi are very apparent.

Their caterpillars are elongated, almost cylindrical. Their chrysalis are almost always angular, sometimes united, but inclosed in a large cocoon.

Some among them, the *hexapodes*, have all the feet fitted for walking, and nearly the same in both sexes *. Their chrysalis, besides the common posterior attachment, is fixed by a silken thread, forming a buckle or half ring round the body. That of some of them is enclosed in a large cocoon. The central partition cellule of the under wings is closed underneath †.

The latter have the internal edge of the under wings concave or folded. Such are

The BUTTERFLIES, properly so called. P. EQUITES, *Lin.*,

Which have the lower palpi extremely short, scarcely reaching the chaperon, by their upper extremity, with the third articulation indistinct.

Their caterpillars, when alarmed or distressed, thrust out from the upper part of the neck, a soft forked horn, which in general emits a penetrating and unpleasant odour. Their skin is naked. The chrysalis is attached by a silk thread, and uncovered.

* The butterflies, properly so called, or those of the division *Equites*, of Linnæus, are allied, on the one hand, to the *Danai festivi*, and on the other to the *Parnassii*. From the latter we pass to *Thais*, and finally to the *Parnassii*. The preceding Danai are allied to the *Helicorni*. Hence it follows that we ought to begin the series of the diurnal lepidoptera, by the tetrapods, as the *Satyri*, the *Pavarii*, the *Morphi*, and the *Nymphales*, so as to arrive through *Argynnis* and *Cethosia*, to the *Helicani*. The Diurna are divisible into two principal sections, those whose chrysalis is suspended vertically, and simply attached by the extremity of the tail; and those in which it is fixed not only by the extremity, but also by a silken thread traversing the body like a buckle or half ring. The first are always tetrapods. We should commence with those whose caterpillars are naked or nearly so, and generally bifid at the posterior extremity, and those whose caterpillars have spines should follow.

† I have used this character in my genera of Crustacea and Insects. M. Dalman and Godart have generalized the relative application of it in this family.

The species of this subgenus are remarkable for their shape and the variety of their colours. They are found more particularly in the intertropical countries of both hemispheres. Those which have red spots on the breast, form the division of *Trojan knights*, of Linnæus. He has distinguished by the name of Greeks, those which have not these spots. Many have the under wings elongated, and such is that of France, which is called by Linnæus, *Papilio machaon*. Wings yellow, with black spots and stripes; the under wings elongated into a tail, and having, near the posterior edge, some blue spots, one of which is like an eye, with red at the internal angle.

The caterpillar is green, with black rings, dotted with red, and lives on the leaves of carrot, fennel, &c.

P. Podalirius, God., and *P. Alexanor*, are also found in France *.

Zelima, *Fab.* These differ from the last only by the club of the antennæ, which is shorter and more rounded. I know two species of them, one of Senegal, and the other of Guinea, and which form part of the fine collection of the Count Dejean.

Parnassius, *Lat.*, Doritis, *Fab.*, whose lower palpi are elevated sensibly above the chaperon, go to a point, and have three very distinct articulations. The bud of their antennæ is short, almost ovoid and straight. The females have a species of corneous pouch, in the shape of a small boat, at the posterior extremity of the abdomen. Their caterpillars have also under the neck a retractile tentaculum, like that of the butterflies proper; but they make with leaves, tied by silken threads, a cocoon, in which they change to a chrysalis.

* For the lepidoptera of Europe, consult the excellent work of Ochsenheimer, continued by Treitschke.

These species are found only in the alpine or subalpine parts of Europe and the north of Asia. Such is

Pap. Apollo, Lin., God. II., B. 11. 1. White, spotted with black; four white eye-like spots, edged with a red circle and a black circle on the under wings. Its caterpillar lives on *Sedum, telephicum, saxifrage*, &c. It is of a velvety black, with a row of red spots on each side and one on the back. The chrysalis is round, of a blackish green, slightly powdered with white or blueish *.

THAIS, *Fab.* These have the palpi of the parnassii, but the bud of the antennæ is elongated and bent. The abdomen of the female has no corneous pouch.

Their caterpillars have not, it seems, the retractile tentaculum. These species belong to the south of Europe, some of them are found only in the mountains, (*Pap. hysipyle, rumina*, Fab.)

In the former the under wings advance under the abdomen, and form a cornice. Their caterpillars have no tentaculum. Many of them live on cruciferous plants.

These lepidoptera (*P. danai candidi*, Lin.) form two subgenera.

PIERIS, *Schr.*, PONTIA, *Fab.*, with the lower palpi nearly cylindrical, little compressed, with the last articulation nearly as long as the preceding, and with the club of the antennæ ovoid †.

COLIAS, *Fab.* In these the club is conical, elongated, and reversed; the inferior palpi much compressed, with the last articulation much shorter than the preceding. *P. hyale*, Lin. *P. rhammi*, Lin. *P. cleopatra*, Lin. &c.

* See God. ibid.

† Here come the lepidoptera, under the general name of Brassicæ, as *P. brassicæ*, Lin., *P. rapæ*, Lin., *P. napi*, Lin., *P. daplidice*, Lin., *P. sinapis*, Lin., *P. cardumines*, Lin. &c, nearly all vernal species.

The other butterflies of the same division (the tetrapods) have the two anterior feet evidently shorter than the others, folded, by no means ambulatory in either sex, and sometimes only in the male. The chrysalis is merely attached by its posterior extremity, and suspended with the head downward.

Sometimes the anterior feet, although the smallest and folded, differ little from the others. These under wings, whose central cell is always closed beneath, extend but slightly, in general, on the abdomen. The lower palpi are distant from each other, slightly cylindrical, and generally very short. All the subgenera of this subdivision are exotic.

We may distinguish *danais, euploea,* Fab., *P. danai festivi,* Lin., in part by their triangular wings, and their antennæ terminated in the manner of an elongated and bent club *.

Idea, *Fab.*, are known by their nearly oval elongated wings, and their antennæ nearly filiform †. In these two subgenera the lower palpi are scarcely elevated above the chaperon, and their second articulation is scarcely as long again as the first. In the two following subgenera the wings resemble the others, but they are commonly narrower and larger, and the abdomen is also proportionally longer than that of the majority of the preceding; the second articulation is much longer than the first, and its extremity reaches beyond the chaperon.

Heliconius ‡, *Latr., Mechanitis, Fab., P. Pelicani, Lin.,* have the antenne as long again as the head and the thorax, and enlarging gradually toward the extremity. In Acræa ‡, Fab., they are shorter, and terminated abruptly in a club.

Sometimes (*P. nymphalis*, Lin.) the two anterior feet are decidedly folded, whether apparent and very hairy, or very small and hidden. The under wings, whose central cellule is

* Latr. Gener. Crust. and Insect. IV., 201.

† Latr. ibid.

‡ Latr. loc. cit.

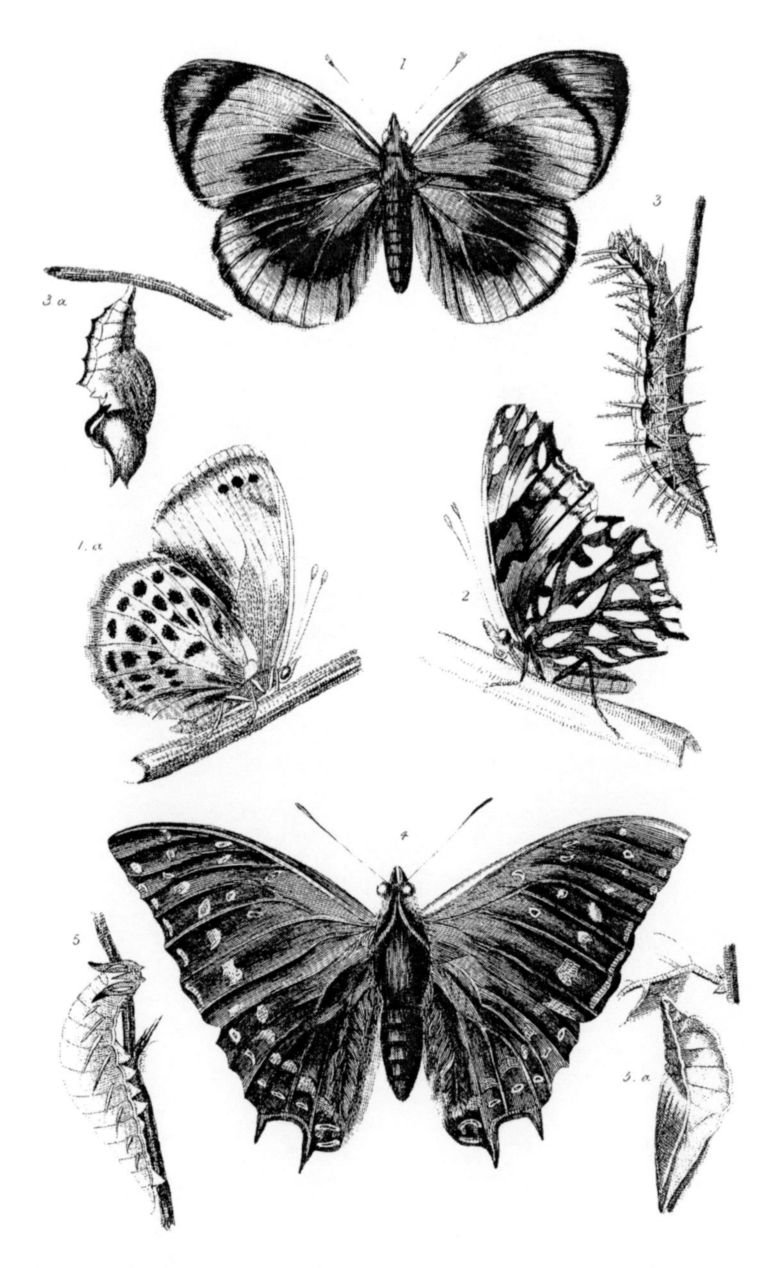

1. Argynnis *callithea*. *God.* 2. Argynnis *moneta*. *Geyer.* 3. *Caterpillar & chrysalis, of* Argyn *paphia*. *God.* 4. Nymphalis *etheta*. *God.* 5. *Caterpillar & chrysalis of* Nymph *ilia*. *God.*

London. Published by Whittaker & C.° Ave Maria Lane. 1831.

open in many, embrace the abdomen very obviously below. The lower palpi are proportionably larger, and often thicker and nearer to each other. Here the central cellule of the under wing is open.

Those whose lower palpi are little compressed, distant from each other, for their whole length, or at least at the extremity, and terminated abruptly by a thin articulation; whose wings have often underneath silvery or yellow spots on a fulvous ground; whose caterpillars are always covered with spines or fleshy and hairy tubercles, compose the subgenera.

CETHOSIA, *Fab.*, and ARGYNNIS, MELITÆA, *Fab.* In the first, many species of which have the wings elevated and elongated, the lower palpi are distant their whole length, the hooks of the tarsi are simple, and the club of the antennæ is oblong. In the second it is short and abrupt; the hooks of the tarsi have a single indentation; the lower palpi are separated only at their extremity. The under wings are often round.

Some (*Argynnis,* Fab.) have pearly spots under their wings. Their caterpillars have spines, two of which on the neck are the largest. Those of the others (*Melitæa,* Fab.) have small hairy tubercles; the wings are checkered; the pearly spots are replaced by yellow, which sometimes takes place in the preceding.

Those which have the lower palpi contiguous their whole length, terminated almost insensibly in a point, and very much compressed, compose five other subgenera.

VANESSA, *Fab.*, differ from the following by their antennæ terminated abruptly by a short knob in an ovoid form. Their caterpillars are covered with numerous spines.

V. morio, (*Pap. antiopa,* Lin.) God. Hist. Nat. des Lepid. de France, t. v. 1., has angular wings of a deep purplish black, with a yellowish or whitish band on the posterior edge, and a row of blue spots above. Its caterpillar is blackish, spiny, with a series of red square divided spots along the back. It feeds on the leaves of the birch, osier, and poplar, and lives in societies. They appear at two separate periods.

Pap. Io. Lin., God., *ibid.* t. v. 2., has angular indented wings, reddish fulvous above, with a large occellated spot on each wing; those of the upper wings reddish in the middle, surrounded by a yellowish circle; those of the under wings blackish, surrounded by a gray circle, and inclosing some blueish spots; under the wings blackish. The caterpillar is black, dotted with white, with simply villose spines; it lives on nettles.

Pap. cardui, God., *ibid.* t. v. 2. Wings indented, red above, varied with black and white; underneath marbled with gray, yellow and brown, with five occellated blueish spots on their edges. The caterpillar lives solitarily on thistles; there are some brownish with yellow stripes, or reddish with transverse yellow bands. It is spiny. This butterfly appears only at the end of the summer.

Pap. atalanta, Lin., God., t. vi. 1. Wings indented and a little angular, black above, crossed by a bright red band, with white spots on the upper wings; beneath marbled with divers colours. The caterpillar is black, spiny, with a range of saffron lines on each side. It lives on nettles, and prefers the berry, and remains hidden at the top between the leaves, which it rolls up and fastens with silk.

The same division includes some other species very common with us, such as *P. polychloros,* Lin., *P. urticæ,* Lin., the *Gramma,* or *Robert the Devil, P. C. album.* The chrysalis of the last represents rudely a human face, or the mask of a satyr. (*God. ibid.*)

Pl. 21

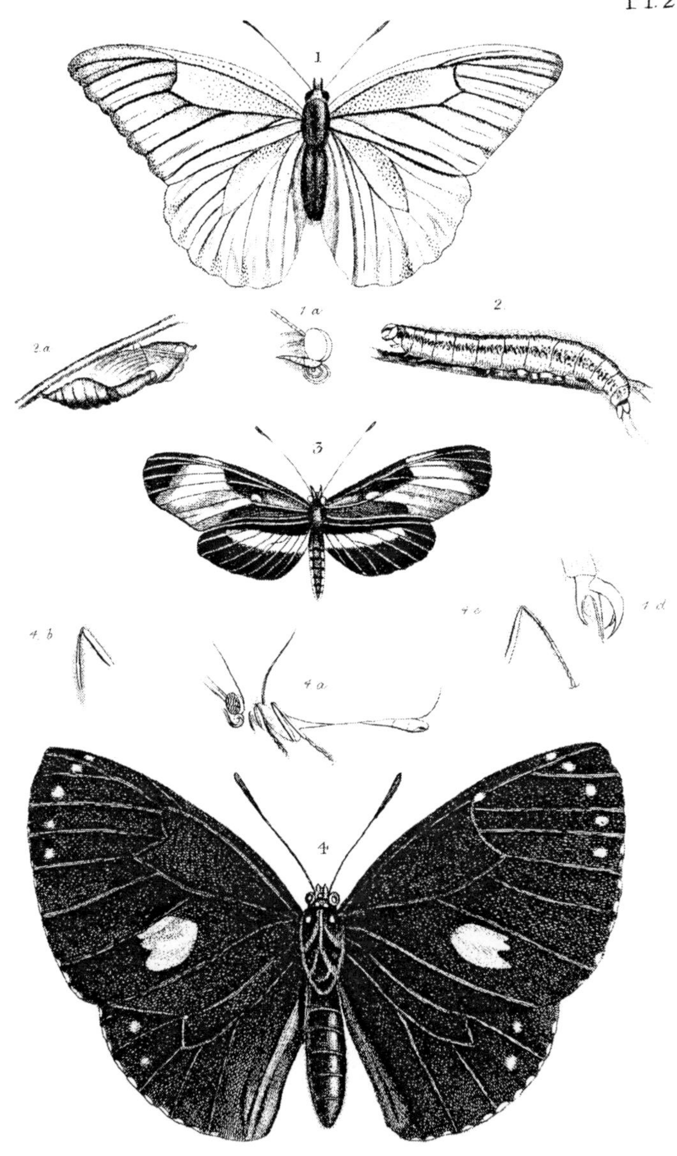

1. Pieris *thiria*. *God*. 2. *Caterpillar & chrysalis of* Pieris *brassicæ*. *L*. 3. Heliconia *Langsdorfii* *God*. 4. Denaida *eunice*. *God*.

London, Published by Whittaker & C°. Ave Maria Lane. 1837.

In the four following subgenera, the antennæ are terminated in an elongated mace, or are nearly filiform. The caterpillars are naked, or have a small number of spines.

Libythea, *Fab.*, in which the males only have the two fore claws very short and broad. Their lower palpi advance sensibly like a beak. The upper wings are very angular (see the work before quoted).

Biblis, *Fab.*, *Melanitis, Ejusd.* In which these palpi are larger than the head, but more obtuse, and a little bent at the extremity, the two fore claws are short and folded in both sexes, and the antennæ have a much smaller terminal mace. The wings are in proportion larger and simply indented. It has been observed that the nervures of the first were much swollen at their origin.

Nymphalis, *Lat.* Similar as to the claws to Biblis; but the lower palpi are shorter. It is only by the elongation of the mace of the antennæ that this subgenus is distinguishable from Vanessa. Nevertheless the caterpillars are different, for besides having but a few spines or some fleshy knobs, they become thin towards the posterior extremity, which is slightly forked.

These butterflies are generally highly ornamented, and have a rapid and elevated flight. Many pretty species of them are found in France, such as those which the amateurs mark by small groups, under the names of *Sylvanis* and *Mars*. The males of the latter have changeable colours. To this subgenus belongs also another pretty species, also indigenous, *P. Jason*, Lin. The form and size of the mace of the antennæ varies a little, as well as the relative proportions of the wings, which has induced some other subgenera, but they have very equivocal characters. The species which approximate the most to Biblis, and which, as *Sylvanus cœnobites*, of Engrammelle, form the genus Neptis, of Fabricius. Among those which differ most from the preceding, both by the antennæ and the under wings, and have tails like certain species of the division of Knights, of Linnæus, we will cite *P. Jason,* mentioned above.

Morpho, *Fabricius,* differ from nymphalis, by having the antennæ more filiform, slightly and gradually thicker toward the end. All the species are peculiar to South America, and very remarkable in shape, in colours, and in the eyed spots of the under side of the wings. Linnæus has united many of them to his *Papiliones milites* *.

Godart has separated, under the generic name of

Pavonia, the species which have the central cellule of the under wings closed, and the inner nerve of the upper wings bent into an S, instead of being straight or slightly bent. One species proper to India, which has the anal angle of the under wings elongated like a tail, *P. phidippus,* is the type of the genus Amathusia, of Fabricius. All the others are from the new continent. The cut of the second articulation of the lower palpi of *Pavonia morpho,* and of the previous subgenera, is rather wide, or the palpi are not strongly compressed, while they are much so in *Satyrus,* a subgenus very analogous to the preceding.

The following have also the discoidal cellule of the under wings closed behind.

Brassolis, *Fab.* In this the antennæ are terminated abruptly in a thick mace, formed like a reversed cone, and the lower palpi are short, not extending beyond the chaperon. The males have, near the internal edge of the lower wings, a longitudinal cleft covered with hair.

Eumenia, *Fab., God.,* whose lower palpi are longer, and the antennæ at a slight distance from their origin gradually enlarge and form a much elongated club †.

* See Godart's Hist. Nat. des Lepid. de France, and the Genus Nymphalis, in the Ency. Methodique.

† Ency. Method. Insect. IX. 826. Godart saw only specimens deprived

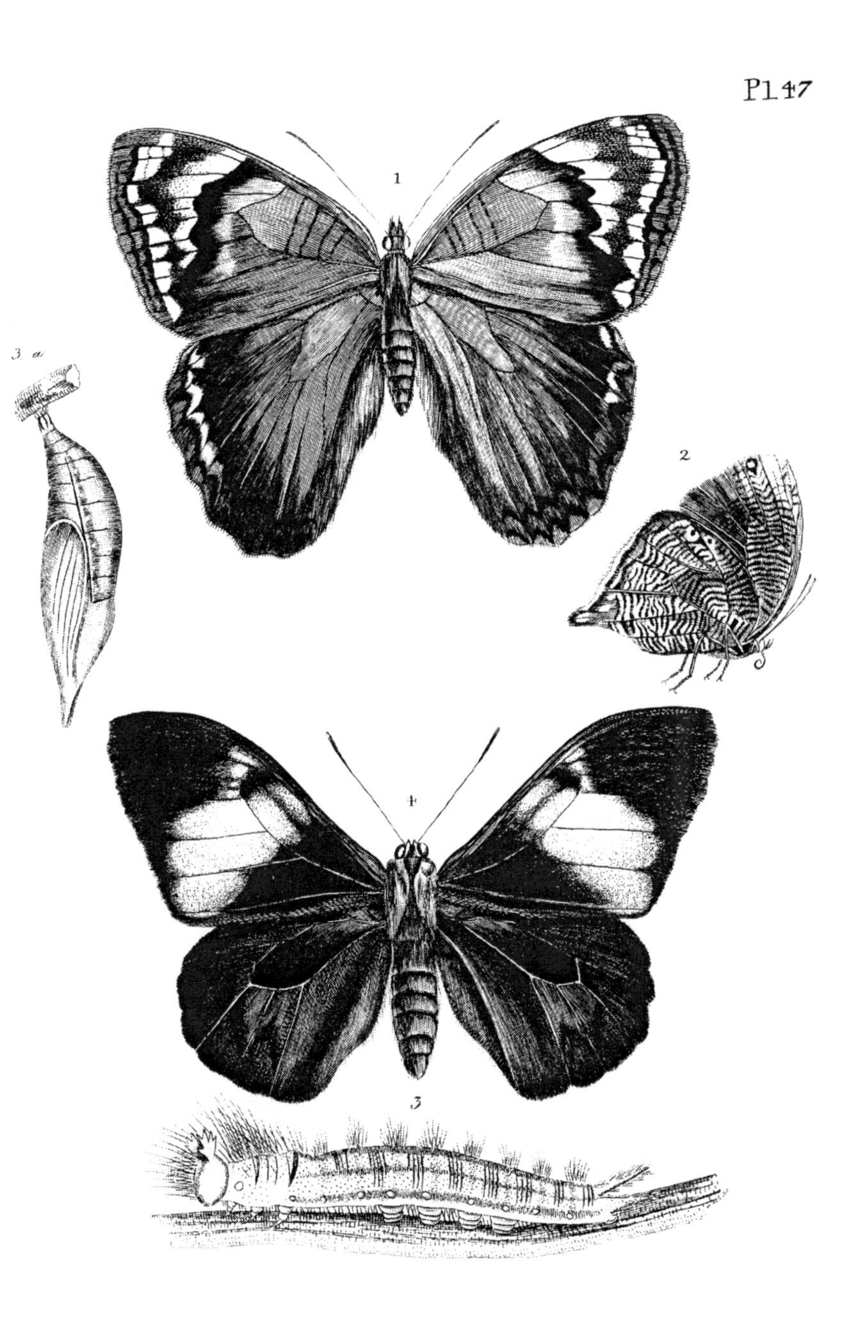

Pavonia *acadina*. God. 2. Morpho *actorion*. God. 3. Morpho *phidippus*. God. 4. Brassolis *astyra*. God.

London. Published by Whittaker & C°. Ave Maria Lane.

1. Satyrus *Balder. Boisd.* 2. *Cat. & Chrys. of* Satyrus *jurtina. God.* 3. Eumenia *toxea. Godard.* 4. Erybia *carolina. Godard.*

London, Published by Whittaker & C°. Ave Maria Lane. 1831.

EURYBIA, *Illig.*, approach BRASSOLIS in the shortness of the lower palpi, but they are in proportion thicker, and the club of the antennæ is formed like a spindle slightly bent.

SATYRUS, *Lat.*, have the lower palpi in the ordinary manner, extending beyond the chaperon, much compressed, with the edge sharp and beset with hairs ; the antennæ terminate in a slight swelling in form of a knob, or in a thin and elongated mace. Godart has observed that the first two or three nervures of the upper wings are very much elongated at their orifice. The caterpillars are naked or nearly shaved close, with the hind extremity of the body contracted into a forked point. The chrysalides are bifid before with two tubercles on the back.

We shall terminate this first section of diurnal lepidoptera, by those whose lower palpi have three distinct articulations, but the last of which is nearly naked, or much less furnished with scales than that which precedes it; and whose hooks of the tarsi are very small, not at all, or very slightly, saillant. The discoidal cellule of the under wings is open behind.

These caterpillars are oval, or formed like the woodlouse. Their chrysalides are short, united, and always attached like those of the butterflies properly so called, by a silken chord, which traverses their body *.

Linnæus included them among his *Papilliones plebei*, the division *ruricolæ*, of Fab., in an harmonious section of his genus *Hesperies*. These form the *Argus* of M. Lamarck.

of their antennæ. M. Poé has shown me some quite perfect which he took at Havannah.

† With reference to this, these subgenera ought to terminate this section, and we should begin it by the satyri. Such was the course we had at first pursued.

Fabricius, lastly (Syst. Gloss.), has divided them into several genera, whose characters however need revision.

Sometimes the antennæ are terminated, as well as in common, by a solid swelling, in form of a bud or knob.

Some, or their males at least, have the two fore-claws much shorter than the others. They compose the subgenus

ERYCINA, *Lat.*, and belong to America.

All the claws are alike in both sexes in the others.

MYRINA, *Fab.*, is distinguished from the following subgenera by the elongation and remarkable projection of their lower palpi *.

The species in which they do not extend much beyond the chaperon, form the subgenus

POLYOMMATUS, so named because they have in general on their wings small eye-like spots.

Many species have moreover been named collectively, *les petits porte-queue*. The most common in the neighbourhood of Paris is

Papilio Alexis. Hübn. t. x. 292—294. *D'Argus bleu.* Geoff. God. Hist. Nat. des Lepid. de France. t. ii. sect. 3. The upper part of the wings of the male is of an azure blue, changing into pale violet, with a slight black stripe pursuing the hind edge, and a very white fringe; that of the wings of the female is brown, with a range of fulvous spots near the posterior edge, and a black stripe in the middle of the upper wings. The under part of all the four wings is nearly the same in both sexes,—gray, with a range of fulvous spots enclosed within two lines of points near the posterior edge;

* Fabricius has established in this division many other genera, but which I have not yet sufficiently studied. Some species of South America resemble Pyrales, in the upper wings being arched exteriorly at the base. The club of the antennæ presents also several modifications, which may serve as grounds of subdivision, but it is necessary to see a great number of species, and especially to know their metamorphoses.

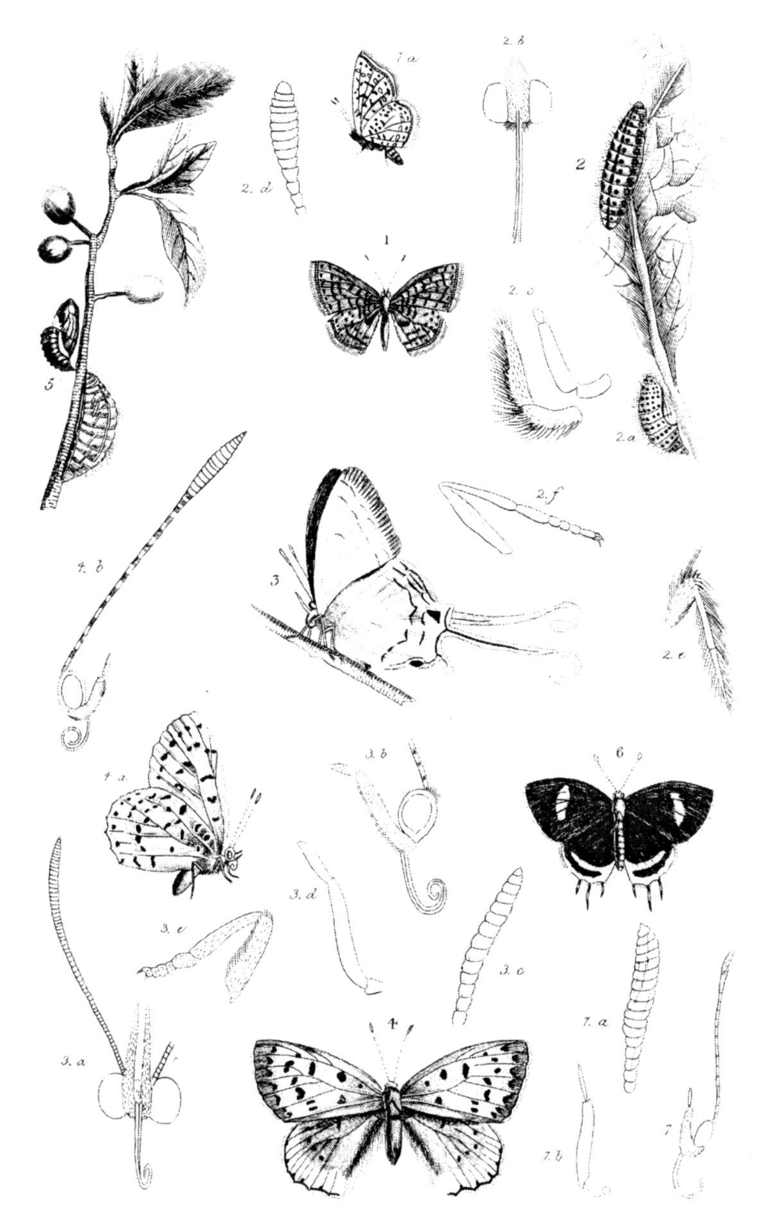

1. Erycina *virginiensis. Boisd.* 2. *Cat. & Chrys. of* Erycina *lucina.* 3. Myrina *Jaffra. God.* 4. Polyommatus *Thoe. Boisd.* 5. *Cat. & Chrys. of* Polyom. *pruni. God.* 6. Zephyrius *amor. Dalm.* 7. *Details of* Polyom. *roxus. God. (g. Lycæna. horsf.)*

London, Published by Whittaker & C.º Ave Maria Lane. 1831.

some black points edged with white are also observable there. The caterpillar lives on the sainfoin, the genista of Germany, &c. Its colours are varied.

Other lepidoptera of the same division have antennæ of a truly unusual form; those of one of the sexes of Barbicornis, God. are setaceous and plumose; those of Zephyrius, Dalm. are terminated by ten or twelve separate globular articulations, or like a chaplet.—[Vide Encyc. Method. Insect. ix. 705. This genus is established on probably false antennæ.]

The second section of the diurnal lepidoptera is composed of species whose posterior legs have two pairs of spines, one at the extremity, and the other above (and so also in the two following families); the under wings are commonly horizontal when at rest, and the end of the antennæ is often terminated in a very bent point. The caterpillars, but a small number of which, however, are known, fold up leaves, spin a cocoon of fine silk therein, and are there transformed into chrysalides, in which the body is united, or destitute of angular projections.

These lepidoptera form the division *Papiliones plebei urbicolæ*, of Linnæus, or the mutilated caterpillars (*Papillons estropiés*), of Geoffroy. Fabricius united them to *Argus*, under the generic name *Hesperia*. But we must also refer to that section some exotic lepidoptera, called Pages by amateurs, and whose natural situation has not been hitherto well determined: such are *Urania*, of Fab. These several lepidoptera conduct us very naturally to the second family: they compose two subgenera.

Hesperia, *Fab.*, or the *Papiliones plebei urbicolæ*, of Linnæus, which have their antennæ terminated distinctly in a bud or mace; and the lower palpi short, broad, and well furnished with scales in front.

Hesperia malvæ, Fab., Rœs. Insect. t. lx. 2. 9. Wings indented, blackish brown above, with white dots and spots;

the posterior edge intersected with similarly coloured spots; the under part of the wings of a greenish grey, with similar irregular spots; its caterpillar is elongated, grey, with a black head, and four yellow points on the neck or first ring, which is narrowed, a character peculiar to the caterpillars of this subgenus; it feeds on mallows, whose leaves it folds up, and in which it is changed; the chrysalis is black, but slightly powdered with blue*

URANIA, *Fab.*, in which the antennæ, at first filiform, diminish into a silken thread at the extremity; the lower palpi are elongated, slender, with the second articulation much compressed, and the last much more slender, nearly cylindrical, and naked†.

The SECOND FAMILY of Lepidoptera.

CREPUSCULARIA.

These have near the origin of the external edge of the under wings a rough scaly bristle, like a spine or horse hair, which passes into a hook of the under part of the upper wings, and holds them when at rest in an horizontal or inclined position‡. This character is found also in the following family; but the crepuscular lepidoptera are distinguishable from them by their antennæ, formed into an elongated mace, either prismatical or spindled.

Their caterpillars have always sixteen tarsi; the chrysalides do not present those points or angles which are to be seen in most of the diurnal lepidoptera, and are commonly inclosed

* For the other species, see Fabricius, Entom. Syst., the division Hesperia urbicolæ; the genus Hespérie, article PAPILLON of the Encyc. Method., and L'Hist. Nat. des Lepid. de France de Godart.

† *Pap. riphæus, leilus, lavinia, orontes,* of Fab.; *Noctua Patroclus,* ejusd. The Urania compose the genera *Cydimon, Nyctalamon,* and *Sematura,* of Dalman: see his prodromus of a monograph of the genus *Castnia.*

‡ Some of the Smerinthi, according to Godart, are deprived of them.

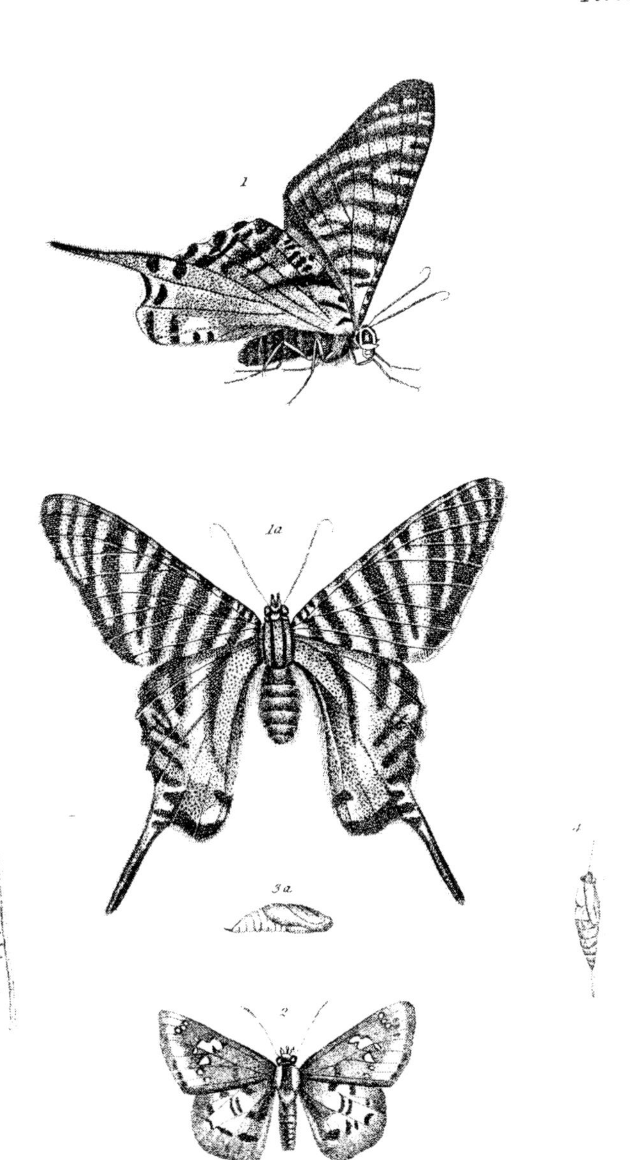

1. *Urania Boisduvalii*. G.
2. *Hesperia sabadius*. Bdv.
3. Cat & Chrys of *Hesperia tages*. Fab.
4. Coc of *Hesperia Linea*. Fab.

London, Published by Whittaker & C°. Ave Maria Lane 1832.

in a cocoon, or hidden either under ground or under some body. These lepidoptera fly but seldom, except at night or in the morning.

This family includes the genus

SPHINX, of *Linnæus*, PAPILLONS BOURDONS, *Degeer*.

So called from the attitude of many of their caterpillars, similar to that of the fabled sphinx. The humming frequently made by the perfect insect in flight has originated Degeer's epithet.

I shall divide this subgenus into four sections, corresponding in the same order to the genera *Castnia*, *Sphinx*, of Fabricius, and to those which he at first named *Sesia* and *Zygæna*.

The first *(Hesperi sphinges)* is composed of lepidoptera, which evidently connect hesperia with sphinx, properly so called. The antennæ are always simple, thick towards the middle, or at the extremity, which forms the hook, diminishing into a point without a tuft of scales at the extremity. All have a very distinct trunk; the lower palpi composed of three apparent articulations; in some the second is elongated, much compressed; and the third is slender, almost cylindrical, and nearly naked: these palpi resemble those of Urania; in others they are shorter but broader, subcylindrical, and well furnished with scales; the antennæ of the latter diminish only at the end.

Those in which the lower palpi are elongated, with the second articulation much compressed, and the last slender and nearly naked, whose antennæ are simply and gradually thicker toward the middle, and become slender after, terminating in an elongated crochet, form the subgenus

AGARISTA*, *Leach*. Those which have the lower palpi

* See the article Papillon de l'Encyclop. Method., genus Agarista. Near this subgenus comes that of *Cocytia* of M. Bois-Duval; the wings

formed in the same manner, but in which the antennæ terminate abruptly in a mace, with a short crochet at the end, compose the subgenus

CORONIS *, *Latr.* Lastly, those which have the antennæ similar to agarista, but in which the palpi are shorter, broad, and cylindrical, are

CASTNIA, *Fab.* All the species known are from the new continent.—(Ency. Method., and the Monograph of Dalman.)

The sphinxes of the second section, *Sphingides*, have the antennæ always terminated by a slight tuft of scales; the lower palpi large, or transversely compressed, very scaly, with the third articulation generally but little distinct.

Most of their caterpillars have the body shorn, elongated, thicker, with a dorsal horn at the posterior extremity, and the sides striped obliquely or longitudinally; they feed on leaves, and change in the earth without spinning a cocoon: such are

SPHINX, properly so called, in which the antennæ, commencing near the middle, form a prism, simply ciliated, or striated transversely like a rasp on one side, and which have a very distinct trunk. They fly with extreme rapidity; hover over flowers, from which they have been called hawk-moths, and hum at the same time. The chrysalides of some of the species have the sheath of the trunk projecting like a nose.

S. euphorbiæ, Lin., Rœs. Insect. t. cl. 1. *Pap. noc.* iii. Upper side of the upper wings reddish grey, with three spots and a large green band; upper side of the under wings red, with a black band and a white spot; antennæ white; upper part of body olive green; abdomen conical, very pointed, and without a terminal brush. The caterpillar is black, with

are vitreous, a character which seems to connect them with Sesia, but the palpi are those of Urania, and the antennæ those of Agarista.

* Formed from a Brazilian species, which I believe to be inedited in the collection of M. le Comte Dejean.

1. Agarista *Pales.* 2. *Cat. & Chrys. of* Agar. *glycinæ.* 3. Coronis *Leachii.* 4. Castnia *acræoides.*

London Published by Whittaker & Cº Ave Maria Lane. 1831.

yellow dots and spots; a line on the back, and the tail and feet red.

Sphinx atropos, Lin., Rœs. Insect. iii. 1. Upper wings varied with deep brown, yellowish brown, and light yellow; under wings yellow, with two brown bands; a yellowish spot with two black points on the thorax; abdomen without a terminal brush, yellowish, with black rings. This is the largest species of this country, the spots on the thorax imitating a death's head; the sharp noise it can make is attributed by Reaumur to a rubbing of the palpi against the trunk *, and by M. Lorey to the rapid escape of the air from two particular cavities in the belly, which have frightened people in certain years in which this insect was more than usually common †. Its caterpillar is yellow, with blue stripes on the sides; and the tail bent back in zigzag. It feeds on potatoes, privet, jasmin, &c., and is turned into a nymph toward the end of the month of August. The perfect insect appears in September

The caterpillars of some species, altogether remarkable for their fine colours (*celerio, nerii, elpenor, porcellus,*) have the anterior end of the body much attenuated, like a pig's snout, whence they have been named *cochonnes,* and capable of being drawn within the third ring; on the sides are many eye-like spots. These species form, with reference thereto, a very natural division.

Other sphinxes have, like sesia, the abdomen terminated by a scaly brush. Scopoli formed them into a genus MACROGLOSSUM. Fabricius at first united them to sesia; afterward (Syst. Glossat.) he separated them, preserving to this generic

* It is shorter in proportion than in the other sphinxes, and it is probable that it was on this peculiarity that the genus *Acherontia* was formed with this species, and are analogous to it, from Java.

† According to M. Passerini, Arm. des Sciences Natur. t. xiii. 332, the seat of the organ producing this noise is in the interior part of the head.

group this denomination, and giving that of *Ægeria* to the primitive genus of the series; but the lepidoptera, which he now designates under the generic name *Sesia*, have the essential characters of sphinx: such is *P. stellatarum*, Lin., and those which have been named *fuciformis, bombyliformis*, &c. The wings of these two last are, in a great part, vitreous or transparent.

SMERINTHUS, *Lat.*, which have the antennæ indented like a saw, and have no distinct tongue.

Sphinx tilæ, but much more common on the elm, *S. ocellata, S. populi*, &c. form this subgenus; they are heavy, and the under wings surpass the upper, as in many species of bombyx.

Our third division (*Sesiades*) of sphinxes includes those whose antennæ are always simple, like an elongated spindle, often terminated, as well as in the other subgenera, by a little bundle of hairs or scales: whose lower palpi, slender and narrow, have three very distinct articulations, the last being pointed: and whose hind limbs have at their extremity very strong spurs: the abdomen is in general terminated by a sort of brush. Their caterpillars gnaw the interior of the trunk or root of vegetables like those of *Hepiales* and *Cossus*, are naked, without posterior horn, and build for themselves in these same vegetables, with the debris of the matters on which they feed, a cocoon, in which they undergo their last transformation.

SESIA.

These have the antennæ terminating by a little tuft of scales; the wings are horizontal, and have certain vitreous spaces; the scales of the extremity of the abdomen form a brush. Many of these insects resemble wasps, or other hymenoptera, diptera, &c.*.

* See the monograph of *Sesia* of Laspeyres, Hübner, Godart, &c.

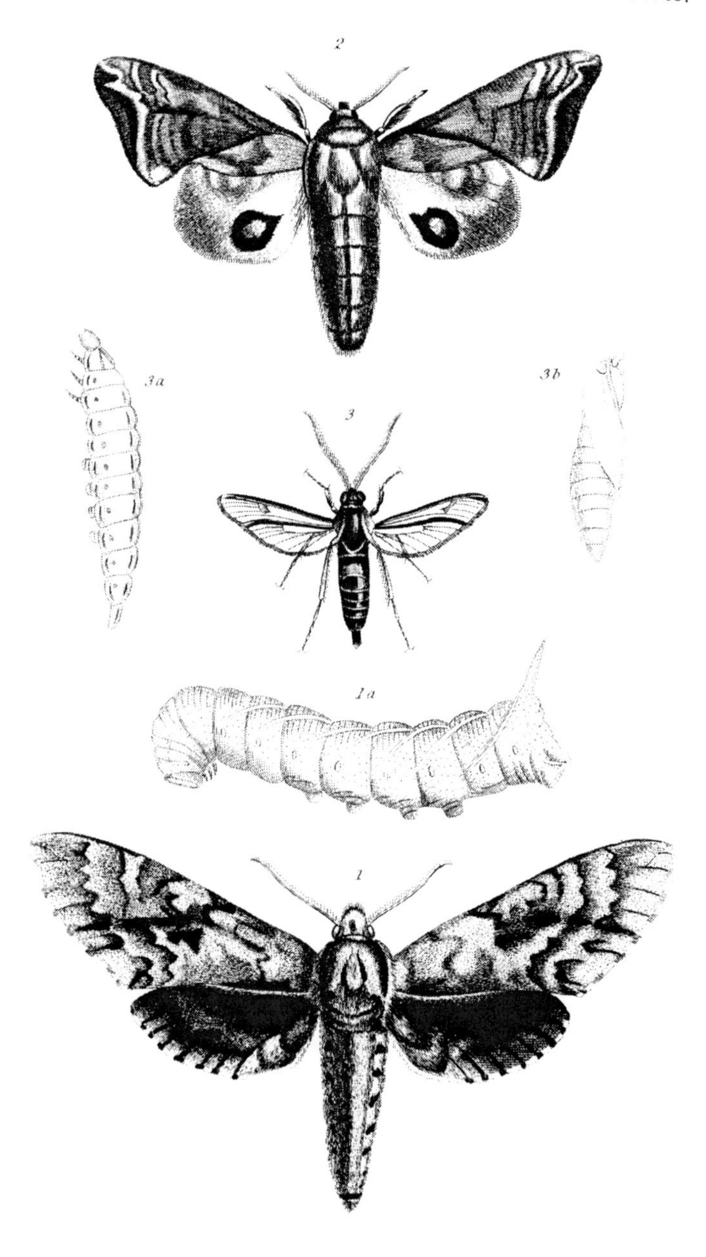

1 Sphinx Jasminearum_Bdv.

2 Smerinthus Io_Bdv.

3. Sesia asilipennis_Bdv.

London, Published by Whittaker & C° Ave Maria Lane, 1832.

THYRIS, *Hoffm*, *Illig.*, similar to Cesia, but with the antennæ much more slender, nearly setaceous, and without a tuft at the end. Their wings are angular and indented; their abdomen terminates in a point *.

ÆGOCERA, having also antennæ without a scaly tuft at the end, but evidently thickened toward the middle, and spindle-shaped; the second articulation of the lower palpi is moreover furnished with a bundle of hairs, advanced like a bird's bill; the abdomen terminates in a simple point; the wings are inclined like the roof of a house, and entirely covered with scales. Their changes are unknown †.

The fourth and last section of the sphinxes *Zygænides* is composed of lepidoptera, whose antennæ always terminated in a point deprived of a tuft, are sometimes simple in both sexes, spindle-shaped, or like a ram's horn, sometimes but little thick, toward the middle almost setaceous, pectinated in both sexes, or at least in the males; the lower palpi moderate or small, subcylindrical, and always formed of three distinct articulations. The wings are always like a roof, and many species have vitreous spots on them: the abdomen has no brush at the end: the spurs of the hind legs are generally small. The caterpillars live nakedly on various vegetables: they are cylindrical, generally hairy, without a posterior horn similar to that of many bombyces, and form a silken cocoon, spindled or ovoid, which they fasten to the stems of plants. The habits of these insects have been well described by Bois-Duval.

ZYGÆNA. Exotic insects from the new continent, whose

* *Sphinx fenestrina*, Fab. Lat. Bois-Duval.

† *Bombyx venulia*, Fab. See Lat., Gen. Crust., and Insect. iv. 210. Dalm. Anal. Entom. 49. It might probably be more consistent with the natural order to place this subgenus near Agarista.

antennæ are simple in both sexes, terminated abruptly in a mace, a spindle, or a ram's horn, and in which the lower palpi are elevated above the chaperon, and are pointed.

Sphinx filipendulæ. Lin. Rœs. Insect. t. class. ii. *pap. noct.* 57. of a black or blueish green; six red spots on the upper wings, the lower red, with the posterior edge of the same colour as the body. Its caterpillar is citron yellow, a little hairy, with five ranges of black spots along the body. It spins on the stems of plants a cocoon of a yellow straw colour, shining, elongated, and spindle-shaped. Its surface is rough, or, as if folded. The perfect insect comes forth in the month of July.

SYNTOMIS, *Illig.*, differ from the last only in having the antennæ less thick, forming a slender, gradual spindle. The lower palpi are shorter and obtuse.

ATYCHIA, *Hoffm.*, *Illig.* In the females the antennæ are simple and bipectinated; in the males the lower palpi are very hairy, and evidently surpass the chaperon; the wings are short, and there are strong spurs at the extremity of the hind legs.

PROCRIS, *Fab.*, approximate to the last in the antennæ; but the lower palpi are shorter and less hairy; their wings are long, and the spurs on the hind legs very small.

Sphinx statices, Lin., De Geer. Insect. ii. 255. Body shining green, and as if gilt; under wings brown; antennæ of the males, with two ranges of black barbs, that of the male slightly serrated.

The other lepidoptera of this division, have, in both sexes, the antennæ furnished with a double range of elongated, or bipectinated teeth. Those which have a distinct trunk form the genus GLAUCOPIS of Fab., and those in which this organ is wanting, or is indistinct, that of AGLAOPE.

A great number of species are found in foreign countries

1. *Thyris sepulchralis.* Bdv.
2. *Ægocera rectilinea.* Bdv.
3. *Zigæna pulchella.* Bdv.
4. *Zigæna filipendulæ.* Lin.
5. *Procris nebulosa* ♀ Klug.
6. *Syntomis mycodes.* Bdv.
7. *Syn. phegea.* Fab.
8. *Atychia pumila.* Och. Lat. ♂. ♀.
9. Cat. & Chrys. of *l'Aglaope infausta.*
10. *Glaucopis Folletii.* Feist.
11. *Aglaope americana.* Bdv.

London, Published by Whittaker & C.º Ave Maria Lane 1832.

belonging to these two subgenera *. These crepuscular moths seem to connect themselves with callimorpho.

The genus *Stygia*, of Draparnaud, which was placed in this tribe, belongs to *Hepialites*.

M. de Villiers, who has given us [Ann. de la Soc. Lin. de Paris. v. 473.) new details of *S. australis* accompanied with good figures, considers it as intermediate between Sesia and Zygenia, but it is without trunk; the palpi are those of *cossus*; the antennæ are short, not at all spindle-shaped, and more analogous to that of certain bombyces, than to that of sesia or zygenia. In the disposition, even of the colours of the upper wings, these lepidoptera approach nearer cossus and zeuzera, than the preceding insects.

The third family of Lepidoptera,

NOCTURNA,

Has, with a few exceptions, the wings tied down when at rest, by means of a corneous sort of horsehair, or bundles of silks proceeding from the outer edge of the under wing, and passing into a ring, or grove, on the under part of the upper wing. The wings are horizontal or leaning, and sometimes rolled round the body; the antennæ, as they advance, diminish in size, from base to point, or are setaceous.

This family composes in the Linnæan system, but one genus, namely

PHALÆNA.

These lepidoptera fly in general only at night, or after sunset; many of them have no trunk; some of the females are wingless, or have them only very small. The caterpillars in general spin a cocoon; the number of their feet varies

* For the species of the above genera, see Lat. loc. cit.; and Hist. Nat. des Lepid. de France.

from ten to sixteen *; the chrysalides are always rounded or without angles or points.

The classification of this family presents considerable difficulties, and our systems are hitherto mere essays, or imperfect attempts †. We will divide them into ten sections. The species with perfectly entire wings, and without fissures, composing a sort of digitations, will fill the first nine. All those which, in the caterpillar state, live almost entirely naked, or in fixed retreats, and many of which have less than sixteen tarsi, and which, in their last state, have the upper palpi very small, or entirely hidden, the wings more or less triangular, horizontal, or sloping, and not folded round the body, will compose the first eight. The last or eighth of these is the only one whose caterpillars have fourteen legs, two of which are anal. If the same number be found in some others, here the two posterior are wanting.

The two divisions, *Attacus* and *Bombyx*, of the genus *phalæna*, of Linn., correspond with the first four sections. The trunk is in general rudimentary, or very small, and its two parts are disjointed; the lower palpi, except in a few, are small and subcylindrical; the antennæ, at least in the males, are pectinated or serrated; the wings are horizontal or sloping, and in many the under wings surpass the upper when at rest, and are even sometimes without the stiff bristle or bundle of hairs, which holds one wing to the other. The

* Degeer has counted eighteen, and all membranaceous, in one species. ii. 245.

† It has been several times attempted to employ characters drawn from the caterpillars. If, indeed, we do not refer to them, a great number of genera must be cancelled. I will cite, for example, that of phælena, properly so called, or the geometricians. If we advert to the perfect insects only, it is impossible to distinguish generically many species, such as the following, *prodromaria*, *betularia*, *hirtaria*, from *bombyx*. It is also evident that we cannot then separate platypterix, and other genera.

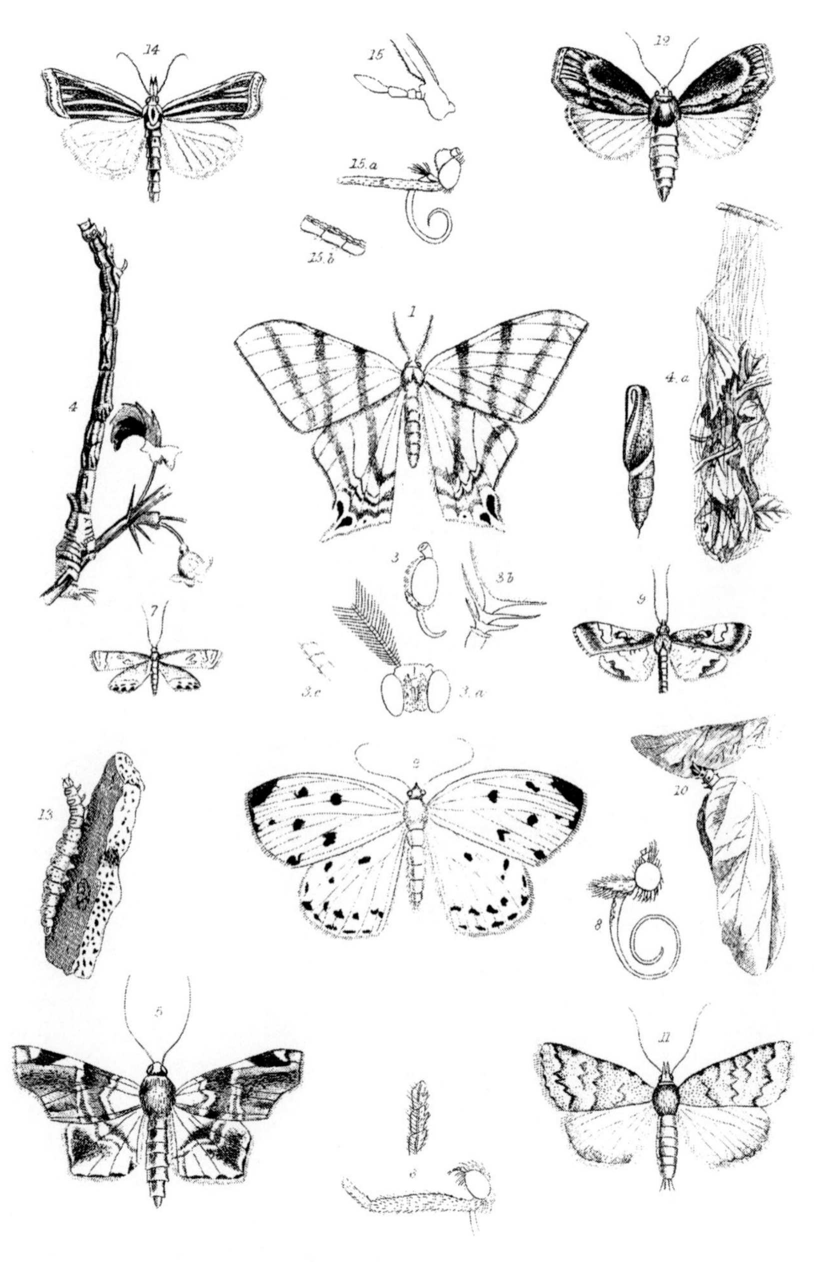

1. Phalæna machaonaria, Bdv.
2. Ph. guttaria, Bdv.
3. Details of Ph favillacea, Hub.
4. Caterpillar of Ph. grossulariata, L.
5. Herminia sidonia. Cram.
6. Head & antennæ of Herm. crassalis. Fab.
7. Botys dilacidalis. Bdv.
8. Head of Botys cingulalis. Hub.
9. Hydrocampe aquatilis. Bdv.
10. Caterpillar of Hydr. nympheata. Lat.
11. Aglossa dilacidalis. Bdv.
12. Galleria cereana. Fab.
13. Cat. of Gal. colonella. Hub.
14. Crambus retusalis Bdv.
15. Details of Cr radiellus. Hub.

London: Published by Whittaker & Co. Ave Maria Lane, 1832.

thorax is always united as well as the abdomen, and is woolly; the latter is generally very voluminous in the females; the cocoon of the chrysalis is, in general, well felted and solid.

Although the nocturnals of the fourth section have many relations to those of the preceding, their caterpillars supply, nevertheless, a character unique in the order: the anal tarsi are wanting, while those of the first three sections have all the sixteen.

The first section, *Hepialites*, has for its type the genus *hepialus* (*hepiolus* of some authors), and *cossus* of Fabricius. The caterpillars are shorn, and remain hidden in the interior of the vegetables on which they feed; the cocoon they form to pass the chrysalis state in, is in great part composed of parts of these vegetables; the edges of the rings of the abdomen of the chrysalis are indented, or spiny; the antennæ of the perfect insect are always short, have in general but one sort of small indentations, short, rounded and serrated; that of some others terminate always in a simple thread, but they are furnished underneath, in the male, with a double range of barbs; The proboscis is always very short and indistinct; the wings are sloping, and commonly elongated; the last ring of the abdomen of the female forms an elongated oviduct, or sort of tail. In the caterpillar state these insects do much mischief to different trees, and to some other useful vegetables.

Sometimes the antennæ, formed alike in both sexes, have only very short teeth disposed in one or two ranges, such are

HEPIALUS, *Fab.*, which are distinguishable by their antennæ, almost granulated, shorter than the thorax. The under wings have not commonly any tie.

Their caterpillars live in the ground and gnaw the roots of plants.

H. humuli, Fab., Harr. English Insects, iv. The male has the upper wings silvery white, without spots; those of the female are yellow, with red spots. The caterpillar devours the roots of the hop, and causes great mischief in hop-grounds *.

Cossus, *Fab.*, have the antennæ as long, or less so, than the thorax; the inner side having a range of small lamellated teeth, short, and rounded at the end.

The caterpillars live in the interior of trees, which they gnaw, and use the saw-dust they make in building their cocoon; their chrysalides, at the period when the insect is about to be disclosed, advance to the exterior opening, which serves as a passage for the insect.

Cossus ligniperda, Fab., Rœs. Insect. tom. i. class ii. Pap. noc. 18. A little more than an inch long, ashy grey, with numerous small black lines on the upper wings, forming little veins interspersed with white. The posterior extremity of the thorax is yellowish with a black line.

Its caterpillar, which is found in spring, resembles a large worm; it is reddish, with transverse blood-red bands. It lives in the interior of the wood of the willow, the oak, and especially the elm; it disgorges an acrid and fetid liquor, contained in peculiar interior reservoirs, and which appears to be used in softening the wood †. Sometimes the antennæ differ considerably in the two sexes; those of the males being furnished beneath with a double range of barbs, and terminated in a thread; those of the females entirely simple, but cottony at the base.

Zeuzera, *Lat.*, Cossus, *Fab.* The caterpillar of a very

* For the other species, see Fabricius, Esper, Engramelle, Hubner, Donovan, Godart, &c.

† *Stygia Australis.* Lat. Genev. Crust. and Insect. iv. 215. God. Hist. Nat. des Lepid. de France. iii. 169; see also the before quoted Memoirs of M. de Villiers, inserted in the collections of those of the Linnæan Society of Paris. tom. v. South America supplies another species. The antennæ differ from that of the cossus, and this subgenus may be preserved. The abdomen terminates in a small brush.

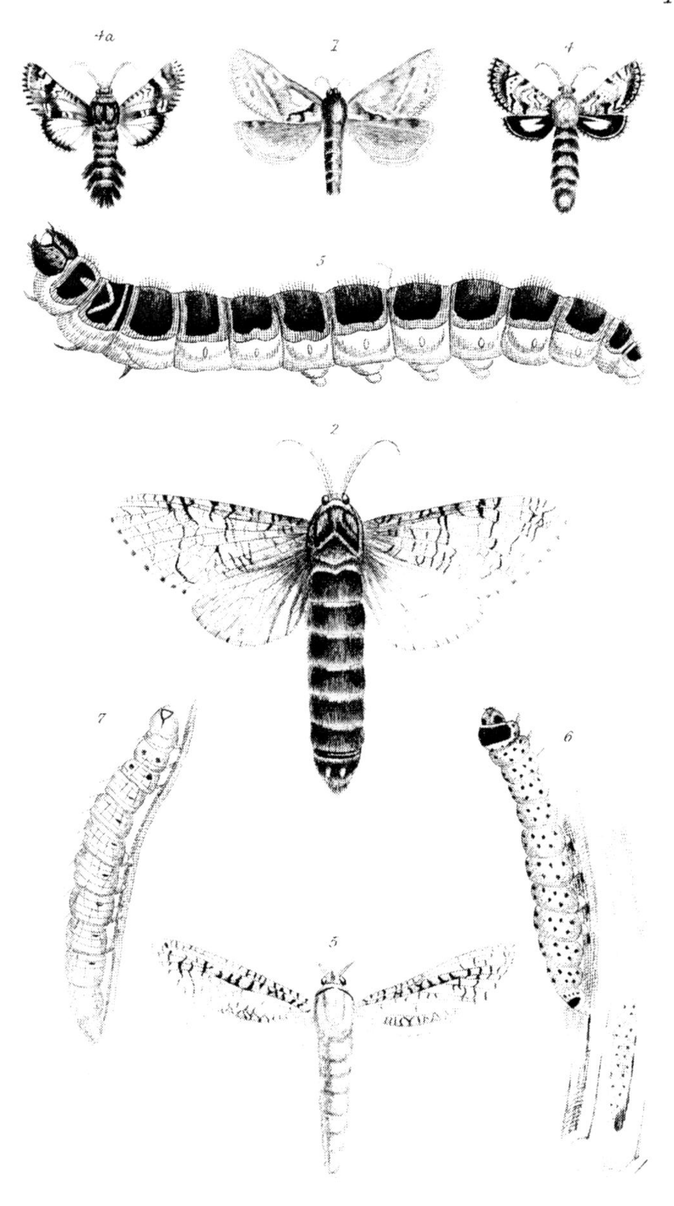

1. Hepialus lupulinus. Fab.
2. Cossus Macmurtrei. Bdv.
3. Caterpillar of Cossus ligniperda. Fab.
4. Stygia australis. Latr.
5. Zeuzera scalaris. Donov.
6. Caterpillar of Zeuzera æsculi. Fab.
7. Caterpillar of Hepialus humuli. Fab.

London, Published by Whittaker & Co. Ave Maria Lane. 1832.

pretty species, *Cossus æsculi*, Fab., whose body is of a fine white, with blue rings on the abdomen, and numerous points of the same colour on the upper wings, lives in the interior of the Indian chesnut tree, the apple and pear tree, and often even in their pith *.

Our second section, that of *Bombycites*, is distinguished from the preceding, and from the third, by these characters: Proboscis always short, and simply rudimentary; wings, whether extended and horizontal, or sloping, with the under always surpassing, laterally, the upper; antennæ of the males entirely pectinated.

The caterpillars live exposedly, and gnaw the tender parts of vegetables; most of them make a cocoon of pure silk; the chrysalides have no indents on the edges of the rings of the abdomen.

We shall form, with these species, whose wings are extended and horizontal, or the *Phalænæ attacæ* of Linnæus, a first subgenus, for which we shall preserve the name

SATURNIA, which M. Schrank has given them, and to which we shall unite the AGLIA, *Bombyx tau, Fab.*, of Ochsenheimer. It comprehends the largest species, whose wings have frequently vitreous spots (*fenestratæ*), such especially among the exotic insects, as the *Atlas* of China, the *B. hesperidæ*, the *B. cecropia*, the *B. luna*, whose under wings are elongated like a tail. They have employed immemorially, at Bengal, the silk of the cocoon of two other species of the same division, *Bombyx mylitta*, of Fabricius, and *Phalæna cynthia*, of Drury. (Ins. ii. 6. 2.; Lin. Soc. Trans. vii. 35.) I am satisfied, by the communication made to me by Mr. Huzard, of a Chinese manuscript upon this subject, that the caterpillars of this bombyx were the wild silk-worms of China. I conjecture that a part of the silks that the ancients procured by their

* Rœsel. Insect. t. iii. 48. 5, 6. *Cossus pyrinus*, Fab., *C. scalaris*, ejusd. *Phalæna scalaris*, Donovan. *P. mineus*, ejusd.

maritime commerce with the Indians, came from the silk of these caterpillars.

Europe furnishes only five * species of this subgenus.

B. pavonia major, Fab. Rœs. Ins. iv. 15—17, is the most common, and the largest of this country, being five inches wide, the wings extended; the body is brown, with a whitish band at the anterior extremity of the thorax; the wings are round, brown, powdered with grey, a white eye-like spot, black intersected by a transparent trait, surrounded by an obscure fulvous circle, a white half circle, another reddish, and finally a black circle in the middle of each. The caterpillar, which lives on the leaves of different trees, is green, with blue tubercles disposed circularly, which are furnished with long hairs, having a mace-like termination. They spin an oval cocoon, contracted into blunt points, with a double neck, the interior of which is furnished partly with elastic and convergent threads, which assist the disclosure of the insect, and which also resists the entrance of all other insects; the silk is very strong and gummy; the bombyx is disclosed in May of the second year †.

The other bombyces have the upper wings inclined sloping; the outer edge of the under wings reaches beyond the upper almost horizontally *(alæ reversæ)*.

Sometimes their palpi advance in the form of a bill, and their under wings are often indented. The insect resembles a packet of dead leaves. These species form the genus

LASIOCAMPA ‡. The species in which the lower palpi do not jut out sensibly, compose the subgenus

* Authors mention but four: but another, perfectly distinct, is lately discovered, which I have seen in the collection of M. Bois-Duval.

† For the other species, see Fab. Entom. Syst. first division of the Bombyces, and Olivier. Ency. Method. first family of the same genus.

‡ *B. quercifolia, populifolia, betulifolia, illicifolia, potatoria,* of Fabricius. This subgenus forms part of the genus *Gastropacha*, of Ochsenheimer.

M. Banon, professor of chemistry at Toulon, to whose friendship I owe

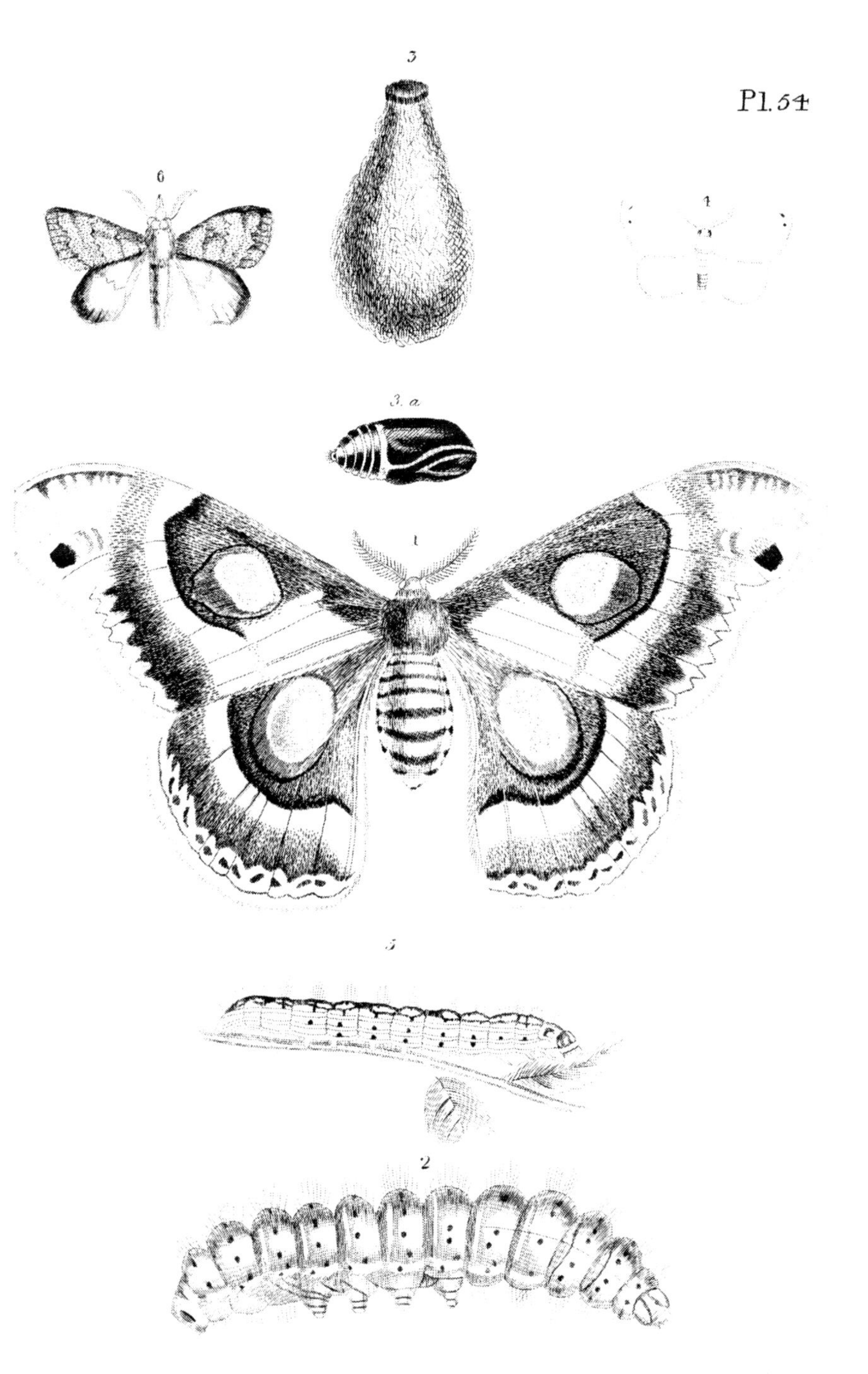

1. Saturnia *bauhiniæ.* 2. *Cat. of* Sat. *luna.* 3. *Cocoon & Chrys. of* Sat. *pavonia minor.* 4. Bombyx *digramma.* 5. *Cat. of* Bomb. *pensylvanica.* 6. Lasiocampa *proboscidea.*

London Published by Whittaker & C.º Ave Maria Lane. 1831.

BOMBYX *, *(proper)*. *B. mori*, Lin. Rœs. Ins. iii. 7. 9. Whitish, with two or three obscure transverse rays, and a spot crossing the upper wings. Its caterpillar is known under the name of silk-worm. It feeds on mulberry leaves, and spins an oval cocoon of a serrated tissue of very fine silk, in general bright yellow, but sometimes white. The variety which gives its silk of the latter colour is generally cultivated in preference to the other.

The bombyx which produces it is originally of the southern provinces of China. According to Latreille, the town of Turfan, in Little Bukharia, was for a long time the resort of caravans coming from the west, and the principal depôt of the silks of China. It was the metropolis of the Seres of Upper Asia. Driven from their country by the Huns, the Seres established themselves in Great Bukharia and India. It is from one of their colonies, *Ser-indi*, that the Greek missionaries transported, in the time of Justinian, the eggs of the silk-worm to Constantinople. The cultivation of them passed at the period of the first crusade, from the Morea to Sicily, the kingdom of Naples, and many ages after, under Sully especially, into France. But the ancients imported their silk by land or sea, from the kingdoms of Pegu and Ava, or from the Eastern Seres, those which are most frequently mentioned in the writings of the earliest geographers. A part of the Northern Seres took refuge in Great Bukharia, and even formed the commerce of the country, as seems

many insects, collected by him at Cayenne, as well as others from the Levant, has communicated to me a lepidopterous insect having all the characters of lasiocampa, but provided with a distinct antlia or proboscis. It appears to be intermediate between this subgenus and *Calyptra*, of Ochsenheimer.

* This generic denomination has been unadvisedly suppressed by Ochsenheimer. We apply it collectively to all the species of his genus, *Gastropacha*, whose lower palpi are not advanced like a bill.

to be indicated by a passage from Dionysius the Periegete. It is known that silk anciently sold for its weight in gold, and is still an important source of wealth to France.

B. neustria, Fab. Rœs. Ins. I., class ii., *Pap. noc.* 6. Yellowish, with a band, or two transverse stripes of yellow brown in the middle of the upper wings. The female deposits her eggs round branches, in form of a bracelet or ring. The caterpillar is striped longitudinally with white, blue, and reddish. They live in society, and frequently do great mischief to fruit trees. They make a cocoon of a thin tissue mixed with whitish dust.

B. processionnea, Fab. Reaum. Ins. II. x., xi. Ash-coloured. Two obscure rays towards the base of the upper wings, and a third blackish a little above the middle: all three transverse. The caterpillars have the body hairy, of an obscure ash colour, with the back blackish, and some yellow tubercles. They live in society upon the oak; spin, in common, when young, a net in which they are covered; often change their domicile until after the third moult, then fix themselves, and make another common habitation of the same material, like a bag, but divided within into many cellules. They generally go forth in the evening in regular order. One individual takes the lead and serves as guide; two others follow and compose the second line; there are three to the third, four to the fourth, and so on to the end, always increasing by one. They follow the movements of the first. These caterpillars spin each a cocoon, one beside the other, with the tissue mixed with the hairs of their body. These hairs, as well as those of several other species, are very fine: they penetrate into the skin, and cause considerable itching and pustules.

B. Pythio-campa is a species analogous to this.

The inhabitants of Madagascar make use of the silk of a caterpillar, which is also gregarious. Its nidus is sometimes

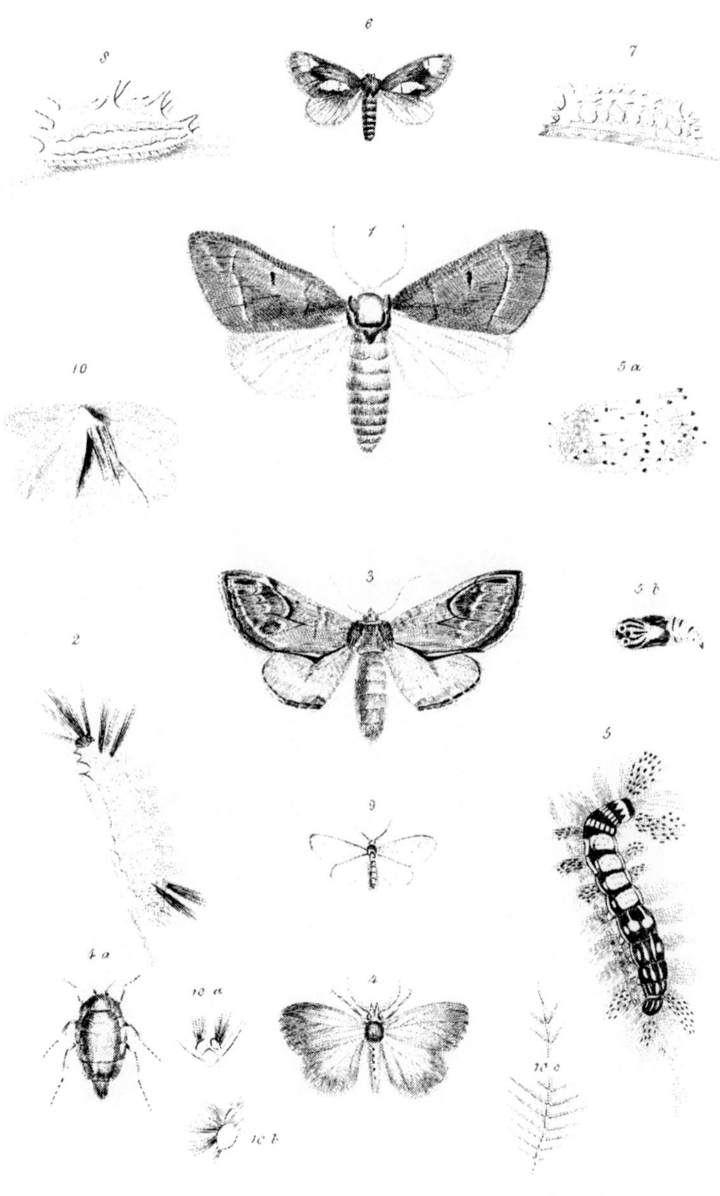

1 Sericaria ranæceps Bdv.
2 Seric tessellata
3 Notodonta ziczac Lin.
4 Orgyia detrita Bdv.
5 Org gonostigma
6 Limacodes delphinii
7 Limac strigata
8 Lim indeterminus
9 Psyche calvella
10 Psyche nitidella

London. Published by Whittaker & Co. Ave Maria Lane. 1832.

three feet high, and the cocoons are so pressed one against the other, that there is no void space: one nidus has sometimes five hundred cocoons. This species, however, belongs to the subgenus *Sericaria* of the following section.

The third section of nocturnal lepidoptera, that of *Pseudobombyces*, has the under wings, as have also, indeed, all the following, deprived of any tie fixing them to the upper when at rest; they are, moreover, entirely covered by the latter, and both are placed sloping or horizontally, but covered to the inner edge. The antlia, toward the end of this tribe, begins to be elongated, and does not even differ in the last subgenus from that of other lepidoptera, except in being a little shorter; the antennæ are entirely pectinated or notched, at least in the males. The caterpillars feed on the exterior portions of vegetables.

We shall first separate the species whose antlia is very short, and not fitted for suction.

In most of these the caterpillars live exposedly, and do not make any portable domicile.

Among these some of the caterpillars are elongated, furnished with common tarsi fitted for walking; the rings of the body are not attached above.

Sometimes both sexes have wings fitted for flight.

SERICARIA, *Lat.*, whose upper wings have no indentations on the inner edge.

B. dispar. Fab., Rœs. Ins. i. 150. ii. *Pap. noc.* 3. The male is much the smallest; the upper wings are brown, with wavy blackish stripes; the female is whitish, with black spots and stripes on the same wings; she covers her eggs with the numerous hairs she carries at the tip of the abdomen. The caterpillars are often very hurtful to our fruit trees*.

* *Bomb. versicolora, bucephala, coryli, pudibunda, abietis, anachoreta,* of Fabricius; or the genera *Endromis, Liparis, Pygæra,* and many species of that of *Orgyia* of Ochsenheimer.

NOTODONTA*, *Ochs*, in which this edge is indented.

This is connected with certain nocturnal subgenera.

Sometimes the females are nearly apterous, as in

ORGYIA, *Ochs*. The caterpillars have tufts and pencils of hairs.

B. antiqua, Fab., Rœs. ibid. xxxix.; the female iii. 150. ii. *Pap. noc.* 13.; the male.

The male has the upper wings yellow, with two transverse blackish rays and a white spot towards the inner angle; the abdomen of the female is very voluminous †.

We then arrive at the *Pseudo-bombyces,* whose caterpillars are rampant, their tarsi being very short, and the scales even being retractile. The body is oval, like the wood-louse, with the skin attached above, from the second ring, so that it forms a vault under which the head is withdrawn.

These species compose the subgenus.

LIMACODES, *Lat.*—Their caterpillars appear to represent, in this division of the nocturnal lepidoptera, those of certain diurnal insects of the same order, such as Polyommates ‡.

If regarded in their early age, the last nocturnal pseudo-bombyces, without an apparent, or at least a useful, proboscis, present another anomaly. Their caterpillars live in the

* The notodonta of the same Ochsenheimer. I have excepted from them a species called *palpina,* which, on account of its large and compressed palpi and of its spiral antlia, ought to form a subgenus of itself, connecting *Notodonta* with *Calyptra* of the same naturalist, and which I put at the head of the noctualites, to pass from them to *Xylena,* to *Cuculia,* &c. Some notodonta have the corslet and crest, a character which appears more proper to this last section: it is the same with those in which the lower palpi are much compressed, like those of the noctualites.—(See hereafter the generalities of this division of night moths.)

† Add *O. gnostigma* of Ochsenheimer. The others are sericariæ.

‡ *Hepiales testudo, asellus, bufo* of Fabricius. See God. Lepid. de France, iv. 2791.

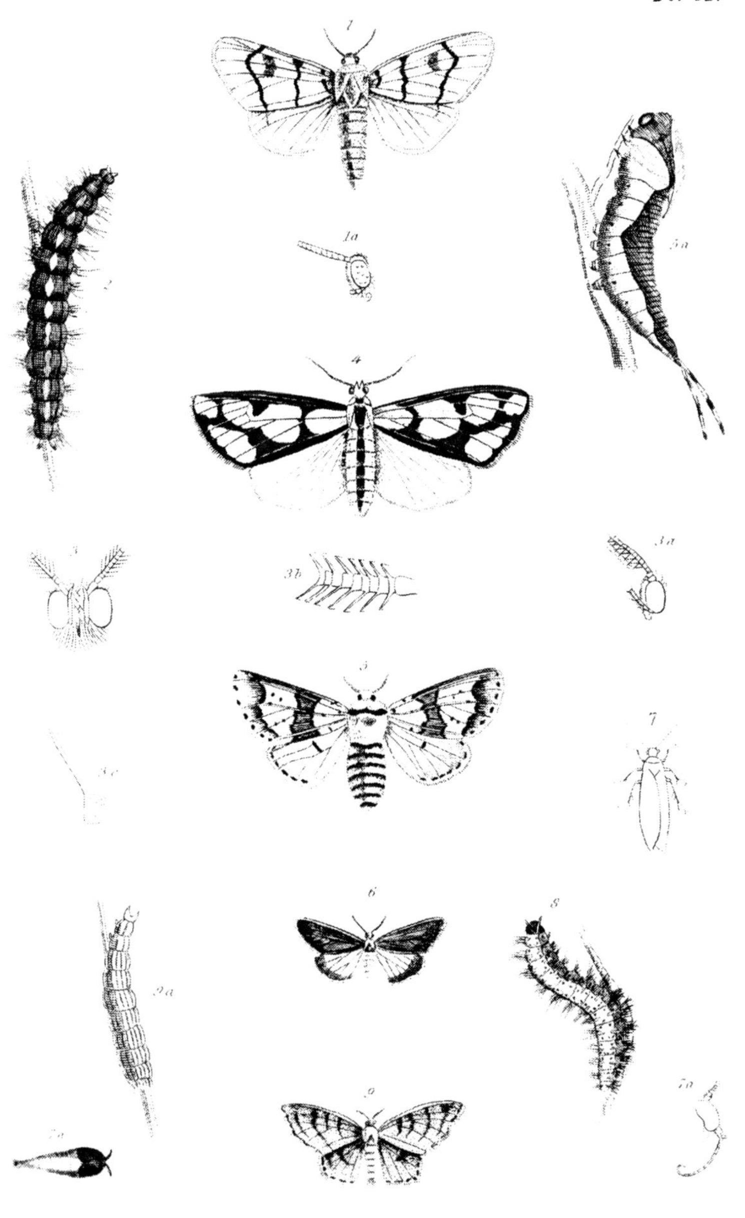

1. *Chelonia evidens*, _ Bdv.
2. *Caterpillar of Chel. nubilis*, _ Bdv.
3. *Details of Chel. chrysorhœa.*
4. *Callimorpha Lecontea.*
5. *Dicranoura borealis*, _ Bdv.
6. *Lithosia læta*, _ Bdv.
7. *Lithosia luteola.*
8. *Caterpillar of Lithosia pulchella*
9. *Platypterix globulariæ*, Bdv.

London, Published by Whittaker & C.º Ave Maria Lane. 1832.

same manner as many tineæ, in portable domiciles, consisting of a silky tube, on which they apply a morsel of a stalk, or small branches of vegetables, forming little bags, placed one above the other. This habitation resembles that of certain larvæ of *Phryganæ*. There are some very remarkable species in the East Indies and in Senegal.

These lepidoptera, united by Hübner to the tineæ, compose the subgenus

PSYCHE, *Schr.* *—The last Pseudo-bombyces, which, in the arrangement of their colours, seem to resemble some of argynuis, have a very distinct proboscis, which is considerably elongated when unrolled beyond the head; such are

CHELONIA, God., *Arctia*, Schr., *Eyprepia*, Ochs., whose wings are sloping, whose antennæ are corded in the males, and who have the lower palpi very velvety, and the proboscis short.

Bomb. chrysorrhæa, Fab. Rœs. Ins. i. class ii. *Pap. noc.* 22. Wings white, spotless, posterior extremity of the abdomen yellow brown. Its caterpillars, in some years, entirely strip the trees of their leaves.

Bomb. caja., Fab., Rœs, *ibid.* i.—Head and thorax brown, upper wings of the same colour, with irregular white stripes, under wings, and upper part of abdomen red, with bluish black spots. Its caterpillar, which feeds on the nettle, lettuce, elm, &c., is called in France hedge-hog, or bear, on account of the long and numerous hairs which cover it. It is blackish brown, with blue tubercles, arranged in rings †.

CALLIMORPHA, Lat., *Eyprepia*, Ochs., has the wings also sloping, but the antennæ are more or less ciliated in the males, the lower palpi are covered only with small scales; the proboscis is long.

* See Ochs., God., &c.

† For the other species, see Latr. Gen. Crust. and Insect. iv. 220. Ochsenheimer and God. Hist. Natur. des Lepid. de France.

Bombyx jacobeæ, Fab., Rœs., Ins., class ii. *Pap. noct.* 49, whose caterpillar is found on the groundsel, is a very common species in France. It is black; the upper wings have a carmine red line and two points. The under wings are of that colour, edged with black. The caterpillar is yellow, with black rings.

LITHOSIA, Fab. The wings are bedded horizontally on the body.

The fourth section of nocturnal lepidoptera, APOSURA*, deviates, as we have stated, in the general divisions of this family, by a unique character drawn from the absence of anal tarsi, in the state of larva or caterpillar. The hind extremity of the body terminates in a point, which, in many species, is forked, or presents two long and moveable articulated appendices, forming a sort of tail. With reference to the proboscis, palpi, and antennæ, these lepidoptera differ but little from the foregoing. Some, such as

DICRANOURA, God., *Cerura*, Schr., *Harpyia*, Ochs., have the exterior appearance of the sericariæ; the antennæ of the males are terminated by a little simple and bent thread. The hind extremity of the body of the caterpillars is forked †.

Some others, such as

PLATYPTERIX, Lacep., *Drepana*, Schr., are much like phalæna, properly so called. Their wings are broad, and the superior angle of the hind extremity of the upper wings is prominent or formed like a scythe. The body is slender. That of the caterpillar terminates in a simple truncated point. They fold down and fasten the edges of leaves in which they inhabit, and on which they feed, with silken threads. The cocoon of the chrysalis is but little clothed. These insects,

* The anus has no tarsi, a character peculiar to the caterpillars of this tribe, which form a lateral ramification to phalæna.

† See Ochsenheimer, Godart, Hübner, and Fische, Entom. de la Russie.

in one word, are connected, in the caterpillar state, with Dicranoura, and in the imago state, with the section Phalæna *.

The fifth section of nocturnal lepidoptera, *Noctualites*, Lat., similar to the last in the shape, relative size of the wings, and the position when at rest, has also as distinctive characters, a corneous, spiral, and generally long proboscis (*Antlia*, Kirby), the lower palpi terminated abruptly by an articulation very small, or much more slender than its predecessor; the latter being much larger, and greatly compressed.

The body of these noctualites has more scales than down. Their antennæ are in general simple; their thorax is often crested on the upper side; the abdomen has the form of an elongated cone; their flight is rapid. Some species appear during day.

Their caterpillars have commonly sixteen tarsi, the others have two or four less; but the two posterior or anal tarsi are never wanting; and in those which have only twelve, the anterior pair is as large as the following. Most of these caterpillars inclose themselves in a cocoon, in which they undergo their change.

These lepidoptera include the division *Noctuæ* of Linnæus; all the genuine divisions which have been lately established, and whose characters are borrowed from the animal in its caterpillar, rather than in its perfect state, are connected with the two following subgenera.

Erebus, Latr., *Thysania*, Dalm., *Noctua*, Fab., whose

* *Ph. falcataria, lacertinaria* of Fabricius, his *Bombyx compressa*. I had at first a notion to form this subgenus into a particular section, which would have been intermediate between the Pseudo-bombyces and the Phalenites. Ochsenheimer places them at the end of the noctualites, passing from *euclidia* to the preceding section; but platypterix appears to us much nearer in the caterpillar state to Harpyia of this naturalist, than to Euclidia and other nocturnals whose caterpillars are pseudo-geometricians.

wings are always extended and horizontal, and whose last articulation of the lower palpi is long, slender, and naked.

These are the largest lepidoptera of this tribe, and which, with the exception of one species, which belongs to the species *Ophiusa scapulosa*, Ochs., are all exotic *.

NOCTUA. In these the last articulation of the lower palpi is very short, and covered with scales like the last †.

* Latr. Gener. Crust. and Insect. iv. 225. Consid. Gen. sur les Crust. &c. The males of all the species have pectinated antennæ, and may form a peculiar subgenus.

† The genus *Noctua*. Fabricius forms of it, in the History of European Lepidoptera of Ochsenheimer, forty-two, beginning with that of *acronicta*, unto that of *euclidia*, inclusive. These are, for the most part, all the divisions established in the systematic catalogue of the lepidoptera of Vienna, transformed into genera, the exposition whereof the nature of our work forbids. That of *Erebus*, being taken from it, appears to us to be divisible into two grand parallel series: one connected with these last mentioned lepidoptera, and the other with *notodonta*. The first is composed of noctuelles, whose caterpillars walk like those which have been called carpenters or geometricians. Some have sixteen tarsi, but the first two or four of the intermediate membranes are shorter; the others have only twelve. Such are the subgenera *plusia* and *chrysoptera*, distinguished from the preceding by the size of the lower palpi, which turn back on the head. The second series will commence by species whose palpi are proportionally larger, whose antennæ are pectinated, and whose proboscis is small; such are *Odonptera palpina*, Nob., and *calyptra* of Ochsenheimer, or the *calpe* of M. Treitschke. In following the genera *Xylena, Cucullia,* the night moths, whose upper wings have the hind edge angular or indented, those whose antennæ are pectinated, and finally those in which these organs are simple. We shall terminate these last species by that whose thorax is united, some of which, of the genus *Erastia*, of this naturalist, appear to lead us to *Pyralis*. All the caterpillars of this second series have sixteen tarsi, with the intermediate membrane of equal size; their walk is rectigrade. The chrysopteron (*Plusia concha*, Fisch. Entom. de la Russ. i. Lepid. iv.) by which we shall finish the other series, have relation to *Herminia* and *Pyralis*. Thus the two series appear to terminate, converging to this last section. Lichenia, or Catocales, of Ochsenheimer, are

Among the night-moths, properly so called, there are some, and that the greatest number, whose caterpillars have sixteen tarsi. We will remark—

N. sponsa, Fab., Rœs. Ins. iv., 19, of an ashen grey, thorax crested, wings overlapping; the upper side of the upper wings of an obscure grey, with black stripes, much undulated, and a whitish spot, divided by some black stripes; the upper side of the under wings is bright red, with two black bands: the abdomen is ash.

The caterpillar lives on the oak. Grey, with irregular obscure spots, and little tubercles; its eighth ring has a bunch, on which is a yellow patch. This species, and some others, are said to be lichenous, because their caterpillars are coloured like the lichens, or the trees. They have the four anterior membranaceous feet shorter, and walk like the carpenter caterpillars.

N. pacta, Fab., is of this number: it is distinguished from the others by the red colours of the upper part of the abdomen, and is found only in the north of Europe *

The caterpillars of some have only twelve tarsi. The perfect insect often has gold or silver spots on the upper wings; such as the two following species †:

N. gamma, Fab., Rœs. Ins. i. class iii. *Pap. noc.* 5, has

large species, with the wings nearly horizontal, which appear naturally to approach Erebus, as well as Ophiusia, Brephos, &c. If they are placed in the other series, they will disturb the harmony of it.

Bombyx cyllopoda, of Dalman, (Analect. Entom. 102.) ought to form a new and remarkable subgenus, inasmuch as the two hind feet of the males are shorter than the others, imperfect, and almost useless in walking. These insects having the antennæ pectinated, the proboscis distinct, and the palpi twice as long as the head, appear to be near the genus *Calyptra*, of Ochsenheimer, or near our *Herminiæ*.

* These two species belong to the genus *Catocala*, of Och.

† Genus *Plusia*, of the same.

the thorax crested; the upper side of the upper wings brown, with brighter parts, and a golden spot, representing a lambda or gamma, in the middle. When the hind extremity of the abdomen of the male is pressed, two tufts of hair are protruded. The caterpillar feeds on various culinary plants.

N. chrysitis, Fab, Esp. noct., 109, f. 1—5. Upper wings clear brown, traversed by two bright brass-coloured bands.

Some of the caterpillars, as that of *N. verbasci*, *N. artimisiæ*, *N. absinthii*, &c., have a peculiar habit of feeding on certain particular flowers *.

Other species of night-moths have the antennæ pectinated, as *P. graminis*, Lin., whose caterpillar sometimes ravages the meadows of Sweden.

The sixth section of nocturnal lepidoptera,

PHALÆNÆ TORTRICES, *Lin.*, have the nearest relation to the two preceding sections. The upper wings, whose exterior edge is arched at the base, and then becomes narrower, and their short and broad shape, like a truncated oval, gives these insects a peculiar appearance. They have been said to be *large-shouldered Phalænæ*. They all have a distinct proboscis, and the lower palpi are in general nearly like those of noctua, but a little advanced.

The lepidoptera are small, prettily painted; carry their wings in a compressed roof, or nearly horizontal, but always lying the upper wings across each other a little along their inner edge.

Their caterpillars have sixteen tarsi; the body is commonly

* They belong to the genus *Cucullia* of Schrank, and other writers on lepidoptera. For the other species, see Olivier, art. Noctuelle, Encyclo. Method., and Latr. Gener. Crust., and Insect. iv. 224. See especially the work of Ochsenheimer on the European Lepidoptera, and the Natural History of those of France by Goddart, continued by Duponchel, well known to entomologists, by his interesting monograph on the genus *Erotylus*, already quoted, and by several memoirs.

shorn, or but little velvety; they twist and roll the leaves; they fix successively, and in one direction, several points at the surfaces of the leaves, by layers of silken threads, and thus make a tube, in which they are covered, and there tranquilly devour the parenchyma of these leaves. Others make a retreat of many leaves or flowers tied together with silk. There are some who prefer fruits.

Many have the hind extremity of the body narrower, and Reaumur calls them *fish-formed* caterpillars. Their cocoon is boat-shaped. These cocoons are sometimes of pure silk; sometimes made of several materials.

The Tortrices include the subgenus

PYRALIS, *Fab.* *

P. Pomana, Fab., Rœs., Insect. i. class iv. *Pap. noc.* 13. Ashy grey; the upper wings finely striped above with brown and yellow, with large golden red spots. The caterpillar feeds on the pips of apples. The perfect insect has deposited her eggs on their germ.

P. vitis, Bosc. Mem. de la Soc. d'Agric, ii. iv. 6. The upper wings deep greenish, with three oblique blackish bands,

* Some divisions established in our Gener. Crust. and Insect. iv. 230, div. 2 and 11, have appeared to us (Fam. Nat. du Regne Animal, 476), to form distinct subgenera.

Some species (*Tortrix dentana*, Hüb.) which carry their wings in a peculiar manner, the upper rising a little at the outer edge and inclining toward the opposite edge, and whose caterpillars have membranaceous tarsi of a peculiar form, which Reaumur compares to wooden legs, compose the subgenus *Xylopoda*. Other species. *P. rutana, umbellana, heracleana,* whose lower palpi bend back over the head like horns, and go to a point, form that of VOLUCRA.

Others also, having the wings narrow and elongated, and the lower palpi longer and advancing, species very nearly allied to *crambus* of Fab., near which, probably, we ought to place them, constitute a third subgenus, PROCERATA, having for its type *P saldonana*, of Fab.

For the other species, see Fabricius and Hübner.

the third of which is terminal. The caterpillar does great mischief in vineyards.

P. Prasinaria, Fab., Rœs. Ins. iv. 10. The largest of the known species; upper side of upper wings soft green, with two oblique white lines. On the elm and the oak: its caterpillar is one of those Reaumur compares to a fish. Its cocoon is boat-shaped.

The seventh section of nocturnal lepidoptera, *Phalænites*, Lat., *P. geometræ*, Lin., includes lepidoptera whose body is commonly slender, with the proboscis either merely rudimentary, or but little elongated, and submembranaceous. The lower palpi are small and subcylindrical; the wings are ample, stretched, or like a flat roof. The antennæ of many of the males are pectinated. The thorax is always united. The caterpillars have commonly but ten tarsi, but others have two more; the anal tarsi are always found. Their name of carpenters or geometricians originates in their mode of walking. When they advance, they fix themselves first by the fore tarsi, or scales; then they elevate the body into an arch, bringing the posterior or extremity of it to the part, or that which is fixed; they then fasten themselves by the hind tarsi, disengage the anterior, and carry the body forward, fixing anew the anterior scaly feet, and repeating the same operation. Their attitude when at rest is very extraordinary: fixed to the branch or stem of a vegetable only by the hind tarsi, their body is suspended in the air in a straight and perfectly immoveable line. It so much resembles the branches and stems, by the colours and inequalities of the surface of its body, as to be mistaken for them. The animal remains many hours, and even entire days, in this strange position. The chrysalides are almost naked, or their cocoon is very thin, and has but little silk.

This section includes, if we abstract the caterpillars from the consideration of it, but one subgenus,

Phalæna, properly so called. The caterpillar of *P. margaritaria* *, Fab., has twelve feet; the others have but ten.

P. sambucaria, Lin., Rœs., Ins. i. class iii. *Pap. noct.* 6. are of the largest species, of this country, of a sulphur colour; the wings are stretched, and marked with two transverse brown stripes; the under wings are elongated at the outer angle like a tail, and one may observe there two small blackish spots. The caterpillar is brown, and resembles, in form and colour, a little club; the head is flat and oval. Dr. Leach (Miscel. Zool.) forms this phalæna, and some other species whose under wings have the same figure, into a genus *Ourapteryx.*

We shall quote, moreover,

P. syringaria, Lin., Rœs., *ibid.* x., whose antennæ are pectinated in the male; the wings are angular and jaspered, with a mixture of yellowish brown and reddish. The caterpillar has four great tubercles on the back, and a little horn or crochet on the eighth ring.

P. grossulariata, Lin., Rœs., *ibid.* ii., whose wings are white, dotted with black; there are two light yellow bands on the upper side of the upper wings, one toward the base, and another near the middle. The caterpillar is bluish grey above, dotted with black, with the lower sides of the belly yellow, sprinkled with black.

The female of *P. brumata*, Lin., as well as those of some other analogous species, have only rudiments of wings. These species appear in winter †.

Degeer describes a species, *Ph. à six ailes*, whose males appear to have six wings, the under wings naving, on the inner side, a small appendix, which is hidden under them ‡.

* The type of my subgenus *Metrocampa.*

† These species form my subgenus *Hybernia.*

‡ For the other species, see Fabricius and Hübner.

The eighth section of nocturnal lepidoptera, DELTOIDES, *Lat.*, furnishes some species very analogous to Phalena, properly so called, but whose caterpillars have fourteen tarsi, and are rollers and sewers of leaves. In the perfect insects, the lower palpi are elongated and bent back; the wings form, with the body, on the sides of which they stretch horizontally, a sort of delta, whose posterior edge has in the middle a re-entering angle, or appears forked. The antennæ are commonly pectinated, or ciliated *

The deltoidinian lepidoptera compose the subgenus

HERMINIA, *Lat.*, which belongs to the division *Phalenæ pyralides*, of Linnæus, and is made up of the genera *Hyblæa* of Fab., and of many of his *Crambi* †.

The ninth section of nocturnal lepidoptera, TINEITES, *Lat.*, *Ph. Tineæ*, Lin., and the greater part of his *pyralides*, include the smallest species of this order, whose caterpillars are always shorn, provided with at least sixteen tarsi, and rectigrade, and live concealed in habitations of their own making, whether fixed or moveable. In these the wings make a kind of elongated triangle, almost flat, and terminated by a re-entering angle. Such are *Ph. pyralides*, of Lin., which may form a section of themselves. These species have four distinct palpi, commonly uncovered. In them the upper wings are long and narrow, sometimes moulded on the body, and giving the shape of a rounded ridge; sometimes they incline almost perpendicularly close to the sides, and are often elevated or ascending behind like a cock's tail. In both cases the

* This section included, in the first edition of this work, all the *Phalenæ pyralides* of Linnæus; but a complication of characters resulted from it, which disappears, if we include it in Herminia only. That of Tineæ will be then composed exclusively of tineæ and pseudo-tineæ of Reaumur.

† Latr. Gener. Crust. and Insect. iv. 228.

lower wings are always large, and folded. These species, moreover, often have the four palpi uncovered.

All the caterpillars whose habitations are fixed or immoveable, are the *pseudo-tineæ* of Reaumur; those which make moveable retreats, and which they carry about, are the *tineæ*, properly so called.

The substances on which they feed, or on which they constantly remain, furnish the materials for these retreats.

Among those composed of vegetable substances, there are some extremely curious. Some of these are covered on the outside with bits of leaves, applied one upon the other, forming a kind of flounce. Others are formed like a cross, and sometimes indented on one side only. There are some, the component matter whereof is transparent, as if cellular, or divided by scales.

The caterpillars of the *tineæ*, properly so called, commonly called worms, clothe themselves with pieces of woollen stuff which they cut with their jaws, and on which they feed, on hair, furs, and the skins of stuffed animals, which they unite with silk. They know how to lengthen at one end their covering, or to augment its bulk, by dividing it, and adding a new piece. They undergo their metamorphosis there, after having first closed the openings of it with silk. We must consult the works of Reaumur, Rœsel, and Degeer, to learn the manner in which they make these habitations, as well as the differences of their composition and figure.

The false tineæ merely excavate the interior of the animal and vegetable substances on which they feed, and thus form simple galleries, in which, if they construct a case, whether with these substances or with silk, such habitations are always fixed, or used as a mere place of retreat.

The caterpillars, which dig in various ways into the parenchyma of the leaves on which they feed, have been called miners. They produce those dried spot-like spaces of undulated lines observable on many leaves. Buds, fruits, seeds, and often that of corn, and even the resinous galls of cariferous trees, supply food and a domicile to others. These lepidoptera are often ornamented with very brilliant colours. The upper wings, in many species, have gold or silver spots, or points, sometimes even in relief.

Some which have the four palpi always distinct *, uncovered, or merely partly hidden (the superior) in part by the scales of the chaperons, advanced, and of middling size, resemble *P. pyralides* of Linnæus; their wings disposed in a ridge, in general flat, or but little elevated, form an elongated triangle, or sort of delta.

Sometimes the spiral tongue is very apparent, and serves common usages. The caterpillars of these species live on various plants.

Botys, *Lat.* Their caterpillars are leaf rollers, and do not differ superficially from the others as to the respiratory organs.

P. urticata, Lin. Rœs. Insect. i. Phal. xiv. The thorax and end of the abdomen are yellow, and the wings white, with blackish spots, forming bands. The caterpillars sew the leaves of the nettle, and remain nine months in the cocoon which they spin, before their transformation to the nymph; it is shorn green, with a deeper stripe along the back.

The same plant supplies food for the caterpillar of another

* The Yponomeut, with one or two exceptions, the Œcophoræ and Adel, are almost the only Tineæ whose upper or maxillary palpi are not very apparent; but as they may be hidden by the lower, and as it is very difficult to establish, in this respect, a fixed and determined line of demarkation, we have not thought it necessary to divide this group by the number of these organs. M. Savigny, in his Memoirs on Invertebrated Animals, has given figures by which they are represented under various degrees of proportion. The new genera, which he has only named, are unknown to us.

species of the same subgenus *. *B. verticalis*, Lin. Rœs. ibid. i. Phal. iv. 4. The perfect insect is pale yellowish, and shining, with some obscure transverse stripes more marked beneath.

HYDROCAMPE, *Lat.* These are species very analogous to the last, but their caterpillars are aquatic, and have, in general, appendices in the form of threads or long hairs, the innermost of which have some tracheæ. They make for themselves, with the leaves of several aquatic plants, tubes in which they remain covered †.

Sometimes there is little or no spiral tongue.

AGLOSSA, *Lat.* The four palpi are exposed, and the wings form a flat triangle; the upper wings have no emargination at their extremity.

P. pinguinalis, Lin. Deg. Insect. ii. vi. 4. 12. Reaum. Insect. iii. xx. 5. 11. The upper wings are grey, with blackish stripes and spots. They are found in houses on the walls. The caterpillar is shorn, of a blackish and shining brown, and feeds on greasy and buttery substances. Reaumur calls them false tineæ of leather, because they also gnaw that matter as well as the bindings of books. They build a retreat in the shape of a long tube, which they apply to the bodies on which they feed, and which they cover with grains, composed, for the most part, of their own excrements. According to Linnæus, they are found, though but seldom, in the stomach of mankind, where it produces more alarming effects than the intestinal worms. A physician, whose veracity I cannot doubt, sent me some caterpillars of this species which had been vomited by a young woman.

That of another Aglossa, *P. farinalis*, Lin., feeds on farina. The perfect insect is also often found on walls, where it re-

* *Ph. forficalis, purpuraria, margaritalis, alpinalis, sanguinalis*, &c. of Fab.

† *P. potamogata, stratiolata, paludata, lemnata, nympheata*, &c.

mains motionless, with the abdomen elevated. The under side of the upper wings is reddish, edged with white behind; the posterior extremity is also reddish, but this colour forms there an angular spot, edged above by a white stripe, also angular; the space included between these spots, or the middle, is yellowish.

GALLERIA, *Fab.* The scales of the chaperon form a projection, covering the palpi; and the upper wings, shorter in proportion than those of aglossa, and emarginated at the lower edge, are, as well as the under wings, strongly inclined, and elevate themselves behind like a cock's tail, as in many species of the following subgenera.

G. cereana, Fab. Hübn. Ins. iv. 25., is about five lines long, ash, with the head and thorax lighter, and little brown spots along the inner edge of the upper wings. Reaumur names the caterpillar—the *wax false tinea.* They do great mischief in bee-hives, piercing the combs, and building, as they proceed, a silken tube, covered with their excrements, and formed of the wax on which they feed. The cocoon of their chrysalides are sometimes found in heaps.

G. alvearia, of Fabricius, approaches nearer to *tinea* than to this subgenus.

His *Crambus erigatus,* and *Tinea tribunella,* and *colonella,* of Hübner, approximate to the preceding by the arrangement of their wings, but their lower palpi are much longer, and these insects have in this respect more affinity to *crambus:* they may form particular subgenera.

Others, whose upper palpi are not always very distinct, have the upper wings long and narrow, sometimes rolled on the body, and sometimes applied perpendicularly to the sides. In this state, the insect has always a narrow and elongated form, approaching a cylinder or cone.

Here, the lower palpi, always large, are carried forward; the last articulation at most is elevated; the upper palpi are apparent.

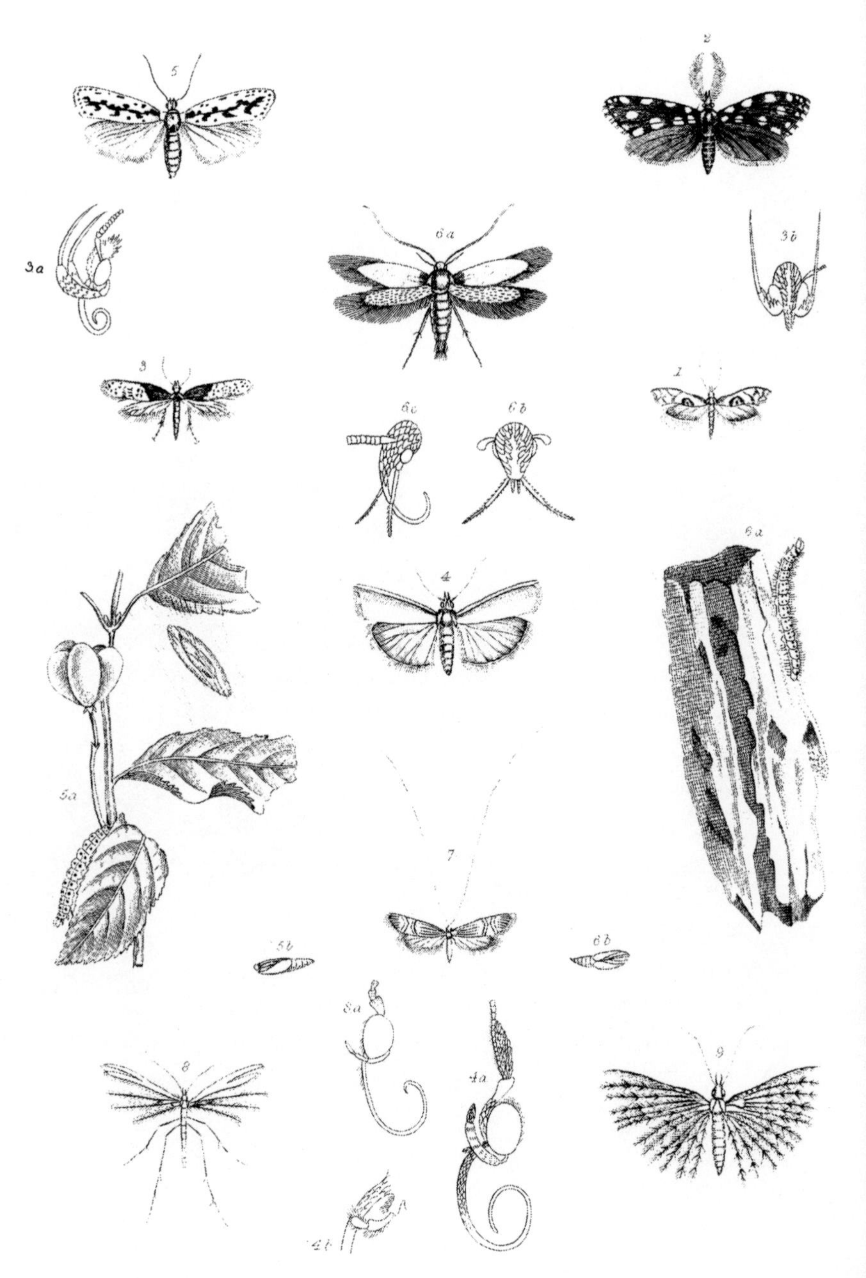

1. Alucita asperella. Hubn.
2. Euplocamus anthracinus. Hubn.
3. Tinea tapezana, Fab. 3. a. Head of Tin. longicornis. Curtis.
4. Ilithyia carnea. Latr. 4. a. b. Details of Il. pinguis. Curtis
5. Yponomeuta pusiella. Hub. 5. a. b. Cat. & Chrys. of Yp. plumbella. H.
6. Æcophora Linneella. Clerk. 6. a. b. Cat. & Chrys. of Æc. majorella. Hub.
7. Adela Degeerella. Fab.
8. Pterophorus ptilodactylus. Hub. 8. a. Head of Pter. spilodactylus. Curtis
9. Orneodes hexadactylus. Latr.

London, Published by Whittaker & Co. Ave Maria Lane. 1832.

CRAMBUS, *Fab.*, have a distinct proboscis; the lower palpi advance, like a narrow bill, to the end. These lepidoptera are found in dry pastures, on many species of plants *.

ALUCITA, *Lat.*, *Ypsolophus*, Fab., have also a distinct proboscis, but the last articulation of the lower palpi is elevated. The antennæ are simple †.

EUPLOCAMUS, *Lat.*, *Phycis*, Fab. with a very short, and scarcely visible proboscis, having, moreover, the last articulation of the lower palpi elevated; the scales of the preceding form a bundle. The antennæ of the males have a double range of little barbs ‡.

PHYCIS, *Fab.* Altogether like the last, but with the antennæ entirely ciliated ||.

There the lower palpi are entirely elevated, and even bent back above the head in many.

Sometimes the lower palpi are very apparent, and of middle size. The antennæ and the eyes are directed laterally.

In the two following subgenera the lower palpi scarcely pass the forehead.

TINEA

Have the proboscis very short, formed of two small membraneous and disjointed threads. Their head is crested.

Pyralis tapezana, Fab. Reaum. Insect. iii. xx. 2. 4. The upper wings black; their posterior extremity, as well as the head, white.

The caterpillars gnaw woollen drapery, and other similar

* Fab. Entom. Syst. Sup. and Latr. Gener. Crust. and Insect. iv. 232. See Hüb. Tin. v. viii. *Crambus carnellus* belongs to another subgenus, ILITHYIA.

† Lat. ibid. 233. Unite to the same subgenus the crambus of div. ii. ii. p. 232.

‡ Lat. Gen. Crust. and Insect. iv. 233.

|| *Phycis Coleti*, Fab.

stuffs; they are concealed under a vault, which they form as they proceed. This is a false tinea of Reaumur.

Tinea sarcitella, Fab. Reaum. Ins. iii. vii. 9, 10. Of a silvery grey, and a white point on each side the thorax. The caterpillar is found on woollens. They make their immoveable tubes by weaving with silk the morsels they detach; they elongate it at the end in proportion as they grow, divide it to enlarge it and insert a piece; their excrements are of the colour of the wool on which they have fed.

T. pellionella, Fab. Reaum. Insect iii. vi. 12—16. Upper wings silvery grey, with one or two black points on each. The caterpillar lives in a felted tube on furs, whose hairs they cut at the root, and thus quickly destroy them.

T. flavifrontella, Fab., destroy natural history collections in the same manner *.

T. granella, Fab. Rœs. Insect. i. class iv. Pap. noct. 12. The upper wings are marbled grey, brown, and black, and are raised behind. The caterpillar ties several grains of corn with silk, and so forms a tube, whence they issue, from time to time, to gnaw the grains and do much mischief.

Ilithyia, *Crambus* †, Fab., have a very distinct ordinary sized proboscis, and the last articulation of the lower palpi is evidently shorter than the preceding.

Yponomeuta, *Lat.*, have also a distinct common sized proboscis, but the last articulation of the lower palpi is at least almost as long as the preceding.

These insects appear connected with Lithosia.

T. evonymella, Fab. Rœs. Insect. i. class iv. *Pap. noct.* 8.

* All authors who have described or figured Tineæ, and other analogous lepidoptera, not having studied them in a strict manner, it is impossible to refer the species mentioned by them to our several subgenera.

† *Crambus carneus*, Fab., and some other species. The antennæ of the males have underneath a swelling like a knot.

Upper wings bright white, with very numerous black points; the lower wings are blackish.

T. padella, Fab., Rœs. ibid. vii. Upper wings lead grey, with a score of black points.

The caterpillar, as well as that of the last, live in an extensive society under one nidus *. They sometimes multiply prodigiously on our fruit trees, devouring the leaves; the branches appear to be covered with crape. In the following subgenus, that of

Œcophora, *Lat.*, the lower palpi are bent back above the head like horns, going to a point, and reaching to the back of the thorax.

The *corn tinea*, which is often so destructive in the south of France, and which is uniformly of the colour of coffee with milk, belongs to this subgenus. I also refer to it *Tinea harisella*, whose caterpillar, according to the observations of the younger Hübner, forms a kind of hammock †.

Sometimes the lower palpi are very small and velvety, the antennæ are almost always very long, and the eyes are very near.

Adela, *Lat.*, *Alucita*, Fab. These insects are found in woods, and many species appear as soon as the oak leaves begin to shoot out. Their wings are generally brilliant.

A. Degeerella, F. de G. Ins. i. xxxii. 12. Antennæ three times the length of the body, whitish, with the lower part black; the upper wings of a golden yellow, on a black ground, forming longitudinal stripes, with a large band of gold yellow, traversed, and edged with violet.

* See Lat. Gener. Crust. and Insect. iv. 222. and Hist. Nat. de Lepid. de France de Godart.

† *T. majorella*, *geoffroyella*, *rufimitrella*, &c. of Hübner. With reference to this and the last subgenus, see the monograph of *Phycis*, inserted in the third volume of the Entomological Magazine of M. Germar.

A. Reaumurella, Fab., is black, with the upper wings golden and spotless *.

The tenth and last section of nocturnal lèpidoptera, FISSIPENNES (*pterophorites*, Lat.,) have considerable relations with the last, as well as to the narrow and elongated form of the body, and the upper wings, but departs from them as well as all others of the same order, in this, that the four wings, or two at least of them, are cleft all along like branches, or fingers barbed on their edges, and are similar to feathers: these wings imitate those of birds.

Linnæus includes these lepidoptera in his division *Phalanæ alucites*. Degeer calls them *Phalènes tipules*.

We shall form of them, with Geoffroy and Fabricius, the subgenus

PTEROPHORUS.

Their caterpillars have sixteen tarsi; feed on leaves or flowers, and do not build retreats.

Sometimes the lower palpi bend back from their insertion, are entirely covered with small scales, and are not longer than the head. These compose the genus *Pterophorus*, properly so called, of Latreille. Their chrysalides are naked, beset with hairs on small tubercles, sometimes suspended by a thread, sometimes fixed by means of crochets at the lower end of the body, to a bed of silk, on leaves, &c.

P. pentadactylus, Fab., Rœs. Ins. i. class iv. *Pap. noct.* 5. Wings white as snow; the upper divided into two slips, and the lower into three †.

Sometimes the lower palpi are advanced longer than the

* See Fab. Entom. Syst. Sup.; Latr. Gener. Crust. and Insect iv. 223, and Hübner, Tineæ xix.

† For the other Pterophori of Fab., with the exception of *P. hexadactylus*, see also Hübner and Degeer.

head, with the second articulation very scaly, and the last nearly denuded and elevated. The chrysalis is enclosed in a silken cocoon. Latreille distinguishes these species under the generic name of ORNEODES*.

* *P. hexadactylus*, Fab. See Lat. Gen. Crust. and Insect. iv. 234, 235.

SUPPLEMENT

ON THE

ORDER LEPIDOPTERA.

THE character of this order will oblige us to deviate a little from the plan which we have hitherto pursued in our supplementary additions to the text of the "Regne Animal." The habits of lepidopterous insects, in their perfect state, afford but slender materials for popular disquisition, however attractive and interesting they may be for their splendour and beauty; but in their first, or caterpillar state, these habits are objects of great interest and curiosity. We shall, therefore, embrace in one general observation, all that we have to say concerning the order, without following the subdivisions in succession, as we have before done. Our supplement will thus naturally divide itself into three parts,—first, some general brief observations on lepidoptera; secondly, the history of caterpillars more detailed; and, thirdly, a notice of such lepidoptera in their perfect state, as may present any thing, in their habits or manners, worthy of the attention of the general reader.

It forms no part of our present object to say much respecting the structure of these insects; but it is impossible for us to pass over in silence the peculiar covering of the perfect insects of this order, and the singular conformation of the mouth.

The order was named by Linnæus from λεπις ἰδος, and

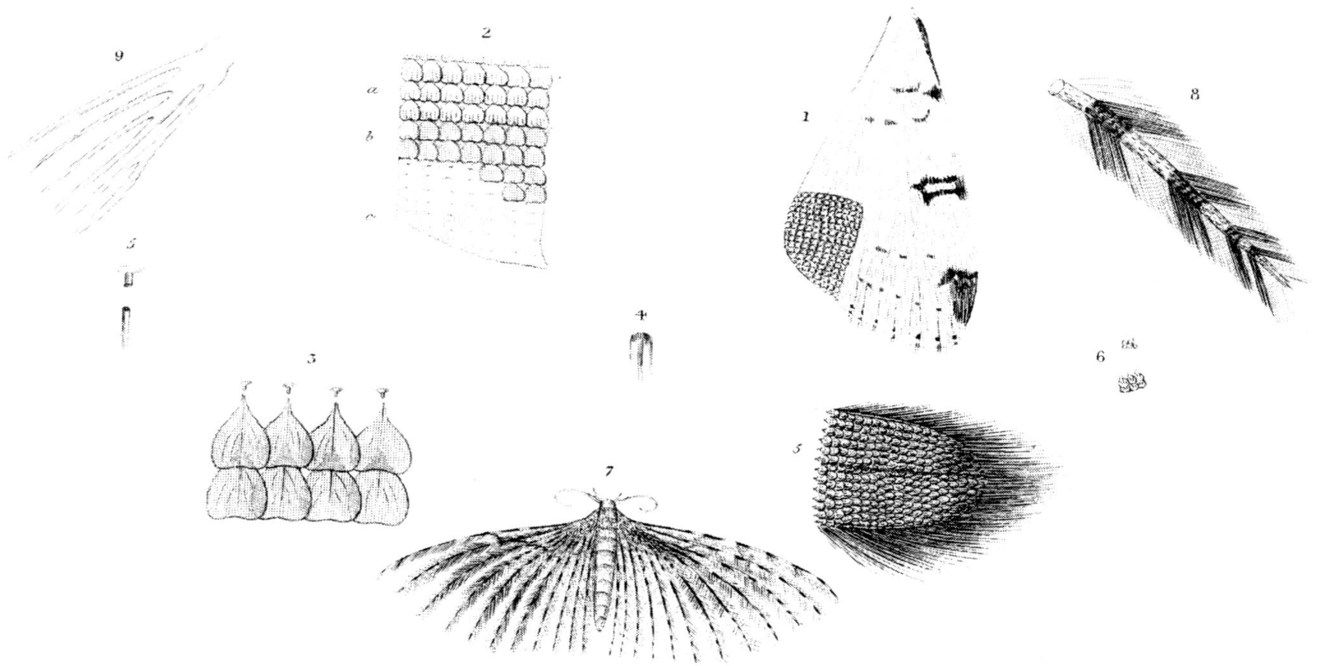

Scales or Feathers of Lepidoptera.— Pl. 2.

Published by Whittaker & Co. Ave Maria Lane, London 1837.

Pl. 81

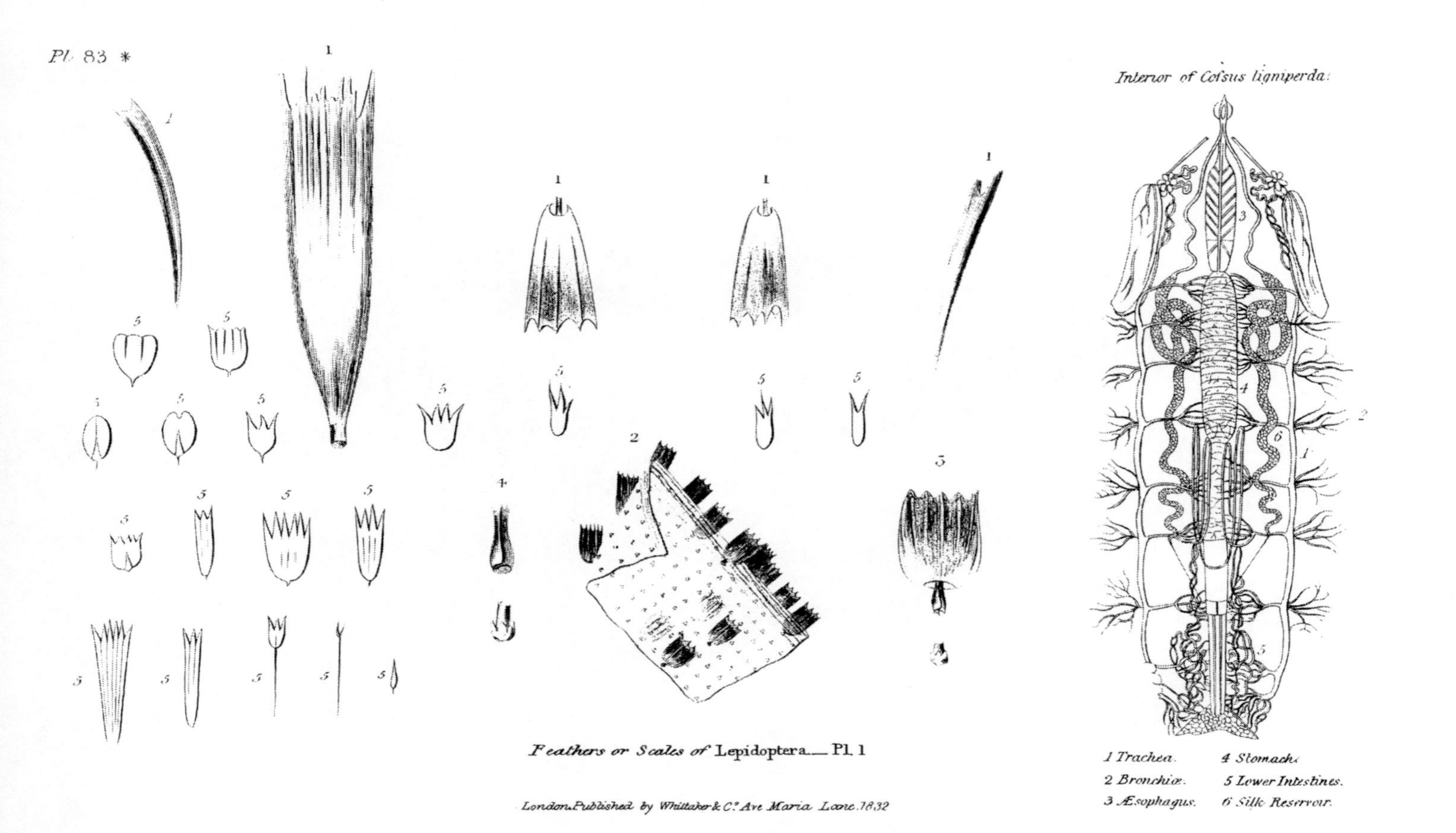

Feathers or Scales of Lepidoptera — Pl. 1

London, Published by Whittaker & C°. Ave Maria Lane, 1832

πτερα, or winged with scales, a character peculiar to the insects proper to it alone, whose wings are covered on both sides with imbricated scales or feathers, presenting to the unassisted eye nothing but the appearance of mere particles of dust or powder, but exhibiting under the microscope a mathematically regular series of scales or feathers, peculiar, as to shape or size, to each species.

It has been much doubted whether these little splendid appendages to the wings are most analogous to feathers or to scales; as they are terminated by a sort of quill, or stem, they seem in that respect much to resemble feathers, but by the nature of their substance, their shape and indentation, they seem to approximate to scales. Four hundred thousand of these have been counted by the patient Leeuwenhoek, on the wings of the silk-worm moth; but this must be a small number, compared with that on the wings of some others of a larger size, as the *Atlas*, though in this and other allied genera, it will be observed, that there are portions, or a portion, of each wing entirely denuded of these scales, or feathers, presenting a semi-transparent appearance; indeed, in some, (*nudaria*) the wings have none of this sort of covering, though in all other respects, the insects evidently belong to this order.

These scales vary infinitely in size, and consequently in number and in shape; some are long and slender, others precisely the reverse: they are to be found round, oval, oblong, rhomboidal, parabolical, triangular, and, in fact, of almost all intermediate figures. Some terminate in a point, others have several, indented, lanceolate, &c.; the surfaces of most, if not all of them, seem to be divided into longitudinal ridges, or furrows, and in some instances may be found, by a strong magnifying power, to have transverse ridges also. We insert, Pl. 83. f. 1. figures of various feathers taken from *Papilio urticæ*, magnified by the compound microscope about 160,000 times. When so much magnified, they assume a hollow form, but it may be hardly safe to conclude that to be their actual conformation, as in the use of the highest powers of the compound microscope, we are liable to much deception, so varied and so casual is the degree of light to be obtained. Fig. 2. represents a part of the membrane of the wing, with a few feathers remaining on each side of it: fig. 3. is one feather greatly magnified, with the petiolus a little removed from its stalk, which stalk and petiolus are still more magnified at fig. 4. displaying the manner in which the former is convoluted, and embraces the latter by its elasticity. Fig. 5. represent the outlines of various other scales.

In some (*P. cratægata*) the membrane of the wing has first a series of orbicular feathers, which are covered by long setaceous feathers. Pl. 81. fig. 1. is the upper wing of this moth, magnified with a part of the upper series of setaceous feathers removed on one side to show the under or orbicular feathers. Fig. 2. is a portion of this wing deprived of the setaceous feathers altogether, with the orbicular feathers remaining on the upper side in part *a,* on the under side only in part *b*, and the membrane of the wing itself deprived of all its feathers on both sides in part *c.* The mode of imbrication of these orbicular feathers, the upper series being a little removed from the stalks, is depicted at fig. 3.; and fig. 4. shows the convolution of the petiolus. Fig. 6. is a few of the feathers taken from the brown spot on the wing. Fig. 7. represents *Ph. hexadactyla;* 8. a single plume more magnified, to display its tubular form; and 9. the radii of the left anterior wing. Fig. 5. shows the lip of the under wing of *Pha. Tinea semiargentella,* with its orbicular and setaceous feathers.

In the formation of the mouth of the lepidoptera, nature appears to have had the intention of developing the jaws, at the expense of the other parts. She has connected their jaws into a kind of sucker, naked, and considerably elongated.

But it is not easy to conceive the mode in which she has changed them into so many tubes. The examination of the proboscis of bees, can alone conduct us to the solution of this difficulty. In them the jaws and tongue are very long, elbowed, and curved underneath. These first parts are so many valvules, like a half-sheath. Let us suppose, that the tongue, which has the form of a tube, be divided longitudinally into two portions, that each of them be incorporated with the jaws, and form their internal paries; that the lower sheath of this tongue, or the chin, be shortened, widened in the form of a membrane, and fixed; that the radical part of the jaws, as far as the origin of their palpi, also undergo analogous changes. In fine, let us suppose that the mandibles and labrum are very much diminished, and become, as it were, nullified; we shall then have transformed the mouth of the bee into that of a lepidopteron. Let us observe, that in the hemiptera, the parts representative of the grinding organs are considerably elongated, and the palpi disappear; in the diptera, the part corresponding to the chin receives the greatest degree of development, so that nature presents, with respect to the mouth of sucking insects, three different combinations:—First, naked jaws, forming of themselves a sucker, whose action is independent of the other parts,—*lepidoptera;* secondly, mandibles, jaws, tongue and chin, greatly elongated, this last part serving as a case to the preceding, which compose the sucker,—*hemiptera;* thirdly, chin also performing the office of sheath, but much more developed; sucker usually short, and varying in the number of its pieces, —*diptera.*

It is to be observed, that the spiral tongue varies in the minute construction of the parts in every species possessing that organ. Our figures of the details of the tongue in *Papilio urticæ,* while they display the peculiarities incident to that species, may serve to illustrate the general character of the organs in all the others. Fig. 1. at Pl. 28. represents, of course greatly magnified, the head, with the eyes and spiral tongue having the left palpus removed, to show the convolution of the tongue. Pl. 140. fig. 1. shows the tongue taken out and partly unrolled, displaying the two segments, as united with the rugæ, of which they are composed, and the papillæ or absorbents toward the end; 2. is a front view of the mouth without the palpi, the clipeus being raised to show the opening of the mouth and the insertion of the tongue; 3. is the under part of the mouth and tongue, with the articulations of the palpi omitted in the last figure; 4, 5, and 6, display the termination only of this organ, with the papillæ; the first is a front view; the next has the segments separated, showing the filaments and membranes which connect the segments; and the third is a lateral view, with the papillæ. Fig. 7 shows the manner in which the tongue, after the extremity of it has been applied to the nectary of a flower, and has absorbed the juices, is conveyed to the mouth, and the papillæ cleansed by scraping on the under part of it.

We never find, in this order, more than two kinds of individuals, males and females. Some amateurs, in breeding caterpillars, have obtained, but very seldom, individuals which they have considered as hermaphrodites, in consequence of the dissimilarity of their antennæ; but these are aberrations analogous to what we observe sometimes in domestic animals, or real monstrosities. The males of the lepidoptera make their appearance first, and are very ardent in pursuit of the females. Those of many nocturnal lepidoptera discover the places of their retreat, by means of a very fine sense of smell, for they penetrate into our houses, for the purpose of fecundating such females as have been taken captive, or have been reared up from the caterpillar state. The Chinese fix on rings with threads the females of two species of wild *Bombyx*, whose caterpillars produce

Pl. 140.

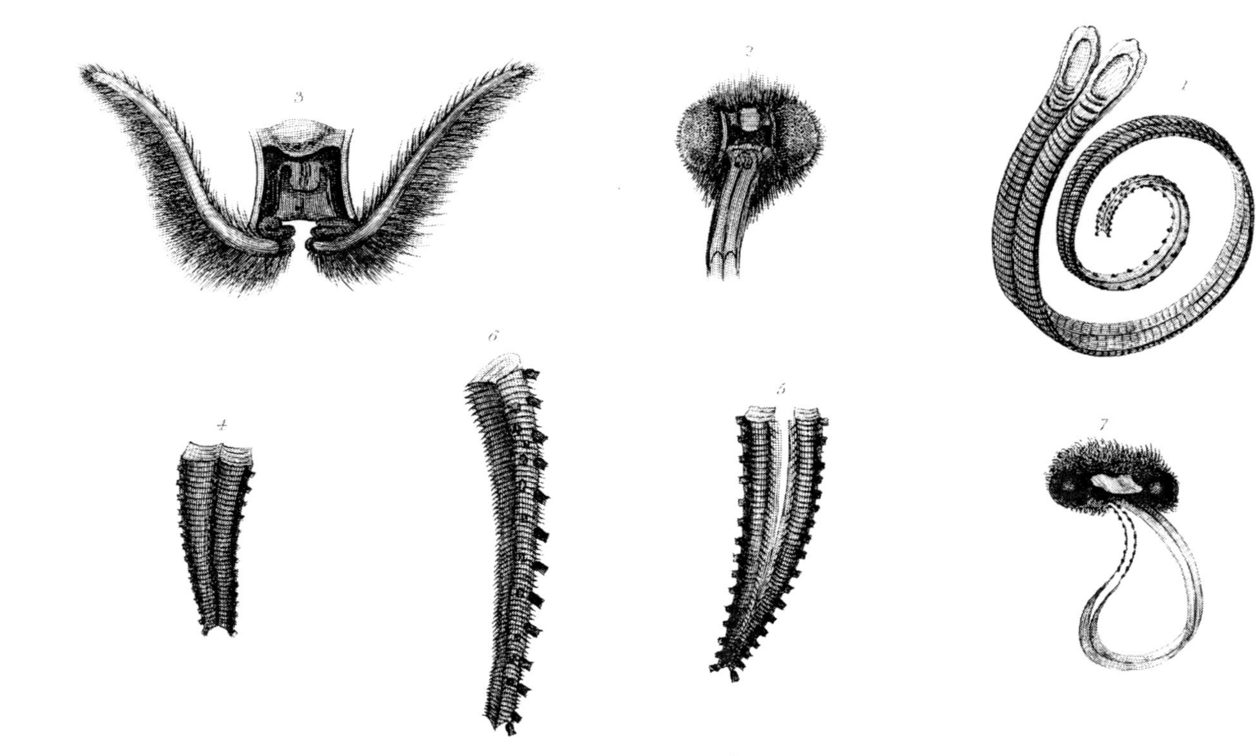

Tongue or Proboscis (Antlia Kirby) of Lepidoptera.

London. Published by Whittaker & C°. Ave Maria Lane. 1832.

silk, place these insects on a tree, or on some body situated in the open air, and the males, guided by their scent, visit the females, to the advantage of the proprietor.

The sexual union lasts a considerable time. The female frequently carries along with her into the air, the male, who is always smaller, and otherwise differs from his companion, in the narrower form of his abdomen, and also often in the antennæ, the tint of the wings, or some modification in their pictorial design.

The females lay their eggs, often very numerous, and of regular or of very varied forms, usually upon vegetable substances, on which their larvæ are to feed. They are fixed there by means of a peculiar viscosity, and arranged sometimes side by side, with considerable art, or even covered with small silken hairs, which the female, for this purpose, detaches from her abdomen. The deposition of eggs generally takes place but once a-year. Many diurnal lepidoptera, nevertheless, deposit twice, first in Spring, and afterwards towards the end of summer, or in autumn. When their amours are over, these insects, like almost all others, very speedily perish. A few females, however, of the number of those produced in the latter part of the season, occasionally survive the rigours of the winter.

We shall now speak of *caterpillars* in general, the history of which we shall find to embrace by far the major portion of the history of the entire order. We are necessitated to commence with some notice of their structure, on which subject, however, we shall study brevity as much as possible.

The caterpillar is a lepidopterous insect in its first state, or from the time in which it issues from the egg, to the period of its transformation into a chrysalis. Under this form, it is one of the most destructive animals with which we are acquainted, an object of hatred to the agriculturist, but a subject of the most curious observation to the natural philosopher.

Its body is composed of twelve membranous rings, which assist it in locomotion. It is this membranous composition of the rings which distinguishes the caterpillars from the larvæ of other insects, whose body is composed of the same number, not of membranous, but of scaly rings.

There are six small black points observable on each side of the head, similar to small simple eyes, to each of which a branch of the optic nerve conducts, but it seems doubtful whether the power of vision resides in them. The eyes which the perfect insect is to have, are covered, in the caterpillar, by the two spherical, hard, and scaly caps of the head. The mandibles are trenchant and powerful, well adapted for cutting food. At the upper extremity of the mouth is a cylindrical nipple, pierced with a small hole, through which the silk spun by the caterpillar issues, and which has received the denomination of *spinneret*.

On the two sides of the caterpillar, are small oblong apertures, called *stigmata*, which are regarded as the organs of respiration. They are eighteen in number, nine on each side. The maximum number of the feet is sixteen, and the minimum eight. Six of these feet are scaly, and serve as cases to the feet of the future butterfly; the others are membranaceous.

There is no kind of animal, the species of which are formed upon so many different models, as the caterpillars. The most remarkable variation is in the number of feet, which, however, is not referable to the scaly, but the membranaceous feet, and also in their distribution. This gives rise to a variety of modes of locomotion among them. Those which have eight intermediate feet, walk parallely to the plane of position, and proceed by short steps. Those which have four or five rings in succession, without feet, curve this vacant space into an arch, and then stretch the body out again. From this peculiar motion, as the animal appears to measure the ground before advancing, these caterpillars are called in French, *geometricians*, or surveyors. Their body is oblong, stiff, and in some species, of the colour of wood, so that it might be mistaken for a small stick. They are also further mistaken, in consequence of the motionless attitudes in which they remain, and which presuppose an astonishing power in the muscles. Some may be seen embracing a little twig, or the tail of a leaf, with the two posterior feet and the two intermediate ones which are next to them, and which they hook on. The rest of the body raised vertically, remains stiff and motionless for hours together; others sustain themselves for an equally long time, in an infinity of other attitudes, which require an incomparably greater degree of force; for we observe them with their body in the air, in all positions. They even keep their body motionless after having given it a variety of fantastic curves. The muscles which have sustained the living caterpillars in these singular attitudes, keep them in them even after death. The caterpillars which have but eight feet altogether, are the smallest of all. Most of these appertain to the moths, and usually lodge in garments which they have formed for themselves, but of a variety of substances, in the interior of leaves, flowers, &c. These have no need of intermediate feet.

Another remarkable variety in caterpillars is that of size. This exists in various degrees amongst them; but we may take, as a mean standard, those of twelve, or thirteen lines in length, and about three in diameter. Such as much exceed this standard, may be considered as very large, while those much below it are very small.

The size of the full-grown caterpillar is, however, prodigious, in comparison to that of the egg and the young one. When we compare a caterpillar just born, and scarcely a line in length, with another at its full growth, about three inches and a half, this augmentation of volume in the same animal must appear considerable, though not equal to what we remark in fish. It has been calculated that twenty-six eggs are necessary to make up the weight of one caterpillar.

Some caterpillars are covered with a smooth skin, quite destitute of hairs. In some, the skin is so thin and transparent, that part of the interior of the animal is perceptible through it. In others, it is thicker and more opaque. Among the latter, some have it smooth, shining, and as it were varnished, while in others, it has a dead, dull appearance. Those caterpillars in which the skin is tender, transparent, and approaching to flesh-coloured, have been most frequently confounded with larvæ; while the latter, whose skin is opaque, and yellow, green, or brown, or striped with these colours, have been named *caterpillars,* though without the scaly head, or any of the distinctive characters of these last animals.

Colours are, no doubt, a very distinguishing attribute of caterpillars. We see, on their bodies, all known colours, and a variety of shades, of which it would be very difficult to find examples elsewhere. Some are but of one colour; others are adorned with different colours, very lively, and very strongly contrasted. Sometimes they are distributed in longitudinal stripes or bands; sometimes in stripes, which follow the contour of the rings; sometimes they are in waves, or in spots of regular or irregular forms; sometimes they are in points, or so varied, as to defy a general, or even a particular description.

Among the smooth caterpillars, some are more so than others; for this name is given to those which have very few hairs, as well as to those which have none. The skin of most of these smooth or *shorn* caterpillars is soft to the touch; but others have it bristled or *shagreened* with an infinity of small hard grains. These are arranged in some order, and appear to be of an osseous or corneous substance. Observed through the microscope, they look like little nipples, proceeding from a circular base.

Many of these last-mentioned caterpillars have a horn on the eleventh ring, directed backwards and a little curved. The caterpillars of the Sphinx are thus furnished. This, from its figure and direction, has been supposed an offensive or defensive weapon; but no one has observed the insect make use of it for such purposes. Besides, though called a horn, it is of a fleshy substance, and too soft to inflict any injury. It is not necessary to suppose that nature has formed every thing for a determined use; nor is it by any means in accordance with observation. There are often many parts which may be considered as kinds of off-shoots of organization, (if we may use such a phrase,) of no apparent utility to the animal, and which it can part with without detriment. It would appear that when certain powerful agents in nature have been set in motion, for given purposes, they have sometimes gone further than was intended, like generals or ambassadors who exceed the limit of their instructions. They produce superfluities, which do not interfere with the general plan of nature, though they answer no particular end. They are like ornaments to a building, which contribute nothing to its strength or solidity; or, like the colours with which nature has invested the variety of her productions, and which cannot modify their internal character. One colour, as far as utility is concerned, appears to be as good as another; and nature, had she pleased, instead of presenting creation as a *picture* to our eyes, might have made it an *engraving*, all of one colour, and characterized by variations of form alone.

Among the *shorn* caterpillars are also placed those which are remarkable enough for rounded tubercles, regularly distributed on each ring. Many large species, and which produce the handsomest lepidoptera, are provided with these. They are truly ornamental. In some, they are of a fine blue, which produces the most beautiful effect on a skin of somewhat a clear brown. Some caterpillars have a green, or a rather yellowish skin, with tubercles of a turquoise colour; others, also green, have them of a lively flesh-colour, contrasting finely with the hue of the skin.

Other caterpillars have hairs, or rather bristles, very thick and hard. These hairs resemble the thorns of plants in their form, and sometimes one constitutes a stem, from which others diverge in various directions. The form, size, colour, and number of these spines, vary in the different species.

In fine, the most common caterpillars, among which are some of the ugliest and handsomest that we can find, are the *hairy*, properly so called. By the quantity, length, and disposition of the hairs, they are distinguished from each other. They are also characterized by the different colours of their hairs. But it would be tedious for us, and indeed of small utility, to enter into any further details on this subject.

The mode of life and the habits of caterpillars are almost as various as the species. Some are solitary, others are social. There are species which live in the earth, the interior of plants, the trunks and roots of trees. The majority delight to lie on leaves, trees or plants, within reach of the aliment which is necessary for their sustenance. Some protect themselves by rolling the leaves, so as to form at once a garment and habitation. Others of a very small species, excavate the interior of leaves, and dwell therein, unperceived of their enemies; and, finally, there are some that form a kind of tunnel in a very exact manner, which renders them invisible, and accompanies them wherever they go.

A particular consideration of the mode of life in caterpillars, is essentially necessary to their just inter-distinction. Those that, though their exterior is pretty similar, yet show very characteristic differences in the mode of life, must certainly be ranged in different species. Thus, there are some which remain solitary during the whole period of their existence, and seem to have no sort of commerce one with another. Others pass the greater portion of their lives in a social state, nor do they separate until, having acquired their full growth, they are about to undergo their first transformation. There are some that even remain together in the state of chrysalis, and are not disunited until they have assumed the final form. The various substances on which they feed, should also be a consideration to guide us in the discrimination of the species.

The first law which nature has imposed upon all living beings is that of self-preservation. They are possessed of the means of living, from the moment in which they commence to exist, and their right to existence is proved by the very possession of these means. Let us, therefore, cease to repeat the absurd dogma, that all things have been created for our use alone, and to blame nature for the production of beings who are to live at our expense. Let us be no longer astonished that those caterpillars, whose multiplication is so prodigious, and whose growth is so rapid, should carry on such ravages, and be at once the scourge of our orchards, gardens, and forests. When the caterpillars are numerous, there are very few plants indeed, which they do not attack and despoil of their leaves. In some years they are so very common, that scarcely any herb escapes their devastations. By gnawing the leaves of trees, they reduce them to a state almost as deplorable as that in which we observe them in the depth of winter ; with this difference, however, that the loss of their leaves in that season, does them no harm, nor is it at all injurious to the principle of vegetation. But when they are despoiled of them in the spring and summer, they suffer very materially. When the caterpillars have devoured the verdure of a tree, they do not always abandon it altogether, though it appears to offer them no further means of subsistence. They often wait for the second germination, that they may feed upon the tender shoots and buds. Some species, however, proceed elsewhere, in search of nourishment. Among animals of the largest species, we find no example of voracity, comparable to that of the caterpillars. There is none that can eat, in the space of twenty-four hours, a quantity of leaves more heavy than itself. Some of them even eat more than double their own weight. But we are so much accustomed to see caterpillars feed on nothing but leaves and herbs, that when we find the trees themselves filled with holes, dried to the very roots, and even broken and overturned, we little suspect that all this mischief is the work of caterpillars.

It has been believed, and it is still very usually imagined, that each plant nourishes its own peculiar species of caterpillar; for many rather doubt if there be a single species of caterpillar, to which nature has assigned but a single species of plant, or a single substance of any kind, as its food. This can only be the case with those species whose diminutiveness conceals them entirely from our sight, and permits to live only, wherever they are found. There is a hairy and red caterpillar, which lives generally on the leaves of the vine, but will nevertheless eat other leaves with great avidity. It derives its nourishment both from leaves that appear to us perfectly insipid, and from leaves which possess an aromatic character. We find species which gnaw indifferently the leaves of the oak, the elm, the thorn, the pear-tree, the plum, and the peach-tree. Other species are found which feed indiscriminately on the leaves of the mallow, the helianthus, the pimpernel, the gilly-flower, lavender, and all kinds of pot-herbs. It appears, however, to be true, that there are but a certain number of plants or trees, of an analogous nature, which are suited to each species of caterpillar. What would become of our crops, if the caterpillars that ravage the woods could at the same time subsist upon the green ear? Accordingly, the plants upon which the caterpillars live, may serve as another foundation for the distinction of species. Thus we may well suspect, that caterpillars, though of similar form and colour, which are found upon an oak and on a cabbage, are not of the same species.

It might be thought strange, that nature has assigned to some caterpillars, as their aliment, not only plants whose bitterness appears insupportable to us, but also plants replete with an acrid and caustic juice, if we did not reflect, that the qualities of bodies act upon each other only in proportion to their respective relations. Thus some caterpillars live on the leaves of certain *tithymali*, notwithstanding the corrosive quality of the milk which they contain. The conduits through which this juice passes in the insect, small and delicate as they may seem to be, are not vitiated by a fluid, which acts very differently on our tongue. It must also appear extraordinary, that there are caterpillars that live upon the nettle; many species found on this plant are in truth armed with long spines, which may seem necessary to keep those of the leaves away from their skin. But many species of the smooth or *shorn* catterpillars are found upon the nettle, and whose skin even appears more tender than that of a number of other caterpillars which live on plants whose leaves are very soft to the touch. These caterpillars of the nettles eat leaves armed with prickles, which, when they come in contact with our skin, cause very troublesome and even painful excoriation. Can the palate and œsophagus of these caterpillars, which one would naturally suppose to be so extremely delicate, be more proof against those prickles than our skin? It may be, indeed, that these caterpillars take these prickles into their mouths by the base, and in a direction so as to do them no injury. But this is merely conjecture.

The majority of caterpillars live on trees and plants, for the purpose of eating their leaves alone. Some, however, gnaw the flowers, and others spare neither the fruits nor roots. There is also a number of them living in the interior of the different parts of trees and plants. Their skin is transparent, and generally more tender than that of others, and less qualified to resist the action of the external air. By exposure, it would dry up too much, and therefore these caterpillars never quit their obscure retreats. Some remain continually in the interior of the branches and stems, especially in the white hazel. The saw-dust which we daily observe issuing through a hole, whose aperture is at the external surface of the bark, indicates the presence of an insect, gnawing away the interior fibres. Among the caterpillars which live on wood, there are some to which the wood of different species of trees is most suitable, in the same manner as different species of plants are most suitable to others. The fruits which we find the most succulent, and the sweetest, have not been accorded to us alone—Nature has thought proper that insects of different genera should partake them with us; the pears, apples, plums, &c., which arrive at maturity sooner than other fruits of the same species, fall in our gardens every year; and these same fruits owe both their precocity and their fall to some insect, which has grown up within them. The most important of our fruits, those which form the basis of our aliments, are not in safety, even after their harvest has been gathered in. It is too well known that the different species of corn, such as wheat, rye, barley, &c. are sometimes entirely consumed in our granaries. Besides many species of larvæ, and perfect insects of different genera, there are a great number of caterpillars that attack fruits of various kinds. As among the caterpillars which live on leaves, some gnaw those of certain plants and trees, on which other caterpillars would die of hunger, so certain species of caterpillars eat fruits which would not be at all suitable to other species. Those which are brought up in pears, would in all probability perish in nuts, and *vice versâ*. The different species of fruits, however, are not so frequently attacked by caterpillars as the leaves. We are ignorant, if the leaves of any plant whatever are spared by caterpillars, but there are species of fruits in which they are never, or at least very seldom, found. It would not be more easy to give a reason why certain species of fruits are spared, while others are injured, than to tell why the leaves of the cabbage are more attacked by caterpillars than those of the beet; or why so many more insects live on the oak than on the linden. Plums are greatly subject to be maggoty;—a kind of small caterpillar grows in their interior. There is neither larva nor caterpillar brought up in the peach or apricot. We know that the lepidoptera do not deposit their eggs at hazard: their principal attention is to deposit them in such places as may afford an abundant supply of suitable food to the future caterpillar, from the moment of its birth. Thus the lepidoptera whose caterpillars are to feed on fruits glue their eggs on these fruits, which are often so young that the petals of the flower have not yet fallen; and they sometimes leave them between these very petals, against the pistil which is the embryo of the fruit. These caterpillars, which are not long in being disclosed, find themselves placed, from their birth, on some tender fruit, which they easily pierce, and introduce themselves into its interior. There they are at once in the midst of such aliment as they are fond of, and placed in shelter. The part even where they have entered sometimes closes up, so that it is difficult, or even impossible to find the little hole which has given a passage to them. The caterpillars that live in fruits are commonly small, very much under the middle standard. The small caterpillars which live in husks or cods, do not seek to conceal themselves in the fruit on which they feed, but have part of their bodies outside. But those which eat fruits not enclosed in shells, always remain in the interior of the fruit. A remark, which should not be omitted, and which has long since been made, respecting larvæ, is, that each fruit contains but a single caterpillar, at least it is scarcely ever otherwise. If two inhabitants are sometimes found in a single fruit, one of them is a larva and the other a caterpillar. There are small caterpillars that lodge in grains : heaps of wheat or barley may be thus entirely filled by them, without its being perceptible from the external appearance of the grain, as the skin is always spared. But on pressing the grain between the fingers, those which have been attacked are easily distinguishable from the others ;—nay, it is possible in this way to distinguish the age of the caterpillar which is lodged there. If, on being pressed, the grain yields on all sides, the caterpillar has acquired its full growth, or has become a chrysalis. If only a portion of the grain gives way, its interior substance has not been altogether gnawn, and the caterpillar is yet young. A grain of wheat or barley contains the just quantity of provision which is necessary for the support of this caterpillar from its birth to its transformation. If a grain be opened containing a caterpillar on the point of metamorphosis, its whole interior substance will be gone, and nothing but the skin remaining. Drink does not appear to be necessary to the caterpillars, or, to speak more correctly, they derive all their fluid nourishment from the substances on which they live.

There is one fact, which must not be forgotten in the history of caterpillars, though it does not present them in the most amiable point of view. The maxim, so often cited against us, that man is the only animal that makes war upon his own species, has been advanced by persons who have paid no attention to the study of insects. Carnivorous insects, as we have had occasion to remark more than once in the course of this work, very often devour their own consimilars. But this trait is still more exaggerated in some caterpillars, which, though born to feed upon leaves, and delighting in that food, yet find the flesh of their companions a preferable viand, and devour each other whenever they have an opportunity.

The time in which the caterpillars take their food, is another distinguishing specific mark amongst them. Some eat at all hours of the day, and some only in the evening or morning, and remain quiet during the heat of noon. There are some, also, that only eat during the night. Thus among the smooth caterpillars there are some brown and green ones which live on cabbage, but quit it in the morning, to conceal themselves in the earth during the day, which never issue from their retreat but in the evening, and gnaw the leaves only during the night. Thus the gardener who desires to rid his cabbages of these caterpillars, and the naturalist who wishes to observe them, must both choose the night for their operations, and seek these insects by candle-light. There are numerous other species which conceal themselves at certain periods both by night and day, and which cannot be discovered but at the moment of their issuing forth. There are other caterpillars fond of the roots of plants, and which remain continually under ground. Gardeners are well acquainted with the species that eats the roots of lettuce, and the cultivators of hops, with that one which gnaws the same parts of that plant, and often occasions infinite loss and damage.

The attitudes assumed by different caterpillars, when we attempt to catch them, may also assist us in establishing many new distinctions between many species. Some roll themselves into a ring the moment they are touched, and remain motionless as if dead. Those which are hairy, in turning in this manner, assume then the appearance of a hedgehog; others suffer themselves to fall to the ground, the moment the leaf on which they are placed is touched; others try to save themselves by flight, and some among these run remarkably fast; many, more courageous, seem desirous of defending themselves: they fix one portion of their body, and agitate the other in various directions, as if to strike the person who disturbs them; some of these put in motion the anterior, some the posterior part of their body. Finally, there are some, which, when touched, give inflections to their bodies similar to those of serpents, and twist and twine themselves in various directions with very great agility.

Though all caterpillars are injurious to vegetation, all are not equally so. There are some species so small, and so little multiplied, that the mischiefs which they do are very trivial. There are others that live on plants, in the preservation of which we are not interested. But there are species so exceedingly pernicious to us, that we extend, in consequence, our hatred to the entire class. They have rendered the very name of caterpillar odious, and stirred up all our energies to the destruction of the entire race. Nor have the ravages of caterpillars been the only motive of prejudice against them. For a long time these insects were considered venomous; an error which had no other foundation than the aversion created by ugliness of exterior, and the timidity of ignorance and folly. Birds feed on caterpillars, and the repast is not only innoxious, but nutritive. Children have been known to eat silk-worms without inconvenience. Caterpillars are given to poultry without doing any harm. Some larger species will, indeed, when touched, excite an irritation of the skin, but without any dangerous result. This irritation is caused only by the hairs with which they are furnished.

But when prejudice and fear are laid aside, when animated

only by the desire of knowledge, we cast our eyes on these despised insects, and examine their properties, their habits, their instinct, nay, even their utility, we not only forget the evils which they cause, but are penetrated with sentiments of admiration and delight. No wonder, then, that they have attracted and fixed the attention of the profoundest observers of nature, and the most worthy of the name of philosophers.

Nature employs, in the preservation of these insects, from year to year, and during the rigours of winter, four different means, but all of equal efficacy. Some pass the winter in the envelope of eggs; some in the caterpillar form; others in the state of chrysalis; and others as the perfect insect.

Those which have passed the winter in the egg, live, after they have quitted it, in the caterpillar form, during a portion of the summer. The shell of the egg, and the place where it is deposited, protect the embryo from destruction.

Others which have left the egg before autumn, and the fall of the leaves, feed upon these as long as the season will permit; and some of them, before the arrival of winter, have attained about half of their complete developement; others, born later, grow but little in the same year. On the approach of winter, the young caterpillars adopt the proper means of sheltering themselves from the cold. In the following spring they quit their asylum, and proceed in search of food. At the commencement of the fine season, we are surprised to find so many caterpillars advanced in growth; but they have existed under this form during a part of the preceding autumn. The winter retreats, which these caterpillars select, or fabricate with great industry, are of various kinds. The solitary caterpillars conceal themselves under stones, inside the bark of old trees, or retire into the earth, to a degree of depth where the cold cannot reach them. The social form nests which are very remarkable, constructed of many leaves which they connect together with silk, and attach to the tops of trees. This is the case with several caterpillars, and especially the smaller species.

Those which pass the winter as chrysalids, are the most numerous. It is towards the end of summer and autumn, that they cease to eat, and prepare for their transformation. Some of them enter the earth for this purpose; others seek retreats in the holes of walls and trees, or under stones; others form cocoons of silk, or other materials in which the chrysalids are enclosed: some have no need of shelter, take the chrysalis form in the open air, and are perfectly impervious to the cold.

The caterpillars which live together, all come from the eggs of a single lepidopteron, which have been deposited one after another, or heaped together, so as to form a sort of nest. The young caterpillars, for the most part, quit the egg all in the same day, and as they are born, so they continue to live together, for the term which their instinct prescribes. These societies of brothers and sisters, commonly consist of two or three hundred caterpillars, and sometimes even of six or seven hundred; some do not separate until they have quitted the exuvia of chrysalis, and even almost form societies for life; others live together only until they have arrived at a certain growth.

Among the latter must be placed the caterpillar which is named *common*, because, in truth, it is most frequently met with, well known from its ravages, its middle size, and hairy body, of a brown colour, and furnished with sixteen feet. The female of the *Bombyx*, to which it belongs, deposits her eggs on a leaf about the middle of summer, and envelops them in a sort of yellow silk. These eggs are four or five hundred, and from each of them, in a few days, issues a small caterpillar. They all remain assembled on the leaf on which they were born; they are scarcely disclosed when they begin to eat and spin in concert; they construct a nest, where they retire during the night, and which serves as a retreat during bad weather, or winter. But too many of these nests are to be seen on fruit trees in autumn, and still more in winter. These nests are thick packets of white silk and leaves, and their form is neither agreeable nor constant. In proportion as the insects grow, they enlarge their lodging, by new layers of leaves and silk. Each nest consists of many enclosures of webs, which form so many apartments, and each enclosure has its doors, which allow the caterpillars to pass from one to the other. The webs, composed of a prodigious number of threads, extended one over the other, render these nests capable of resisting all attacks of wind and weather. The rain cannot enter, because all the apertures are below, and it slips off from the silken tissue without penetrating it. The period in which these nests might be most deranged, would be in spring, if the stems which they envelope cease to be covered with new leaves, or to grow themselves; but the caterpillars take good care to prevent this, by gnawing the principal buds of the stem, and thus hindering it from germinating. These are admirable productions, and it is pleasant to see the caterpillars, in fine weather, issuing forth from their little doors, to enjoy on the outside of the web, the air, or the sun. Some extend their promenade farther, but they never proceed beyond the length of the branch which sustains the nest. In walking, they carpet all their path, and do not proceed beyond the spot where these traces of silk are seen to terminate. Though they do not seem to observe a very strict polity, they are not altogether undisciplined. None of them ever fail to re-enter their habitation on the approach of night or bad weather. It is a very amusing spectacle to see these little caterpillars going and coming, some on one side, some on the other, without confusion, and when they meet, saluting each other, like the ants; also descending the branch in great numbers, and ranging themselves side by side, on the top of a leaf, for the purpose of foraging. The sound of the voice, or of an instrument, appears to annoy them, and the slightest motion in the neighbourhood of their dwelling, or on the leaves where they have established themselves, soon determines them to regain their retreat. In fine, after several moultings, the period of their dispersion arrives; the society is dissolved, and each caterpillar retires to pass the rest of its life in solitude.

The pine forests nourish caterpillars of another species, which pass a great portion of their life in society, and seem more worthy of attention than the preceding, from the quantity and quality of the silk, of which their nest is composed. These nests are sometimes larger than a man's head, and the silk is strong and white. The ravages of this caterpillar can neither excite nor deserve our vengeance. Its gnawing the narrow and pointed leaves of the pine, which is the only tree that it attacks, is of little consequence. So far from injuring us, its silken cocoons might be of great utility, if proper pains were taken in preparing them for manufacture. These caterpillars, like the preceding, compose a nest in common, whither they retire for the night. They come forth on the return of day-light, spread themselves over the tree, and gnaw the leaves. They march in the same order as those called *processionary*. A little time after their birth, they work in concert at the nest, which is at first very small, but as they grow, they increase its circuit by spinning new webs. All the interior of the nest is filled with webs, directed variously, which form different lodges, with apparently the same communication as those of the common caterpillar. The principal entrance is not constantly in the same place, and other smaller ones are observable. These caterpillars walk very fast, and at first separate but a little from each other, to gnaw the leaves in their immediate neighbourhood. When they have let themselves down, they make use of a very slender thread of silk, as of a ladder, to remount to their nest. Although they appear to issue forth more willingly by night than day, and appear to avoid light, some are observed to come out at all hours of the day. They walk in procession in file, and in the finest order. They all defile one by one, with an equal and rather slow step. The file, which is often very long, is continuous throughout, the head of the caterpillar which follows touching the tail of that which precedes. Sometimes they defile on a straight line; sometimes they describe a variety of curves, many of which representing festoons or garlands, are very agreeable to the eye. These kind of evolutions are still more amusing when many societies are in the same neighbourhood, and the processions set out from different nests. They sometimes proceed to very great distances from the nest, and then the files are very long. While one procession follows the same right line, others turn in different directions; some are mounting, some descending. All the caterpillars of the same procession march with an uniform and grave step, none ever attempting to advance before another, none ever remaining behind in the interior of the file. The caterpillar which is at the head of the procession, determines the evolutions of all the troop. Each preserves his place, and directs his march by that of the caterpillar immediately preceding. When the first caterpillars in a procession have made a halt, they usually assemble one near the other, and over each other in a heap, and enclose themselves in a sort of pouch of filigree work, very similar to a fishing net. When our processionaries return to the nest, it is by the same way that they came. They often remove from their domicile to some distance, and make many turns, but are sure to find their way back again. It is not the sense of sight that directs them so certainly in their marches. This is a point clearly proved. Nature has accorded them another means of regaining their lodging. We pave our ways: these caterpillars carpet theirs, and never walk but on silken tapestry. All the paths which lead to their nest, are covered with threads of silk. These threads form traces of a lustrous white, at least two or three lines in breadth. By following these traces in file, they never fail to find their nest, how tortuous soever may be the windings in which they engage themselves. If the finger be passed across the trace, the way will be broken, and the caterpillars thrown into the greatest embarrassment. They will be observed to stop all on a sudden at this place, and give all the marks of fear and distrust; their march will remain suspended, until some one of them, bolder or more impatient than the rest, shall have dared to cross the gap. The thread which it spins in crossing, becomes a bridge for another. The latter, in passing, spins another thread, his followers each another, and the path is soon repaired.

The industrious processes of insects, and indeed of animals in general, are apt to lay hold of our imagination. We are fond of according to them our own reasonings and our own views. But this is by no means a correct mode of viewing the question. These caterpillars, no doubt, do not carpet their paths for the purpose of preventing themselves from losing their way; but they do not lose their way, because they carpet their paths. They spin continually, because they have continual need of evacuating the silky matter, which is reproduced by nutriment, and contained in their intestines. In satisfying this want, they secure their march without thinking on the subject; and for that very especial reason they do so better. The construction of their nest is connected with the same want. Its architecture is agreeable to the form of the animal, to the structure and play of its organs, and to the peculiar circumstances in which it is placed. This is one of those operations which is directly traceable to the organization of the animal; and though most of those, which we are pleased in our ignorance to call *instinctive*, cannot, from our limited powers of investigation, be traced so clearly, perhaps the hypothesis which refers them to the same source, is as good as any other; at all events, if instinct be a mystery, our own reason is not less so; and man can no more point out the cause of the one than that of the other.

When these caterpillars have acquired their full growth, and the time of their metamorphosis is approaching, they abandon their nests, separate, and go to construct in the earth cocoons of pure silk.

The caterpillar of the *bombyx neustria*, of Fabricius, which, from the longitudinal bands, of various colours, which adorn its body, has something of the appearance of a riband, is common in gardens and orchards. The leaves of fruit-trees, and those of many others, are its food, and it sometimes does infinite mischief. These caterpillars are also social during their infancy, and spin in concert a sort of cloth, which serves them as a tent, under which they have some leaves to feed on. When their provision is out, the family remove to another part of the tree, where they can find more, and where they establish themselves afresh, as before. This economy, which continues while the caterpillars are young, is sufficient to strip a tree entirely, when there are two or three of these numerous families. During the night, they return into the nest, but by day-time they parade upon its surface, withdrawing, however, immediately on the approach of rain. They walk like the preceding, in procession, but not in so continuous a file, nor in equal range. The procession is often interrupted in its march by caterpillars returning to the nest, and by others halting. Having gone a certain distance, the processions often stop, and the caterpillars flock together. Some then return by the same road, others pursue their course, always at a slow pace, and without the least confusion. They find their way by the same means as the last. If any of them that go first be touched slightly with the finger, it is amusing to see them shake their heads several times, and go back quickly, without being stopped in their flight by any of their companions, who follow the first route with a tranquil step. If their way be divided, like the others, their embarrassment and inquietude are extreme. The cordons which they form by their various evolutions have a beautiful effect, appearing, at a distance, like traces of gold; some in a right line, and others in curves of various inflexions. What renders this spectacle more agreeable is, that the cordon of gold is couched upon a silken riband, of a bright and silvery white. After the second moulting, these caterpillars no longer observe the same order, but wander about without any fixed direction, and some become completely solitary.

Tufts of grass may be seen in autumn, in the meadows, covered with white webs, which might be taken for those of spiders; these, however, are tents under which some caterpillars live. There is nothing regular in their arrangement, and the interior is divided by many partitions into different lodges, which are wider as they approach the base. When the insects have consumed all the provision under one of these, they remove to another situation. Thus they live during summer, but for the winter they form a more solid lodging, in the interior of the principal tent, in the form of a purse; they are there one upon the other, and each rolled up. On the return of the fine season, they make new tents of silk, which serve to defend them against the rain: they leave several oblique apertures for egress and regress. These societies do not consist merely of caterpillars of the same family; those of several nests often join in making a tent. About the middle of spring they separate, each to live in solitude, and prepare for its metamorphosis.

We must now speak of social caterpillars, which remain together even under the chrysalis form.

Of all caterpillar republics, the most numerous are those of a species usually living on the oak, and which have particularly been named *processionary*. Each of these consists of seven or eight hundred individuals. While young, they have no fixed establishment. The different families encamp, sometimes in one place, sometimes in another, on the same tree where they were born. They spin their nests together. In proportion as they change their skin, they form new establishments; but at the period of their full growth, their habitation is fixed. These nests must be large to contain the numbers, but have nothing singular or invariable in their form. The walls are formed by many layers of webs applied upon each other. Between the trunk of the tree and these walls is a cavity, in which the caterpillars occasionally enclose themselves, and near the trunk is the aperture of entrance. These nests are rarely found in the middle of the forest, and though placed within reach, are not casually perceptible, as they nearly resemble the lichens with which the stems of the oak are covered.

When these insects quit the nest, they go in procession, but in an order somewhat different from what we have already described: a single caterpillar opens the march, the others follow in file, and sometimes two, three, or four a-breast. They observe so perfect an *alignement*, that the head of one never passes that of another. When the conductor halts, the troop which follows does not advance, but awaits the signal from the leader. In this order they traverse the paths, and pass from one tree to another in search of food. When they find a branch covered with fresh leaves, the caterpillars spread over the leaves, and are so contiguous to each other that their bodies touch. When they have terminated their repast, they regain their nest in the original order. Sometimes, indeed, this little army makes a number of singular evolutions, but it is always conducted by a single caterpillar. The van always constitutes one angle, but the main body is gradually enlarged, so that some ranks consist of fifteen or twenty caterpillars. In a wood where there are several nests of these processionaries, to see them forming their battalions, is an admirable spectacle for the lover of nature. One issues forth from a nest whose aperture is scarcely sufficient for two to pass abreast; it is followed by others in file, and having arrived at the distance of about two feet from the nest, it makes a pause, during which those which are in the nest continue to issue forth. They each take their rank; the regiment is formed; at length the conductor marches, and the whole troop follows, entirely subordinate to every movement of its chief. The same scene is passing in the neighbouring nests:—they are all vacated at once. The hour has arrived when the caterpillars are to take their food, for it is towards night that they commence these processions. During the heat of the day, they usually repose in the nest. In commencing the nest, which is to serve them as a last retreat, they give it as much thickness and breadth as is necessary, but they sometimes lengthen it afterwards if they find its capacity insufficient. The exuviæ from their last moultings strengthen the envelope, being connected together with new threads, and the tissue, at first transparent, in the course of a few days becomes perfectly opaque. In this nest they assume the form of chrysalids, each spinning a particular cocoon The nests, the hairs of these caterpillars, and even the air which is charged with them, produce violent irritation in the skin, followed by swellings.

In spring, on the leaves of apple-trees, is found another species of caterpillar, which remains social in the chrysalis state. These caterpillars remain in a sort of hammocks, which they construct, and not only repose therein, but find their food. They eat only the parenchyma of the surface of the leaves, and, what is singular, their bodies never touch the leaf which they are gnawing, as if it were too delicate to support the touch. It is covered by a very soft skin, endowed with great sensibility. Let them be touched ever so slightly, these caterpillars advance, or draw back in their hammock, with great quickness. We are surprised to find that they turn neither to the right nor left while they are executing movements of so much promptitude. In fact, each caterpillar is lodged in a sort of long sheath of filigree work, spun by itself, and imperceptible to the eye. The whole nest or hammock is an assemblage of these sheaths, couched parallelly one upon the other, in each of which is enclosed a caterpillar. The nest envelopes a certain number of small shoots or leaves, and when the parenchyma of these is gone, the caterpillars proceed to hang another hammock on some neighbouring leaves. They do this successively several times in the course of their lives. These would be taken at first sight for the webs of spiders. Nothing is perceptible but a confused assemblage of webs, of irregular forms, and very transparent. The nest begins at certain leaves, and ends at others more or less remote; when they abandon it, the new one which they construct is always at but a small distance from the first: all occupy it at once, and each furnishes a great number of threads. At one of the ends of their last nest each constructs a cocoon, of a very white silk, in which they enclose themselves, in order to undergo their transformation into a chrysalis.

Caterpillars in general are not to be considered as very sociable beings: the greater number live without having any communication with their consimilars; and as for those that live in society, they rather provoke us to destroy them, than excite our curiosity to observe them. Having given some notions concerning the mode of life of the most common species which are social, we shall now invite the attention of the reader for a brief space to the habits of some which are solitary, and which are not less worthy of admiration.

Some of these, though living in great numbers on the same tree, must yet be considered as solitary, because they perform no works in common, and the labours of one have no influence on those of the other: they live in common as though they were alone. But others are still more solitary: they make, in succession, several habitations, without even placing themselves within the reach of communication with any of their own species. Such are almost all those that curve or roll the leaves, for the purpose of lodging in them, and all those which connect together several leaves, to unite them in a packet, towards the centre of which they remain.

We see almost every day, on our fruit-trees, and various plants, some leaves which are simply curved, others folded in two, others rolled several times over, and others, in fine, collected together into a shapeless packet. It will be easily perceived that these leaves are held together in these different states by a great number of threads, and that the cavity within these leaves contains a caterpillar. If we look at the oak-leaves particularly, towards the middle of spring, we shall find many of them, folded and rolled in different ways, and with a most surprising degree of regularity: the upper part of the end of some appears to have been brought back towards the under part of the leaf, there to describe the first turn of a spiral, which has afterwards been covered with many other turns, produced by successive rollings, carried sometimes as far as the middle of the leaf, and sometimes beyond it. The centre of the roll is hollow; it is a tunnel, the diameter of which is proportioned to the body of the caterpillar which inhabits it; other leaves are rolled towards the upper part; others, very numerous, are rolled towards the under part like the first, but in directions totally different. The variety of forms, however, which these rolling caterpillars use in their habitations, may be better comprehended by inspection, or from figures, than from description *.

Sometimes many leaves are employed in a single roll. Such works would not be very difficult to our fingers; but the caterpillars possess no instrument equivalent to them. Moreover, in rolling the leaves, it must retain them in a position, from which their natural spring is perpetually inclining them to diverge. The mechanism by which the caterpillars perform this part of their work is easy to observe. We see packets of threads attached by one end to the exterior surface of the roll, and by the other to the flat of the leaves. We can very well conceive these little cordages sufficient to preserve the rolling form to the leaf; but it is not so easy to divine how the caterpillars produce this particular form, how, and in what time they attach the connective threads. All this can be ascertained only by observing the insect at work.

This is not so easy; it would be difficult to seize the exact moment of their operation, and the presence of a spectator might tend to derange them. We may choose a

* Language is comparatively inadequate to a clear and minute description of the objects of sight; language is the symbol of mind, of thought, of sensation, of what is internal, rather than what is external. Writing has taught us

"To paint in mystic colours, sound and thought;"

but painting and sculpture alone can embody to our view the outward forms of things. Verbal description may assist our recollection of what we have seen, or convey a notion of what is perfectly analogous thereto; but if every one would be as candid as myself, I am sure they would confess, that an object which they had never seen, of which from analogy they could form no previous notion, cannot be clearly conveyed to their apprehension by words alone. The *concrete* is as ill expressed by language as the *abstract* by painting. E. P.

more facile method, by sticking in a large vessel full of humid earth, some branches of oak newly broken, by distributing on their leaves a certain number of caterpillars, taken out of the rouleaus, which they have already made. They are very impatient at being uncovered, feeling (for all the rolling caterpillars are smooth) the necessity of being sheltered from the impressions of the open air. Accordingly, they soon go to work, under your inspection, just as they would do in the forest. It is, in general, the upper part of the leaf which they roll towards the under, but some commence the rouleau at the very end of the leaf, and others at one of the denticulations of the sides. The head of the caterpillar is applied against the under part of the leaf, quite near the edge, and from there, as far as it can extend near the side of the principal nervure. It returns directly to the place from which it had at first set out, and comes back in the same manner, to retouch the most distant part of the edge. Thus it continues to give itself successively more than two or three hundred alternate movements. Each movement of the head, each going, produces a thread, and each return produces another, which the caterpillar attaches, by each end, to the places where its head appears to apply itself. All these threads form a sort of link, and having given a sensible curvature to the leaf, towards the under part, the caterpillar begins another at two or three lines distance from the preceding. The part which is between the first link and the second, now begins to curve more, and that which is beyond, already curved, will become more so by a third link. The extent of the part which is to form the first turn of the roll, is not great. It is just like a piece of paper that we begin to roll from the angles. Accordingly, three or four packets of thread are sufficient to give the necessary curve to this first turn. It is by means of similar threads, and similar links, that the second turn is to be formed. Nevertheless, though the leaf curves more and more, in proportion as each link is finished, we cannot yet perceive the cause of this rolling. After having considered each link as composed of thousands, very nearly parallel, to form a more exact idea of it, we must regard it as composed of two sets of threads placed one below the other. All the threads of the upper set, cross those of the lower. The packet is wider at both its ends than it is in the middle; nevertheless, the number of threads in the middle is equal to that of either end. Wherefore, then, do they occupy less space? It is, that they are more crowded one against the other in their crossing. If we choose to follow the caterpillar while it is spinning the threads of each of these sets, we shall discover the double use of both these sets. The threads of the first set, being all attached pretty nearly parallel to each other, the caterpillar passes to the other side, to spin those of the second set. While it is spinning, it cannot go from one to the other extremity of this second set, without passing over the threads of the first; and far from trying to avoid them, it applies its head and a part of its body to them. The threads of this set form a kind of cloth, capable of supporting this pressure; they consequently draw the two portions of the leaf one towards the other. That which is near the edge gives way, approximates, and the leaf curves. The only question now is, to preserve the curve which the leaf has assumed, which is done by a new thread attached by the caterpillar. A caterpillar which has to roll a thick oak-leaf, could not spin threads sufficiently strong, to hold against the stiffness of the principal nervures, and especially of the middle ones; but it knows how to render them supple. It gnaws away, in three or four different places, those nervures where they are thicker than the rest of the leaf. When, after having rolled a portion of the leaf, it finds a large denticulation which considerably out-edges, instead of rolling it, it bends it by the threads, which it attaches to the end, and in the sequel, it forms a tube of a proportioned diameter, and very well rounded. Besides the links which are all along the last turn of the roll, the insect has often occasion to put some at both ends, or at least at one of the ends. But these are so disposed, that they do not deprive it of the liberty of issuing forth from the interior of this roll, and re-entering. This is its domicile; it is a species of cylindrical cell, which only receives the light at the two ends, and its walls furnish nutriment to the animal that inhabits it.

The different species of caterpillars which roll the leaves of the oak, elm, and other trees, proceed exactly on the plan we have now described. It would, therefore, be superfluous to dwell upon them. Plants have also their rolling caterpillars. In general almost all the rollers are remarkable for their vivacity. A caterpillar that rolls the leaves of sorrel, exhibits this peculiarity in its habitation, that it is placed perpendicularly on the leaf. This position will not allow the caterpillar to roll the leaf as it finds it. It cuts a strip of the leaf, but without detaching it altogether. The broadest part of this strip constitutes the height of the roll, and its length furnishes all the necessary turns.

There are other leaf-rollers which also vary the forms of their habitations, and pursue different plans; but our limits will not permit us to go into further details concerning these processes. We must say a word, however, respecting the leaf-binders, which are more numerous than the leaf-rollers. Their works are more simple, but nevertheless do not exhibit less industry. These kind of works are to be seen on the oak, as well as the habitations of the rollers. The end of the leaf has been brought back, almost brought down flat, and no sensible elevation left, except at the place of the fold. The direction of these curves is very various, some downwards, and some upwards; but the leaf-binders may be said generally to inhabit a kind of flat box. They have no great space, but it is proportioned to their body, as they are usually small. There is an aperture at each end, but almost imperceptible. They only eat into a portion by the thickness of the leaf, for were they to gnaw it altogether, their habitation would soon be open. They only attack the larger nervures and fibres, detaching merely the softest substance, the *parenchyma*. It is worthy of remark, in some of the leaf-binders, that they have a particular place for the deposition of their excrements, which is the part where they have first commenced to gnaw the interior of their habitation.

Many caterpillars, still smaller than the last, unite several leaves in one and the same packet. These leaves are very variously and irregularly arranged, and attached in such places as are most easy of union. Here the caterpillar is comfortably placed in the midst of the packet, and surrounded with plenty of provision.

All the *geometer* caterpillars, with but ten feet, live solitary, and for the most part, neither bind, roll, nor connect, several leaves together. Their plan is more simple. They remain between two leaves applied flatly together, and retained so by threads of silk at the two surfaces which touch. They let themselves fall when the leaves are agitated, or an attempt is made to seize them. They do not, however, let themselves fall to the ground, but are retained by a thread, like spiders, in the air. The mechanism of this is curious enough, but our limits oblige us to refer to Reaumur and M. Latreille for the details of it.

We must now dismiss all further consideration of the habitations of caterpillars, though much has been discovered, and doubtless more yet remains to be discovered concerning them, of high interest to the lover of nature. But in fact, a complete history of caterpillars would constitute a work of no inconsiderable extent, while we can afford but comparatively a few pages to its discussion.

Among the facts deserving of examination in these insects, there are few more so than their changes of skin. Most of them undergo this change but three or four times previously to their final transformation; in some, however, it takes place eight or nine times. These moultings are called diseases in the silk-worm, and are so in fact, for they often perish in consequence.

The exuvia rejected by the caterpillar is so complete, that it has all the external appearance of the insect itself; the coverings are detached from every part, which we may easily conceive to be no trifling operation; to perform it the social caterpillars retire to their nests, and hook themselves by the claws of their feet to the silken tissue. The solitary ones spin light tissues for this purpose. It is more easy to detach the covering when they have obtained a hold of this kind.

As the moulting approaches, the colours of the caterpillar begin to fade; the skin dries up, and becomes cleft on the second or third ring of the back. The insect continues to swell that part of its body which is opposite the cleft, until the aperture is sufficient to enable it to draw the body out. Having disengaged the upper part of the body, it elongates the lower, to disengage it likewise. This operation, though laborious, does not take a moment in performing.

The colours of the caterpillar with the new skin are fresher and finer, and sometimes seem altogether different from what they were before. The new skin, even when furnished with hairs, is perfect under the old; and we may suppose such caterpillars as change several times are furnished with so many of these skins, one above the other. The insect is always weak after each moulting; the parts are tumid, but this moisture speedily evaporates.

These changes are necessary to the growth of the caterpillar; for the solid and external parts do not grow, and therefore, as the internal growth advances, a new coat becomes necessary.

The final transformation into chrysalis now approaches, and the caterpillar forms its cocoon. Most of these cocoons, as is well known, are spun of silk, but there are many varieties in their structure, mode of suspension, &c. &c. Some caterpillars cannot make their cocoons of pure silk alone: they mix it with earth, or make the cocoon of earth alone. When the time of their transformation approaches, they conceal themselves under ground, and there the chrysalids remain until they are ready to appear winged. Many, however, neither form cocoons, nor retire under ground; but in general, they remove from the places where they have lived, and choose some retired place for their transformation. Their positions are different; some are hung vertically in the air, the head downwards; the end of their tail alone is attached to some elevated body;—some are attached against the walls, having the head higher than the tail; and others present a variety of different attitudes; some, although attached by the tail, are further fastened by a little cincture embracing the back, which is composed of a great number of silken threads closely approximated, and much stronger than is necessary for the support of the insect; others are attached with less art, but even the most simple suspensions presuppose modes of operation not very easy to be divined.

As the metamorphosis approaches, the caterpillars often quit the plants or trees in which they have lived. After having ceased to take food, they evacuate copiously: some totally change colour; but it more usually happens that their colours become effaced. In those that have a horn in the hinder part of the body, that appendage, from having been opaque, becomes transparent. The stigmata also appear to close at the moment of tranformation.

The cocoons of the silk-worms are, doubtless, the handsomest of any made by caterpillars, if we speak in reference to the substance of which they are composed, and its capacity of being manufactured. Nevertheless, other caterpillars form cocoons, which, though less useful, are more remarkable in their form, and seem to presuppose a greater degree of intelligence in their constructors; some apecies content themselves with filling a certain space with threads, crossing each other in various directions, but leaving between them several vacancies. The insect occupies the centre of this space; the threads support it, without concealing it. Others, though they form the cocoon better, yet its tissue allows the animal to appear. Nevertheless, the majority of those which use few and separated threads, are by no means fond of being seen, and choose a shelter under the leaves of trees, &c.; and even some of those that arrange their cocoons with more exactitude and closeness, will cover them with the leaves of the tree or plant on which they have lived.

The cocoons which are most frequently exposed to view, are those which are composed of pure silk. They are generally of an elliptical, though sometimes of a cylindrical form. The web of some is very fine and close, while that of others is coarser. They are all composed of a single continuous thread, unless it happens to break, which is very seldom the case. It is fortunate that these threads in the case of the silk-worm, are not strongly cemented together, as it would be impossible to unwind the cocoons, which is very easily done, when they are held in warm water. The gummy matter with which the silk is impregnated, dries up instantaneously, and retains no more viscosity than is sufficient to attach it very slightly to the thread. In the cocoons of many species of caterpillars, the threads are attached by stronger and more tenacious gluten. These cannot be unwound, and the only resource is to card them. But others are so closely cemented, that even this operation would not succeed, but would reduce them into small fragments. Six different layers are distinguishable in the cocoon of the silk-worm ; and a single thread may be unwound from it more than a thousand feet in length.

The most usual colours of the cocoons of different species are white, yellow, brown, or red ; but they exhibit shades of all these colours extremely varied. The silk-worm sometimes employs two or three days to finish its cocoon ; others take but a single day, and others but a few hours. But the forms, the materials, the industry exhibited by these insects in the construction of their cocoons, would even, if imperfectly described, carry us far beyond our limits. Nothing, indeed, can be fitter to interest the curiosity of a philosophical observer, than such remarkable varieties of architecture in insects of the same class. These variations take place not only among different species, but also among different individuals of the same species, when placed under different circumstances ; when deprived of their usual materials, they will employ others presented to them, and arrange them according to the necessity which locality, &c. may impose.

That insects sometimes commit mistakes in the execution of their works, is a fact, which might, if necessary, be alleged in proof that they are not mere machines. Many examples of such mistakes (perhaps we should say, *irregularities,*) occur among the caterpillars. Thus we sometimes find two or three silk-worms shut up in a single cocoon, in which they nevertheless undergo their metamorphosis perfectly well.

We must now not only dismiss all further consideration of cocoons and chrysalids, but deny ourselves the pleasure of investigating a subject of not less interest, we mean the *vital economy* of caterpillars. The reader who is curious on this subject, will find ample details in the works of Lyonnet, Kirby, &c.

It is much to be regretted, that amidst so many caterpillars that spin silk, we have been able to avail ourselves of the labours of but three or four species. The silk produced by the silk-worms, is not always equally good. That of China is renowned for its fineness. In some countries, the silk is very coarse and indifferent, which doubtless depends upon the different qualities of food.

There can be no doubt but that we might derive great benefit from the productions of other caterpillars, if not in the manufacture of silk, at least in that of many other articles of equal utility, such as fine cotton, cloth, paper, &c. Some experiments which have been made, clearly prove the practicability of this. It may also be mentioned by the way, that the fluid with which the caterpillars moisten the silk, might be turned to useful purposes in the making of varnish, &c.

The enemies of caterpillars are very numerous, especially among birds, which annually destroy prodigious numbers of them. There are also many insects which destroy them, and some of their own species attack each other.

We must now conclude our supplement on Lepidoptera, with reference to the plate representing the progress of one species through its different states of existence, and with a very brief consideration of the habits of some of the perfect insects, and descriptions of several new species of which we present figures.

The metamorphosis of a *Pap. Urticæ,* are depicted by a series of sketches, taken from observation. *See* Pl. 115. Fig. 1. represents the larva just issued from the egg. 2. The same in progress to maturity. 3. The same at its full growth. 15. Its head magnified. 4. Its appearance when about to

Transformations of the Butterfly.

London Published by Whittaker & C° Ave Maria Lane, 1832.

change into the pupa state. 5. In the act of changing on the following day. 6. The pupa come forth, the change having been performed in about four minutes. The exuvia of the larva being raised up to the point of suspension, was thrown off, and is represented afterwards at fig. 7. After remaining twelve days in the pupa state, the perfect insect began to appear as at fig. 8. 9. Represents the butterfly in the act of escape; and 11 a, and 11 b, the pupa case, and the insect totally extricated, with the wings folded, collapsed, and humid. At 12, it is gradually expanding the wings, during which operation it voids a sanguineous-looking excrement; and at fig. 13, the perfect insect is seen with the wings expanded for flight. The whole of this wonderful process from fig. 8, was performed in about seven minutes.

It is probable, that the BUTTERFLIES were the first insects which attracted attention, and furnished the first materials of entomology. Nature, in investing them with such brilliant and various colours, and giving them such facility of flight, seemed desirous of reproducing to our view, in the class of insects, the beauties of the humming-birds and colibris. The analogy holds good even in the organs with which they take their nutriment, and the nature of that nutriment itself.

The plan of nature, in creating these sylph-like inhabitants of air, is wonderful indeed. Who could ever pre-suppose that so lively, delicate, and brilliant an insect as a butterfly, so airy in its habits, and so fastidious in its food, should be derived from a crawling, sombre, and voracious worm? The butterfly, on issuing from its cocoon, is entirely formed. Nothing of its prior state remains. Its figure, its habits, all, in a word, is so changed, that it can no longer be recognized. The butterfly is agility itself, and grace personified. It appears to disdain the earth, and in its magnificent robe, to seek the skies; while it is sustained with nectar, like the fabled divinities of old. Issuing from its dark cradle, it seems to rejoice in its new-born existence, to court the sunbeam, or delight in recognizing the groves or fields where its laborious infancy has been passed. Its life is now a scene of perpetual enjoyment. It wanders from flower to flower; and, like a gay gallant, flutters from female to female, continually in pursuit of the pleasures of novelty and change.

But a life of this description affords no materials for the historian: however pleasant to contemplate, it is barren of facts. The historian of nature, like that of man, must not look to peace and pleasure, for the most abundant interest of his subject. He must find it in labour, bustle, and industry, amid the excitement of war, and the vicissitudes of peril.

The female butterfly deposits her eggs on such vegetables as are proper to nourish the caterpillars which proceed from them. She contents herself with simply agglutinating them; and we find no example in her, of the singular foresight which belongs to some of the lepidopterous species.

The mode of flight in those diurnal lepidoptera varies as considerably as in birds, and affords to the experienced naturalist a criterion of the distinction of species. Inclement weather would not suit these tender insects; their wings would be rendered useless, and they would not propagate their species. Perfect summer, or at least the whole of that season when nature reposes in the bosom of calm and sunshine, and the fields are enamelled with flowers, is the period of their happy but transitory existence. They pass the winter in the caterpillar or chrysalis form, or, but very rarely indeed, as perfect insects, in a lethargic slumber. The woods and forests are most favourable to their reproduction. Some species, however, haunt rocky and stony places; other diurnal lepidoptera delight in the neighbourhood of streams and waters; but in general their localities are determined by the

I.O. Westwood del.t

Papilio Cleotas. G.R. Gray.

London. Published by Whittaker & C.o Ave Maria Lane. 1832.

Pl. 38

1. Papileo *Childrenæ.* G.R. Gray. 2. Pieris (*Euterpe.*) Swainsoni. Gray.

London. Published by Whittaker & Co. Ave Maria Lane. 1837.

presence of such vegetables as are peculiarly adapted for their own nutriment, and that of their offspring. After the impregnation of the female, the male soon perishes, and the latter survives only to deposit her eggs, and then undergo the same destiny.

It seems needless to repeat the minute differences which distinguish the several subgenera of diurnal lepidoptera, or butterflies: they are sufficiently indicated in the text. We proceed, therefore, to a description of our figures of two new species of true *Papilio*, from the splendid collection of lepidopterous insects belonging to Mrs. Children. The first is named, in honour of the above-mentioned lady, *Childrenæ*. It has the upper wings black, with the middle of a fine bright green, divided by the black nervures, also a white oval spot near the exterior border; the lower wings black, with a longitudinal spot of virescent crimson near the anal angle; beneath the upper wings, black, with a white spot, which is divided by a nervure; the lower wings with a transverse interrupted band of crimson; the body black, the sides of the thorax spotted with red. It is from Brazil.

The other species is named *Cleotas*. It is of a bright black, with two rows of bright yellow spots, the first series commencing from the centre to the posterior angle of the wing; these are the largest, and rather square; the other row is near the exterior margin; these last are small and of a triangular shape. Between these two series is a row of spots formed of rather scattered yellow scales; the lower wings dentated and black, with a band near the centre of rather large yellow square spots, then a row of lunated spots formed of scattered yellow scales, then a series of very small similar spots; those near the anal angle are red, and the other three yellow; beneath the upper wings brownish black, with a large spot of yellowish in the costal nervure, also some large yellow spots near the centre; those nearest the posterior angle are the smallest, some of them in a lunar form; beneath the lower wings brownish black, with the band as above, white; each spot formed of red scales, and some scattered, then a narrow lune of black, next a line of blue scales, and a row of red lunes on the posterior margin. This species is from South America.

We have figured two species which belong to the genus Pieris: the first we name *Pieris Swainsonii.* It is a greyish black, with a mascular band of pinkish white spots, the inner margin greyish white; the lower wings, from base to the centre, greyish white; beneath the upper wings, brownish black, with a broad oblique pink band cut by the nervures; the lower wings with a long narrow yellow spot on the anterior margin, and spotted at the base with crimson, also a bluish grey transverse band near the centre. This Brazilian species belongs to the subgenus *Euterpe,* of Mr. Swainson, and is in the collection of Mrs. Children.

The second, *Pieris Nemesis,* of Latreille. It is black; the upper wings with six spots of yellow, the three nearest the base the largest; beneath these wings, brownish black, with two small spots on the side; the lower wings with the anterior part ash virescent, then a transverse band, narrowing towards the exterior border; the posterior part of the wings of a fine jonquil yellow, beneath brownish black, with a transverse broad yellow band near the posterior margin; the body black, with the under side of the abdomen yellow. This South American insect belongs to Mr. Swainson's subgenus *Lieinix.*

Of *Heliconia* we figure a· species which M. Latreille names *Cyrene.* It is black, with ten transparent spots placed in the following manner, (from base) 1, 2, 4, 3,—those nearest the base the largest; the lower wings with a transverse

1. Pieris *Nemesis*. *Lat.* 2. Heliconius *cyrene*. *Lat.* 3. & 4. Nymphalis *Pitheas* *Lat.*

London. Published by Whittaker & C°. Ave Maria Lane. 1832.

1. Nymphalis *Thamyris. Lat.* 2. Nymphalis *chrysites. Lat.* 3. Nymphalis *leucophthalma Lat*

London. Published by Whittaker & C°. Ave Maria Lane. 1837.

transparent band in the centre, not touching the exterior margin; beneath the wings reddish brown, bordered with black, and spotted white; the antennæ black; the feet and palpi are part black and part white. This species is from Peru.

The genus Nymphalis we illustrate with the following species, viz. *Nymphalis Pitheas,* Latr. It is violet black, with a transverse band, and the space at the base of a fine crimson; the lower wings violet black, with the base and interior margin crimson; beneath these wings pale rose, with two large black spots, with bluish pupil; the posterior and interior margins bordered with black, which has a line of yellow on each side, also two small lines of blue at the anal angle.

The second species is named *Nymphalis Thamyris.* The four wings above are of a velvety black, with the base of a brilliant blue, which is less extended in the lower wings. The upper wings have two spots of the same colour near the summit of each wing, the posterior one the largest; the lower wings have the same tinge along the exterior border; their tails are speckled near the tip with bluish grey, the tip is black; the interior border of the same wing is for the most part bluish; beneath the four wings are of a ferruginous brown, rather deep, shining, and speckled with grey; the upper wings have the posterior border, and a ray of the same colour; this ray crosses from the summit to the middle of the interior border, and this is bordered with a line of deep brown; the lower wings have on each, three small spots, rather distinct, round, and disposed in a transverse line, with two black, the centres of which are occupied by a small spot, rather circular; the upper part is grey, and the lower of a deep bluish ash,—the interior spot is much the smallest; beneath the tail is speckled as above.

The third species, *Nymphalis Chrysites,* above the fore wings is black, with a violet reflection; the fore wing with three transverse bands, and two bands nearest the base of upper wings, continuing on the lower wings, of a bright yellow; beneath the wings, like the upper surface, but much paler; and the centre of the bands with a small spot, oval and bluish; the extremity of the wing is of a reddish brown, or rather bay, with a tinge of bluish ash; in the place of the third band there are four small eyes, the two upper are very small, particularly the second, which is black, with the pupil and iris of a bluish ash; the third is whitish, surrounded with two circles, one of brown and orange, rather larger; the fourth is much larger, very distinct, black, with a circle of pale grey, beneath which are two spots of bluish ash; beneath the lower wings, bay, with a tinge of bluish ash, particularly towards the base and interior margin; in the middle of the exterior border is a spot of pale golden yellow, shining, in the form of a large triangle, with two eyes at the summit of the posterior angle, with black pupil, with a bluish point in the middle. These three last species are from South America.

The fourth species we name *Nymphalis Annetta:* it is black on the upper surface, with a narrow whitish band across, near each anterior angle; and the base of a fine crimson colour towards the posterior angle; the lower two wings with a spot near the centre of each wing, and the nervures of a fine crimson; beneath the fore wings, like the upper surface, but of a paler colour, and spotted in the costal nervure; the lower wings black, with four yellowish brown bands, the two nearest the margin connected between each nervure, so as to form rings, with a violet lune in the centre of each, except the anal one, which has two lunes and some undulated bands; interior margin brownish black, the base spotted with crimson; the body above black, and beneath brownish. This species is from Brazil.

We have figured a species of the genus Erycina, which we

I.O. Weswood del.t

Nymphalis Annetta. _ G.R. Gray.

London. Published by Whittaker & C.o Ave Maria Lane 1832

1 Erycina (Chorinea) Xanthippe G.R.Gray
2 Erycina (Dorila) asteris G.R.Gray
3 Barbicornis (Chroma) basalis Lat. God
4 Erycina (Lamproptera) Curius G.R. Gray.

London Pub.d by Whittaker & C.o Ave Maria Lane 1852.

J. O. Westwood del.t

Eumenia Childrenæ. — G. R. Gray.

London. Published by Whittaker & C.o Ave Maria Lane. 1832.

consider as new, *Erycina Xanthippe*. The wings are transparent, with the margin and cross bands black, also with two crimson spots at the anal angle; tail black, tipped with white; beneath, the same, except that there are two white spots beneath the crimson; it is from Brazil.

We also figure a new species of *Eumenia*, dedicated to Mrs. Children, in whose collection it is contained. It is of a deep black, the anterior margin of a rich violet blue, also with a row of triangular spots on the margin of the wings, which are also fringed with white; beneath the wings are black, with a violet blue spot in each costal nervure of the fore wings, and also very much spotted with yellowish green. This splendid species is from Mexico.

The *Barbicornis basalis*, of *M. Godart*, is black, with two fulvous transverse bands; the lower wings black, with a longitudinal band near the anterior margin; also a spot at the base of each tail, fulvous. This is from Brazil.

We now come to the *Crepuscular* family of lepidoptera, and all we have to say concerning them in the perfect state, may be comprised under the head of SPHINX.

These insects, like the butterflies, are very beautiful, their wings being adorned with the most lively and most agreeably varied colours. Although they are in general rather large and heavy, they fly more lightly than the *bombyces*, and they are discovered by the noise which they make. At the setting of the sun, they proceed to take a little nutriment from flowers. They pass with rapidity from one to another, pumping the juices which they contain, with their long proboscis, and hover round these flowers without resting on them.

There is nothing particular in the coupling of these insects. The males finish their career with that of their pleasures. The females lay a great number of eggs.

The *Sphinx Atropos (Death's Head Moth)*, which is deep brown, varied with yellow, makes a singular noise, resembling a plaintive cry, which Reaumur considers to be produced by the friction of the palpi against the proboscis. This stridulation, however, is differently explained by a French surgeon, named Lorey. According to him, this noise which the moth makes when touched, or when, having entered an apartment, it fears to be unable to retreat, is entirely owing to the air. This air escapes from a trachea, which exists at both sides of the base of the abdomen, and which, in a state of repose, is closed by a bundle of very fine hairs, united by a ligament which originates on the lateral and internal parietes of the upper part of the abdomen, which bundle is dilated by the divergence of the rays which compose it, forming a little sun, or very pretty asterisk. M. Lorey, to convince himself of the truth of this, successively amputated the palpi, the trunk, and the head, without putting a stop to the stridulation.

This plaintive sound, and the peculiar figure of the corslet, with three spots upon it, representing, in some degree, a death's head, have rendered the sphinx atropos an object of superstitious terror. A number of them made their appearance on a certain year in the province of Bretagne, to the great terror of the peasants. They believed them to be the cause of certain epidemic maladies, which then prevailed in that country, and regarded them as the fore-runners of death. Their lamentable and funeral cry is certainly more than enough to frighten the ignorant and superstitious.

We shall now proceed to the nocturnal division of lepidoptera, beginning with PHALÆNA.

These insects come from the caterpillars which we have mentioned under the name of *geometers*, and respecting which, it is unnecessary to add any thing here. They remain for a greater or less time under the form of chrysalis. They then couple and die after the oviposition. These last become perfect insects towards the end of summer.

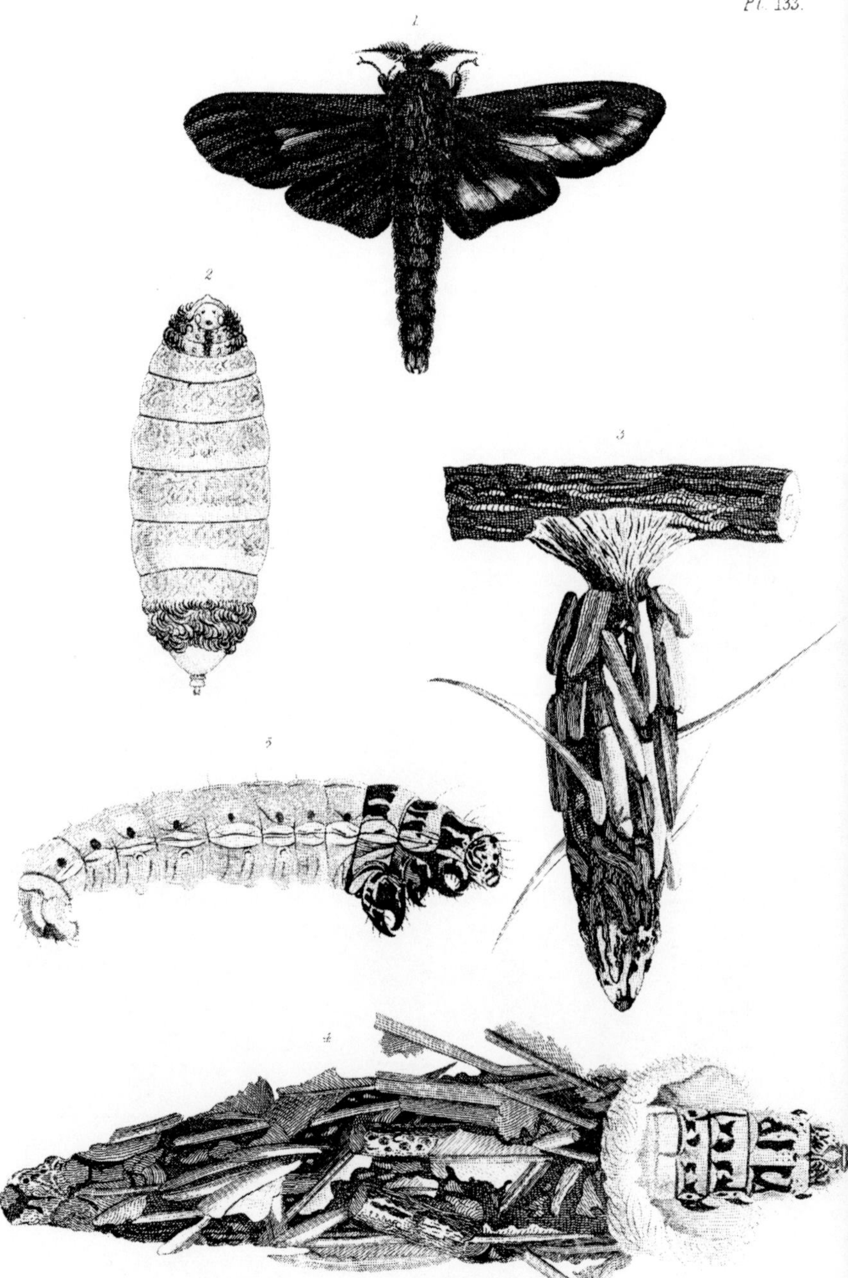

1 *Oiketicus Kirbii male.* 3 *Habitaculum of the male before his metamorphosis*

2 *Female of same.* 4 *Larva of female.*

5 *The same withdrawn from the habitaculum*

London, Published by Whittaker & C.º Ave Maria Lane, 1832.

But those whose caterpillars do not undergo their metamorphosis until autumn, pass the winter in the chrysalis form, and do not become perfect insects until the following spring.

Like all the nocturnal lepidoptera, the phalenæ seldom fly but after the setting of the sun. They remain during the day tranquil upon the leaves. They inhabit gardens, meadows, and especially woods and forests. The males of three or four species appear to have six wings, the lower ones having, near their origin, a sort of flat, oval appendage, folded in two, and couched on the upper part of these wings.

The Bombyces, in their perfect state, live even a shorter time than the other nocturnal lepidoptera. Incapable of taking nutriment, since they have neither proboscis nor tongue, they come into the world only to propagate their species. The same day that they have quitted their cocoon, they are in a state for coupling. The males are very lively and ardent. They fly rapidly, often in a zig-zag course, and sometimes even in the open day.

Respecting the silk-worm, which is the caterpillar of *Bombyx mori*, as we have already so much extended our generalities on caterpillars, we must be excused from entering into any details here.

As to *Tinea* (*Moths proper*) there is nothing in the habits of the perfect insect, to arrest us here. The ravages made in furs, &c. by their caterpillars, are too well known, and the mode of checking them, properly belong to a subject different from ours.

The females of many lepidoptera, even in the perfect state, are nearly apterous, as is the case with *Phalæna antiqua* of our own country; but the most decidedly apterous female is to be found in a species described and figured by the Reverend Lansdown Guilding, in the fifteenth volume of the Linnean Transactions, *Oiketicus Kirbyi*. The reverend gentleman states that the larvæ are very common on many of the trees in the West Indies. When the pupa has slept the appointed time, the female still resident within the habitaculum formed by the larva, opens the carina at the upper part by the motion of its head, and prepares to receive the winged male; and, by the extraordinary extensive power of the male organ, copulation is effected, without the female quitting the habitaculum, at the bottom of which the eggs are subsequently deposited, and the parent dies.

Certain species of minute fungi have been found on the dead larvæ and pupæ, and even on the perfect insects of this class;* and recent observations seem to lead to the conclusion, that these fungi grow also upon them in the living state. We have copied the figure of a moth from Cramer, with a parasitic fungus, a species of *Isaria* growing upon it. The insect is the *S. Achemenides* of the above named entomologist, and is of a nearly uniform yellowish-brown colour.

* *See* Kirby and Spence, iv. 215.

1 Sphinx Achemenides. *3 Phalæna Tilea.*

2 Phalæna Orosea. *4 Sphinx Atropos.*

London. Published by Whittaker & C.º Ave Maria Lane, 1832.

1 Noctua Ancea. 3 N. Damonia _ female.

2 N. Damonia _ male. 4 N. Euristea.

5 N. oculata.

London, Published by Whittaker & C.° Ave Maria Lane. 1832.

THE ELEVENTH ORDER OF INSECTS.

The Rhipiptera.

Stresiptera of Mr. Kirby. On the two sides of the anterior extremity of the trunk, near the neck, and the exterior base of the first two feet, are inserted two small mobile, crustaceous bodies, like small elytra, narrow, elongated, dilated into a knob, curved at the end, and terminating at the origin of the wings. The proper elytra covering always the whole, or the base of these last organs, and springing from the second segment of the trunk. These bodies are not true cases, but pieces analogous to those observed at the base of the wings in lepidoptera. The wings of rhipiptera are large, membranaceous, divided by longitudinal nervures, forming rays, and fold up like a fan in their entire length; their mouth is composed of four pieces, two of which, shorter, appear to be so many palpi, with two articulations; and the others inserted near the internal base of the preceding, have the form of small linear laminæ, pointed and crossing at their extremity, like the mandibles of many insects. The head has two large and hemispherical eyes; two antennæ, approximating at their base, on a common elevation, almost filiform, short, and composed of three articulations, the first two of which are very short, and the third, very long, is divided from its origin into two branches, long, compressed, lanceolate, and applied one against the other; the simple eyes are wanting; the abdomen is almost cylindrical, formed of from eight to nine segments, and terminates in pieces which have some analogy with those of the *Cicadariæ psylus* and *Chrysis*, of the order Hemiptera. The feet, six in number, are almost membranaceous, and terminated by filiform tarsi, with four membranaceous articulations, the last without hooks. The four anterior feet are close together, and the last two thrown backwards; the breast is ample, and divided by a longitudinal furrow; the hinder extremity of the metathorax is prolonged like a scutellum over the abdomen: the sides of the hinder portion of the trunk are strongly dilated, and form a sort of swelled buckler.

They are composed of two genera, XENOS and STYLOPS.

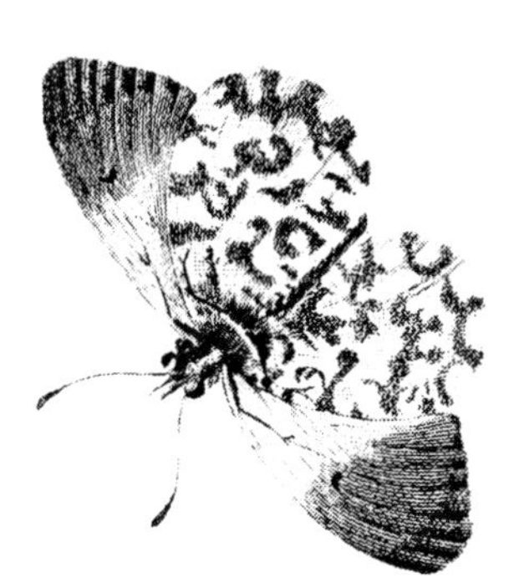

London. Publish'd by G & W. B. Whittaker Feb.y 1824.

SUPPLEMENT

ON

THE RHIPIPTERA.

This little order, the characters of which have been given in the text, will exact a very small portion of our attention. The insects of which it is composed live, when in the larva-state, some of them (*Stylops*), between the segmentary scales of the abdomen of the *Andrenæ*; the others, as *Xenos*, between these same parts of *Polistes*, a genus of the tribe *Vespiariæ*. One of these larvæ, observed by Mr. Kirby, is oval, annulated, with the anterior part of the body dilated into the form of a head, and the mouth composed of three tubercles. These larvæ are metamorphosed into nymphs in the place where they were born, under their own skin. The Rhipiptera may be thus considered, with reference to certain animals, as a kind of *Œstri*.

These larvæ (*Xenos*) are often to be found in the interior of the abdomen of the wasp, as many as six in number. The body is milk-white, cylindrical, and divided into nine semi-rings; the head at a fixed period is disengaged, appears between the scales of the wasp's belly, and is coloured more strongly by the contact of the air. These larvæ have no organs proper for ambulation, and when taken from the body of the wasp, they have a marked undulatory movement, which may be instantaneously increased by irritation. When exposed to the air they soon perish. The insect is not easily procured in a perfect state, as it is difficult to preserve the wasps alive that have the larvæ in their interior. M. Jurine convinced himself that the larva of Xenos spins no cocoon for its transformation, but that its own skin, dried up, and grown hard, answers the same purpose.

We are enabled, by the kindness of Mr. Children, to insert the figure of a new and curious species of Stylops, which Mr. Gray has dedicated to that gentleman, and named it Stylops Childreni. It is velvety black, with the antennæ and palpi piceous; the abdomen is pitchy brown, and the legs are rufescent; the wings are slightly tinged with a sooty irridescence, with the anterior margin darker.

THE TWELFTH

AND

LAST ORDER OF THE CLASS OF INSECTS.

The Diptera, *Antliata*, Fab.

Six feet; two membranaceous extended wings, with two mobile bodies above them, called balancers; a sucker formed of scaly pieces, like hairs, from two to six in number, inclosed in the upper gutter of a sheath, like a proboscis, terminating in two lips, or covered by one or two inarticulated laminæ, like a case. The simple eyes, when existing, are always three. The antennæ are in general inserted on the front, and approximated at their base. Those of the diptera of our first family are very analogous to those of the nocturnal lepidoptera; but in the following families, they are composed but of two or three articulations, the last of which is usually fusiform, furnished with a small appendix, like a stylet, or a thick hair, sometimes simple, sometimes barbed. The pieces of the sucker perform the office of lancets, and pierce the envelope of the substances on which these insects feed, causing the fluid to ascend by pressure, to the pharynx, which is at the base of the sucker. The sheath seems merely as their support, and is usually refolded, while they are in action. The base of the proboscis often bears two filiform or claviform palpi; sometimes with five articulations, but more usually with one or two. The wings are simply veined and generally horizontal. The use of the balancers is not very well known. Many species have, besides them, two membranaceous pieces, named *winglets.* The prothorax is always very short, and often only its lateral portions are discoverable. In some diptera, these are prominent like tubercles. The meta-thorax composes the major portion of the trunk, and in front or behind the prothorax, are two stigmata; two others are visible near the origin of the balancers. The abdomen is often attached to the thorax only by a portion of its transverse diameter. It has from five to nine apparent rings, and is usually pointed in the females. In those which have the fewest number of rings, the last are oviduct, presenting a series of small tubes reentering into one another, like a telescope. The feet generally long and slender, terminate in a tarsus of five articulations, the last of which has two hooks, and very often two or three membranaceous cushions. The larvæ have no feet.

We shall divide this order into two principal sections, in the first of which the head is always distinct from the thorax, the sucker inclosed in a sheath, and the hooks of the tarsi simple. The transformation of the larva into nymphs, takes place out of the belly of the mother. The first subdivision consists of diptera, in which the antennæ are divided into a great number of articulations. They form our first family.

The Nemocera.

The antennæ most frequently with fourteen or sixteen articulations, and six to nine or twelve in the others. They are filiform, often hairy, and much longer than the head. The body is long, the head small and rounded, eyes large, proboscis projecting, sometimes short and terminated by two lips, sometimes long and like a siphon. There are two external palpi inserted at its base; usually filiform or setaceous, and composed of four or five articulations. The thorax is thick and raised, the wings oblong, the balancers entirely uncovered, and not accompanied with ringlets; the abdomen long, most usually with nine rings pointed in the females, thicker at the end in the males, and provided with pincers in the males; feet very long and slender.

Some, whose antennæ are always filiform, of the length of the thorax, bristling with hairs, and with fourteen articulations, have a long, advanced filiform proboscis, inclosing a sharp sucker, composed of five hairs; they constitute the genus

CULEX, *Lin.*—CULICIDES, *Lat.*

Body and feet very elongated and hairy; antennæ much furnished with hairs, forming a sort of plume in the males; eyes large and convergent at their posterior extremity; palpi advanced, filiform, hairy, of the length of the proboscis, and with five articulations in the males; shorter and less articulated in the females. Proboscis, a membraneous, cylindrical tube, terminated by two lips, and also composed of a sucker of five scaly threads. Wings horizontal, one over the other, above the body, with small scales.

In the work of M. Meigen, on the European diptera, the genus *Culex*, of preceding writers, forms three. Those in which the palpi are in the males longer than the proboscis, and very short in the females, compose that of CULEX, proper.

The species, whose palpi are in the males of the length of the proboscis, form that of ANOPHELES.

Those in which they are very short in both sexes, comprehend that of ÆDES. Hoffmannegge.

M. Robineau Desvoidy has added three others: those whose palpi are shorter than the proboscis, intermediate legs and tarsi dilated and much ciliated, are distinguished by the name of SABETHES. Those whose proboscis is elongated, recurved at the end, palpi short, but first articulation thicker, second shorter, and the other three cylindrical, compose the genus MEGARHINUS. The *Culex ciliatus*, of Fabricius, forms a third, PSOROPHORA; principal character, two small appendages, one on each side of the prothorax.

In the other Nemocera, the proboscis, whether terminated by two lips, or, like a siphon, is perpendicular, or curved upon the breast. The palpi are curved underneath, but then with one or two articulations at most. Linnæus comprehends them in his genus

TIPULA. *Tipulariæ, Lat.*

Which we divide in the following manner:—

In the first section, the antennæ are sensibly longer than the head, at least in the males, slender, filiform or setaceous, more than twelve articulations, and the feet long and slender.

Some of them, all winged, never have simple eyes; wings roof-like, palpi always short; head not, or very little prolonged in front; eyes crescent-formed; legs without spines. This subdivision is composed of small species.

Sometimes the antennæ are entirely furnished with hairs, but much longer in the males, and forming a triangular plume.

Those whose antennæ are composed in both sexes, of fourteen ovaliform articulations, the last little differing from the preceding, and the wings couched horizontally one over the other, compose the subgenus CORETHRA, *Meig.*

Those whose wings are inclined, antennæ composed of thirteen articulations in the males and six in the females, compose the subgenus CHIRONOMUS, *Meig.*

TANYPUS, *Meig.* Antennæ fourteen articulations in both sexes; last but one very long in the males; all the others nearly globular.

Sometimes the antennæ, always of thirteen articulations at

least, and usually grained, have but short hairs, or but a small bundle of them at the base.

CERATOPOGON, *Meig.* *Culicoides, Lat.* Antennæ with simply a bunch of hairs at the base. Proboscis, as in the two following subgenera, has the form of a pointed bill.

PSYCHODA, *Lat., Meig.* No bundle of hairs to the antennæ; wings roof-like, and a great number of nervures.

CECIDOMYIA, *Meig.* Wings couched on the body, and but three nervures.

Other species, still of the division of those whose antennæ are manifestly longer than the head, and slender, are also without simple eyes, but the usual ones are entire, oval, or round. The wings, sometimes separate, have always membranaceous nervures united transversely. The anterior extremity of the head comes out like a muzzle; palpi usually long; extremity of the legs spiny.

In many the wings are always extended; the antennæ of the males usually barbed, pectinate or serrated. The palpi are composed of five articulations, the last very long, and seemingly formed of others very small.

CTENOPHORA, *Meig.* Filiform antennæ, pectinate in the males, grained or serrated in the females.

PEDICIA, *Lat.* Antennæ almost setaceous, simple, with the first two articulations larger, elongated; the three following top-like; the next three globular, the last seven slender, and almost cylindrical.

TIPULA (proper), *Lat.* All the articulations, except the second, which is globular, almost cylindrical.

NEPHORTOMA, *Meig.* The first and third articulations elongate and cylindrical, and the rest arched; nineteen in the males, fifteen in the females.

PTYCHOPTERA, *Meig.* Sixteen articulations, third much longer than the others, and the following oblong.

In all the following subgenera, the last articulation of the palpi is scarcely longer than the others, and has no appearance of annular divisions. The antennæ have more than ten articulations.

Those in which they are for the most part grained, of nearly the same thickness, and furnished with verticillæ of hairs, compose, in M. Meigen, divers genera.

RHIPIDIA, *Meig.* The only tipulariæ of this subdivision, with pectinate antennæ in the males.

ERIOPTERA, *Meig.* Many nervures in the wings, but furnished with hairs.

LASIOPTERA, *Meig.* Hairy wings, but only two nervures.

LIMNOBIA. Smooth wings, and antennæ simple.

POLYMERA, *Wiedmann.* Antennæ with twenty-eight articulations.

In the other subgenera, the antennæ terminate in many articulations, more slender and almost cylindrical.

TRICHOCERA, *Meig.* First articulations of the antennæ almost ovaliform, the following more slender and pubescent.

MACROPEZA, *Meig.* Distinguished by the great length of their last feet.

DIXA, *Meig.* First articulation of the antennæ very short, second almost globular, and the following proportionally more slender.

Sometimes the antennæ have but ten or six articulations: those in which they have but ten, form the genus MÆKISTOCERA, of M. Wiedmann; the wings are separated.

Those where they are six, are HEXATOMA, *Lat.*, comprehending *Anisomera* and *Nematocera,* Meigen. In the first, the third articulation of the antennæ is much longer than in the second.

Other tipulariæ present an anomaly very rare in this order, namely, the want of wings; antennæ filiform, but a little slender towards the extremity, and but little hairy; feet long;

abdomen of the females pointed, ending in a bivalve oviduct. This sub-division comprehends the genus *Chionea*, Dalman.

Another division of tipulariæ, *Fungivora*, have two or three simple eyes; the antennæ are much longer than the head, slender, and have fifteen or sixteen articulations; the last articulation of the palpi without apparent division; wings always couched on the body, and the nervures are usually less numerous; eyes either entire or emarginated; feet long and slender, and extremities of the legs spiny.

Some have the palpi curved, and four articulations at least, very apparent; antennæ filiform or setaceous.

In some the anterior extremity of the head comes out like a bill, and then the head is almost entirely occupied by the eyes; there are always three simple eyes; antennæ and their articulations short.

Those in which the head is almost entirely occupied by the eyes, simple eyes of equal size and muzzled, advanced, but not longer than the head, form the subgenus Rhyphus, *Lat.*

Those in which the eyes occupy only the sides of the head, and the anterior simple eye larger than the two others, and the muzzle prolonged under the breast, like a proboscis, compose the subgenus Asindulum.

Gnovista, *Meigen.* Palpi seem inserted near the end of the proboscis, and not near the base.

The head, in none of the following subgenera, presents any anterior prolongation. Eyes always lateral.

Sometimes the antennæ are longer than the thorax, like hairs, with the first two articulations thicker; three simple eyes. In Bolitophila, *Hoffmans, Meig.*, they are in a transverse line. In Macrocera, *Meig.*, they form a triangle.

Sometimes the antennæ are of the length of the head and thorax.

Some subgenera, in which the eyes are always entire, are

remote from the others, by having their four hinder legs externally furnished with small spines—as

Mycetophila, *Meig.* But two very small simple eyes, inside apart.

Leias, *Meig.* Three simple eyes approximated.

Sciophila. Articulations of antennæ more distinct; their wings have, beside the usual closed cell extending from the base to the middle, another small complete cubital cell.

Of the subgenera without spines on the legs, and with three approximating simple eyes, we shall first separate those whose antennæ have sixteen articulations. Eyes not emarginate.

Platyura, *Meig.* Antennæ rather thick and compressed, and a little perfoliate.

Synapha, *Meig.* Wings with but one cubital cell.

The next have the eyes clearly emarginate on the internal side.

Mycetobia, *Meig.* The wings have a large closed cell from base to middle.

Molobrus. Cell from the base of the wing to its posterior edge.

Campyolmyza, *Wied.*, *Meig.*, whose antennæ (females) have but fourteen articulations; wings hairy, and without nervures internally.

The last fungivorous tipulariæ,

Ceroplateus, *Bosc.*, *Fab.* Palpi raised; but one ovoid articulation; antennæ fusiform, and compressed.

The last general division of these, called *floral*, has the antennæ but little longer than the head, thick, eight to twelve articulations, in form of a perfoliate knob, nearly cylindrical in most, fusiform in some, and in others terminated by a thick and ovoid articulation. Body short and thick; head entirely occupied by the eyes in the males. Such are, especially,

CORDYLA, *Meig.* Fusiform antennæ, composed of twelve articulations; eyes round, entire, and apart. No simple eyes. Feet long, and legs spiny at the end.

Now come the subgenera, whose antennæ have eleven articulations, composing almost a cylindrical knob; eyes of the males, are contiguous; no simple eyes; eyes of the females emarginate internally, and crescented.

SIMULIUM, *Lat.* Antennæ a little crooked at the end.

In the following, three simple eyes are distinct.

SCATHOPSE, *Geoff.*, *Meig.*, *Fab.* Crescented eyes in the females; palpi very small, and with but one distinct articulation.

PENTHETRIA, *Meig.* Eyes entire and apart; feet long and without spines.

DILOPHUS, *Meig.* Eyes contiguous in the males; a range of small spines at the extremity of the anterior legs.

The last floral tipulariæ have but eight or nine articulations to the antennæ; those which have nine, forming a cylindrical and perfoliate knob, compose the subgenus BIBIO, *Geoff.*, *Meig.*

ASPISTES. Only eight articulations, the last an ovoid knob.

All the following diptera, with very few exceptions, have the antennæ with three articulations, the first sometimes so short, that it need not be reckoned; the last, in many, is annulated transversely, but without distinct separations. It often has a thread, lateral in some, at the top in others, presenting at its base one or two articulations, and sometimes simple, sometimes silky. If it be terminal, its length sometimes diminishes, and thickness augments, so that it forms a stylet. The palpi have never more than two articulations.

Some have the sucker always composed of six or four pieces, and the extremity of the proboscis always projects; the palpi, when they exist, are external, and inserted near the edges of the mouth; the sucker originates near this cavity.

There are three families in this sub-division,

First,

Tanystoma.

The last articulation of the antennæ, not comprehending the stylet which may terminate it, presents no transverse division. The sucker is composed of four pieces.

In the first division, the proboscis, always entirely or nearly salient, advances more or less like a tube or siphon, cylindrical, conical, or filiform; the palpi are small.

Some have the body oblong, thorax contracted in front, and wings couched on the body; the proboscis is most usually short; the antennæ are always approximate, and the palpi apparent.

Asilus, *Linn.* Proboscis directed forwards.

Some (*Asilici*, Lat.) have the head transverse, the eyes lateral and apart; proboscis as long as the head, and a complete triangular cell near the internal edge of the wing.

Sometimes the tarsi terminate in two hooks, with two intermediate cushions.

Here the stylet of the antennæ is not very perceptible, or when it is so, its second and last articulation is not prolonged like a thread.

There are some among them whose antennæ are but little longer than the head; the stylet scarcely sensible, or very short, conical, and pointed.

Laphria, *Meig.*, *Fab.* The stylet of the antennæ fusiform, scarcely sensible, and the proboscis straight.

Ancylorhyncus, *Lat.* The stylet of the antennæ is scarcely projecting, and pointed; proboscis compressed, arched, and hooked.

Dasypogon, *Meig.*, *Fab.* Stylet very distinct and conical, and proboscis straight.

In the two following subgenera, the antennæ are manifestly longer than the head, and often on a common peduncle; stylet elongated, of the same thickness as the antennæ, with two articulations, the second longer, cylindrical or ovoid, and terminating in an obtuse point.

CERATURGUS, *Wied.* Antennæ not on a common elevation.

DIOCTRIA, *Meig.*, *Fab.* Antennæ on a common peduncle.

Sometimes the stylet is prolonged like a thread; those in which the thread is simple, form the subgenus ASILUS proper.

Those in which it is plumose, form the subgenus OMMATIUS, *Illig.*, *Wied.*

Sometimes the tarsi terminate in three hooks instead of the intermediate cushions.

GONYPUS, *Lat.* Stylet terminating in a short thread.

The others (*Hybotini*, Lat.) have the head more rounded, proboscis very short, fewer nervures to the wings, and no complete triangular cell.

Sometimes the last articulation of the antennæ is large, formed like an elongated knob, and terminated by a very small stylet. ŒDALEA, *Meig.*

Sometimes it is short, ovoid or conical, with a long thread.

HYBOS, *Meig.*, *Fab.* Hinder thighs large and swelled.

OCYDROMIA *Hoff.*, *Meig.* Of the usual size.

EMPIS, *Linnæus.* Proboscis perpendicular, or directed backwards.

Some have the antennæ with three articulations; sometimes the last is like an elongated cone.

Here the proboscis is much longer than the head; stylet with two articulations always short, EMPIS proper.

RAMPHOMYIA differ from Empis only by the absence of a small nervure at the end of the wings.

In HILARA the antennæ are terminated by a small stylet, with two articulations. In BRACHYSTOMA, *Meig.*, it is a long thread.

Sometimes the last articulation, terminated by a thread, forms with the preceding a spherical body, GLOMA, *Meig.*

Others present distinctly but two articulations in the antennæ: the last is ovoid or globular, proboscis generally short, and palpi couched upon it.

HEMERODROMIA, *Hoff.*, *Meig.* Haunches of the two anterior feet remarkably long.

SICUS, *Lat.* Thighs of the first or second pair of feet swelled.

DRAPETIS, *Meig.* Last articulation of the antennæ almost globular, and proboscis scarcely projecting.

The other tanystomata of our first division have generally the body short and broad, the head exactly applied against the thorax, the wings apart, and the abdomen triangular; proboscis often long.

CYRTUS, *Lat.*

Wings inclined on each side of the body; winglets very large, covering the balancers; head small and globular; thorax very much raised or humped; abdomen cubic; and proboscis directed backwards, or not existing.

Those which have the proboscis extended backwards, form the genus PANAPS, of *Lamarck*, and that of CYRTUS, proper, of *Latreille*. ASTOMELLA, of *M. Dufour*, is distinguished by antennæ with three articulations, the last like an elongated button, and without stylet. In the genus HENAPS, of *Illiger*, and that of ACROCERA, of *Meigen*, the antennæ are very small, with two articulations, and terminal thread.

BOMBYLIUS, *Lin.*

Wings extended horizontally on each side of the body, balancers naked, antennæ much approximated, abdomen triangular or conical, and proboscis directed forwards; antennæ always composed of three articulations, the last of which is elongated, almost like a compressed spindle, and terminated by a very short stylet; palpi small, slender, and filiform; the proboscis very long, and more narrow towards the end; feet long and much attenuated.

Some have a proboscis clearly longer than the head, very slender, and pointed.

TOXOPHORA, *Meig.*, are distinguished from the rest by their antennæ as long as the head and corslet, advanced, filiform, ending in a point, with the first articulation longer than the others; body elongated.

Among those whose antennæ are much shorter, XESTOMYZA, *Wied.*, approximate to the preceding in the length of the first articulation: it and the last are almost fusiform.

Another subgenus, in which the first articulation is also very long, but cylindrical, is APATOMYZA, *Wied.*

In the following, of the same division, the last articulation is the longest.

Sometimes the first two are short, and of nearly an equal length.

LASIUS, *Wied.* The head in one of the sexes is almost entirely occupied by the eyes; the last articulation of the antennæ is linear, compressed, and without any sensible stylet; the abdomen is voluminous.

USIA, *Lat.* Last articulation of the antennæ conical, truncated at the end, and terminated by a stylet. No apparent palpi.

PHTHIRIA, *Meig.* Distinct palpi.

Sometimes the second articulation is evidently shorter than the first, as in

BOMBYLIUS (proper), *Meig.*

GERON. A more remarkable elongation of the last articulation of the antennæ, terminating like an awl.

The other species have the proboscis of the length of the

head, swelled at the end; first articulation of the antennæ is the largest; in PLOAS, *Meig.*, it is much thicker than the following. The rest are CYLLENIA, *Lat.* and *Meig.*

ANTHRAX, *Scop.*, *Fab.*, MUSCA, *Lin.*

Body depressed, not gibbous; head equally high and broad; antennæ very short, and except in *Stygides,* apart, terminated awl-like; proboscis generally short, often withdrawn within the oral cavity; palpi usually concealed, and filiform; abdomen partly square.

Some have antennæ closely approximating at the base; STYGIDES, *Lat.*

ANTHRAX, proper, *Meig.* Head almost globular. Three simple eyes closely approximate.

HIRMONEURA. The anterior simple eye removed from the others.

Sometimes the head is proportionally shorter; almost hemispheric, and compressed transversely; antennæ very much apart; proboscis longer than the head; palpi sometimes external; extremity of the wings often reticulated. Those in which they are always so, whose proboscis is but a little longer than the head, palpi not apparent, first articulation of antennæ cylindrical, last in an elongate cone, form the subgenus MULIO, *Meig.*

Those in which the summit of the wings is most frequently reticulated, proboscis longer than the head, palpi external, first two articulations of antennæ very short, last in a short cone, with a sudden stylet, are NEMESTRINA, *Lat.*

Our second general division of Tanystomata has a membranaceous proboscis, short stem, but little projected, and terminated by two distinct lips. Some (*leptides*) have the wings apart, and several complete cells. The antennæ do not end in a palette; the palpi are filiform, or conical.

Sometimes these palpi are withdrawn within the oral

cavity. The antennæ terminate like a spindle, or elongated cone, with a small articulated stylet. THEREVA, *Lat.*, *Meig.* BIBIO, *Fab.*

Sometimes the palpi are external; the last articulation of the antennæ is either globular or reniform, ovoid or conic, and always terminates in a long silken hair: the tarsi have three cushions. Such are,

LEPTIS,

Divided into many subgenera.

ATHERIX, *Meig. Fab.* First articulation of the antennæ larger than the second; third lenticular and transverse.

LEPTIS, proper, *Fab.*, *Meig.* Last articulation always globular or ovoid, never transverse.

Some have the antennæ shorter than the head, with the three articulations almost of equal length, such are the *leptis* of M. Macquart, when the palpi are advanced, as in the preceding.

Sometimes they are raised perpendicularly. These are CHRYSOPILUS, *ibid.*

In the others, the antennæ are as long as the head; first articulation elongate and cylindrical; second short; third conical.

The other Tanystomata of our second division, have the wings couched upon the body, and have but two complete cells; antennæ terminated in a palette, almost always accompanied with a thread; the palpi generally are like laminæ, couched on the proboscis.

These characters, a body laterally compressed, triangular head, somewhat advanced, abdomen curved underneath, and long slender feet, furnished with spines, characterize the genus

DOLICHOPUS, *Lat.*, *Fab.*,

Now forming a small tribe. DOLICHOPODES.

In the first instance the proboscis is elongated, like a little bill.—Ortocheile, *Lat.*, *Meig.*, *Macq.*

Sometimes, as in all the others, the proboscis is very short, as Dolichopus proper, where the third articulation of the antennæ is almost triangular.

Sybistroma. Last articulation of the antennæ like the blade of a knife; sometimes the third articulation of the antennæ, whether oval or triangular, or very long and narrow, is almost lanceolate, as in Rhaphium, *Meig.*

It is triangular, and with a hairy thread in Porphyrops, *Meig.*

The thread is simple in Medeterus, *Fisc.*, *Meig.*

Sometimes the third articulation of the antennæ is almost globular; the thread always hairy: it is terminal in Chrysotus; inserted a little underneath in Psilopus; near the base in Diaphorus.

The Megacephali have the body depressed, and the head hemispherical. The feet are short, and the posterior tarsi often broad and flat. Some have a thread to the last articulation of the antennæ; when this is terminal, and the first articulations of the posterior tarsi broad and flat, we have the subgenera Callomyia, *Meig.*, and Platypeza, *Meig.*

In the genus Pipunculus, *Lat.*, the thread is inserted on the back of this articulation, and the tarsi not dilated.

The others have no thread. Scenopinus, *Lat.*, *Meig.*

The third family of Diptera, that of

Tabanides,

Has projecting proboscis, usually terminated by two lips, and advanced palpi; last articulation of the antennæ annulated. Sucker of six pieces. It comprehends the genus

Tabanus, *Lin.*

Body generally hairy. Head of the breadth of the thorax,

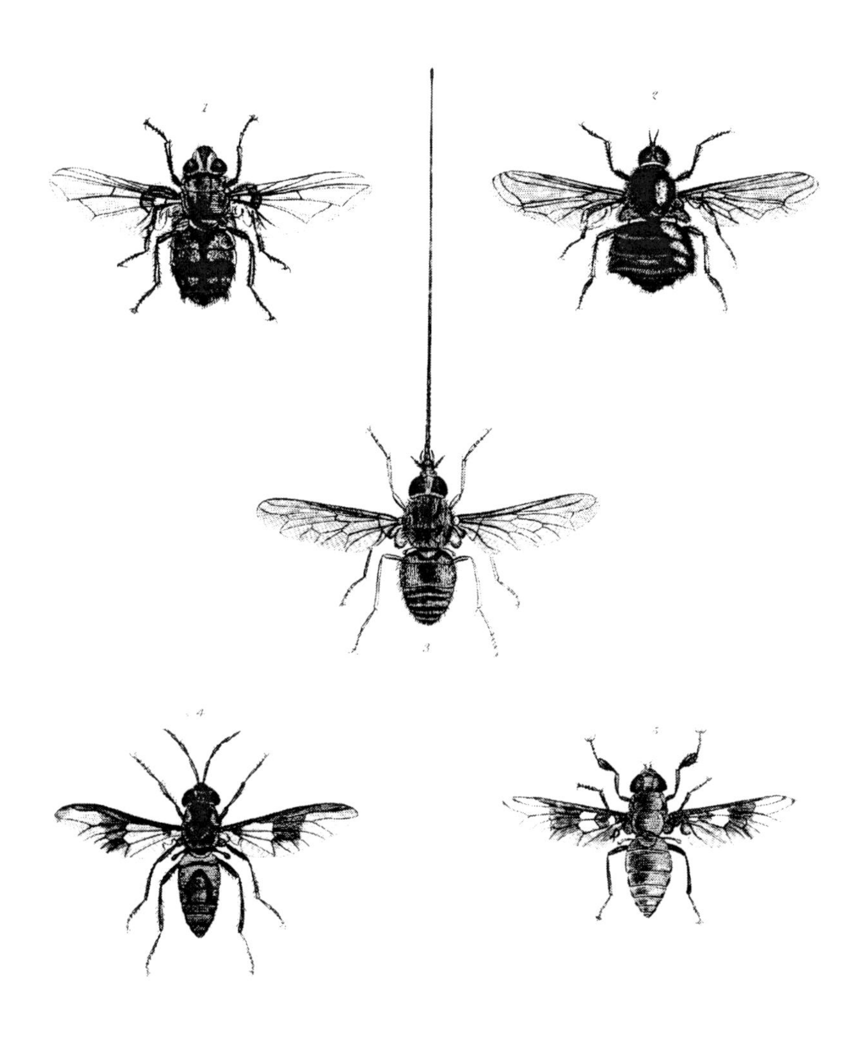

1. *Rutilia Australasia*. G.R.Gray
2. *Beris? violaceus*. G.R.Gray.
3. *Pangonia longirostris*. Hardw.
4. *Hæmatopota lunata*. G.R.Gray.
5. *Tabanus Africanus*. G.R.Gray.

London, Published by Whittaker & C° Ave Maria Lane, 1837.

and almost hemispherical. Antennæ about the length of the head; three articulations, the last the longest and without thread; proboscis generally membranaceous, about the length of the head, and almost cylindrical. Two palpi usually couched upon it, thick, hairy, conical, and of two articulations. Sucker inclosed in the proboscis, and composed of six small pieces like lancets. Wings extended horizontally on each side of the body; winglets nearly covering the balancers; abdomen triangular and depressed; tarsi with three cushions.

Some have the proboscis much longer than the head, siphon-like, scaly, and pointed; palpi very short; last articulation of antennæ divided into eight rings. PANGONIA, *Lat.*, *Fab.*

The others have the proboscis shorter, membranaceous, and terminated by two large lips; length of the palpi, one half of that of the proboscis; last articulation of the antennæ with four or five rings.

Sometimes the antennæ are but little longer than the head; the last articulation crescented, and terminated like an awl, and with five rings. TABANUS proper.

Sometimes the antennæ are clearly longer than the head, last articulation in an elongated cone, and often with but four rings: simple eyes often wanting.

Some with the last articulation of the antennæ awl-like, and in five rings, have three simple eyes.

Those in which the first articulation is manifestly longer than the rest, are SILVIUS, *Meig.*

CHRYSOPS. *Meig.* First two articulations of nearly equal length.

The others have no simple eyes. Last articulation of antennæ but four rings. Sometimes, as in HÆMATOPOTA, *Meig.*, it is subulate.

HEXATOMA, *Meig.* Antennæ longer than in the preceding; last articulation very long.

Fourth family of Diptera, that of

NOTACANTHA.

Sucker formed only of four pieces; proboscis with very short stem, almost entirely within the oral cavity. This organ membranaceous, its lips raised, palpi terminating in a knob, and raised; wings crossed; scutellum often denticulate.

There are three principal sections. In the first (*Mydasii*, Lat.) scutellum not denticulate; body oblong; wings apart; antennæ sometimes with five distinct articulations; last two in some claviform, in others subulate. Sometimes but three articulations; the last large, cylindrical, pointed, and of three rings. *Mydas* excepted, there is no appearance of stylet.

Some have the antennæ much longer than the head, with five articulations, and a very short thread. These compose the genus

MYDAS,

Thus divided:

CEPHALOCERA, *Lat.* Proboscis long, siphon-like.

MYDAS proper, *Fab.* The proboscis short, and terminated by two large lips.

In the others, the antennæ are scarcely longer than the head, cylindrical, and pointed at the extremity. Tarsi with three cushions. Hinder cells of the wings closed by the posterior edge.

CHIROMYZA, *Wied.* Antennæ with five articulations much apart.

PACHYSTOMUS, *Lat.* Three articulations, the last in three rings.

In the second section (*Decatoma*, Lat.) the antennæ have always three articulations, the last the longest, without appendage, and divided into eight rings, claviform in some, and nearly cylindrical in the others. Tarsi with three cushions. They constitute our genus

XYLOPHAGUS.

In some the antennæ are much longer than the head. First two articulations very short; third long, compressed, a little bent in the middle, its lower portion an elongated cone; the other an oval palette. Scutel unarmed. HERMETIA, *Lat.*, *Fab.*

The antennæ of the others are never much longer than the head; last articulation an elongated cone.

In these the scutellum has no spines.

XYLOPHAGUS, proper. Body narrow and elongate, antennæ a little longer than the head. The latter short and transverse.

ACANTHOMERA, *Wied.* Antennæ of the length of the head, and terminated by an articulation like an elongated cone. Head hemispherical, and eyes large. Interocular space with a sort of horn.

Sometimes the scutellum is armed with spines. These have simple antennæ.

CŒNOMYIA, *Lat.*, *Meig.* Antennæ scarcely longer than the head; third articulation conical.

BERIS, *Lat.*, *Meig.* Antennæ a little longer than the head; first two articulations of equal length.

CYPHOMYIA, *Wied.* Antennæ still more elongated; first articulation longer than the second.

In some, the antennæ, near the middle, throw out three or four linear threads, and the upper articulations are silky; near the end they are almost setaceous. PTILODACTYLUS, *Wied.*

The third section (*Stratyomydes*, Lat.) has also the antennæ with three articulations; the last stylet or thread not reckoned, has five or six rings. Stylet or thread in almost all; when not existing, third articulation long, and fusiform. Wings couched one upon the other, in many species. This second section comprehends the genus

STRATYOMYS, *Geoff.*

Some have the third articulation of the antennæ long, fusiform, and usually terminated by a stylet, with two articulations. Scutellum generally with two spines.

Sometimes the proboscis is very short. Front of the head not advanced.

STRATYOMYS, proper. Antennæ much longer than the head. Five rings, but no stylet.

ODONTOMYIA, *Meig.* Antennæ scarcely longer than the head. The last articulation conical, and curved inwards, represents the extremity of the stylet.

EPHIPPIUM, *Clitellaria, Meig.* Antennæ also not longer than the head; the third articulation forms a cone, and the fourth ring a truncated cone. Stylet with two articulations; the second longest and bent.

OXYCERA, *Meig.* Like the last, but the third articulation shortest, and subovoid, and the fourth ring shorter, without a sudden narrowing at the end.

There the proboscis is long and slender, bent at the base and lodged in a lower cavity of a projection before the head, which carries the antennæ, the shape and proportions of which are the same as in the preceding subgenera.

NEMOTELUS, *Geoff., Fab.* In the others the third articulation of the antennæ forms, with the preceding, an ovoid or globular knob, terminated by a long thread. The scutellum is seldom spiny.

CHRYSOCHLORA, Lat. *Sargus,* Fab. The third articulation conical, terminated with a thread.

SARGUS, *Fab.* The corresponding articulation is subovoid, with the thread inserted on the back, near the junction of the third and fourth ring. The first articulation subcylindrical.

VAPPO, Lat., Fab. *Pachygaster*, Meig., differs from Sargus only in having the antennæ shorter, with the first two articulations transverse.

Our second general division of Diptera having the sucker enclosed in a case, and the antennæ with two or three articulations, includes those whose proboscis, commonly membranous, being bent, with two lips, and carrying two palpi a little above its elbow, is generally entirely enclosed in the cavity of the mouth, and has only two pieces in the sucker, where it is always protruded. The last articulation of the antennæ, always accompanied with a stylet or a thread, has never any annular divisions. The palpi when at rest are hidden.

This division will form our fifth family, that of

ATHERICERA.

The proboscis is in general terminated by two large lips. The sucker never has beyond four pieces, and often only two.

This family includes the genera *Conops*, *Œstrus*, and the greatest part of *Musca*, of Lin.

We should naturally separate from the last, the species in which the sucker is composed of four pieces instead of two. These will form the first tribe, that of SYRPHIDÆ.

Their proboscis is always long, membranous, and bent near the base; the upper piece of the sucker inserted near the bend is large, vaulted, and emarginated at the end; the other three are linear and pointed, or thread-like; to each of the two lateral, which represent the jaws, is annexed a little membranous palpus; the lower thread is analogous to the tongue. The head is hemispherical, and occupied principally by the eyes, especially in the males; the anterior end of it is often elongated like a muzzle, and received underneath the trunk when folded on itself. Many species are like the flesh-flies, and others the wasp.

This tribe includes only one genus,

Syrphus.

The first general division shall be composed of those whose trunk is shorter than the head and thorax.

Then will follow those with a prominence in front of the head.

At the head of which will be those whose antennæ, shorter than the head, have a feathery thread. Their body is short, velvety, with the wings wide. These will include three sub-genera.

Volucella, *Geoff.*, *Lat.*, *Meig.*, *Fab.*, with the third articulation of the antennæ, or the pallet, oblong.

Sericomyia, Meig., Lat., *Syrphus*, Fab., with the pallet of the antennæ semi-orbicular.

Eristalis, *Meig.*, *Fab.*, with the outer and closed cell of the posterior limb with a strong emargination, rounded at the outer side; it is straight in the others.

To these will succeed others with the thread simple, or without apparent feathers.

Some, like Eristalis, have the last external cellule of the wings strongly unisinuated.

Mallota, Meig., *Eristalis*, Fab. The last articulation of the antennæ forms a kind of transverse trapesium.

Helophilus, Meig., *Eristalis*, Ejusd., Fab. The pallet of the antennæ demi-oval.

Others differ from the last by the exterior and closed cellule of the posterior limb, the external side being straight, or slightly sinuous. The anterior projection of the head is very short. The abdomen is in general narrower and more elongated than in the other.

SYRPHUS proper, Lat., Meig., *Scæva,* Fab. The abdomen runs to a point.

Another neighbouring subgenus, with the abdomen still longer, and terminated with an elongated knob, is BACCHA, *Meig., Fab.*

To which should be united, as I think, *Scæva,* Fab., *Conopseus,* Meig., although the pallet of the antennæ is less orbicular than in *Baccha.*

We shall pass to similar subgenera, but which have the length of the thread of the antennæ at least equal the length of the face of the head. In these, the antennæ are not placed upon a common pedicle, and they are not longer than the head.

PARAGUS, Lat., Meig., *Mulio,* Fab. Here the antennæ start from a common elevation, and are not longer than the head.

Sometimes the thread is lateral, SPECOMYIA, *Lat.,* in which it is inserted on the second articulation; the last is shorter and subovoid.

PSARUS, *Lat., Fab., Meig.* The thread of the antennæ is inserted on the back of the third articulation, which is nearly oval. The common pedicle is more elevated than in the others.

CHRYSOTOXUM, Meig., *Mulio,* Fab. The thread of the antennæ is also inserted on the third and longest articulation. The wings are wide apart.

Sometimes the thread, always thick, and stylet formed, terminates the antennæ.

CERIA, *Fab.* Body like a wasp, second and third articulations form a spindle-shaped knob, with a short stylet. The wings are very wide, and the outer cell of the posterior limb has on the outer side a decided re-entering angle.

CALLICERA, *Meig*. The body shorter, wider, and silky; the second and last articulation of the antennæ, form an elongated, compressed, spindle-shaped knob, with a long stylet; no re-entering angle, as in the last.

In the following, the nasal prominence is not formed, the thread of the antennæ is simple, the wings are hidden one upon another.

In the first, the antennæ, as in the former, are near each other at the base.

CERATOPHYA, *Wied.* The scutellum is unarmed; the third articulation nearly as long again as the first.

APHRITIS, Lat., *Mulio,* Fab., *Microdon,* Meig. The scutellum has two teeth; the first articulation is as long as the other two.

In this subgenus, in Ceratophya, and in Ascia, the first two closed cellules terminate in an angle.

The antennæ of the following are shorter than the head; the posterior claws are often wide, especially in one of the sexes. Sometimes the pallet of the antennæ is oblong, or an elongated triangle; the posterior thighs are thick and indented; the wings are hidden on one another.

MERODON, Meig., Fab., *Milesia, Eristalis,* Lat. Syrphus, *Fab.* The abdomen triangular, or conical.

ASCIA, *Meg., Meig.* The abdomen narrowed at the base, and club formed.

Sometimes the pallet of the antennæ is short, suborbicular, or ovoid. Here the abdomen is shaped as in the last.

SPHEGINA, *Meig.* The pallet orbicular; the posterior thighs club-shaped, and spiny beneath.

There the abdomen is either triangular, conical, or subcylindrical. In some, the wings are scarcely longer than the abdomen.

We shall separate those whose posterior thighs are enlarged and spiny within.

Such are ECOMERUS, *Zeig.* and XYLOTA, whose abdomen is nearly linear.

In the two following the thighs are merely unidentated at most.

MILESIA, Lat., Fab., Meig., *Tropidia,* Meig. Posterior feet suddenly broader than the others; the abdomen conical or subcylindrical.

PIPIZA, Meig., *Psilota, Eristalis,* Fab., *Milesia,* Lat. Abdomen semi-elliptical, and rounded at the end, eyes pubescent.

BRACHYOPA, *Hoff., Meig.,* are distinguished from all the foregoing, by the wings being longer than the abdomen. These seem to conduct us to Rhingia, the last subgenus of the tribe.

The Syrphides that we have seen, had a proboscis shorter than the head and thorax, and the projection on the face short and perpendicular; this proboscis is nevertheless obviously longer, sublinear, and the projection of the head more elongated and directed forward, like a pointed beak. Such are RHINGIA, *Scop., Fab., Meig.*

PELECOCERA, *Hoffman,* has the thread of the antennæ short, thick, silky, cylindrical, and divided into three articulations. The pallet is like a reversed triangle.

The sucker of all the other athericera is composed of two threads only, the upper of which represents the labrum, and the lower the ligula. These will form three other small tribes, corresponding with the genera *Œstrus* and *Cenops* of Lin., and *Musca,* of Fab., such as he at first made it.

Stomoxys and Bucentes being allied to the last genus, we shall commence this tribe with the ŒSTRIDES, composed of the genus ŒSTRUS, of Lin.

These are very distinct, inasmuch as, instead of a mouth, we see only three tubercles, or slight rudiments of a proboscis and palpi. Their antennæ are very short, inserted each in a fosset under the forehead, and terminated by a rounded pallet, having on the top a simple thread. Their wings are commonly wide. The alulæ are wide, and hide the balancers. The tarsi are terminated by two hooks and two pellets.

The third tribe of ATHERICERA (CONOPSARIÆ) is the only one in which the proboscis is always out, forming a cylindrical, conical, or setaceous syphon. The reticulation of the wings is the same as in our first division of Muscidæ. They compose the genus.

CONOPS, *Lin.*

Some have the body straight and elongated; the abdomen in a knob bent underneath, with the male sexual organ projecting; the second and third articulation forming a spindle-shaped knob. Here the proboscis is advanced and bent only near the insertion. Sometimes the antennæ are much longer than the head.

SYSTROPUS, *Wied.*, *Cæphenes*, Lat. The last articulation alone forms the knob without a stylet. The abdomen is long and slender.

CONOPS (proper), *Fab.*, *Lat.*, *Meig.* The last two articulations form the knob, with a stylet at the end.

Sometimes the antennæ are shorter than the head, and terminate in an ovoid knob. The wings cross each other on the body. ZODION, *Lat.*, *Meig.*

There the proboscis is bent towards the end, and near the middle; the short antennæ terminate in a pallet with a stylet. MYOPA, *Fab.*

The others resemble common flies.

STOMOXYS, *Geoff.*, *Fab.* The proboscis is bent only near the base.

BUCENTES, *Lat.*, *Stomoxys*, Fab., *Syphona*, Meig. The proboscis is bent twice, as in Myopa.

G. caruns, of M. Nitzsch, has only rudimentary wings.

A visible proboscis, always membranous, and with two lips generally supporting two palpi (*Phora* only excepted), capable of being withdrawn into the buccal cavity, and a sucker of two pieces, distinguish the fourth and last tribe, that of Muscides, from the other three. The antennæ always terminate in a pallet with a lateral thread. These include the old genus *Musca*, of Fab., which, however, has been greatly modified; nor are all the difficulties which attend the study of it as yet removed, and some even of the several new genera into which it has been split, may still be considered as magazines for further subdivisions.

Musca.

The antennæ are inserted near the forehead; the palpi implanted on the proboscis, and withdrawn with it into the buccal cavity, and transverse nervures to the wings, will form the first section of the winged muscdæ, including eight principal groups and sub-tribes.

Our first division, Creophilæ, have large alulæ nearly covering the balancers. The wings wide, each of the two terminal and exterior cellules of the posterior cell closed by a cross nervure.

Among these may be distinguished those in which the epistoma is not advanced like a beak, and in which the sides of the head are not elongated like horns. Some have the thread of the antennæ simple, or without visible hairs. In one subgenus, that of

Echinomya, Dum., *Tachina*, Fab., Meig., the second articulation of the antennæ is the longest of all; the last, or pallet, is wider, compressed, like a reversed triangle, *M. grossa*, Lin.

In the other creophilæ, the third articulation is longer, or as long as the preceding.

Sometimes the anterior face of the head is nearly smooth,

or with short hairs disposed in two longitudinal ranges. The abdomen convex ; the rays distinct, more or less triangular.

In these, the thread of the antennæ is bent in the middle.

GONIA, *Meig.* But in these, as in the other creophilæ, it is not bent.

MILTOGRAMMA, *Meig.* The third articulation of the antennæ longer than the preceding.

TRIXA, *Meig.* In which it exceeds but little that of the preceding.

There the abdomen is sometimes much swollen, and the separation of the wings is indistinct; sometimes it is very flat. The wings of the last are wide, and often slightly arched on the outside.

GYMNOSOMIA, Meig., *Tachina,* Fab. Abdomen like the last; second and third articulations equal and linear.

CISTOGASTER, *Lat.* Abdomen as last; antennæ shorter; third articulation longer and wider.

PHASIA, Meig., *Thereva,* Fab. Abdomen flat, semicircular; wings simply furnished with little hairs.

TRICHIOPODA, Lat., *Tachina,* Fab. Abdomen flat, oblong; hinder legs with a fringe of lamelliform ciliæ.

Sometimes the fore face of the head has two ranges of long hairs. Some of these have the wings vibratile; the abdomen narrow and long. They form three subgenera. The wings of the first two like those of the last, and most others have the two cellules external and closed from their posterior end, nearly equally elongated behind; the outer one goes beyond the other, and its posterior angles are sharp; the antennæ are as long or scarcely shorter than the face of the head.

LOPHOSIA, *Meig.* The last articulation of the antennæ forms a very broad triangular pallet.

OCYPTERA, *Meig., Fab.* The corresponding articulation, scarcely wider than the preceding, forms a long square.

In the subgenus Melanophora, *Meig.*, now suppressed by him, and united to *Tachina*, the antennæ are shorter. The exterior cellule is more advanced than the other.

The abdomen of the other creophilæ is but little elongated, triangular, and the wings are not vibratile.

Phania. The posterior of the abdomen is elongated, narrowed, and folded underneath. The third articulation of the antennæ long and linear.

Xysta, *Meig.* has five or six rings in the abdomen; last two articulations nearly equal; the hind legs slightly arched, compressed, and ciliated. This appears to be intermediate between Gymnosoma and Phania, and approaches also to Trichiopa.

Tachina, *Fab.*, *Meig.* The abdomen is not bent underneath at the posterior end, and has outwardly but four rings; the antennæ as long as the head or nearly so, terminate by a longer articulation than in the last.

Some species, forming a particular section, live in the larva state, in the bodies of certain caterpillars, and destroy them; but we shall pass on to the Creophilæ, with the thread of the antennæ visibly feathery. The third articulation always forms an elongated pallet.

Dexia, *Meig.* Appear like Ocypterus, the abdomen being narrow and long, especially in the males.

Musca proper, Lin., Fab., Meig., *Mesembrina*, Meig., have the abdomen triangular; the eyes contiguous behind, or very near each other in the males.

Here is placed most of the flies whose larva feed on flesh, vegetables, &c. *M. vanitoria*, Lin., *M. cæsar*, Lin. *M. domestica*, Lin.

Sarcophaga, Meig., *Musca*, Lin., Fab., differ from the last only in having the eyes wider apart; the young sometimes break from the egg in the body of the parent, and hence they are said to be viviparous. *M. carnaria*, Lin.

We shall terminate these Creophilæ by certain subgenera, which contrast with the last either in the head, the situation of the wings, or the cells of the posterior extremity. The thread of the antennæ is velvety in most of them.

In some, such as the two following, the wings terminate as in the last, or have at their posterior end, between the middle and the side, two complete cells.

ACHIAS, *Fab.* Very singular from the horn-like elongati on on the sides of the head, like Diopsis, but they have the antennæ inserted at the top of the forehead, and like those of flies in form, &c.; the wings are wide.

IDIA, *Meig.*, *Wied.* The anterior extremity of the head projects like a horny beak; the wings are bent on the body.

In the other two and last subgenera of Creophilæ, the terminal cells of the wings are closed at the outer edge; the eyes are wide; the abdomen is flat.

LISPA, Lat., Fab., Meig., *Musca,* Deg. The body oblong; the antennæ inserted near the forehead, nearly as long as the face of the head; the last articulation longer; linear and furnished with a plumose thread; the wings hide one another; the palpi are dilated above.

ARGYRITES, *Lat.*, resemble Phasia, but the antennæ are inserted under the forehead, short, and the last articulation furnished with a simple bent thread; the palpi terminate in a subovoid, short club.

In all the other Muscidæ, the alulæ are small, or nearly obliterated; the balancers are uncovered; and the principal longitudinal nervures of the wings reach to the posterior edge, which, with a few exceptions, close the posterior cellules at least.

A second general division of Muscides, ANTHOMIZIDES, is composed of species like common flies; the wings are generally hidden, and not vibratile; the antennæ short, terminated by a long square or linear pallet, with the thread feathery; the head hemispherical, with hairs in front; the eyes near each other in the males; the abdomen exteriorly with four rings. Some have the antennæ nearly as long as the face of the head, with the thread feathery; sometimes the abdomen of both sexes terminates in a point.

ANTHOMYIA, Meig., *Musca*, Lin., Fab. Eyes separated in both sexes; proboscis not terminated like a hook or by a sudden angle, *M. pluvialis*, Lin.

DRYMELA, *Meig.* Proboscis as last; eyes united behind in the males.

Sometimes the abdomen is enlarged at the end, forming a knob. CŒNOSIA, Meig., *Musca*, Deg. *M. Fungorum*, Deg.

The others have the antennæ shorter, with a simple thread; the eyes of the males united behind; the mouth very downy. ERIPHIA, *Meig.*

Our third division, HYDROMYZIDES, has the head subtriangular; the eyes very prominent; a swollen vaulted muzzle; a small arched lamina edging the top of the buccal cavity, which is large; proboscis thick; sides of the face without hair; antennæ very short, generally plumose; wings incumbent one on the other; strong thighs, generally enlarged.

ROPALOMERA, *Wied.* Thighs enlarged; an elevation or tubercle on the face.

OCHTERA, Lat., *Musca*, Deg., *Tephritis*, Fab., *Macrochira*, Meig. Anterior thighs very large, compressed, indented underneath, terminated by a strong spine.

Others of this section have not the thighs enlarged.

EPHYDRA, *Fall.* Like the last, but the thread of the antennæ is merely thickened underneath, and simple; the pallet round at the end; the vertex is behind a small elevation.

NOTIPHILA, *Fab.* Head more round, without anterior projection; the eyes less prominent; the thread of the antennæ plumose; no elevation to the vertex. *M. cellaria*, Panz.

The three following divisions have the body oblong; the wings bedded, and not vibratile; the head running to a point, commonly truncated or obtuse at the anterior and upper end; face covered with a white membrane, furrowed longitudinally on each side; the head is often compressed under the antennæ, and the lower or buccal part advanced like a truncated muzzle; in others, the face forms a very inclined plane not elevated below; the antennæ inserted at the top of the forehead, shorter in all the other Muscidæ than in these.

The fourth division, SCATOMYZIDES, and the following, is distinguished from the sixth, by having the head seen above, never longer than it is wide, and the form subspherical or triangular; the posterior tarsi never longer than the body, or very slender; the body narrow and large.

These are distinguishable from DOLICHOCERES, by having the third articulation of the antennæ longer than the preceding, *Lexoceres* alone excepted; the head seldom advances at its anterior and upper end beyond the eyes, appears in general, seen above, nearly hemispherical, rather wider than long.

Sometimes the posterior tarsi are broad, wide apart, with the thighs thick and compressed, and the articulations of the tarsi enlarged; antennæ short; last articulation lenticular; sides of the face hairy.

THYREOPHORA, Lat., Meig., *Musca,* Panz. Antennæ longer, in a subfrontal cavity; pallet lenticular, but not transverse; head sloping from the top to the mouth; posterior thighs thick; second and following articulations of the tarsi nearly alike; all the terminal cellules of the wings closed by the posterior edge; palpi form a spatula at the end. *M. cynophila,* Panz.

SPHÆROCERA, Lat., *Borborus,* Meig., *Copromyza,* Fall. Antennæ projecting; pallet sub-hemispherical, transverse;

head under the forehead suddenly concave; posterior thighs compressed; first two articulations of the tarsi wider than the following; the last of the two cells which occupies the middle of the length of the wings closed before the posterior edge; proboscis thick; body depressed.

Sometimes the posterior tarsi differ little or nothing from the others; the antennæ often as long as the face of the head, with the thread downy; sides of the face sometimes glabrous; some such have the abdomen of the males at least elongated, terminating in a knob or a stillet, and these have the sides of the face hairy.

DIALYTA. Abdomen with four segments; thread of the antennæ simple.

CORDYLURA. Abdomen with five segments at least, ending in a knob in the males; wings as long as the abdomen.

SCATOPHAGA, Lat., Meig., *Musca*, Lin., Fab. Wings longer; abdomen not enlarged at the end. *M. stercoraria.*

These are without hairs on the face; body linear.

LEXOCERA, *Lat.*, *Fab.*, *Meig.* Antennæ longer than the head; they resemble small ichneumon flies.

CHYLIGA, *Fall.*, *Meig.* Antennæ shorter; thread stylet-formed.

The others have the antennæ shorter than the head, advanced wide apart; pallet never longer than it is wide, sub-ovoid or sub-gobular; some of these have the thread downy, body elongated, and the abdomen in a point.

Some again have the face naked and the pallet ovoid, as

LISSA, *Meig.* Upper part of the head with an elevation; abdomen linear, not ending in a stylet.

PSILOMYIA, Lat., *Psila*, *Meig.* Body less elongated; abdomen with an articulate stylet. GEOMYZA, of *Fallen*, may be united to these.

Titanura and *Tenypeza*, of Meigen, seem allied to some of the preceding subgenera, only the legs are longer and more slender; the first outer nervure of the wings is simple, without cells; the exterior terminal cells are wide.

Others have the sides of the face hairy; the first articulation of the antennæ the most slender; the two following forming a rounded knob.

LONCHOPTERA, Meig., *Dipsa*, Fall. The eyes smooth, on an elevation; wings without transverse nervures; the third longitudinal nervure from the outside bifurcate.

The body of the others is more like that of common flies. One subgenus only, that of

HELEOMYZA, *Fall.*, has no hairs on the face.

Two others have the thread downy, or plumose, DRYOMYZA, with the face concave, with a short truncated muzzle; and

SAPROMYZA, *Fall.*, *Meig.* Face straight without a muzzle.

The last Scatomyzides have the thread of the antennæ simple, very short, wide, and straight, with the last articulation semi-ovoid. These diptera are very small, nearly smooth, dark, varied with yellow, the tarsi strong, and the eyes large; the upper part of the head is flat, with a small triangular brown space, in which the eyes are placed. The two transverse nervures approximate each other near the middle. These compose our genus OSCINIS, *Lat.*, *Fab.*, to which we shall refer *Chlorops*, of Meig.

The fifth division, DOLICHOCERA, includes Dumeril's genus, *Tetanocera*, and is near the last, but the second articulation of the antennæ is as long, or longer, than the third. These organs are wide and advanced, and with few exceptions, as long, or more so, than the head, and end in a point. The upper surface of the head forms an obtuse triangle.

Some have the antennæ shorter than the head. OLITES, *Fab.*, with the thread simple, and without a buccal projection.

EUTHYCERA, *Lat.* Second articulation of antennæ the largest, thread plumose, a small truncated muzzle.

Others have the antennæ as long, or longer, than the head, as SEPEDON, *Baccha*, Fab. Second articulation longest, thread simple.

TETANOCERA, Dum. Lat. *Seatophaga*, Fab. Second articulation compressed and square, with the thread sometimes plumose.

The sixth division, LEPTOPODITES, is remarkable for the tenuity and length of the legs, the last two being twice the length of the body. The head is spherical, ending in a point; the abdomen pointed in the females, but with a knob in the males; antennæ short.

MICROPEZA, Meig. *Calobates*, Lat. Head ellipsoidal; last articulation of antennæ semiorbicular, thread simple. *M. filiformis*, Fab.

CALOBATA, Fab. *Micropeza*, Lat. Head spheroidal; last articulation more elongated; thread often plumose.

The CARPOMYZÆ form the seventh division of Muscidæ; wings elevated, or wide apart, when at rest, and then capable of a vibratory motion, spotted black, or yellow, like the common flies, but with the eyes always wide apart; the balancers visible, the abdomen with four or five outer wings, often terminated in the females by a hard pointed oviduct; antennæ pallet-formed, short, and the thread rarely plumose. Many approach the last subgenus, as

DIOPSII, *Lin.*, *Fab.*, called also spectacle flies, the eyes placed at the extremity of two lateral elongations, the scutellum with two spines. They are all exotic.

CEPHALIA, *Meig.* The pallet elongated, and the thread pubescent. Fore part of the head elongated, and the palpi spatuliformed.

SEPSIS, Fall., Meig., *Tephritis*, Fab., *Micropeza*, Lat. The pallet shorter, semi-elliptical, thread simple, palpi filiform, thickening at the end. *M. cynipsea*, Lin.

The other Carpomyzæ have the appearance of common

flies. Sometimes the upper surface of the head is nearly horizontal, or slightly inclined, so that the antennæ, seen in profile, seem to be inserted near the level of this plane. The palpi and proboscis can be withdrawn into the buccal cavity. The wings, when at rest, elevated, and the abdomen apparently composed of five rings.

ORTALIS, Fall., *Seatophaga, Tephritis Dictya*, Fab., *Tephritis*, Lat. Abdomen of the females without a pointed ovipositor. Many species have the body more elongated than the other subgenera. The pallet of the antennæ is sometimes long and linear (*O. paludium*, Fall.), sometimes shorter and broader (*M. vibrans*, Lin.), *M. Cerasi*, Lin.

TETANOPS, *Meig.* Female with a visible ovipositor, head seen above, subtriangular, and as long as it is wide.

TREPHRITIS, Lat., Fab., Fall., *Trypeta*, Meig., *Dacus*, Fab. Ovipositor; head transverse, and rounded, (*Dacus*) Fab., *M. Cardui*, Lin.

Sometimes the head is more compressed transversely. The proboscis very thick, and partly projecting; the wings wide apart horizontally, the abdomen outwardly with four segments. PLATYSTOMA, Meig., *Dictya*, Fab.

This last conducts us to the *Timiæ*, of Wiedemann, which compose our eighth division, GYMNOMYZIDES. Small flies, with a short, arched, smooth body, shining black: head compressed laterally, without any projection, and the buccal opening large; the wings incumbent on the body, and extending beyond it; the scutellum advanced; the abdomen compressed, short, with a point at the end, tarsi smooth, a little downy.

Some have the antennæ as long, or less so, than the head, and wide apart, CELYPHUS, *Dalm.*, distinguished from all other Diptera by the scutellum, covering the whole upper part of the abdomen. But one species, *C. obtectus*, Dalm. of Java.

LAUSCANIA, *Lat., Fab., Meig.* The scutellum of common dimensions; the thread of the antennæ plumose.

The others have the antennæ shorter than the head.

Here they are always very short, inserted on a sort of edge, traversing the face. The first cellule of the posterior edge of the wings is generally nearly closed; the antennæ are lodged in fossets. The species whose first cellule of the posterior edge is nearly closed, form, according to Meigen, two genera, *Timia* and *Ulidia,* which we unite in one subgenus, MOSILLUS, *Lat.*

The species with the corresponding cellule entirely open and longitudinal, include two genera of Meigen, *Homalura,* with five segments, and *Actora* with six.

The Gynomyzides, with short antennæ, inserted as in the last, compose the genus GYMNOMYZA, of Fallen. Those with the antennæ inserted higher, with an elongated pallet, compose the genus LONCHÆA, *Fall., Meig.*

HYPOCERA, our second section, forming the ninth subtribe, includes but one subgenus. Palpi exterior; antennæ inserted in the oral cavity, short, ending in a subglobular, thick articulation; wings furnished with numerous ciliæ on the side, with a strong oblique nervure from the base to the side, from which nervure three others run nearly parallel and longitudinally. Body arched, tarsi strong, spiny, thighs wide, compressed. They form the genus PHORA, Lat., *Trineura,* Meig.

The Diptera of our second family compose a distinct order, *Omaloptera,* according to Dr. Leech, those of which deprived of wings and balancers, have a considerable relation to the apterous hexapod insects, composing our order of Parasites, *Pediculus,* of Lin. This second section

PUPIPARA,

Will form our sixth and last family of Diptera. Reaumur

calls them *Nymphipara.* The head, viewed above, appears divided into two distinct portions; the posterior carries the eyes, and receives, in an anterior emargination, the anterior portion; this also is divided into two, the posterior being the broadest, coriaceous, carrying the antennæ, the other portion constituting the manducatory apparatus. The lower and buccal cavity of the head is occupied by a membrane, from the end of which proceeds a sucker, springing from an advanced pedicle, composed of two nearly approximated threads, and covered by two coriaceous laminæ, narrow, elongated, and downy, performing the office of a sheath. Whether these laminæ represent valves, as I have conjectured, or are the parts of a proper sheath, as M. Dufour thinks, who has discovered two little bodies, which he takes for palpi, it is not the less true, that this proboscis differs from that of other Diptera, and that the case in these is more like the proboscis of a flee, though it is without articulations.

The body is short, broad, flat, and covered with a tough, leathery skin, the antennæ, at the sides of the fore part of the head, sometimes form tubercles with three threads, sometimes little downy laminæ. The eyes in some are very small. Whether or not they possess simple eyes, is not determined. The wings are wide apart, and with balancers; the sides of the wings are edged with ciliæ, but in some the wings are rudimentary, and the balancers disappear; the tarsi have two strong nails, with one or two teeth underneath. The abdomen can extend itself considerably, which is quite necessary in the females, as the larvæ live there till their period of change into nymphs.

These Diptera, sometimes called spider flies, live exclusively on quadrupeds or birds; run very fast, and often sideways.

Some have the head very distinct, and articulated with

the anterior extremity of the thorax. They form the genus

HIPPOBOSCA, *Lin.*, *Fab.*

HIPPOBOSCA, proper, have wings; the eyes very distinct, occupying all the sides of the head; the antennæ in form of tubercles, with three threads, *H. Equina*, Lin.

ORNITHOMYIA, *Lat.*, differ from the last, by having the antennæ laminous, downy, and advanced, the longitudinal nervures of the wings very distinct, and receding to the edge.

These insects form, in Dr. Leech's monograph, four genera, 1. FERONIA, *Nirmomyia*, Nitzch. Antennæ in form of tubercles, nails with two teeth. 2. ORNITHOMYIA. Simple eyes, nails with three teeth; antennæ laminous; wings almost equally broad and round. 3. STENEPTERYX. Like the first, but the wings narrow. 4. OXYPTERUM. Wings acute: antennæ dentiform: eyes small, and without the simple eyes, like Hippobosca and Feronia.

STREBLA, *Dalm.*, differ from Ornithomyia, by having the wings crossed on the body, longitudinal nervures united by transverse nervures, eyes very small at the posterior angle of the head.

MELOPHAGUS, *Melophila*, Nitz. Wingless, and with the eyes indistinct, *H. ovina*, Lin.

One species of these lives on the stag; the thorax is a little larger than the head, it forms the subgenus LIPOTENA, of Nitzsch. Near these also should be placed his genus BRAULA, the only species of which lives on the common bee: it is quite blind. The thorax is divided into two transverse portions, the last articulation of the tarsi has underneath a transverse row of prickles.

The other Pupipara, *Phthiromyia*, Lat., have the head very small; it forms near the anterior and dorsal end of the thorax a little vertically elevated body. This composes the genus

NYCTERIBIA, Lat., *Phthiridium*, Herm.

Which have neither wings nor balancers, and are even still more than the last like spiders. They live on the bat. Linnæus placed the only species he knew with the lice.

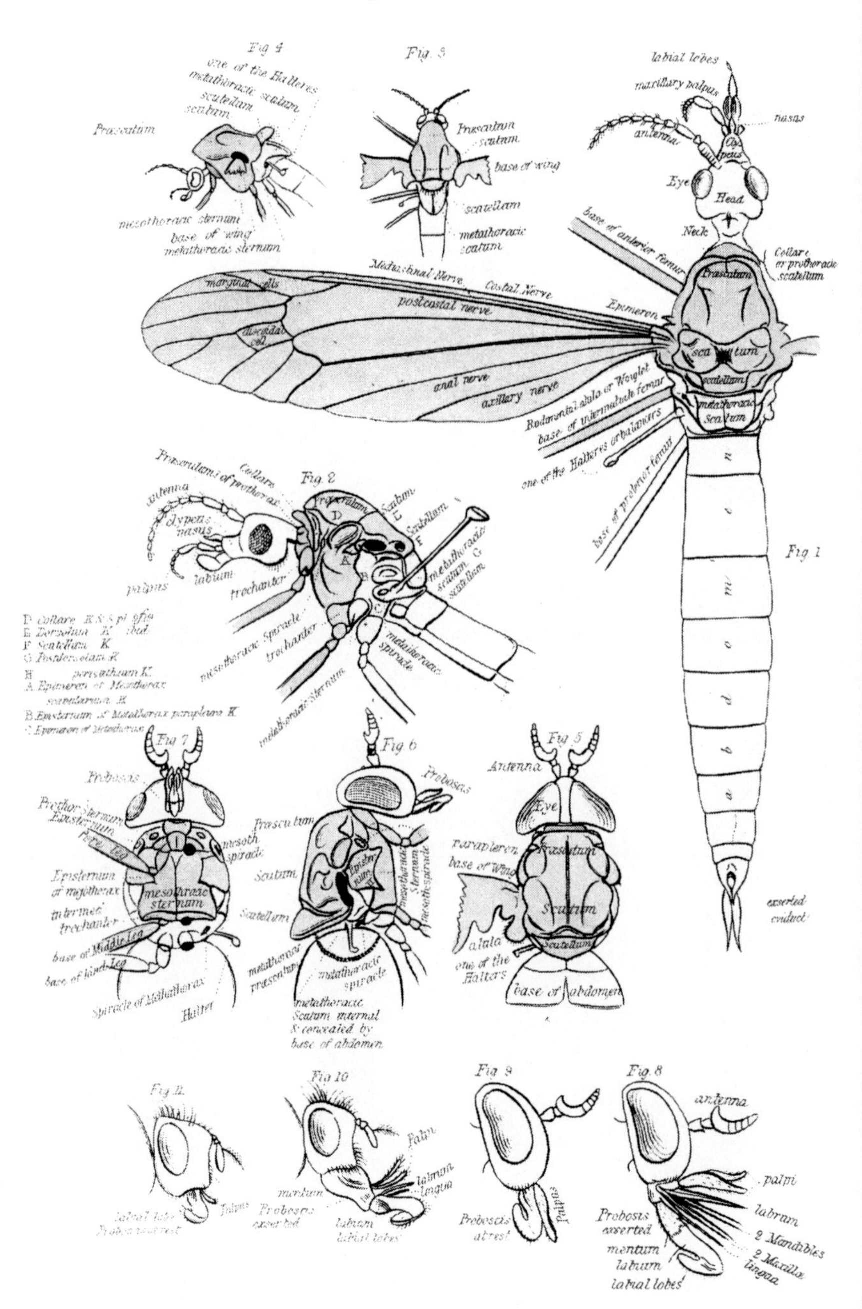

J. O. Westwood F.L.S. &c del.t

Fig. 1 & 2 Tipula ♀
— 3 — 4 Cheronomus ♀
— 5 — 9 Tabanus bovinus
— 10 — 11 Musca Cæsar

London Published by Whittaker & C.o Ave Maria Lane, 1832.

SUPPLEMENT

ON THE

ORDER DIPTERA.

On the characteristics of this order, as given in the text, it will be necessary to enlarge a little here. The body is covered with a membranaceous elastic skin, incapable of resisting a strong pressure. Like that of other hexapod insects, it is composed of three principal parts. The head, which is usually compressed, and turns upon itself, as on a pivot, from right to left, and from left to right, presents two composite eyes, often very large, and contiguous in the males; two antennæ and a proboscis, or sort of bill, sometimes projecting, sometimes more or less expansible, and retractile into an interior cavity, when it is not in action. The head likewise exhibits in the majority, three small simple eyes, triangularly disposed.

The antennæ are usually inserted on the forehead, and approximated at their base. Those of the first family, the *Nemocera*, have very great relations in their setaceous form, the number of their articulations, and often in the hairs or barbs, with which they are furnished, with the antennæ of the nocturnal lepidoptera. But in the subsequent families, which united, compose the major part of the diptera, they are short, most generally inclined or couched on the space comprised between the eyes, and simply formed of from one to two articulations, the last often fusiform. The mouth of these insects is fitted only for the conduction and extraction of fluid materials. In many, the two palpi are attached to two pieces of the sucker, which is a proof that those two pieces correspond to the jaws of the grinding or masticating insects.

Above the wings, are two small and very mobile bodies, composed of a line or stem, and terminated by a swelling like a button or knob; those are called the *halters* or *balancers*: their use is unknown. Those two papyraceous pieces, which are above the balancers, and named *winglets* in the text, often conceal the balancers, and their size is in an inverse proportion to that of these latter parts.

The final tarsus often has two or three membranaceous articular cushions, which serve the purpose to these insects, of fastening themselves to the most polished bodies, and that even in a vertical position. Sir Everard Home has published in the Philosophical Transactions, 1816, some very curious observations on the form of these parts. Whether spongy or membranaceous, whether fleshy or vesicular, which exhibit to us the tarsi of divers insects, attached or walking in a direction contrary to that of their gravitation. He explains to us in the same place, with similar detail, the nature and disposition of the glands and scales, with which the under part of the toes of the geckos is so abundantly clothed. To the anatomical observations collected on these insects, by Swammerdam, Reaumur, and Ramsdorff, M. Dutrochet, a most able and ingenious naturalist, has added some that are new and very curious. He has found that the stomach of many diptera was accompanied by an especial organ which he names paunch (*la panse*) and where a part of their aliment is deposited. Swammerdam had seen in the lepidoptera a similar appendage.

Many diptera, such as the culices, tabani, and stomoxys, incommode the human species by their bites, and most cruelly torment many domestic animals. Others, such as the œstri, deposit their eggs in the bodies of animals, that when the larvæ come forth, they may obtain there their proper nutriment. Others of them again, for a similar reason, infect the meat and other viands, which we preserve for the purposes of food, whenever they can obtain access to them. There are some of them that are still more pernicious, since, in the larvæ-state, they destroy the young corn, and other fruits of the earth, and have the power in the localities where they are exceedingly multiplied, of annihilating the hopes of the agriculturist. But if such diptera are extremely injurious to man, there are others which afford a compensation for our losses, by destroying hurtful insects, and by consuming animal and vegetable substances, whose putridity would otherwise corrupt the fluid which we respire. Many larvæ of diptera also serve to accelerate the dissipation of the stagnant and infectious waters of pools, marshes and ditches.

The duration of the life of the diptera, even reckoning from the moment in which they issue from the egg, is, generally speaking, extremely short. It is often limited to but a few months, or some weeks. All of them undergo a complete metamorphosis, but one which is modified in two principal ways. The larvæ of several of them change their skin, when they pass into the state of nymph; but the others do not moult: their skin most frequently becomes hard and contracts. It assumes a brown tint, exhibits no external signs of irritability, and becomes a tolerably solid shell or cocoon, which has the appearance of a grain or an egg. The body of the larvæ is detached at first from this skin, leaving on its interior parietes the external organs which were peculiar to it, such as the hooks of its mouth, and soon it presents itself under the form of a soft or gelatinous mass, which is by some writers denominated an *elongated ball*, and at the exterior part of which, we cannot yet distinguish any of the parts which characterize the perfect insect. Finally, at the end of a certain time, these organs become pronounced and determined. The insect is then really in the state of nymph, properly so called. It issues forth, causing the anterior and upper extremity of its envelope to come off like a cap.

The larvæ of the diptera have no feet, but we observe in some, particularly in those of the aquatic diptera, divers appendages which resemble them, and which even contribute to locomotion. Many of these larvæ have a soft and variable head, and which in fact can only bear this name, by reason of the organs of manducation, which form the most apparent part of it. This character is exclusively peculiar to the larvæ of those diptera, which are transformed under their skin, and is not observed in any of the preceding orders. The mouth is generally provided with two hooks, which serve to cut in upon the substances from which they derive their nutriment. The principal orifices of respiration in almost all the larvæ of the same order, are situated at the posterior extremity of their body. Many present besides, two stigmata on the first ring, or that which comes immediately after the head.

The characters of the five great families of this order are founded on the general relations of forms and habits, and on the principal differences presented by the metamorphosis of these insects. In the first four families, the larvæ have the body ringed, elongated, and with very distinct stigmata. They proceed from eggs deposited by the mother, on the substances on which they feed. The larvæ of the diptera of the last family have the figure of an egg, or rather of a bean-seed, the skin of which is continuous, without sensible stigmata, or annular divisions, and presents only at one of its extremities, a small scaly plate. These larvæ do not come forth from the belly of the mother, but pass at once into the nymph-state, and even preserve their primitive form. Those of the first two families change their skin when about to undergo this second metamorphosis. Their body, with the exception of the last species of the second family, has a scaly head, and of an invariable form; but in the larvæ of the two following families, the head is always soft and susceptible of assuming divers sorts of figures. These larvæ are metamorphosed under their own skin, and present us, before passing to the nymph-state, with that mode of existence which we mentioned above, to be sometimes denominated *elongated ball*. The skin of the nymphs of the diptera of the fourth family, experiences a contraction which has so modified its external form, that it no longer preserves any of its primitive traits. But the nymphs of the third family still retain the form that was peculiar to them in the larva-state. Their skin is merely something firmer.

The first family of diptera, the Nemoceræ, is composed of the genera *Culex* and *Tipula*, of Linnæus. For its characteristic features, independently of what we have already observed, we must refer to the text. Many of the insects which compose it, especially the young, assemble in numerous troops in the air, and form there, in flying, sorts of dances. Some are to be found at almost every season of the year. They are placed end to end, in the act of sexual intercourse, and often fly while in this situation. Many females lay their eggs in the water, which is the habitat of their larvæ and nymphs; others place them in the earth, in tow, or on vegetables.

The larvæ are always elongated, similar to worms, with a scaly head, of an invariable figure, and the mouth presents parts analogous to jaws and lips. They all change skin to become nymphs; these nymphs are sometimes naked, sometimes enclosed in cocoons which their larvæ have constructed. They approximate in their figure, to the perfect insect, exhibit the external organs, and conclude their metamorphosis in the ordinary manner. At the anterior part of the corslet we often observe two respiratory organs, in the form of tubes, of horns, or of earlets.

The CULICES, or gnats, which are but too well known to us by their annoyance and their bites, are small insects, with membranaceous wings, elongated body, and mounted upon long legs, or to use the expression of Reaumur, sustained as it were, on stilts. The head is small and rounded, with two large net-work eyes, but no vestige of the small simple eye. The anterior extremity of the head gives birth to the palpi and proboscis. Reaumur has considered the palpi, in reference to the individuals of the male sex, as serving as a sort of sheath for the case of the sting, or for the external body of the proboscis. The palpi of the females are much shorter, less articulated, and the same naturalist has given them the denomination of *barbs*. He applies the same name to one of the lower articulations, and the thickest of the palpi of the males. The proboscis forms a sort of advanced, long, slender cylindrical syphon, and terminated by a swelling like a button. This external part is only the case of the sting or sucker. It consists of a membranaceous, very flexible, long, and narrowed piece, rolling on itself to form a funnel, and which leaves above or between its edges, a furrow, in which the sucker is inserted; it is, though under another form, analogous to the proboscis of the common flies. Its button represents the two lips which terminate the latter. The sucker is composed of five small, scaly lancets, similar to silky hairs, or very fine drawn threads, very sharp, and two of which, at least, have denticulations at their extremity directed backwards. Reaumur has not been able to distinguish what is the respective position of each of these pieces, when they are assembled. But it appears to M. Latreille, that they are disposed in three ranks thus: one upper, and isolated, the other four situated in pairs underneath. The first and the two lower ones probably unite to form a sort of sheath to the intermediate two, which sheath Reaumur compares to a canula. When the gnat intends to sting, it at first puts forth the point of all these threads united, out of the aperture of the end of the case, and applies it to the skin. It then sinks in this point by degrees, resting the end of the sheath or the button, on the part where the sting has penetrated, and as it enters farther, the sheath bends in proportion, and forms a more acute angle, so that if the sucker is almost totally sunk in the body where it has penetrated, the sheath is almost entirely bent double. For more minute details on this curious subject, the reader may consult an article by Roffredi, in the *Acad. Collect. Part. Etomg.*, t. xiii. *p.* 412.

Although the wound produced by the sucker of the gnat be but slight, it nevertheless causes a tumour in the skin, and at the same time, an insupportable itching, as the part is bedewed with a poisonous fluid, capable of creating considerable irritation. This caustic liquor, it has been supposed, was given to the insect, to mix with the blood, so as to impart to it fluidity, and render it more easily passable through the proboscis. This at least is the opinion of the advocates of final causes.

It has been remarked, that there are some persons not liable to be stung by these insects, although their skin may appear to be extremely delicate, while others cannot go into the country without being almost literally devoured by them. We shall notice presently the remedies which may be employed against these bites; it also appears, that it is invariably the female gnats alone which attack and torment us, for the purpose of sucking our blood.

When the gnats cannot find a sufficient quantity of blood to satiate their rapacity, they suck the juices of plants. Some of them are found on flowers, particularly on those of the willow, and on the buds of this tree. In the hot days, and in places enlightened by the sun, they remain tranquil until evening. Sometimes, however, they commence their pursuits from the middle of the day, especially in the woods. Placed upon the leaves, they give a sort of balancing motion to their bodies, up and down, bending and straightening their limbs alternately, and with a tolerable degree of quickness, as is done by many *tipulæ.*

The corslet of the *culices* is thick, raised, and as it were, *humped.* The wings, which they carry inclined horizontally, are extremely pretty, when seen through the microscope. Their surface is punctuated and furnished along the nervures and the internal edge, with small scales in the form of oblong palettes, and which hold by a pedicle. The exterior edge of these scales has, from distance to distance, sorts of prickles in the place of scales. The winglets are wanting, or at least, not very distinct; but the balancers are discernible without difficulty. The abdomen, which is long and cylindrical, is also covered with scales and hairs, which are more numerous on the sides, and form there a species of fringe. It is terminated in the males by some hooks, sometimes four in number, simple, and two of them smaller; sometimes but two in number, and armed at the end with a small mobile claw. They answer the purpose of seizing, in sexual intercourse, the posterior part of the abdomen of the female, which instead of these parts, has two small palettes, capable of being applied one against the other. There is a little pellet at the end of the feet.

There are but few insects so greedy of our blood, and which pursue us with so much avidity as the culices. Garments even of a thick and close texture, are insufficient to protect us from their sting. The inhabitants of almost all country parts, especially of those which are aquatic, or well-wooded, are tormented with them during the summer, and those of towns have no small difficulty in defending themselves against their assaults. In some southern countries of France, their attacks during the night are only avoided by placing an envelope of gauze over the beds, which is called in French, *cousinière*, from *cousin*, gnat. "In this country," says Mr. Kirby, "they are justly regarded as no trifling evil, for they follow us into all our haunts, intrude into our most secret retirements, assail us in the city and in the country, in our houses and our fields, in the sun and in the shade; nay, they pursue us to our pillows, and either keep us awake, by their incessant endeavours to fix themselves upon our face, or some uncovered part of our body; or, if in spite of them, we fall asleep, awaken us by the acute pain which attends the insertion of their oral stings, attacking with most avidity the softer sex, and trying their temper by disfiguring their beauty. But although with us, they are usually rather teasing than injurious, yet, upon some occasions, they have approached nearer to the character of a plague, and emulated with success the mosquitos of other climates. Thus, we are told, that in the year 1736, they were so numerous, that vast columns of them were seen to rise from the air, from Salisbury Cathedral, which, at a distance, resembled columns of smoke, and occasioned many people to think that the Cathedral was on fire. A similar occurrence, in like manner giving rise to an alarm of the church being on fire, took place in July, 1812, at Sagan, in Silesia. In the following year, at Norwich, in May, at about six o'clock in the evening, the inhabitants were alarmed by the appearance of smoke issuing from the upper window of the spire of the cathedral, for which, at the time, no satisfactory account could be given, but which was most probably produced by the same cause. And in the year 1766, in the month of August, they appeared in such incredible numbers at Oxford, as to resemble a black cloud, darkening the air, and almost totally intercepting the beams of the sun. One day, a little before sunset, six columns of them were observed to ascend from the boughs of an apple-tree, some in a perpendicular, and others in an oblique direction, to the height of fifty or sixty feet. Their bite was so envenomed that it was attended by violent and alarming inflammation; and one, when killed, usually contained as much blood as would cover three or four square inches of wall. Our great poet, Spencer, seems to have witnessed a similar appearance of them, which furnished him with the following beautiful simile:

> "As when a swarm of gnats, at even-tide,
> Out of the fens of Allan doe arise,
> Their murmuring small trumpets sounden wide,
> Whiles in the air their clust'ring army flies,
> That as a cloud doth seem to dim the skies;
> Ne man nor beast may rest or take repast,
> For their sharp wounds and noyous injuries.
> Till the fierce northern wind, with blust'ring blast,
> Doth blow them quite away, and in the ocean cast."

In Marshland, in Norfolk, as I learn from a lady, who had an opportunity of personal inspection, the inhabitants are so annoyed by the gnats, that the better sort of them, as in many hot climates, have recourse to a gauze covering for their beds, to keep them off during the night.

But these insects are infinitely more redoubtable in other climates, especially in the neighbourhood of the frozen regions of the north, and the arid countries near the equator. There they become, indeed, a plague, and one of the severest scourges of human life, "We may be disposed to smile," (to use again the words of the last cited writer) "perhaps, at the story which Mr. Wild relates from General Washington, that in one place the mosquitos were so powerful as to pierce through his boots (perhaps they crept within the boots); but in various regions, scarcely anything less impenetrable than leather can withstand their insinuating weapons, and unwearied attacks. One would at first imagine that regions, where the polar winter extends its icy reign, would not be much annoyed by insects; but, however probable the supposition, it is the reverse of fact, for no where are gnats so numerous. These animals, as well as numbers of the *Tipulariæ* of Latreille, seem endowed with the privilege of resisting any degree of cold, and of bearing any degree of heat. In Lapland, their numbers are so prodigious, as to be compared to a flight of snow, when the flakes fall thickest, or to the dust of the earth. The natives cannot take a mouthful of food, or lie down to sleep in their cabins, unless they be fumigated almost to suffocation. In the air you cannot draw your breath without having your mouth and nostrils filled with them; and unguents of tar, fish-grease, or cream, or nets steeped in fetid birch oil, are scarcely sufficient to protect even the case-hardened cuticle of the Laplander from their bite. In certain districts of France, the accurate Reaumur informs us, that he has seen people, whose arms and legs have become quite monstrous from wounds inflicted by gnats; and in some cases, in such a state, as to render it doubtful whether amputation would not be necessary. In the neighbourhood of the Crimea, the Russian soldiers are obliged to sleep in sacks to defend themselves from the mosquitos; and even this is not a sufficient security, for several of them die in consequence of mortification produced by the bite of these furious blood-suckers. This fact is related by Dr. Clarke, and to its probability, his own painful experience enabled him to speak. He informs us, that the bodies of himself and his companions, in spite of gloves, clothes, and handkerchiefs, were rendered one entire wound, and the consequent excessive irritation and swelling, excited a considerable degree of fever. In a most sultry night, when not a breath of air was stirring, exhausted by fatigue, pain, and heat, he sought shelter in the carriage; and, though almost suffocated, could not venture to open a window, for fear of the mosquitos. Swarms, nevertheless found their way into his hiding-place; and in spite of the handkerchiefs with which he had bound up his head, filled his mouth, nostrils, and ears. In the midst of his torment, he succeeded in lighting a lamp, which was extinguished in a moment, by such a prodigious number of these insects, that their carcases actually filled the glass chimney, and formed a large conical heap over the burner. The noise they make in flying cannot be conceived by those who have only heard gnats in England. It is to all that hear it a most fearful sound. Travellers and mariners, who have visited warmer climates, give a similar account of the torments there inflicted by these little demons. One traveller in Africa complains, that after a fifty miles journey, they would not suffer him to rest, and that his face and hands appeared, from their bites, as if he was infected with the small-pox in its worst stage. In the East, at Batavia, Dr. Arnold, a most attentive and accurate observer, relates, that their bite is the most venomous he ever felt, occasioning a most intolerable itching, which lasts for many days. The sight, or sound, of a single one, either prevented him from going to bed all night, or obliged him to get up many times. This species is distinct from the common gnat, and appears to be nondescript. It approaches nearest to *C. annulatus*, but the wings are black and not spotted. And Captain Stedman, in America, as a proof of the dreadful state to which he and his soldiers were reduced by them, mentions, that they were forced to sleep with their heads thrust into holes made in the earth with their bayonets, and their necks wrapped round with their hammocks.

"From Humboldt, we also learn, that, between the little harbour of Higuerote and the mouth of the Rio Unare, the wretched inhabitants are accustomed to stretch themselves on the ground, and pass the night buried in the sand, three or four inches deep, leaving out the head only, which they cover with a handkerchief. This illustrious traveller has given an account, in detail, of these insect plagues, by which it appears, that amongst them, there are diurnal, crepuscular, and nocturnal species, or genera; the *mosquitos*, or *Simulia*, flying in the day; the *Temporanceros*, probably a kind of Culix, flying during twilight; and the *Zancudos*, or *culices*, in the night: so that there is no rest for the inhabitants from their torments, day or night, except for a short interval between the retreat of one species and the attack of another.

" It is not, therefore, incredible, that Sapor, king of Persia, as is related, should have been compelled to raise the siege of Nisibis by a plague of gnats, which attacking his elephants and beasts of burden, so caused the rout of his army, whatever we may think of the miracle to which it was attributed; nor that the inhabitants of various cities, as Mouffet has collected from different authors, should, by an extraordinary multiplication of this plague, have been compelled to desert them; or that, by their power to do mischief, like other conquerors, who have been the torment of the human race, they should have attained to fame, and have given their name to bays, towns, and even considerable territories."

A number of remedies have been proposed to dissipate the itchings occasioned by the bite of gnats. Some employ alkalies, both fixed and volatile; others lavender-drops, *theriaca* of Venice, mixed with sweet oil, and also vinegar, has been tried. Some apply leaves of the green elder, to the juice of which are added vinegar and common salt; others find plantain leaves better, or sweet basil. It is probable, that nothing is better than strongly to compress the wounded part, so as to cause a drop or two of blood to come forth, and even to enlarge the wound, and wash it with water.

It is towards evening that the coupling of these insects takes place. The males assemble, flying continually from one side to the other, without going any distance, and the females repair to them. As soon as a male sees one appear, he approaches, joins himself to her in an instant, and hooking on, suffers himself to be drawn along in the air, where they fly along together. Their union lasts but a very short time, and as soon as it is terminated, they separate. The female being fecundated, proceeds to deposit her eggs. She places them on the water, so that the larvæ, when born, may enter into the element where they are destined to reside. During the time of laying, the attitude of these females is very singular. They hook their four anterior feet on some object that floats upon the water, either a leaf, or a little piece of wood, and cross their two hinder feet. These two feet form an angle, and it is in this angle that they deposit their eggs, side by side. In proportion as the mass of eggs increases, it is elongated, and assumes the form of a little boat, more elevated at its two extremities. The feet which support it are removed by little and little, and when the insect has finished her deposition, she abandons it, and leaves it to float upon the water, at the mercy of the wind.

These insects produce many generations in the year, from the return of spring to the end of autumn. Each brood contains from two hundred and fifty to three hundred and fifty eggs, which are oblong, more pointed at the upper extremity, considered in their vertical situation, more thick and rounded below, and terminated abruptly by an edged neck, resembling that of some flagons; there we perceive a small circular aperture, which appears closed by a little membrane, through which the larvæ issues forth.

The boat, which is formed by the eggs, must always float upon the water, and their germ would perish, if they were submerged. Their neck alone enters into this fluid. Those which are just laid are all white, but they afterwards assume greenish shades, and in less than half a day, they become greyish. The laying usually takes place in the morning, about five or six o'clock, and the larvæ are born in about two or three days.

The larvæ swarm, in spring and summer, in the stagnant waters of marshes and ponds. They are even to be seen there in abundance from the time that the ice is melted. They usually remain at the surface of the water, or a little above it, in an inverted position, with the head down, respiring the air through the aperture of a funnel, which terminates their body. They are very lively, and when the slightest impression is made upon the water, they dive or swim, but return almost immediately to resume their original position. Their body is without feet, elongated, greenish at first, then greyish, transparent, and composed of three parts, the head, the corslet, and the abdomen. The head is detached from the corslet, and holds to it by a sort of neck. It is more brown than the rest, almost heart-formed, flatted from the upper to the under part, and presents, on each side, the appearance of a blackish eye, and a short antenna, with a single articulation, curved like an arch, with two aigrettes of hairs, one on the exterior side, and the other at its extremity. Around the mouth are small barbles furnished with hairs, which the larva causes to play with quickness, and which thus produce little currents, by means of which microscopic insects, little plants, and terrene bodies, are carried towards its mouth, and become its food. Among the tentacula, there are two more considerable, having the figure of crescents, and the internal side of which, on the concave part, is furnished with a fringe of very close hairs, like two tufts. The corslet is thick, rounded, with three small aigrettes of hairs on each side. It is followed by the abdomen, which is much more slender, in the form of an elongated cone, and divided into eight rings, which have each a tuft of hairs on each side. The eighth, and last, is terminated by two tubes; one upper, forming an angle with it, and the extremity, or aperture of which, has the form of a star with five rays; the other lower, more thick, but shorter, almost perpendicular to the length of the body, with two articulations, the last of which is longer at its upper contour, edged with long hairs, which are arranged like a funnel when they float in the water. From the end of this same tube proceed four slender, oval, transparent laminæ, as it were scaly, placed in pairs, capable of separation one from the other, and look like fins all round. Its aperture gives issue to the excrements, which are of a greenish colour. The other, or superior tube, is destined for respiration, and susceptible of contraction and dilatation. The transparence of the skin even permits us to discover the two tracheæ, which repair thither, after having traversed the entire length of the body. The intestines and the aliments are also perceptible.

The larva is subject to several moultings, and undergoes three at least, in the space of two or three weeks. To despoil itself of its skin, it places itself horizontally at the surface of the water, having the back upwards. The change then takes place in the usual way, by means of a cleft which is made over the corslet, and which afterwards extends to the rings of the abdomen.

The nymph swims very well, here and there, just like the larva, but without taking any nourishment. When in a state of repose, it has a shortened and lenticular form, its abdomen being folded underneath, and applied against the breast. It is placed vertically in the water, but differently from the larva, the respiratory organs being situated on the corslet, they resemble two species of horns or asses' ears. Their upper extremity is cut obliquely, and the insect always keeps it above the surface of the water. The eyes are distinct; under the corslet is found a thick mass, enclosing the antennæ, the mouth and the limbs. The feet are rolled up there. The abdomen is elongated, divided into rings, and is terminated by two oval palettes.

If the nymph is desirous to descend to the bottom of the water, it unfolds its body, elongates itself, and gives strokes with the tail. But when it ceases thus to operate, it is speedily brought back to the surface. After having passed eight or ten days in this state, the insect has arrived at the final term of its metamorphosis. But this period is very critical, in consequence of the danger it then incurs, of being overturned in the water, and perishing there, its new organs not suffering it to live any longer in the same element in which it had existed until that time. This moulting takes place in the same manner as the preceding. The gnat, on issuing from its spoil, through the upper cleft which is made there, raises at first its head and corslet, as much as it can above the aperture; the posterior extremity of its body contracts a little, and then elongating, is pushed upwards. It disengages itself more and more, rises, and erects itself in proportion, and finally concludes by being in a perpendicular position. Its old skin becomes a kind of boat, of which the body is the mast. It rests upon it only by a very small portion of its hinder part. After having drawn from their sheaths the two anterior feet, and then the two following, it carries them forward. Soon after it leans towards the water, and places these organs on it. Thus supported, and in safety, it unfolds its wings, and flies off in a few instants. At its birth, its body is whitish, with the corslet greenish; but the colours soon assume the tints which are proper to the insect in its final state. Such is the summary of the history of this exceedingly unpleasant dipteron. Those who are desirous to know more of it in detail, may consult the works of Reaumur and Degeer, not to mention Swammerdam and Kleeman, where they will find ample matter to satisfy their curiosity.

We now proceed to the TIPULARIÆ and their subdivisions. We shall first speak of the tribe in general.

Their body is usually elongated, the head round, and occupied more or less by two large complex eyes, but the small simple eyes they seldom possess. The corslet is inflated and round; the balancers are long; no winglets. The abdomen is long and almost cylindrical; the feet long and slender in the greater number. The tarsi are terminated by two hooks, and a small cushion.

It is very easy to distinguish, at first sight, the tipulariæ, from the other diptera, by the length and trifling bulk of their bodies, by the extent of their wings, and by their long and slender feet, which can scarcely support the body, which the insect balances, and causes to vacillate continually. Many small species have considerable resemblance to the gnats, with which they have been confounded by Swammerdam and Goëdart; but a slight examination of their mouth is sufficient to distinguish them from those insects whose proboscis is long and advanced, whereas that of the tipulariæ most frequently projects but little, and is bilabiate.

It is in the meadows that the larger species are most commonly to be observed, which in most countries have their peculiar name. Goëdart and Leuwenhoek have named them *tailors;* some French authors have called them *tipules couturières;* the small species are called *tipulæ culiciformes.* Among the first, there are some that are twenty lines in length. We call the largest species in this country, by the popular name of *long legs.*

From the commencement of spring until the end of autumn, we observe the larger tipulariæ in meadows, &c., but especially in this last season. Although they rise pretty high, they fly but a short distance. At certain times they make no use of their wings but to assist them in walking, and reciprocally their feet to assist them in flying. They make use of them to sustain their body above plants, and to push it forward. Some of the smallest species remain almost continually in the air. In all seasons of the year, at certain hours of the day, clouds of them are observed rising and falling, pursuing a vertical line. They make a trifling noise, which could be but little heard, if it were not produced by the innumerable quantities of those which fly at the same time and altogether

The larvæ of these insects vary much in their form, and the nature of their habits. In general, they resemble elongated worms. Their head is of an invariable figure, and their body divided into rings. Some have pediform appendages, others are destitute of them. Those of the large species have the head small, usually concealed under the first ring. This head is provided with two fleshy horns, and in front with two hooks, below which are two immoveable scaly pieces. These pieces serve to cut and bruise the aliments on which they feed. On the last ring of their bodies is a depression, which contains the two stigmata through which they respire the air. These larvæ live in the humid grounds of meadows, where they remain at one or two inches in depth. They feed on earth and soil. Although they do not eat plants, yet nevertheless, they do them much injury, because, as they often change place, they raise and detach the roots, which they expose to be dried up by the soil. These larvæ also live in the cavities of half rotten trees, where they find a mould tolerably similar to that of the dunghill. They undergo their metamorphosis in the earth, and change there into nymphs of a greyish colour, the rings of which bristle with tuberosities and spines, simple or forked, and inclined backwards. It is upon their head at this time that the organs of respiration are seated, which consists of two horns more or less long, according to the species. A little time before their final metamorphosis they make use of the points of their rings to push and raise themselves above the surface of the earth, to the height of about half their body, and there they remain until the skin which holds them, as it were swaddled up, cleaves to give them passage, at the moment when they become perfect insects. Almost immediately after their last metamorphosis, the tipulariæ couple, and during the coupling, the male remains hooked to the female with the pincers which terminate his abdomen. Their junction lasts nearly four and twenty hours, without interruption, and they often fly without separating.

When the females are fecundated, they deposit their eggs in the earth, employing for this operation the scaly pieces in the form of pincers, which they have at the extremity of the belly. During the laying their attitude is very singular. They keep their body elevated vertically, and sink the upper part of their pincers in the earth, as far as the organ of the lower piece, which is the conduit through which the eggs pass. After having left one in the first hole, they go away to form another, and so on to the last. These eggs are oblong, a little curved, and of a shining black. Each female lays a tolerably large number.

As to the larvæ of the smaller tipulariæ, some live in cow-dung, others in different kinds of mushrooms, and some others in the water. The agaric of the oak breeds rather a singular species, which does not penetrate into the substance of this plant, but remains underneath it. This larva, which is without appendages in the form of feet, and whose skin is humid and gluey, like that of snails, never crawls upon the agaric naked. It carpets all the places where it passes, with a gluey coat, which it draws from its mouth. When it wishes to fix itself any where, it applies this liquor against one of the points of the place which it is to inhabit, and spins it into fine laminæ, of which it applies several, one against the other, and attaches the ends to an opposite point. It forms also a sort of little roof in the same manner, and remains in shelter between this matter, which serves it both for bed and tent. We find little more than eight or ten of these larvæ on the largest *agarics*. Arrived at their full growth, towards the end of summer, they enclose themselves in a cocoon with large meshes, which they construct with a fluid similar to that of which they compose their nest, and it also serves them to fill the vacancies of those meshes. These cocoons are of a conical figure, and rough on their surface. The perfect insect issues from them in about fifteen days after the larva has been changed into a nymph.

The larvæ which live in the water differ much from each other in their conformation. They have nothing in common but the stigmata, the number of which is the same in all, though diversely figured. Some swim with very great agility, others inhabit holes which they make in the earth, at the edges of streams, where the water penetrates. Many inclose themselves in cases, which they form with the fragments of rotten leaves, of grains, and other materials which they find within their reach. The nymphs of these larvæ scarcely differ less amongst each other than the larvæ themselves. Some remain motionless at the bottom of the hole which the larvæ had inhabited; others swim and run with swiftness in the water. All are provided with organs, by means of which they respire, and they frequently apply them to the superficiés of the water to pump in the air. The tipulariæ which these larvæ produce are rather small. They are commonly called *culiciformes*. Their resemblance to the gnats causes them to be dreaded by those who do not know them; but they do no harm. Those who are desirous to know more particularly the habits of this genus, may consult Reaumur, and especially the Memoirs of Degeer.

All these insects multiply considerably, and in spite of their enemies, the species are very numerous. In their final form, the tipulariæ are pursued by birds, which destroy a great quantity of them; and those whose larvæ live in the water, serve to feed the fish, and carnivorous aquatic insects. Some species of those are to be found in the middle of winter.

In the subgenus Corethra of this tribe, Degeer has particularly observed the metamorphoses of *C. culiciformis.* Its larva, which he found in the month of May, in the water of ponds and marshes, resembles much, in its form and colour, that of the common gnat. Its body is a little smaller, of a clear brown, with some spaces deeper and more transparent. It is composed of a head tolerably thick, rounded, presenting two eyes, and some barbles; of a corslet very large, raised, and in the form of a rounded ball, and of an elongated abdomen, growing slender by little and little towards its extremity, and composed of eight rings. The last forms a sort of conical tail, curved underneath, furnished with hairs, and pierced at its extremity for the passage of the excrements. Near its lower junction with the preceding ring, is a sort of fin, which consists of an assemblage of long, black hairs, placed like rays, and forming together a circular plate. There is visible, at the end of the eighth ring, a piece of a conical figure, perpendicular, having also an aperture at its extremity, but which serves the larva for the purposes of respiration. We distinguish, in the interior of the corslet, two bodies tolerably voluminous, oblong, brown, and from each of which proceeds a vessel which repairs to the posterior extremity of the body. These two vessels approach each other very nearly, through a good part of their length, and form undulations. They dilate at the seventh ring, and afterwards contract. Degeer supposes, with considerable foundation, that these are tracheæ.

These larvæ hold an attitude different from those of the gnats. Their body is in an horizontal position, in the middle of the water, and always maintains itself there in *equilibrio.* It descends slowly, and by its own proper weight, to the bottom of this fluid. But a movement of the tail restores its equilibrium. Thus it is that these larvæ reascend to the surface of the water; but they are seldom seen there, and they swim as it were by jumps.

Their nymphs have also a great conformity with those of the gnats. They swim in the same manner, equally suspend themselves at the superficies of the water, by two sorts of horns or earlets, situated on the corslet, and which are the organs of respiration. Being lighter than the fluid which they inhabit, they always naturally proceed towards its surface, and cannot dive, but by giving some strokes with the tail. Their head, corslet, and the pectoral prominence in which the antennæ, and the organs of locomotion are enclosed, form an irregular mass. The abdomen is elongated, curved like an arch, so that its extremity is placed near the under part of the head. It is divided into eight rings, the sides of which have angular inequalities. The nymphs, however, occasionally extend it in consequence of its great flexibility, in a right line. At other times, it applies itself so exactly against the breast, that its body has the figure of a flatted bean. We see, on each side of the head, a large, oval, black eye. The abdomen is terminated by two oval fins, similar to two small leaves, having nervures, and by a point, which appears to correspond to the tail, or the last segment of the abdomen of the larva. The interior of the body also presents those two tracheæ, of which we have spoken, and its colour is not changed. The two respiratory tubes of the corslet may be easily detached, and the insect will not perish. Degeer, however, remarked, that an individual on which this operation was performed, did not suffer its last metamorphosis. This takes place at the end of about eighteen days, reckoning from the passage of the larvæ to the nymph-state.

The Chironomi, like the last, belong to that division of

tipulariæ which have been named *culiciformes*. The front feet, which are very long, they hold in the air, and agitate like antennæ, when they are in a state of repose. The larvæ of these diptera are also aquatic, and have great affinity with those of *culex*. The larvæ of one species (*C. plumosis*), which have often been confounded with those of the gnat, come in great quantities into stagnant waters, and into buckets, &c. filled with this fluid, and exposed to the air. They there exhibit small terrene masses of irregular figures placed against the sides, and especially at the bottom of these vessels, many of which, pierced with little holes, have some resemblance to honey-combs, but with round apertures. There are some, however, which are oblong, and turned in a vermicular form: they are the entrances of the cells, or habitations of these animals, which are sometimes observed to put forth their head or the anterior part of the body. The body is long, cylindrical, composed of twelve rings, and a scaly head, of one unvarying figure. It presents, at this part, two appendages in the form of inarticulate stumps, membranaceous, and resembling sorts of legs. From the middle of its penultimate ring, and from its junction with the last, hang two cords (four in all), usually waved and interlaced. From this is derived the name of *vers polypes*, given by Reaumur, to these larvæ. The aperture through which they reject their excrements, is situated at the end of the last ring, and forms by its contour, a square, having at each angle a little oblong body, similar to an olive; two of these bodies are directed towards the head, and two others backwards. From the origin of each of these, proceeds another appendage, but as large again, oblong, bellied at the base, with the end flat, and crowned with stiff and prickly hairs; these two parts are probably respiratory organs.

These larvæ sometimes come out of their dwellings, and

swim pretty near the surface of the water, turning in a circle in various directions, or giving their bodies all the movements necessary to carry them whither they are desirous to go.

They even remain thus, out of their retreats, for whole days together, assembling in great numbers round some leaf or some little moss, and fix themselves there by the posterior extremity of the body. Some hundreds of these larvæ are often to be seen agitating themselves at the same time, and making contortions which appear extremely forced. Some of them rest upon the bodies of others; each of them constructs its tube, with whatever it meets most light and spongy; and though Reaumur was unable to perceive any thread escaping from the mouth of these animals, he conjectures, nevertheless, from the mode in which they then execute their movements, that they really spin, and thus connect the different molecules of which their cells are composed. The two anterior false feet even appear to serve under these circumstances, to retain the materials. Fixed by the posterior extremity of their body, they curve it, bring their head close against this resting point, then deposit the little grains, and renew the same manœuvre, until they have concluded the formation of their tubes.

It is there also that these larvæ are transformed into nymphs remarkable for the handsome white tufts, which furnish the two ends of their body. The anterior is composed of several little feathers, which extend even on the sides of the corslet, where some of them form sorts of stars, with five branches. The abdomen is narrow and very long. It is terminated by two hooks, and its tuft is disposed like a fan.

These nymphs are very lively when they are drawn out of their cases, and when they are put into the water, they agitate and torment themselves. When the period of their final metamorphosis is arrived, ten or twelve days after the preceding, they repair to the surface of the water, there change place, cause their body to assume different inflexions, and even sometimes pass an entire day in this situation before they complete their metamorphosis. Their spoil, which retains its tufts, preserves them from the submersion to which they are exposed.

The most remarkable species of TANYPUS is that which Degeer calls *tipula variegata*, and of which he has given us the history. Its larva is found in the month of May, in the waters of ditches and marshes. It swims there like a little serpent, bending its body on one side and the other. It also walks at the bottom of the water, and over the aquatic plants. Its body is cylindrical, about three lines in length, not thicker than a horse-hair, transparent, and of the colour of a dead leaf, with many small black spots, or deep blue. It is divided into ten rings, separated by well marked incisions. Its head is oval, scaly, tolerably large, furnished with two little filiform antennæ, with some small setaceous barbles, making part of the mouth, and two small eyes, one on each side, in the form of black points. Its interior presents two black opaque bodies, having the form of kidneys. The first ring is much longer and thicker than the others. It represents a sort of corslet, in the interior of which are distinguished two granular bodies, which, according to Degeer, may be reservoirs for air. To the anterior and lower extremity of this ring, are attached two false feet, similar to wooden-legs, or crutches, proceeding from a common stem, which afterwards divides into two branches, and the extremity of each of which is crowned with long mobile hooks, curved externally and superiorly. The animal causes those organs to move in various directions, and can even draw them within the body, so far, that externally, they present only the form of stumps. At the extremity of the last ring, are two other feet, almost similar, but entirely separate, in their whole length, always hanging in a perpendicular manner, and always stiff, or extended.

Their direction may change, when the larva, to walk, bends like some caterpillars, the hinder part of its body downwards, but it can neither bend nor shorten them. Immediately above their origin, the last ring is terminated by four triangular, and very diaphanous plates, and by some hairs. A little higher, or towards the back of the same ring, are two small cylindrical perpendicular stems, having at their extremity from five to six long hairs, disposed like an aigrette. These appendages, as well as the preceding laminæ, are respiratory organs.

The nymph resembles that of other aquatic tipulariæ. Its body is of a clear and transparent brown, formed, 1. of a rounded head, a little emarginated, and pointed superiorly, and with two oval eyes; 2. of a thick, and as it were, humped corslet, having two raised pieces above, in the form of earlets, oval, terminated by a little point, serving for respiration, and on each side one oval plate, which encloses the wings; 3. of an elongated cylindrical abdomen, composed of eight rings separated by deep incisions, curved like a buckle, and terminated by two small elongated conical points, with aigrettes of long hairs.

This nymph always remains perpendicularly in the water, the head upwards, and the curve of the abdomen underneath. It is most frequently placed in the middle of the water, fixes itself, by means of the points of its tail, and sometimes for a considerable time, to aquatic plants, and to different bodies. It also comes to the surface of the water, and swims with quickness, moving its abdomen. At this time it erects the abdomen, and beats the water with reiterated strokes; it often allows itself to be carried along by the motion of this fluid. The metamorphosis of the larvæ into nymphs takes place towards the end of May; a few days after, these nymphs become perfect insects: for this purpose, they place themselves at the surface of the water, and the skin of which they despoil themselves, forms, as in the metamorphosis of the gnats, a sort of boat, which protects the new-born insect from drowning.

We pass to Cecidomyia. This genus is composed of very small diptera. The abdomen of the females is terminated by a very sharp point, sometimes in the form of a dart, which serves them to sink their eggs into the buds of the leaves or flowers of many plants. These parts acquire an extraordinary development, and form a sort of gall, which serves both as habitation and food to the larvæ of these insects. The extremity of the young shoots of the juniper, of the willow, of the lotus, of the vetch, &c. often exhibit those singular excrescences, or sorts of vegetable monstrosities. Degeer has compared that which he observed at the extremity of the young branches of a sort of willow, to a double rose, but green, or to the fruit of the hop. Swammerdam and Frisch had observed it before.

The young shoots of the same tree also present some irregular tuberosities, forming elongated masses, or rounded, or of various figures, and which are ligneous excrescences, produced by the wound of diptera of the same genus. They are the cradle of their posterity, which pass the winter there.

The gall, or scurf, of the juniper, is composed of six leaves, three of which are smaller, interior, and enveloped by the others. Degeer observes, that the leaves of this shrub are always placed three by three, in opposition one to the others, around the same point of the branch, and that the end of the young shoots is also constantly terminated by three leaves, between which are often seen three others smaller. He concludes from this, with reason, that the gall or scurf of this tree is produced by the germ itself, which without this alteration would have produced a new shoot. The larva eats and destroys the interior leaves, and spares only the three external ones, which, thus receiving a part of the nutritive juice destined for the preceding, grow to a preternatural size. In Sweden, the peasants and country people generally, use them as a remedy for the hooping cough, and name them in consequence of this use *kik-bar*, which means berries for the hooping cough. They boil them in milk, which they then give to the patient to drink. These juniper-galls are found at all seasons of the year; but it is only from the month of September to the May of the following year, that they contain the larva. It is placed there vertically, with the head upwards. Its body is but a line in length, of a very lively orange colour, without feet, divided into twelve rings, and less thick in front than behind; its head is rounded with a small pointed eminence, which is apparently the sucker. The larva is transformed in the month of May, or at the commencement of June, into a nymph, which preserves the same colour, and is of an oval figure, with two small conical points in the form of straight horns to the head. Degeer suspects that these are the organs of respiration. It cannot undergo its final metamorphosis if the gall or scurf in which it is enclosed should be dried up. The perfect insect leaves its spoil at the pointed opening of the interior leaves: small ichneumons attack these larva. The nymph of the rose-like gall of the willow is of a red, more or less deep, according to its different parts, and does not present the two horns which we have mentioned, as belonging to the preceding species. It glides, like it, between the leaves which compose its habitation, to get rid of its skin at the entrance. We may observe on the ligneous gall of the willow, the remains of different germs, which would have shot forth leaves, if the gall had not absorbed all the nutritive juice destined for the growth of these leaves. These vestiges of germs form holes or cavities, which have a communication with the interior of the gall. The nymphs of the cecidomyiæ, which inhabit it, glide by little and little into these cavities, as far as their mouths, from which they even issue forth in part. It is then that they become perfect insects. The larva of the galls of this tree spin themselves a cocoon, in which they differ from those of the juniper.

Degeer has found on the leaves of the pine the cocoons of two other species of cecidomyia: one of these cocoons was made of pure silk, of a yellowish white, and fixed by threads of the same material; but the other cocoon had, besides the silky envelope, an external coat of resinous and white matter. The larva which forms this singular cocoon is distinguished from the preceding by the presence of two ranges of pointed nipples, and cleft at the end, resembling feet. Degeer counted fourteen of these. To issue from its cocoon, the nymph detaches, in the form of a cap, a small portion of one of its extremities.

On the lote-tree, named *corniculata*, we very often see some flowers, not yet blown, altogether swelled or monstrous, and which resemble bladders pointed in front. Their petals are never open. The larva of another species of the same genus live in a state of society at the bottom of flowers; their body is of a yellowish white, a little flatted, pointed in front, and rounded behind. At the time of their passage to the nymph-state, they abandon their primitive dwelling, enter into the ground, and undergo their metamorphosis there.

In Tipula (proper) the species *oleracea* is worthy of a slight notice. It is found in great abundance in meadows during the autumn. Reaumur has given a very detailed description of the sexual organs of this species. The attitude of the female at the moment of laying is very singular: she holds herself upright, and walks from time to time, and always in this vertical direction; the two posterior feet placed beyond the back, and tolerably behind are, with the scaly point in the form of a tail at the end of the abdomen, the only parts which she places on the ground. The animal makes, by means of this point, a hole in the earth, and after having introduced a single egg, and perhaps one or two more, she makes a step forward to re-commence a similar operation. Her tail or anal point is formed of two valves, the inferior of which serves to conduct the eggs. It is in the meadows, and in flower-beds, &c. of gardens, newly worked, that she deposits them, and the plants which are there are not useless to her, as the tipula rests upon them her anterior feet.

The body of the larva is in the form of an elongated cylinder, a little attenuated at both ends, greyish, and without feet. The head is scaly, small, provided with two small antennæ, and has for a mouth two corneous hooks, fixed, convex without, concave within, denticulated at the upper edge, with a fleshy and triangular piece in the intermediate space. The last ring presents six rays of unequal size, disposed in its contour, along with six stigmata ranged on two lines, and the upper two of which are the largest. The interior of the body presents two large longitudinal tracheæ, which, near the posterior extremity, are divided, each into a bundle of small threads, leading to the stigmata.

When these larvæ are very much multiplied in a meadow, they prove hurtful to its vegetation; not that they gnaw the roots of the plants, but, by dint of working, they hinder them from receiving the nutritive juices from the earth.

The nymph is elongated, cylindrical, with two small horns to the head, proper for respiration, and small spiny tubercles on the rings of the abdomen. On the point of completing its final metamorphosis, this nymph makes use of them to raise itself, and get near the surface of the earth, when it becomes a perfect insect.

We have already, under the head of the gnats, noticed the manners of the *mosquito*. We could not well separate the consideration of these insects, in that point of view, as their habits are identical, and they have been constantly confounded together both by travellers and naturalists; M. Latreille, however, places this formidable insect in his genus Simulium. To what we have said above, we shall only add, that insects of one species of *simulium* come in immense quantities in the spring and the end of the summer into the countries of Servia and Bannat. They attack cattle, penetrate into their parts of generation, and cause them to perish in the course of four or five hours. They are destroyed or driven away by smoke. The species to which we now allude is with Linnæus *Culex reptans*, and has been made a *tipula (Erythrocephala)* by Degeer. It is to be found in France, in the neighbourhood of Paris, and in the southern departments. M. Latreille tells us of having once been bitten on the hand by one of these insects. The *mosquito* of travellers has all the comparative characters belonging to the present genus, as M. Latreille had occasion to verify on some which were shown him by Michaux, the celebrated botanist, who brought them from North America. This species differs from the French one only in being entirely black.

Passing over the intervening sub-genera of tipulariæ, which afford little to arrest our attention, we proceed to the family of Tanystoma, which will not detain us long. The larvæ resemble long worms, almost cylindrical, without feet, with a head either scaly and uniform, or soft and variable, and always provided with hooks, or retractile appendages, which serve to gnaw or to suck the substances on which they feed. The majority live in the earth: they change skin to undergo their second transformation. Their nymphs are naked, and present many of the external parts of the perfect insect, which issues from its spoil through a cleft on the back.

The division Asilici embraces the genus Asilus of Linnæus. These insects are found in fields, gardens, and meadows, especially towards the end of summer and in autumn: they fly with rapidity, particularly when the sun is very hot;

in flying they make a humming noise, tolerably loud. All are carnivorous, and feed exclusively on insects which they catch on the wing: they seize with their anterior feet, humble-bees, tipulæ, flies, and even coleoptera; they kill them by pricking them with one of the pieces of their sucker, which is a true sting, in the form of a stylet, very pointed at the extremity, and then they suck them.

The larva of *Asilus Forcipatus* lives in the earth; its body is elongated, a little flatted, without feet, and divided into twelve rings: the head is scaly, armed with two mobile hooks, curved underneath, and furnished with some hairs; its skin is smooth and shining. It presents four stigmata; two are anterior, and form so many points; the other two are placed on the penultimate ring, and consist of two small tubes, cylindrical, and inclined towards the hinder part. It is also in the earth that the larva is transformed into a nymph, quitting its skin like those of the tipulariæ, and without forming a cocoon.

The Empides in general are of small size, live on prey, and not unfrequently on the juice of flowers: they seize small insects with their feet, and then suck them with their proboscis: they are frequently observed to couple, the male being on the back of the female, and sometimes occupied in sucking a fly. Their larvæ are unknown.

The Bombylii are very agile, and fly with great rapidity. They hover about flowers without settling on them, and introduce their long proboscis to extract the honied fluid, on which they feed. In flying they make a noise similar to that of the humble-bees. They are found in summer. Their larvæ and metamorphosis are unknown: it may be remarked that they usually frequent hot and sandy soils.

The larvæ of Anthrax are unknown. The perfect insect is found during the fine season in places furnished with flowers, or near walls situated to the south. These diptera fly very lightly, especially when the sun is brilliant; they are seen to hover in the air, then to fix themselves on plants, and it is only by a considerable degree of address and celerity that they can be caught. Some have transparent wings, and without colours; others have them opaque, and highly coloured.

The Nemestrinæ also fly with the greatest lightness, transport themselves to very considerable distances, and do not repose a long time on the same flower; they extract from them with promptitude, by means of their long proboscis, the melliferous juices which they contain, and pass hurriedly from flower to flower: they only rest upon those the nectar of which has not been exhausted by any other insect. These diptera are peculiar to the southern countries of Europe, of Asia, and to Egypt.

The Leptides (formerly *Rhagio*, of Fab.) are found on trees and leaves; it would seem that they live on rapine. The *Leptis Scolopacea* is found throughout all Europe. The larva is long, cylindrical, of a yellowish white: its head is small, scaly, brown, and furnished with two small antennæ; the upper part of the body is furnished with ten fleshy nipples, which perform the office of feet, and serve the purpose of locomotion. This larva lives in the earth, where it undergoes all its metamorphoses. Arrived at the term of its growth, it changes into a nymph, which has several ranges of short spines on the body, and which becomes a perfect insect at the end of the month of April.

The habits of the larva of *Leptis Vermileo* are very curious. It is elongated, cylindrical, and of a yellowish grey: the body is divided into eleven rings; the head is of a fleshy substance, conical, and furnished anteriorly with a sort of scaly dart; the last ring of the body is terminated by four fleshy appendages, tolerably long, in the form of nipples, furnished with long and stiff hairs. The anus is placed upon the back, like that of the larva of the *Crioceris*, already noticed, which covers itself with its excrements. This larva, which has been designated under the name of *Lion-worm*, lives on insects; it establishes its dwelling like those of the ant-lions, and is often found in society along with them. It is at the foot of old walls, in sandy soils, that it forms a sort of tunnel, sheltered from the rain; it places itself in the middle, and there remains in ambush to seize and devour the little insects which have the misfortune to fall into the snare which is laid for them: after it has seized its prey it encompasses it with its body, pierces it with its dart, and very speedily kills it, then it sinks under the sand, and drags in the insect, to suck it at leisure, and having drawn out all the substance it flings away the carcase. Towards the end of the month of May this larva, which has then acquired its full growth, is changed into a nymph in the sand, without making any cocoon, and it becomes a perfect insect about fifteen days after its metamorphosis.

The DOLICHOPI, so named in consequence of the length of their feet, have the body ornamented with very brilliant colours. These insects are universally dispersed; some most frequently remain near humid places, running on the ground, over leaves, and sometimes even on the surface of the water; others have the habit of frequenting walls and the stems of trees. Elevated on their long legs, they walk with rapidity in search of their food, which consists of small insects. M. Latreille has seen an individual of the species *Rostratus* dilate most remarkably the lips of its proboscis to swallow a living acarus.

Degeer has noticed the metamorphosis of the species *Ungulatus*, but as he says nothing relative to their habits, we shall pass them over. The nymph appears to be of a very unquiet nature, being continually in a sort of rotatory motion.

Our notice of the little family of TABANIDES shall be confined to the TABANI. These insects resemble large flies; they are sufficiently known by the torment which they cause to horses and oxen during summer, by sucking their blood with the utmost pertinacity. They usually appear at the commencement of this season: it is in low meadows and humid woods that they are to be found in the greatest abundance; they fly in the open day with rapidity, and uttering a humming sound, especially when the weather is very warm, and the sun shines bright. They fasten on the oxen and horses which they pursue, and sometimes also attack men, but more rarely: it has been remarked that it is the females alone which are so eager for blood, and the same observation has been made upon the gnats, the males of which never come to sting us; the male tabani, therefore, must derive their nutriment from the flowers, on which they are often occupied in sucking the melliferous fluid with their proboscis. They fly most usually in a small space, making many turns, and appearing to invite the females to approach them.

We know nothing concerning the coupling of these insects, and their larvæ are but little known. Degeer is the only naturalist who has observed that of the *Tabanus Bovinus*. According to this author, the larva is of a yellowish white, without feet; its body is cylindrical, slender at its anterior part, and divided into twelve rings. Attached to its head are two large, mobile, scaly hooks, curved downwards, and which appear to answer the purpose of digging into the earth, where it lives, and in which it buries itself. It is also in the earth that it undergoes its metamorphosis, and is changed into a nymph of a cylindrical form, whose belly is divided into eight rings, which have each at their posterior edge a fringe of long hairs. The last is armed at its extremity with six hard scaly points, which serve as resting points to the nymph when it reascends to the surface of the earth. About a month after the larva has changed form, the perfect insect issues from its skin of nymph, which is cleft over the head and corslet.

These insects are found both in the old and new world.

Many are especially remarkable for their colours and the lustre of their eyes. In the *Tabanus Autumnalis*, when the insect is dead, the eyes are brown, but of a brilliant black when it is living.

The diptera which constitute the family of NOTACANTHA, were placed by Linnæus in his genus *Musca*. The majority of them inhabit marshy places, their larvæ being aquatic, and remain on leaves, and the flowers of plants. Some others frequent woods, and appear to deposit their eggs in the carious part, or the wounds of trees.

The larvæ have the body long, flatted, divided into rings, of which the last, usually the longest, forms a sort of tail, terminated by some barbed or plumose hairs, disposed in a ray, at the point of the union of which is the aperture which gives passage to the air. Their head is scaly, small, oblong, and provided with small hooks and appendages. Such is especially the form of the aquatic larvæ of this family, the only ones, in fact, with which we are well acquainted. They respire by holding the end of their tail suspended at the surface of the water. Their skin becomes the cocoon of the nymph. Their body, when this takes place, does not change figure, but becomes stiff and incapable of motion. It floats upon the water, and its tail often makes an angle with it. The nymph occupies but one of the extremities of its internal capacity. The perfect insect issues forth by a cleft which is made over the second ring. It places itself on its spoil, where its body becomes strengthened, and completes its development.

The BERIS are small insects, and appear in spring, some inhabit woods, and appear to deposit their eggs in the humid caries of trees. Others inhabit marshes, and their larvæ are probably aquatic.

The insects of the genus STRATIOMYS, now forming the division *Stratiomydes* of Latreille, have been known for a long time. Swammerdam has given their history under the name of *Asilus*, and Reaumur under that of *Mouches à corslet armé* (flies with armed corslet). Geoffroy, in establishing this genus, preserved the French name, which he Latinized, or to speak more truly, *Hellenized*, by that of *Stratiomys*.

The larva of this insect lives in the water; it is without feet, and usually of a greenish or yellowish brown. Its body is elongated, flatted, thicker at its anterior than at its posterior extremity. Its head is small, furnished with hooks, which answer the purpose of seizing the little insects on which it feeds, and with a fleshy nipple with which it sucks them. At its last ring it has an aperture, or sort of stigma, through which it pumps the air of which it has need; near this aperture is a species of funnel, formed by a great number of hairs, which hinder the water from penetrating into the stigma. When this larva is desirous of respiring, it raises its last ring above the water, and remains a certain time in this position with the head downwards; but when it wishes to sink in the water, it folds back its hairs in the form of a packet, with which it covers the aperture of the stigma, which by this means continues dry. Arrived at full growth, it undergoes its metamorphosis under its larva skin, which hardens and serves as a cocoon, without changing form. The nymph, much shorter than the skin, only occupies the anterior part, while the last four rings remain empty. Eight or ten days after this metamorphosis, the perfect insect issues from its cocoon, and goes in search of flowers, to suck the honied fluid which they contain, and returns to the water only to deposit its eggs. Most of them, however, frequent aquatic situations, and remain in the plants or the leaves of vegetables which grow there.

We now come to the fifth family of diptera, the ATHERICERÆ.

The body of the larva is very smooth, very contractile, and more narrow and pointed in front. The head is of a variable form, and its external organs consist of one or two hooks, accompanied, in some genera, with nipples, and probably in all with a sort of tongue destined to receive the nutritive juices. The number of their stigmata is usually four, two of which are situated, one on each side, on the first ring, and the two others on as many circular scaly plates, at the posterior extremity of the body; it has been observed that these last are formed, at least in many, of three smaller and closely approximated stigmata. The larva can envelope these parts with the surrounding flesh, which forms a sort of purse. It does not change skin; that which it has from its birth becomes in solidifying a species of cocoon for the nymph; it contracts, assumes an ovoid or spherical form, and the anterior part, which was more narrow in the larva, augments in thickness, or is sometimes more thick than the opposite extremity. The traces of the rings are discoverable there, and often the vestiges of the stigmata, although they no longer serve the purpose of respiration. The body is detached by little and little from the skin or cocoon, presents itself in the figure of an elongated and very soft ball, on which no parts are distinguishable, and passes soon after into the state of nymph. The insect issues from its cocoon by causing its anterior extremity to shoot off, like a cap; it detaches it by the efforts of its head. This portion of the cocoon, besides, is so disposed as to open with facility.

Few of the Athericeræ are carnivorous in the perfect state; they sojourn for the most part on flowers or leaves, and sometimes on the excrements of animals.

The first sub-genus of the first tribe (Syrphidæ) is Volucella. These insects resemble humble-bees, and it is even in the nests of these hymenoptera that their larvæ are found. Some of them are remarkable for the fleshy and radiated spines on the posterior extremity of their body. These diptera frequent woods, and generally appear in spring: they send forth a tolerably strong humming sound.

The larva lives, as we have said, in the nests of the humble-bees; its body is widened from front to rear, has some transverse wrinkles, some little points on the sides, six membranous threads, disposed like rays at its posterior extremity, and presents underneath two stigmata and six pairs of nipples, each furnished with three long hooks, which answer the purpose of walking.

The ERISTALES also bear some resemblance to the humble-bees: their most remarkable species is the *Eristalis narcissi* whose larva is described by Reaumur. It gnaws the interior of the bulbs of the narcissus, and thus causes the plant to perish. Its body is cylindrical, narrowing to a point at both ends, and wrinkled; its anterior extremity, near which are two stigmata, is armed with two parallel hooks, and above each of them is seen a sort of cleft horn; the posterior stigmata are placed on a cylindrical part, covered by flesh: to the anus are attached two nipples. Sometimes the larva undergoes its transformation in the bulb itself, but it most generally issues forth from it, making a round hole. The cocoon of the nymph is thick, wrinkled, and grey, with two horns in front. In the perfect insect the corslet is covered with fawn-coloured hairs, but those of the abdomen are paler, or greyish.

We now come to the genus SYRPHUS, from which the tribe has received its name. The larva of the Syrphi, or as they have been called, *Aphidivorous worms*, remain upon the trees or plants which are peopled by the aphides, which constitute their sole aliment, and of which they destroy an immense quantity. They resemble a membranaceous, conical, elongated worm, a little flatted above, slender, and terminating in a point at its anterior part, with the hinder part thick, and more or less rounded. This part presents two approximating stigmata, placed each on a small elevation, and concealed by the penultimate ring: the skin is green or yellowish, with a band, or a line differently coloured, along the middle of the back; it is rough or unequal, and more or less provided with small, fleshy, and conical eminences, sometimes very apparent, sometimes visible only with the assistance of the microscope, and sometimes terminated by an articulated point in the form of a spine. The transparence of the skin permits us to distinguish the dorsal vessel, the intestinal canal, and some other interior organs. The under part of the body is furnished with nipples representing feet, and the number of which is various; one species (*Musca pyrastri, Linn.*) has no less than forty-two of them, which are disposed on six ranges. The action of these parts, the alternate elongation and contraction of the rings of the body, and a viscous matter which transudes from the belly, give to these larvæ the means of advancing by creeping, and of even climbing to the top of the branches to seek their food, and also affixing themselves there. The manner in which they walk has some relation with that of the *geometrical* caterpillars; they fix themselves at first with the scaly and pointed instrument of the head, then they contract the body, and glue its posterior extremity in the plane of position by means of the viscous substance of the belly, after which they disengage the head, and carry it forward to rest it anew, and recommence a similar manœuvre.

Placed in the midst of a numerous family of aphides, which appear not to know that they are their enemies, and which manifest no mistrust, they wait without stirring until some one of the aphides shall touch them, or walk imprudently over them; immediately they turn their head, quickly dart it on the aphis, so as to seize it, and sink into the body of the animal the sort of dart with which it is armed, and of which we have already spoken; they then raise the anterior part of their body, and thus holding the aphis in the air, they suck it quietly to the very last drop; they afterwards throw away the skin, the only remaining part of the carcase: this operation lasts but a very few minutes.

The larva devours other aphides in succession, according as it is more or less hungry, but always preserving the same attitude. The aphis being thus suspended in the air, cannot hook itself with its feet and disengage itself. In the moment of suction the interior of the larva allows us to see a small black and elongated part, moving incessantly, and which appears to be a sort of sucker, or piston, adapted for attracting into the stomach the fluid and nutritive substance which escapes from the body of the aphis. Some of these larvæ are satisfied with divers species of aphides; but there are others more delicate, which live only upon one: hunger occasionally obliges them to devour one another.

Having acquired their full growth, these larvæ, which never change their skin, prepare themselves for their last metamorphosis by fixing their hinder part on a stem, a leaf, or some other suitable object, by means of the glutinous liquid of the under part of the body; this substance then hardens like gum. Thus arrested, they commence by degrees to shorten the form of their body, so that its length is diminished about one-third; the skin hardens, becomes like parchment, and then forms a hard and solid cocoon, in which the animal at first passes to the state of nymph, and afterwards to that of the perfect insect; it then despoils itself of the thin pellicle which enveloped it in the first of these states, and issues forth through a circular aperture at the thick end of the cocoon, from which there is detached a portion in the form of a cup, by the effect of the pressure which the insect has exercised over this part, in pushing it out with its head. A very remarkable fact is, that in this passage to the nymph-state, the most slender portion of the body of the larva, or the anterior, is now the thickest, or forms the thick end of the cocoon, while the posterior extremity, which was the thickest, has diminished in volume. A few days are sufficient for the development of these insects. The last rings of the abdomen form in the females a sort of tube or auger, which serves them for the deposition of their eggs.

The Chrysotoxa are not without some resemblance to wasps, from the yellow spots which cover their black, and almost smooth body. They are found on flowers, the honey of which they suck, and over which they frequently hover for a long time. Their flight is rapid.

The Ceriæ are distinguished from the other genera derived from that of *Syrphus*, by the length of their antennæ, their much separated wings, and their elongated and almost cylindrical abdomen. There is a small prominence at the anterior extremity of their head.

These insects are found on flowers, but more particularly on the trunks of trees, where they appear to deposit their eggs. In the elongation of their body and antennæ, the separation of their wings, and their black colour, mingled with yellow, they, like the last, also resemble wasps.

Both the habits of Rhingiæ and their larva are unknown; but it would appear that they live in the dung of animals. Reaumur found that of the *Rhingia rostrata*, in a box which he had filled with cow-dung, for the use of some other larvæ which feed upon it. This species inhabits Europe in general.

We have now to speak of a very mischievous and tormenting tribe, the Œstrus, or *Gad-fly*.

These insects are seldom found in their perfect state at the time of their appearance, and the places of their habitation being very limited, as they deposit their eggs on the bodies of many herbivorous quadrupeds, they must be sought for in woods, pasture grounds, &c. Each species of œstrus is generally parasitical upon a particular species of mammalia, and chooses for the deposition of its eggs that part of the body which alone can be suitable to the larva which shall spring from them: the ox, the horse, the ass, the rein-deer, the stag, the antelope, the camel, the sheep, and the hare, are, as yet, the only animals known to nourish the larvæ of œstri. These animals appear to be singularly afraid of this insect when she approaches them to deposit her eggs.

The habitations of these larvæ are of three kinds, which may be termed *cutaneous*, *cervical*, and *gastric*, according as they live in tumours found on the skin, in some part of the interior of the head, or in the stomach of the animal. Our countryman, Mr. Bracey Clark, the first veterinarian in Europe, has treated this subject very amply and satisfactorily. But we must content ourselves here with a few generalities.

The eggs from which the cutaneous larvæ are derived, have been lodged under the skin by the mother, with an instrument like a wimble, such as we have frequently had occasion to mention before. These larvæ have no occasion to change their locality. At their birth, they are in the midst of the purulent matter which constitutes their aliment. The eggs of the other species are simply deposited and glued on some parts of the skin, which are near the natural cavities, into which the larvæ are to penetrate, or placed so as to be licked by the animal, who thus transports the larvæ into its mouth, and thence into the stomach. Thus the female of the *œstrus ovis* places her eggs on the internal edge of the nostrils of this quadruped, which then becomes furiously agitated, strikes the ground with its feet, and flies along with the head downwards. The larva insinuates itself into the maxillary and frontal sinuses, and fastens itself, by means of two strong crooks with which its mouth is armed, to the lining membrane of these sinuses. The gad-fly of the horse deposits her eggs on the internal part of the legs, on the sides of the shoulders, and, but seldom, on the withers. The species which is called *hæmorrhoidalis*, whose larva also lives in the stomach of the same soliped, places her eggs upon its lips. The larvæ attach themselves to the tongue, and proceed through the œsophagus into the stomach, where they live on the humour secreted from the internal membrane. They are most commonly found around the pylorus, and rarely in the intestines; they are often very numerous, and suspended in clusters. Mr. Clark, however, believes that they are rather useful that hurtful to this quadruped. Reaumur was of the same opinion. Vallisnieri, however, attributed to this cause an epidemic malady, which raged among the horses in the Veronese and Mantuan territory in 1713. Dr. Gaspari, having on this occasion dissected some mares that died of this malady, found their stomachs full of the aforesaid larvæ: each larva was lodged in a kind of cell, formed by gnawing the membrane of the stomach. The external membranes were inflamed, and the internal ulcerated. There were more than seven hundred larvæ reckoned in the body of a single female. It was supposed that these insects also introduced themselves into the fundament of horses, to lay their eggs, but Mr. Clark has shown this opinion to be erroneous.

The *Œstrus bovis* deposits her eggs upon the skin of the ox, in a number of small wounds which she makes, and in which the larva finds both food and an habitation. The places where they abide are easily recognized by an external tumour.

The larvæ of the œstri have, in general, a conical form, and are without feet; their body is composed, mouth not included, of eleven rings, charged with small tubercles and small spines, often ranged like cords, and which facilitate their progression ; the principal respiratory organs on a scaly plane of the posterior extremity of the body, which is the thickest. It appears that their number and disposition are different in the gastric larvæ; it also appears that the mouth of the cutaneous larva is only composed of nipples, while that of the interior larva has always two strong hooks.

These larvæ do not undergo their metamorphosis in the place where they have lived; they issue forth, drop on the ground, and seek some retired place for this purpose. Those which have lived in the stomach, follow the course of the intestines, and escape by the anus, assisted, perhaps, by the excrementitious dejections of the animal, whose parasites they have been. It is usually in June and July that these metamorphoses take place.

The œstri, as perfect insects, are incapable of taking nutriment; they couple immediately, and shortly die.

It would appear, that man himself is not altogether privileged against the attacks of these insects. Dr. Latham, according to the report of Mr. Clark, extracted some larvæ from the maxillary sinuses of a woman, which he referred to the *œstrus bovis.* Gmelin mentions a larva, of whathe calls an *œstrus hominis,* which sometimes afflicts the inhabitants of South America ; and Baron Humboldt has seen some Indians covered with tumours, arising from the presence of larvæ. These testimonies are not to be disputed; but M. Latreille thinks that the insect is mistaken, and that the larvæ rather appertain to the *Musca canaria,* of Linnæus, or some other analogous species.

On the little tribe of CONOPSARIÆ, there is not much to say. The CONOPS are insects of exceeding vivacity; they are found in gardens and meadows. Different from the Asili, which are carnivorous, and with which they have some resemblance, they feed on the nectar of flowers alone. The *Conops rufipes* is found in Europe, in the environs of Paris, and elsewhere, towards the middle of summer, on the flowers in meadows; it lives both in the larva and nymph-state, in the interior of the abdomen of humble-bees, and issues forth when it has undergone its last metamorphosis, through the intervals of the rings; of this extraordinary fact, M. Latreille gives us his own personal testimony. He has frequently found one or two individuals of this species in boxes, in which he had shut up some humble-bees.

The Stomoxys much resemble the common fly, from which, however, they are distinguished by their prominently advanced proboscis. These insects are found every where, both in the country and in houses. They are very troublesome, and prick very strongly with their proboscis, both men and animals, especially in autumn, the season in which they are most common. Their larvæ is unknown, or perhaps confounded with that of the common fly.

The well-known tribe of Muscides, will next claim a little of our attention.

We are not to suppose that this tribe, thus named by M. Latreille, from the word *musca* (a fly), embraces the whole genus thus designated by M. Linnæus. It only comprehends a part of it, which corresponds, with some trifling difference, to the genus *musca,* of Fabricius, as originally limited by that naturalist.

The muscides in general, have all the bearing of that insect, so well known under the denomination of the *common* or *domestic* fly. The head is hemispherical, the complex eyes large, and there are three small simple ones, perfectly distinct. The front of the head is commonly more membranaceous than the hinder portion, and of a different colour, with a longitudinal furrow, or fossette, on each side, for the reception of the antennæ. The corslet is cylindrical, and apparently but of a single segment. The abdomen of various figures, but generally long, and the feet with two hooks, and two pellets or cushions.

The larvæ of the muscides are apod, or without elongated feet, and generally cylindrical; they are soft and flexible; the front of their body is pointed and conical, the hinder part thick and rounded; the head, which is soft and fleshy, is furnished with one or two scaly hooks, which serve to mince the substances on which they feed; these hooks, by their retraction or advancement, render the form of the head variable. They are accompanied, at least sometimes, by nipples, and probably in all by a sort of tongue, calculated to receive the nutritive juices; no eyes are perceptible. The parts which might be taken for such, are only stigmata or apertures, for the entrance of air into the trachea. The number of their stigmata is usually four, two of which are situated on the first ring, and which are those just mentioned, and two others placed on the middle of a circular, and often scaly plate, terminating the last ring; the flesh of its contour can envelope these organs like a purse, and hinder the introduction of tumours, or injurious substances. Sometimes each stigma is composed of three small approximating clefts.

These larvæ feed on various substances, both animal and vegetable. Some devour the flesh of dead animals, whose putrefaction they accelerate, others live in excrements, dunghills, and unctuous earth; some species eat cheese; some others inhabit the bodies of caterpillars, and different larvæ, which they gnaw and consume. Among those which feed on vegetable substances, some live in leaves, which they sap internally; others live in galls, mushrooms, seeds of plants, and fruits. The rat-tailed larvæ, which inhabit muddy and marshy waters, and which feed on the fragments of rotten leaves, and other such matters, belong to the *Syrphides.* The use of the carnivorous larvæ of the genus musca appears to be to consume the carcases of animals which are dispersed in the woods and fields, and which have been spared by wild beasts. From their numbers, they are capable of consuming a carcase in a very little time. Those which live on excrements seem to be born for the purpose of purging the earth from aggregations of filth, which might otherwise prove pestilential.

The larvæ of the muscides do not quit their skin for the metamorphosis. This external skin grows hard, becomes scaly, and forms, as it were, an oblong cocoon, of a reddish or moronne brown, which encloses all the parts of the insect. In this sort of cocoon the larva assumes at first the figure of an elongated ball, in which no distinct part is visible; it is but a simple mass of soft flesh. Afterwards this ball becomes developed, and takes the figure of a nymph, in which all the external parts of the perfect insect are to be seen. In the larva the anterior extremity of the body was the most slender part, here, on the contrary, it is the thickest; the other was most bulky in the larva, in the nymph it is the reverse.

Among the diptera of this tribe there is one that deposits its eggs in cheese. From these come forth larvæ, whose external form presents nothing very remarkable, but they exhibit a phenomenon of a surprising kind, namely, the leaps which they execute by rising and shooting into the air, sometimes to the height of more than six inches. These leaps are the more astonishing in so small an insect, inasmuch as it appears to possess no organ which can assist it in making them. To discover how this manœuvre is performed, we must attentively consider a larva which is preparing for a leap. We shall observe it rise upright on its posterior part, and maintain itself in this position by means of some tubercles which are situated at the last ring of its body. Subsequently it bends itself, forms a sort of circle by bringing its head towards its tail, sinks the two hooks of its mouth in the two sinuosities which are at the skin of the last ring, and holds them strongly fastened. All this operation is but the affair of a moment. Then it contracts itself, and rears up so promptly, that the two hooks in springing from the two sinuses in which they were retained make a slight noise. By this quick movement the body strikes the ground with force, and rebounds at the same time to a considerable height. It is to Swammerdam that we are indebted for the first observations on this manœuvre of these larvæ. They are often found in great quantities in old and half rotten cheese; we vulgarly call them *maggots.*

After having remained for a longer or a shorter time in the nymph-form, according as the season is favourable to their development, these diptera issue forth from their cocoons. To effectuate this end they break off and push out a portion with their head, which swells in this operation; on issuing forth their wings are folded, rumpled, and so short, that they appear to be but stumps; but they are soon developed, extend, and become level and smooth, as is the case with other insects.

These insects, like all others to be fecundated, must couple: there is nothing singular in their mode of coupling, excepting in that of the domestic fly; the female of this species, instead of receiving the organ of the male, introduces, on the contrary, into his body a long fleshy tube; we often observe the males jump upon the females to solicit them to couple, but it never takes place except when the latter are so disposed.

This species of fly, and some others, are subject to a malady which is rather singular, and the cause of which is unknown; their belly swells remarkably, its rings are separated, and the pieces moreover which cover them are removed from one another; the skin is extremely tense, and perfectly white. If their belly be opened we find it full of a fat unctuous matter, of a white colour, which penetrates the skin, and accumulates on the surface of the body. In this state these flies hook themselves with their feet upon walls, on windows, and on plants in meadows, where they are found dead.

The flowers of the rose-laurel (*Nerium oleander*), and some others, also often exhibit the carcases of many small flies, and of *anthomyiæ*, which are suspended to the threads of their stamina. But under these circumstances, these flies have not been poisoned. A very viscous fluid has glued the extremity of their proboscis against these parts, so that they are not able to disengage themselves, and consequently perish. We have the authority and personal testimony of M. Latreille, for this explication of the fact. Other diptera may likewise perish, in the corollæ of some flowers (*Dionœa muscipula*), in consequence of the irritation of the corollæ, which obliges them to close, and retain these same insects captive.

In a very short time, often even but in a few hours after their fecundation, the females proceed to lay their eggs, and place them in those situations in which the larvæ are destined to reside. The sense of smell, in a choice so important to the prosperity of their generations, is their only guide. They are sometimes, however, deceived in this particular. Thus the meat-fly will occasionally deposit its eggs to no purpose on the *arum dranunculus*, a plant which exhales a cadaverous odour. Felix d'Azara reports, in his voyage to Paraguay, that a swarm of flies, and probably of a species analogous to the preceding, assailed him, as well as his horse, in one of his journeys, and overwhelmed him with an innumerable multitude of eggs. The other species of fly before mentioned, also sometimes deposits its eggs on the human body, as its larvæ have occasionally been extracted from human subjects.

Some flies, but few in number, present us with this peculiarity, that they give birth to living larvæ. But they are less fruitful than those which lay eggs, since the larvæ occupy more space in the interior of the insect they bring forth but two young ones at a time.

The Ocyptera constitute a genus very much approximating to Musca proper. They are found pretty often upon flowers, and sometimes on the glass of cross-bar windows; they run very fast, agitating their wings. We are informed by Degeer, that one species is viviparous. But it is to M. Leon. Dufour, that we are indebted for a knowledge of two species, the *O. Cassidæ* and *O. bicolor.* That of the first species lives in the visceral cavity of the *Cass. bicolor,* and that of the second in the same cavity of the *Pentatoma grisea.* Both of them feed on the epiploon of those insects. Their body is oblong, soft, whitish, perfectly free of hair, furrowed, and contractile. Its anterior end presents two nipples having each two small cylindrical bodies, terminated like a button, umbilicated at the centre, and two corneous pieces pretty strong, having each, externally, a large hook, or two which makes them appear forked and set back to back by their convexity. It appears, according to the figure given by this naturalist, that there should be one for each nipple and these interior. He considers them as mandibles, and the sort of palpi of which we have spoken, and whose disk is pierced at the centre, should be a sort of *pedipalpi,* performing the office of a cupping-glass, or serving the purpose of tact. The body of these larvæ is terminated by a sort of syphon, of the length of one-third of the body, of a more solid consistence, of an invariable form, and narrowing as it goes on, with the appearance of two hooks at the end. The posterior extremity of this syphon, occupying one of the metathoracic stigmata, and in contact with the air, serves for the respiration of the larva. Neither eyes nor antennæ are discoverable. It is in the same sojourn that the larva passes into the nymph state. This nymph is ovoid, without any trace of rings, and presents at one of the ends (*O. Cassida*) four, and (*O. bicolor*) six tubercles. It quits its dwelling before it becomes a perfect insect, sometimes without the insect in which the larva has resided perishing, and sometimes at the expence of its life. These larvæ have two salivary vessels, four biliary vessels, some tracheæ, altogether tubular, without a mother-of-pearl appearance, or transverse striæ, and disposed into two principal trunks, sending forth a great number of ramifications. These trunks appear to have one common mouth through a single orifice at the base of the caudal syphon. The alimentary tube is about four times the length of the body, and presents a capillary œsophagus, or crop in the form of a turbinated calyx, which insensibly degenerates into a tubular stomach, folded back upon itself, and followed by a flexuous intestine, a rectum but little perceptible, and terminated by an oblong cœcum.

The FLIES (MUSCA proper) are insects, which of all others must be best known, since they are met with every where, both in fields and houses. They fly with rapidity, and then send forth a buzzing sound, which, as is usually believed, is caused by the friction of their wings against the sides of the corslet. They are extremely troublesome, and incessantly torment both men and animals. Those which fly in our apartments, and which may be termed domestic flies, place themselves continually, in crowds, upon our viands, and particularly on pastries, sweetmeats, &c., which they suck with their proboscis, for they are extremely fond of sugar, and of every thing which is sweet. They injure gildings, ceilings, and the frames of pictures, by depositing their excrements upon them, which are in a liquid form. They are very abundant during the whole summer, but more especially in July and August. Many species frequent flowers for the purpose of sucking the honey; but many others go in search of carcases, meat, &c., principally for depositing their eggs there. Those carnivorous habits even form the distinctive character of the diptera of this genus.

We insert a figure of Dalman's *Celyphus obtectus*, which is purplish blue, with the antennæ and legs rufous, and the wings virescent. Of the singular genus *Diopsis* we insert a new species, *D. fasciata*. The thorax is black, the head and anterior legs fulvous, but the posterior pair are partly black; the abdomen is brown, with several narrow white cross bands, and there are four spines on the posterior part of the thorax.

We also insert the figure of a new genus of the Muscidæ, which Mr. Westwood calls *Diateina,* which is allied to Colobata, but differs, by having the third articulation of the antennæ very long, and the first joint very fine and plumose; the legs are long, with the femora of the anterior pair serrated. The species he names *Holymenoides:* it has the head and thorax fulvous, with a large black mark on the centre of the latter; the body is fulvous (partly mutilated in the specimen), the anterior legs black, the four posterior legs fulvous, with the tip of the femora and ciliæ black. This curious insect is from Brazil.

We have illustrated the curious genus Achias with several figures; the first which we will notice is named *Phagiocephalus tubularis* by Wiedemann. The head is ferruginous, beneath whitish; the middle of thorax and base of abdomen whitish; the wings marked with brown, which is trilobed on the anterior margin. This species is separated from Achias on account of the antennæ being naked and simple, the thorax somewhat circular and large, the abdomen small and oval.

The second is figured under the name of *Diopsis brevicornis,* of Say; but Mr. Say has since formed this species into a separate subgenus, under the name of *Sphryracephala,* in his work on American insects.

The third *Trigonosoma perilampiformis* has the thorax subquadrate, the abdomen triangular and short, the general colour is brown.

And the third and last species is the *Zygothrica dispar,* of Wiedemann. It is brown; the abdomen yellow, with a spot of black in the centre; the legs also yellow. The generic characters are—the antennæ short and plumose; the head very broad in the male, and narrower in the female; the abdomen large and vespiform.

The family of Pupipara is composed of the Linnæan genus

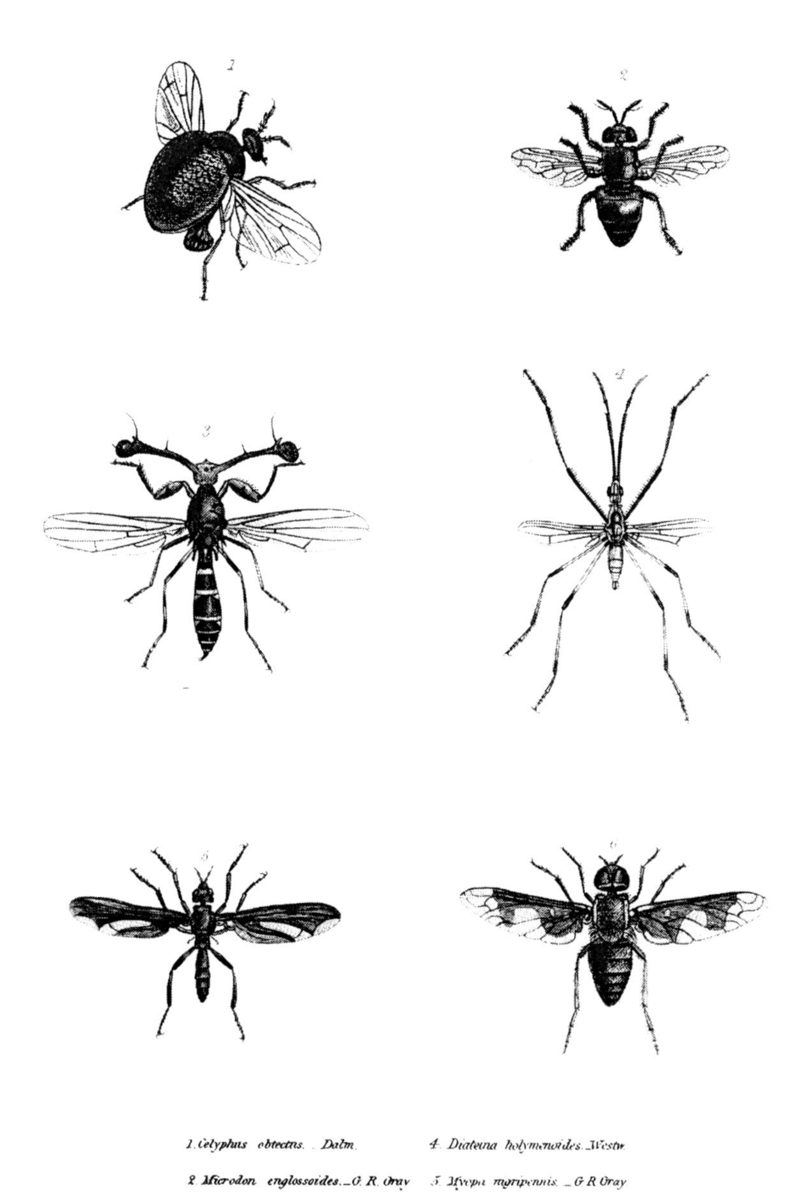

1. *Celyphus obtectus.* — Dalm.

2. *Microdon englossoides.* — G. R. Gray

3. *Diopsis fasciata.* — G. R. Gray

4. *Diateina holymenoides.* — Westw.

5. *Myopa nigripennis.* — G. R. Gray

6. *Anthrax marginicollis.* — G. R. Gray

London. Published by Whittaker & Co. Ave Maria Lane. 1832

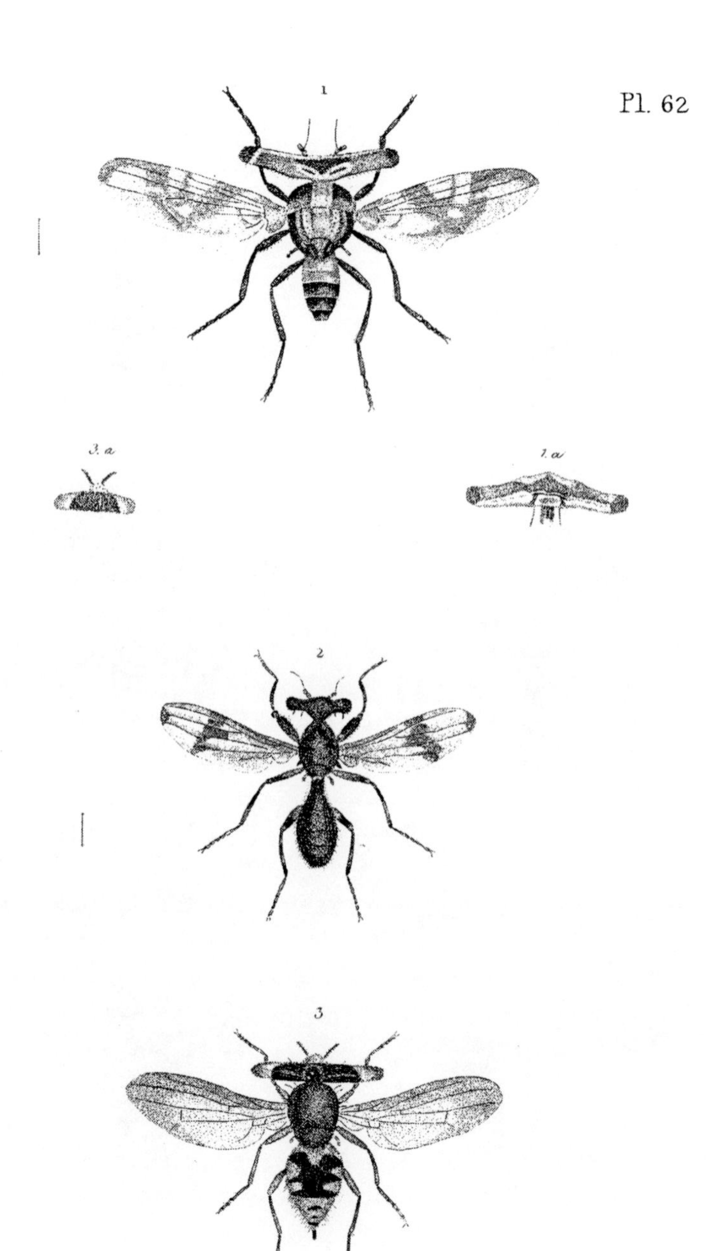

1. Plagiocephalus *lobularis.* *Wiedem.* 2. Diopsis *brevicornis.* *Say.* 3. Zygothrica *dispar.* *Wiedem.*

London, Published by Whittaker & C.º Ave Maria Lane, 1837.

Hippoboscus. To the anatomical researches of M. Leon. Dufour, we are indebted for the knowledge of some very curious facts relative to this family, such as the existence of salivary glands, of a sort of matrix, consisting of a large musculo-membranaceous pouch, intended for a real gestation, analogous to the female uterus in the human species, and ovaries totally different from those of other insects. They are formed of two ovoid, obtuse bodies, filled with a white homogeneous pulp, free and rounded at one end, and leading by the other to a proper conduit. According to him these ovaries, in their configuration and position, approximate very singularly to those of women. Reaumur had a glimpse of their existence. The matrix, at first very small, is dilated by the successive progress of gestation, enormously infringes upon all the viscera, and finishes by invading the entire abdominal capacity, to which it imparts a most considerable amplification.

The *Hippobosci* deserve most particularly to be known, for the state in which they appear at the moment of their birth. They fix themselves on the neck, the shoulders, and other parts of the body of the horse; it is to those parts that are the least defended by hair that they most particularly attach themselves. They often remain under the belly, between the thighs, and sometimes even pass under the tail. It is then that they occasion the greatest annoyance to the animal. Horses are not the only animals which are tormented by their attacks: they are found frequently enough on horned cattle, and in the country they are at times found on dogs, which has caused them to be sometimes named *dog-flies;* but the flatted form of their body, which almost touches the surface on which they are placed, sufficiently distinguishes them from the muscidæ. They hold their feet considerably apart from the body; they rather use their wings for the purpose of getting away, and when an attempt is made to seize them, they betake themselves to a rapid flight.

For our acquaintance with the generation of the hippoboscus, we are principally indebted to the illustrious Reaumur. It was he who discovered that this insect lays a singular egg, almost as thick as its belly, from which issues forth an insect, which, to all appearance, does not pass through the larva state, but which possesses the full size, and all the parts which are peculiar to it, in its final form, at the moment of its exclusion. This egg, on coming out of the body of the female, is as white as milk. At one of its ends is a large black plate, which shines like ebony: it is of a rounded form, flat like a bean, emarginated at the end where the plate is, and forms, in that part, as it were, two horns, or two rounded eminences. This plate is hard, whereas the shell of the egg is soft, and yields a little to pressure. The egg, as we have said, when newly laid, is perfectly white, with the exception of the plate and the eminences, which are black; but this last colour soon becomes general. The skin is then shining, and will resist a tolerably strong pressure of the fingers: thus this envelope, is a species of cartilage or shell, of a sensible thickness, and not easily to be cut, even with a good pair of scissors. The diameter of the greatest breadth of these eggs is more than a line and a half, and that of the greatest thickness a line and a quarter. The dimensions of the body of the female which has not laid her egg, or is not ready to lay, scarcely equals that of one of these eggs; from which it follows, that the interior cavity of the body in the ordinary state is not, within a great deal, capable of containing one of them. But it is with the body of this insect as with a bladder, or a purse, which extend in proportion as they are filled. It would be an immense operation, indeed, for an insect to bring forth from its body an egg whose volume should surpass that of the body itself; but the hippoboscus only lays eggs in proportion to its size. It is not until after they leave the body, that these eggs acquire this monstrous bulk; but their growth is so instantaneous, that the majority of observers have believed that they came out as large as we see them.

Nature, in producing these insects, appears to deviate from the path which she has pursued to conduct the others to perfection. It is under its shell that the insect grows: enclosed in this shell, it undergoes all its metamorphoses; accordingly, this envelope is by no means analogous to that of ordinary eggs. It has even been the skin of the insect before it was metamorphosed into a nymph. Reaumur has proved this by opening one of the eggs which the perfect insect had just quitted; he found in the interior the spoil of the nymph, as we find in the cocoon of a fly that of its nymph, under the skin of the larva which it has quitted, and which becoming hard, has served it as a cocoon.

The hardness and solidity of the shell of each egg render it very proper for defending the insect, which it encloses. This advantage, one might imagine, would turn against the hippoboscus, when with such feeble parts, which have not yet acquired all the consistence which the air will give them, it is obliged to force the walls of its prison. But the same art which has been employed in the construction of the cocoons of flies has also been adopted in that of those of the hippobosci. From the thick end of each of them, where the head is, it is easy with the point of a pen-knife to make a sort of cap shoot off, which, when pressed, will be found divided into two equal pieces. If we observe an entire cocoon with a microscope, we may perceive a slight mark, which indicates the place where this cap unites with the rest of the cocoon. When the proper time is come for the insect to separate it, the animal no doubt has the power of inflating its head, as flies have in similar situations.

Experience has proved that the hippobosci have no more objection to piercing the human skin than that of the horse or the ox; but their bite is not more sensibly felt than that of a flea; it excites a strong degree of itching at the moment of suction, but is followed by no swelling, it only leaves a small red spot, which disappears after the departure of the insect, from which it appears that the hippobosci are not so redoubtable as the gnats, which do not fail to envenom the wounds which they inflict.

We are not acquainted with the number of eggs laid by the hippoboscus, the time which elapses between the coupling and the oviposition, or the interval which passes between the laying of each egg.

"In employing this last term," says M. Latreille, "to indicate the form of these insects at the moment in which they issue from the mother's womb, I merely lend myself to the usual form of speech, that which has been employed even by Reaumur and Degeer themselves. But the facts which I have related, compared with those which have been collected respecting the metamorphoses of the other animals of the same class, clearly evince that this expression is inexact. The egg of the hippoboscus discloses in the belly of the mother, the larva remains and is nourished there, nor is it expelled until the moment in which it is passing to the state of nymph. This egg, therefore, is nothing but an oviform nymph, having a little shell for an envelope, and formed, like the cocoons of many other diptera, by the old dermis; but it presents no appearance of rings, and this character distinguishes it from the preceding. It may be compared to a *bean*, and this denomination would suit it better than it suits the chrysalis of the *Silk-worm*, to which it has been applied."

The Ornithomyiæ are found upon some birds, more especially upon swallows in their nests; they hook themselves very strongly to the plumage and the skin.

The Nycteribiæ are very curious insects of this family, and so closely approximating to the lice, that they would appear at first sight, to constitute a passage from those insects

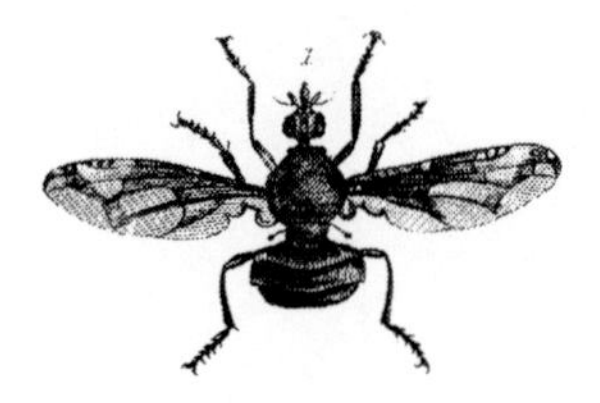

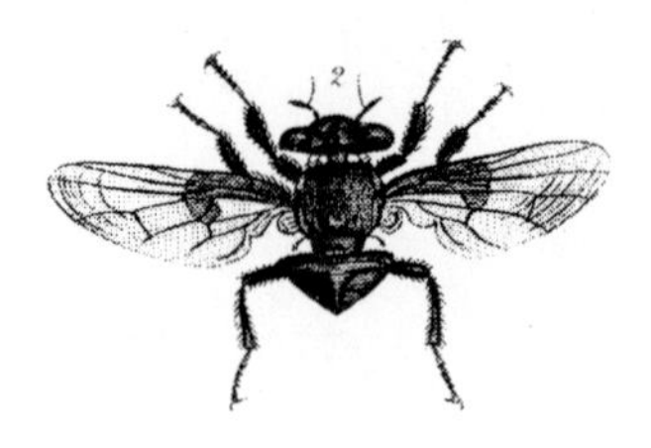

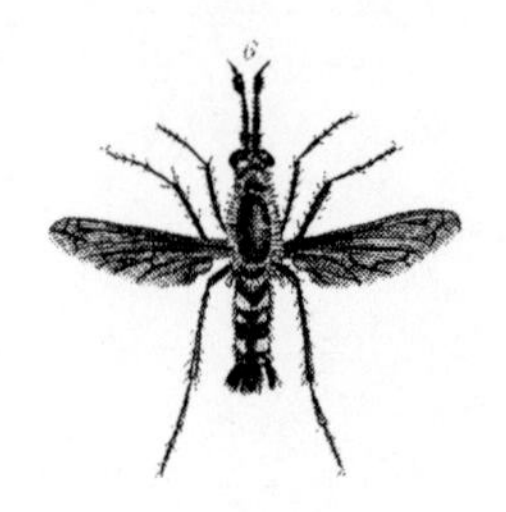

1. *Tephrites Violacea* _ G. R. Gray.

2. *Trigonosoma perilampiformis* _ G. R. Gray.

3. *Pterodontia flavipes* _ G. R. Gray.

4. *Psilodera capensis.* _ G. R. Gray.

5. *Toxophora falva.* _ G. R. Gray.

6. *Ploas? Ægeriiformis.* __ G. R. Gray.

London: Published by Whittaker & C° Ave Maria Lane, 1832.

to the Arachnida. They present, however, all the major characters of the hippobosci, of Linnæus. Any detail of these characters must not be expected here. Linnæus tells us, that the *Pediculus Vespertilionis*, which M. Latreille conceives to be a nyctèribia, has not the power of walking upon a smooth and level plane. This is confirmed by Hermann. These insects, having the head implanted on the back, and small, it is difficult to conceive how they can suck the blood of the bats which forms their exclusive aliment; but, according to the observations of Montagu, they are turned upon their backs in the operation of suction.

We further illustrate this order by the following figures of new species, viz. *Toxophora fulva*. The species is fulvous, with a black mark on the thorax, and black lines across the abdomen. *Ploas œgeriformis*, piceous with a yellow margin to the thorax; the legs and antennæ are black. *Pterodontia flavipes*, a singular insect, which should be placed near Henops of Illiger. It is black, with the scutellum and base of the squamulæ brown, the base and centre of the abdomen black, the rest reddish brown, the wings yellowish, the legs yellow, with the posterior femora black. It differs from Henops in having the wings very truncate, with a spur near the tip, and the scutellum rather longer. *Psilodera capensis* is also a new genus, allied to Bombilius, but differing from it by having the thorax very gibbose, more like that of Henops; the wings are diaphanous. It is fulvous, more or less bright, and the head black. The specimen is from the Cape, and is in the British Museum. *Tephrites violacea* has the thorax pyriform, blue; the abdomen violet blue; the legs black; the wings brown, marked with white. *Microdon englossoides* is glossy green, with the posterior part of the abdomen and legs blue, the wings rather brown. *Miopa nigripennis* has the body and legs light brown, the wings dark brown, with two marks on the posterior border white, the one near the base long and narrow, curving towards the base. Lastly, *Anthrax marginicollis*, has the thorax green, with a white line on each side; the body blue, and the wings diaphanous, with the anterior part and base black.

It is no less a point of duty than of inclination in the Editors, on closing the present portion of their work, to acknowledge their obligations to John George Children, Esq., and to the Rev. Frederick William Hope, for the very kind and liberal manner in which those gentlemen have allowed so many of the new genera and species in their entomological cabinets to be figured and described in this work. An inordinate love of distinction too often induces persons of respectable acquirements, in particular studies, though of contracted intellect in the gross, to pursue their favourite object under the cloak of an abstract love of science, *per fas et nefas*, and to add to legitimate emulation the paltry arts of envy and detraction; this it is which has induced the trite observation, that men of science and of literature are a jealous body; and it is, therefore, the more essential that distinguished instances of liberality, and disinterested love of science, when they do occur, should not be slighted or forgotten.

Mr. G. R. Gray has selected from the above mentioned collections, and has named and described the several species figured. He has also designed the plates intended to illustrate the terminology of the several orders.

ALPHABETICAL LIST

OF

SPECIES OF INSECTS,

FIGURED.

NOTE.—*Many of the species figured did not occur till after the supplementary observations on the order to which they belong were printed; as these therefore are not noticed in the work, a brief description of them is inserted in this list: descriptions are also inserted of all those published by M. Guerin, in his "Iconographie du Regne Animal," which have been copied in this work, no accounts of which have hitherto been published by M. Guerin.*

PL. FIG.

8. 8 ACANTHOCELIS ruficornis, vol. i. page 201
Black, with the antennæ and palpi rufous.

96. 2 Acanthomera gratilla, i. 552.
Brown.

88. 8 Acanthopterus budensis, details of, ii. 103

88. 7 Acanthopterus tripunctatus, ii. 103
Head and thorax red, the latter spotted with black, elytra black, with base and band yellow.

95. 1 Acrocinus Trochlearis, ii. 108.
Brown, spotted with yellow.

113. 9 Acrydium tarsatum, ii. 176
Green; legs banded with black and yellow, tipped with red.

113. 10 Acrydium migratorium, details of, ii. 176

126. 7 Adela Degeerella, ii. 627

80. 2 Adelium cupreum, ii. 22. and pl. lxxiv. f. 2. details.

PL. FIG.

16. 4 Adelocera Chabannii, vol. i. page 317.
Fulvous, marked with black.

56. 10 Adelostoma rugosa, i. 547. and pl. lix. f. 12. details.
Brownish black.

126. 6 Æcophora Linnæella
Reddish brown, upper wings partly red, with three silvery dots.

23. 1 Ægialia cornifrons, i. 460.
Shining black.

120. 2 Ægocera rectilinea, ii. 600
Fulvous, with a transverse band of white, and three black marks, lower wings deeper, with margin and spots black.

51. 2 Æsalus scarabæides, i. 493
Reddish brown.

94. 3 Æshne of Egypt, lower lip of, ii. 296

24. 2 Agacephala furcata, i. 469
Fulvous, thorax green bronze.

L.

M.

R.

S.

U.

THE END OF VOL. XV.

LONDON:
GILBERT AND RIVINGTON, PRINTERS, ST. JOHN'S SQUARE.

CPSIA information can be obtained at www.ICGtesting.com
Printed in the USA
BVOW061349030512

289283BV00001B/18/P